Numerical Modeling of Water Waves

T0199589

Numerical Modeling of Water Waves

Pengzhi Lin

CRC Press
Taylor & Francis Group
Boca Raton London New York

CRC Press is an imprint of the
Taylor & Francis Group, an **informa** business
A TAYLOR & FRANCIS BOOK

CRC Press
Taylor & Francis Group
6000 Broken Sound Parkway NW, Suite 300
Boca Raton, FL 33487-2742

First issued in paperback 2019

© 2008 Pengzhi Lin
CRC Press is an imprint of Taylor & Francis Group, an Informa business

Typeset in Sabon by
Integra Software Services Pvt. Ltd, Pondicherry, India

No claim to original U.S. Government works

ISBN-13: 978-0-415-41578-1 (hbk)
ISBN-13: 978-0-367-86623-5 (pbk)

British Library Cataloguing in Publication Data
A catalogue record for this book is available from the British Library

Library of Congress Cataloging in Publication Data
A catalog record for this book has been requested

Visit the Taylor & Francis Web site at
http://www.taylorandfrancis.com

and the CRC Press Web site at
http://www.crcpress.com

Contents

Preface

Water wave problems are of interest to coastal engineers, marine and offshore engineers, engineers working on naval architecture and ship design, and scientists working on physical oceanography and marine hydrodynamics. The theoretical study of water waves can be traced back to two centuries ago. In the past few decades, research in water waves has been very active, driven by the increasing demand of sea transport and offshore oil exploration. Various water wave theories have been developed to describe different wave phenomena. These wave theories enable us to understand the physical mechanisms of water waves and provide the basis for various water wave models.

Historically, the techniques of water wave modeling were developed in mainly two areas, coastal engineering and offshore engineering (including naval architecture). While the traditional wave modeling in coastal engineering has emphasised detailed wave transformation over a rigid and fixed bottom or structure, the wave modeling in offshore engineering and naval architecture has focused on wave loads on relatively large bodies and the corresponding structure responses. In recent years, there has been a trend to couple the modeling of complicated wave transformation (e.g., wave breaking) with the analysis of body motion. Furthermore, efforts are also being made to develop general-purpose models that can be applied not only to water wave problems, but also to many other types of turbulent-free surface flow problems (e.g., river flows, mold filling).

This book is based largely on the author's teaching and research during the last 14 years. The relevant postgraduate courses include Coastal Engineering and Offshore Hydrodynamics developed at National University of Singapore and Advanced Turbulence Modeling developed at Sichuan University. The relevant research is the modeling of water waves and turbulent-free surface flows at Cornell University, Hong Kong Polytechnic University, National University of Singapore, and Sichuan University. The intended readers of this book include students, researchers, and professionals. The main purpose of this book is twofold: (1) to introduce readers to the basic wave theories and wave models so that readers are able to choose appropriate existing models for different physical problems and (2) to provide adequate details of numerical techniques so that advanced readers can construct their own wave models and test the models against the benchmark tests provided.

This book is organized in the following way. The general background of water waves and water wave modeling is introduced in Chapter 1 and the basic hydrodynamics including turbulence modeling is reviewed in Chapter 2. In Chapter 3, the linear and nonlinear water wave theories based on potential flow assumption are discussed, followed by the introduction of various wave phenomena in oceans. In Chapter 4, different numerical methods are introduced and compared. Readers will be exposed not only the classical numerical methods (e.g., finite difference method, finite element method, and boundary element method), but also the innovative methods (e.g., meshless method, lattice Boltzmann method). Various types of water wave models based on different wave theories and solved by different numerical methods are introduced in Chapters 5. Extensive working examples and benchmark tests are provided to demonstrate the capability of these models. The modeling of wave–structure interaction, which is of primary interest to coastal and offshore engineers, is continued in Chapter 6, again with model demonstration and benchmark tests. The summary for the suitability of different wave models for different wave problems is provided in Chapter 7. In the same chapter, some related subjects that are not adequately covered elsewhere in the book are briefly introduced and the future trend of water wave modeling is highlighted.

Many people have contributed in various ways to this book. The author is particularly grateful to his Ph.D. students, Dr. Liu Dongming, Chen Haoliang, and Lin Quanhong at National University of Singapore, for running the benchmark tests and editing the book draft. Ph.D. student Xu Haihua and former student Lee Yi-Jiat have helped to draw sketches, prepare appendix scripts, and proofread the manuscript. Colleagues at National University of Singapore (Prof. Chan Eng Soon, Prof. Jothi Shankar, Prof. Liong Shie-Yui, Dr. Guo Junke, Prof. Cheong Hin Fatt, Dr. Pavel Tkalich, and Prof. Vladan Babovic) and Sichuan University (Prof. Xu Weilin, Prof. Cao Shuyou, and Prof. Yang Yongquan) have offered the author much help and that help is acknowledged with thanks. The author is grateful to many colleagues who generously provided pictures of their own research for this book and their names are acknowledged in the captions of the pictures. The author would also like to thank Tony Moore at Taylor & Francis who initiated the writing of this book in the summer of 2005, Katy Low at Taylor & Francis who assisted in the preparation of the book draft, and Cherline Daniel at Integra who helped with the typesetting of the final book proof. During the writing of this book, the author was supported, in part, by the research grants through

National University of Singapore (R-264-000-182-112) and Sichuan University (NSFC: 50679046 and 50525926; 973: 2007CB714100). Last but not least, the author would like to thank his family and friends, who have provided continuous support throughout the journey of his life, not only the writing of this book.

Professor Pengzhi Lin
Sichuan University

1 Introduction to water wave modeling

1.1 Introduction to waves

Waves are a common phenomenon in nature. By general definition, a wave is a movement with a certain periodic back-and-forth and/or up-and-down motion. A wave can also be defined as a disturbance that spreads in matter or space, obeying a certain "wave equation." Among the many types of waves, those that we are familiar with are sound waves, light waves, radio waves, and water surface waves. Although light waves and radio waves belong to electromagnetic waves that can travel in a vacuum, other types of waves need a medium to transmit the disturbance.

Waves that propagate through media, depending on the form of wave transmission, may be divided into two major categories, namely body waves and surface waves. In the case of body waves, they can travel in a single medium. A typical example is sound waves, which are caused by a pressure disturbance in a solid, liquid, or gas. Wave propagation is accomplished by the consecutive compression and dilatation that occur in the wave propagation direction. Another example of such a wave is the P-type seismic wave. Besides the above pressure-induced body waves, there are also stress-induced body waves, in which the medium moves in a direction perpendicular to the direction of wave propagation, for example, S-type seismic waves.

Surface waves always appear on the interface of two different media. The restoring force, which attempts to restore the interface to the equilibrium state, plays a predominant role in wave propagation. Surface waves can occur at the interface between a solid and a fluid, e.g., Rayleigh-type seismic waves. They can also occur at a fluid–fluid interface, e.g., ocean internal waves between two fluid layers with different density and ocean surface waves on an air–water interface.

1.2 Ocean surface waves and the relevance to engineering applications

In this book, we shall mainly focus on ocean surface waves. Ocean surface waves, or more generally surface water waves or simply water waves, are mainly generated by wind in deep oceans. Other possible sources of wave generation include astronomical attraction forces (e.g., tide), seismic disruptions (e.g., tsunami), underwater explosions, and volcano eruptions. Like all other surface waves, ocean surface waves require restoring forces to propagate. On most occasions, the gravity of Earth is the major restoring force. It is only when the scale of the wave is very small [i.e., O(1 mm)], that surface tension becomes the main restoring force for the resulting capillary wave.

Water waves share many similarities with light waves. For example, water waves experience refraction when they propagate over a changing bathymetry, similar to light traveling from one medium to another medium. Water waves form a certain diffraction pattern behind a large object or through a narrow gap. This is again similar to what light experiences under a similar condition. In fact, many theories established in optics, linear or nonlinear, can easily find their counterparts in water wave theories. The only exception is probably the wave-breaking phenomenon, which is an important physical process for water waves to dissipate excessive energy in terms of turbulence, but a similar process is not identified for light.

The study of water waves has important applications in engineering design. In deep oceans and deepwater offshore regions, wave height and wave period are two major design criteria. The practical problems include the maneuvering of ships against waves, the safe operation of offshore structures such as floating production storage offloading (FPSO) vessels or floating airports or terminals in extreme waves, and the stability of offshore structures such as bottom-mounted jacket platforms subjected to wave attacks. Because water depth is usually larger than wave length, the main wave mechanism to be considered in design is wave diffraction. The calculation of six-degree-of-freedom (DOF) structure motions, i.e., sway, surge, heave, pitch, roll, and yaw, and the control of the motion are the key considerations in the design. Other related problems are liquid sloshing in a tank induced by different external excitations and the "green water" effect caused by large-amplitude waves impinging on ship decks or offshore platforms.

In nearshore regions, waves can go through complex transformations with the combination of wave shoaling, wave refraction, wave diffraction, and wave breaking. There is also an active nonlinear energy transfer among different wave harmonics when waves propagate over abruptly changed bottom topography. Because of this, it is more challenging to understand wave motions in shallow water than in deep water. The engineering concerns for water waves in nearshore regions are quite different from those in deep seas. The functional performance of various coastal protections, which range from breakwater and groin to seawall and revetment, is one of the most important considerations in coastal engineering. Most of these protections are designed to provide a calm or at least reduced wave environment in the protected areas such as harbors or beaches.

1.3 Wave modeling

There are various types of techniques for modeling a prototype wave system, i.e., analytical wave modeling, empirical wave modeling, physical wave modeling, and numerical wave modeling. With their inherent advantages and limitations, these techniques shall be applied for different purposes.

1.3.1 Analytical wave modeling

A physical wave system in nature can be very complicated. We may find a way to represent the wave system by analyzing the system with a simplified theoretical model that should be able to capture the most important characteristics of the wave system. The model is usually expressed by mathematical equations, which are in the form of partial differential equations (PDEs) or ordinary differential equations (ODEs) governing the space–time relationship of the important variables for the waves. A good theoretical model is always constructed on rigorous mathematical derivations, each step of which has clear physical and mathematical assumptions and implications. When the equations are solved analytically for a specific condition, the closed-form solution can be obtained and used to predict wave system behavior accurately. This procedure of solution seeking is called analytical wave modeling, which is a powerful tool for us to understand the physical phenomenon of a particular wave system. One typical example of this kind of modeling is the study of the fission and fusion of solitons by solving the KdV nonlinear wave equation using inverse scattering transform. Another example is the wave-scattering pattern around a large vertical circular cylinder with the use of diffraction theory. Unfortunately, although most of the wave systems can be formulated theoretically, only very few of them can be solved analytically. This greatly limits the application of this approach to general wave problems.

1.3.2 Empirical wave modeling

An empirical formula is usually a simple mathematical expression summarized from available field data of a prototype system. It can describe the system behavior in terms of simple algebraic equations with important parameters. Since the empirical approach is simple and easy to implement, it is widely adopted in engineering design. In empirical wave modeling, empirical formulas can be used to estimate the maximum wave load on a structure (i.e., the Morison equation), the reflection coefficient from a certain type of structure, the maximum wave run-up on a beach, and the maximum wave overtopping rate over a dike. However, since empirical formulas are established on the known problems and database, when a new prototype system is considered, the existing empirical formulas may not be suitable.

1.3.3   Physical wave modeling

To understand a prototype wave system in nature, we can alternatively build a small-scale physical model in the laboratory as the miniature of the prototype. The wave characteristics (and wave loads on the model structure if wave–structure interaction is investigated) can be obtained by laboratory measurements. The information can then be extrapolated based on certain scaling laws to estimate what will really happen to the prototype system. This approach is called physical wave modeling. It is straightforward and allows us to visualize and understand the important physical processes from the small-scale model study. Typical examples of physical wave modeling include the studies of wave loads on an offshore structure, wave transmission and reflection from a breakwater, etc. However, when a physical system becomes very complicated, a physical model that satisfies all the important scaling laws, on which the physical model is designed and constructed, may not exist. Other approaches may have to be sought under such circumstances or when a physical model study is too expensive and time-consuming. Readers are referred to Hughes (1993) for a more detailed description of the physical models in coastal engineering and to Chakrabarti (1994) for offshore structure modeling.

1.3.4   Numerical wave modeling

A numerical wave model is the combination of the mathematical representation of a physical wave problem and the numerical approximation of the mathematical equations. Compared to theoretical modeling, the difference is only in the means of finding the solution of the governing equations for the wave problems. To model ocean waves numerically, we must start from some existing "wave equations" obtained from theoretical studies. In most cases, we can have more than one wave equation to describe the same wave phenomenon, depending on different levels of approximation made in the theoretical derivations. Similarly, the same wave model can also be applied to many different wave problems, as long as the basic assumptions in the model are valid.

1.4   Numerical models for water waves

In the following subsections, we will classify wave models based on their modeling capability so that readers can have a broad idea of the wave models that are available and the problems that they can solve. In Chapter 5, in addition to providing sufficient theoretical background of wave theories, we will further elaborate on the theoretical assumptions and capabilities of these wave models.

1.4.1 Wave spectral models

The most commonly employed wave model in modeling large-scale wave motion is the wave energy spectral model or wave action spectral model. This type of model is constructed on the assumption that a random sea state is composed of an infinite number of linear waves whose wave height is a function of wave frequency and the direction of wave propagation. For an individual wave train, the rate of change of wave energy (or action) flux is balanced by the wave energy transfer among different wave components in different directions (i.e., wave refraction) and different frequencies (i.e., nonlinear wave interaction) as well as energy input and dissipation. The representative wave energy spectral models are WAM (WAve prediction Model) (Hasselmann *et al.*, 1988) and WaveWatch III (Tolman, 1999), which can be used to model large-scale variations of wave heights in deep oceans. Being connected with an atmosphere model, a wave energy spectral model is able to predict global ocean wave climate. When the current effect is considered, the wave action spectral model can be formulated and used instead to simulate combined wave–current interaction in a large-scale nearshore region. The representation of such a model is SWAN (Simulating WAves Nearshore; Ris *et al.*, 1999).

With the wave phase information being filtered out in the formulation, the wave spectral model can use computational meshes much larger than a wavelength and thus can be applied in a very large area. Without wave phase information, however, the model is unable to represent wave diffraction that is phase-related. For this reason, this kind of model is usually used to provide far-field wave information only. The detailed wave pattern around coastal structures where diffraction is important is left to other types of wave models.

1.4.2 Mild-slope equation wave models

The mild-slope equation (MSE) was originally derived based on the assumptions of linear waves and slowly varying bottoms. The equation can be used to describe combined wave refraction and diffraction in both deep and shallow waters. In most events, the MSE model is employed to study monochromatic waves, though it can be applied to irregular waves by summing up different wave harmonics. The extension of the MSE to abruptly varying topography and larger amplitude waves was also attempted in the last two decades. So far, most of the applications of the MSE models are limited to the region from an offshore location to a nearshore location some distance away from the shoreline before wave nonlinearity becomes strong. One exception is its application in harbor resonance modeling because water depth in a harbor is usually deep even along the boundaries. The MSE has three different formulations, namely the hyperbolic MSE for a time-dependent wave field, the elliptic MSE for a steady-state wave field, and the parabolic MSE for a simplified steady-state wave field that has a primary wave propagation direction.

1.4.3   *Boussinesq equation wave models*

To model nearshore waves with strong wave nonlinearity, a Boussinesq equation wave model (or simply the Boussinesq model in water wave modeling) is often a good choice. The Boussinesq equations are depth-averaged equations with the dispersion terms partially representing the effect of vertical fluid acceleration. Rigorously speaking, Boussinesq equations are valid only from intermediate water depth to shallow water before the waves break. However, in engineering applications, the equations are often extended beyond the breaking point, up to wave run-up in the swash zone. This is possible by adding an artificial energy dissipation term for wave breaking. Besides, efforts are also made to extend the model to deeper water. Unlike the wave spectral model and the MSE model, the Boussinesq model does not have the presumption that the flow is periodic. Therefore, it can be applied to waves induced by impulsive motions, i.e., solitary waves, landslide-induced waves, tsunami, and unsteady undulation in open channels.

1.4.4   *Shallow-water equation wave models*

To model tsunami or other long waves (e.g., tides), a shallow-water equation (SWE) model is more likely to be adopted. Compared with the Boussinesq model, the SWE model is simpler because the flow is assumed to be uniform across the water depth and the wave-dispersive effect is neglected. The SWE model has a wide application range in modeling tsunami, tides, storm surges, and river flows. The main limitation of the SWE model is that it is suitable only for flows whose horizontal scale is much larger than vertical scale.

1.4.5   *Quasi-three-dimensional hydrostatic pressure wave models*

All the earlier discussed wave models can be operated on a horizontally two-dimensional (2D) plane due to vertical integration. When a shallow-water flow is modeled, the depth-varying information may still be needed on some occasions. A typical example is ocean circulation where the horizontal length scale is much larger than the vertical scale, but the vertical circulation, though relatively weak, is still of interest in many events. Thus, in this case, a three-dimensional (3D) model would be necessary. This type of model often solves the 3D Navier–Stokes equations (NSEs) directly under the hydrostatic pressure assumption. Under such an assumption, the solution procedure to the 3D NSEs is greatly simplified and the model is referred to as the quasi-3D model, against the fully 3D model to be discussed later. The model of this type is often solved in the σ-coordinate that maps the irregular physical domain to the regular computational domain for the ease of application of the boundary condition. The representative model of this kind is the Princeton Ocean model (POM).

1.4.6 Fully three-dimensional wave models with turbulence: Navier–Stokes equation models

To model fully 3D wave problems, we must turn to the general governing equations, NSEs, without a hydrostatic pressure assumption. The NSEs are derived from the general principle of mass and momentum conservation that is able to describe any type of fluid flow including water waves. With the inclusion of a proper turbulence model, it is possible for an NSE model to simulate difficult wave problems, e.g., nonbreaking or breaking waves, wave–current interactions, and wave–structure interactions. When breaking waves are simulated, there is the potential to include air entrainment on the surface of the water; when the wave–body interaction is computed, it can treat both the rigid body and the flexible body. With almost no theoretical limitation, this type of model seems to be the best choice of all. However, the main barrier that prevents the wide application of such a model is the expensive computational effort. To solve the fully 3D NSEs, the computation can be much more expensive than all the previously introduced wave models. So far, the application of such models in engineering computation is still restricted to the simulation of local wave phenomena near the location of interest, e.g., the surf zone when the breaking wave and/or sediment transport is considered and the flows around coastal and offshore structure when the wave–structure interaction is considered.

1.4.7 Fully three-dimensional wave models without turbulence: potential flow models

When turbulence is negligible and the bottom boundary layer thickness is thin (e.g., nonbreaking waves), the NSEs can be reduced to the Laplace equation based on potential flow theory. The theory is applicable to most of the linear and nonlinear nonbreaking waves and their interaction with large bodies. The Laplace equation can be solved numerically by many different methods. One of the most commonly used methods is the boundary element method (BEM), which converts domain integration into surface integration with the use of Green's theorem. This type of model is capable of simulating highly nonlinear waves in both deep and shallow waters. It is effective in the study of nonlinear wave transformation over changing topography, linear wave diffraction over a large body, and wave force on a large structure. The major limitation of such a model lies on the potential flow assumption that requires the flow to be irrotational. For this reason, the model is unable to simulate breaking waves as well as wave interaction with small bodies, during which the flow becomes rotational. Furthermore, the computational cost for modeling 3D fully nonlinear waves using BEM is rather high.

1.5   Books on water waves

There are many books on water waves. While some of these books address the general aspects of water wave theories (e.g., Stoker, 1957; Whitham, 1974; Newman, 1977; Lighthill, 1978; Mei, 1989; Dean and Dalrymple, 1991; Dingemans, 1997), other books address either specialized topics related to water waves (e.g., Le Méhauté and Wang, 1996) or the engineering applications in coastal and offshore engineering (Chakrabarti, 1987; Goda, 2000; Kamphuis, 2000; Reeve *et al.*, 2004; Mei *et al.*, 2005). The books dedicated to the numerical modeling of water waves normally cover only a particular type of wave model, e.g., Massel (1993) for wave spectral models and Dyke (2007) for shallow sea models. Only recently, Tao (2005) published a book (in Chinese) on water wave simulations by different wave models. This book is intended to be a different addition to the available literature so that readers can develop a thorough understanding of various water wave theories on which different wave models are constructed and applied to different physical problems.

2 Review of hydrodynamics

In this chapter, we will review the fundamentals of hydrodynamics, especially those closely related to water waves. The review will start from the NSEs, which are the general equations that govern the motion of fluids. The potential flow theory will then be deduced from the equations with the assumption of irrotational flow and negligible viscous effect. This will build up the theoretical background for water wave theories, which will be detailed in Chapter 3.

Under certain environmental conditions (e.g., waves under strong winds and waves in shoaling), waves become unstable and break. It is during this process that strong turbulence is generated. Turbulence can also be generated during wave–structure interaction. The solution to these challenging problems requires appropriate turbulence modeling. In this chapter, we will also review most of the commonly used turbulence closure models whose advantages, limitations, and applicable ranges are discussed.

2.1 Basic equations for hydrodynamics

The governing equations for fluid flow motions are based on the principles of mass conservation and force balance. The equations can be formulated differently, depending on the flow variables used. The most common formulation is based on the primitive variables of fluid flow velocity and pressure. However, other equivalent formulations are also possible by using the derived variables such as vorticity and/or stream function.

2.1.1 Primitive variable formulation: Navier–Stokes equations

2.1.1.1 The momentum equation

Newton's Second Law, which dictates that the net force applied to a body is equal to mass of the body multiplied by the acceleration of the body attained, was originally constructed on the Lagrangian frame that follows the motion of the object. This law is the basis for classic mechanics, of which fluid mechanics is a branch. The momentum equation is essentially the reiteration of Newton's Second Law in the Eulerian (fixed) frame. With the use of the Reynolds transport theorem, a translator between the Lagrangian frame and the Eulerian frame, Newton's Second Law in the Eulerian system can be recast into the following momentum equation:

$$\left[\frac{\partial u_i}{\partial t} + u_j \frac{\partial u_i}{\partial x_j}\right] = \frac{Du_i}{Dt} = f_i \tag{2.1}$$

where D/Dt is called the total derivative (also called substantive derivative, material derivative, or Lagrangian derivative), representing the time rate of change following the fluid particle motion; $i = 1, 2, 3$ represents three orthogonal directions in space (in most of the cases, $i = 1, 2, 3$ are in x-, y-, and z- directions, respectively); ρ is the fluid density; u_i is the fluid velocity in the ith direction; and f_i is the force applied to the fluid system per unit mass in the ith direction. The first term in the bracket on the left-hand side (LHS) of (2.1) the local acceleration of the fluid particle and the second term is the convective acceleration. The force f_i for most of the real fluids is composed of three contributions, namely pressure force, body force, and viscous force, i.e.:

$$f_i = -\frac{1}{\rho}\frac{\partial p}{\partial x_i} + g_i + \frac{1}{\rho}\frac{\partial \tau_{ij}}{\partial x_j} \tag{2.2}$$

where p is the pressure, g_i is the gravitational acceleration in the ith direction, and τ_{ij} is the second-order viscous stress tensor. For Newtonian fluids, the stress is linearly proportional to the rate of the strain of the fluid particle:

$$\sigma_{ij} = \frac{1}{2}\left(\frac{\partial u_i}{\partial x_j} + \frac{\partial u_j}{\partial x_i}\right)$$

as:

$$\tau_{ij} = 2\mu\sigma_{ij} \tag{2.3}$$

where μ is the molecular viscosity of the fluid.

Continuum: It is noted that the above equation is established by assuming that the fluid is a continuum, meaning that all fluid variables vary continuously through the fluid. This assumption is valid for most of the fluids that have the mean free path (the average distance a molecule travels between collisions) of molecules much shorter than the characteristic length scale for a problem.

2.1.1.2 The continuity equation

Under the same Reynolds transport theorem, the statement that the mass of a fluid system is unchanged regardless of its motion (the basis for classical mechanics, but not the theory of relativity) will be translated into the equation in Eulerian system as follows:

$$\frac{\partial \rho}{\partial t} + \frac{\partial(\rho u_i)}{\partial x_i} = 0 \tag{2.4}$$

This equation is often called the continuity equation that ensures the mass conservation in a fluid system. Equations (2.1) and (2.4) are the well-known NSEs. In the case where the fluid is inviscid, the term $\tau_{ij} = 0$ and the equations are reduced to the so-called Euler equations.

Equations (2.1) and (2.4) become complete with the supplementary state equation that relates the local pressure with the local density as follows:

$$\frac{p}{p_0} = (k+1)\left(\frac{\rho}{\rho_0}\right)^\gamma - k \tag{2.5}$$

where p_0 and ρ_0 are the reference pressure and density, respectively, and k and γ are the coefficients that are related to fluid property and thermodynamic status.

The explanation of the Reynolds transport theorem and the derivation of NSEs can be found in many fluid mechanics books (e.g., the Streeter *et al.*, 1998: 115, 201).

Incompressibility: Like solids, all real fluids are compressible. This means that when a force is applied on a fluid particle, the particle will change shape and/or volume. If the force applied is normal to the surface, the fluid will be stretched or compressed. The level of dilatation or compression subject to the normal force is the property of the fluid, which is often referred to as the compressibility of the fluid, and it is characterized by the modulus of elasticity. When the fluid flow is in the low-speed regime where the flow speed U is much smaller than the speed of pressure wave C (and thus we have a Mach number $Ma = U/C << 1$), the compressibility of the fluid can be neglected, meaning that the volume of the fluid particle remains the same regardless of the forces being applied. Under such a circumstance, the density of a fluid particle remains constant and the fluid is called incompressible. Mathematically, this means that following a fluid particle the total derivative of the fluid density is zero, i.e.:

$$\frac{D\rho}{Dt} = \frac{\partial \rho}{\partial t} + u_i \frac{\partial \rho}{\partial x_i} = 0 \tag{2.6}$$

Substituting (2.6) into (2.4), the continuity equation for incompressible fluid has the following new form:

$$\frac{\partial u_i}{\partial x_i} = 0 \tag{2.7}$$

The above equation is simplified compared with the original continuity equation that is time-dependent.

Constant density: If the fluid has a constant density (i.e., fluid density does not change in time and space), the same simplified continuity equation (2.7) can be derived and this often creates confusion between an incompressible fluid and a constant density fluid. One should realize that although the final continuity equation takes the same form, the way of imposing it is different between an incompressible fluid and a constant density fluid. The latter has a stronger restriction on density variation, namely the individual derivative of density with respect to time and space is zero, i.e., $\partial\rho/\partial t = \partial\rho/\partial x_i = 0$. For an incompressible fluid, however, only $\partial\rho/\partial t + \partial\rho/\partial x_i = 0$ is imposed and we can have $\partial\rho/\partial t = -\partial\rho/\partial x_i \neq 0$. This implies that an incompressible fluid can be at the same time *stratified* (i.e., density varies in time and space). This has special implication in ocean wave theories because the sea water in many occasions is stratified but can still be assumed to be incompressible.

To summarize, the continuity and momentum equations for incompressible or constant density fluids are as follows:

$$\frac{\partial u_i}{\partial x_i} = 0 \tag{2.8}$$

$$\frac{\partial u_i}{\partial t} + u_j \frac{\partial u_i}{\partial x_j} = -\frac{1}{\rho}\frac{\partial p}{\partial x_i} + g_i + \frac{1}{\rho}\frac{\partial \tau_{ij}}{\partial x_j} \tag{2.9}$$

In case the viscosity is constant, the diffusion term with the second-order tensor τ_{ij} can be simplified to $\nu\partial^2 u_i/\partial x_j^2$, where $\nu = \mu/\rho$ is the kinematic viscosity. The momentum equation then becomes, with the use of the continuity equation:

$$\frac{\partial u_i}{\partial t} + u_j \frac{\partial u_i}{\partial x_j} = -\frac{1}{\rho}\frac{\partial p}{\partial x_i} + g_i + \nu\frac{\partial^2 u_i}{\partial x_j^2} \tag{2.10}$$

Now all variables in the equation are either scalar or vector. This enables us to write NSEs in the equivalent form using vector operators:

$$\nabla \cdot \mathbf{u} = 0 \tag{2.11}$$

$$\frac{\partial \mathbf{u}}{\partial t} + \mathbf{u} \cdot \nabla \mathbf{u} = -\frac{1}{\rho}\nabla p + \mathbf{g} + \nu\nabla^2\mathbf{u} \tag{2.12}$$

where the operators $\nabla\cdot$, ∇, and $\nabla^2 = \nabla \cdot \nabla = \Delta$ represent divergence, gradient, and Laplacian, respectively, and the vector is represented by boldface. In this book, both notations are mixed, depending on whichever is more convenient.

Note that the NSEs introduced above are established on an inertial reference frame and Cartesian coordinate system. However, there are cases where the problem can be more conveniently solved in a noninertial reference frame (e.g., liquid sloshing in an excited tank). The virtual forces will then have to be added in the momentum equation (see Section 6.5.1.4). In contrast, sometimes it may be more convenient to solve the equations in another coordinate system (e.g., cylindrical coordinate and spherical polar coordinate). In this case, the governing equations originally established on the Cartesian coordinate must be converted using mathematical identities (see Appendix I).

2.1.2   Vorticity–velocity formulation

Definition of vorticity: The vorticity ω is the measure of the local rotational rate of a fluid particle, and it is defined as the curl of local fluid velocity:

$$\omega = \nabla \times \mathbf{u} = \begin{vmatrix} \mathbf{i} & \mathbf{j} & \mathbf{k} \\ \dfrac{\partial}{\partial x} & \dfrac{\partial}{\partial y} & \dfrac{\partial}{\partial z} \\ u & v & w \end{vmatrix} = \left(\frac{\partial w}{\partial y} - \frac{\partial v}{\partial z} \right)\mathbf{i} + \left(\frac{\partial u}{\partial z} - \frac{\partial w}{\partial x} \right)\mathbf{j} + \left(\frac{\partial v}{\partial x} - \frac{\partial u}{\partial y} \right)\mathbf{k}$$

$$(2.13)$$

where **i**, **j**, and **k** represent the unit vectors in x-, y-, and z- directions, respectively.

Momentum equation: Taking the curl ($\nabla\times$) on both sides of the momentum equation (2.12), we have:

$$\frac{\partial \omega}{\partial t} + \mathbf{u} \cdot \nabla \omega = \omega \cdot \nabla \mathbf{u} + \nu \nabla^2 \omega \qquad (2.14)$$

This is the equation describing the transport of vorticity. To arrive at the above equation, the fluid density is assumed to be constant and the vector product identity $\nabla \times (\nabla \varphi) = 0$ (φ can be any scalar) is used. The first term on the right-hand side (RHS) of (2.14), $\omega \cdot \nabla \mathbf{u}$, can be dropped for a 2D problem because the two vectors are orthogonal to each other.

Continuity equation: Similarly, if we take the curl on $\omega = \nabla \times \mathbf{u}$ and make use of the continuity equation, we have:

$$\nabla^2 \mathbf{u} = -\nabla \times \omega \qquad (2.15)$$

This equation describes the relationship between velocity and vorticity under the constraint of the continuity equation. The equation is obtained by using Lagrange's formula of vector cross-product identity $\nabla \times (\nabla \times \mathbf{u}) = \nabla (\nabla \cdot \mathbf{u}) - \nabla^2 \mathbf{u}$ (the complete vector operation identities are given in

Appendix II). Equations (2.14) and (2.15) can be proven to be equivalent to the original NSEs expressed in terms of primitive variables (2.11) and (2.12). The benefit of using the above equations lies mainly on the easier treatment of the solid boundary, the source of vorticity, compared to the primitive variable formulation.

Once both velocity and vorticity are obtained in the flow field, the pressure can be obtained by solving the pressure Poisson equation (PPE) obtained by taking the divergence on both sides of the original momentum equations of NSEs:

$$\text{Tr}(\nabla\mathbf{u}\otimes\nabla\mathbf{u}) = -\nabla\cdot\left(\tfrac{1}{\rho}\nabla p\right) \tag{2.16}$$

where Tr( ) is the trace of a matrix and $\otimes$ represents the cross-product of two tensors.

2.1.3   Stream function–velocity formulation for 2D problems

Continuity equation: Consider another vector product identity $\nabla\cdot(\nabla\times\psi) = 0$ for any vector field; we realize that the divergence-free velocity field for incompressible fluid flow actually allows the introduction of a vector potential ψ that is defined by $\mathbf{u} = \nabla\times\psi$. Taking the curl of the above definition, we have:

$$\nabla^2\psi = \nabla(\nabla\cdot\psi) - \omega \tag{2.17}$$

This equation in combination with the vorticity transport equation (2.14) forms another equivalent formulation of the NSEs in terms of vector potential and vorticity.

However, such a formulation does not seem to have any immediate advantage over the primitive variable formulation, unless it is reduced to the following form for 2D problems [e.g., on the (x,y) plane]:

$$\nabla^2\psi = -\omega \tag{2.18}$$

where both ψ and ω become scalars [by considering the single vector component perpendicular to the (x,y) plane and thus in the z- direction]. The term $\nabla(\nabla\cdot\psi)$ equals zero because the gradient in the z- direction is always zero for 2D problems.

Momentum equation: Similarly, the vorticity equation can be simplified as the convection–diffusion equation:

$$\frac{\partial\omega}{\partial t} + \mathbf{u}\cdot\nabla\omega = \nu\nabla^2\omega \tag{2.19}$$

and the velocity above is related to ψ by:

$$u = \frac{\partial\psi}{\partial y} \quad\text{and}\quad v = -\frac{\partial\psi}{\partial x} \tag{2.20}$$

The scalar ψ is often referred to as the stream function because the contour line of this function is also the streamline, the tangent of which defines the local fluid velocity direction. The advantage of using the stream function–vorticity formulation is that instead of working on the vector transport equations, we now work on two governing equations for scalars. Furthermore, the stream function along a solid surface is a constant, which makes the implementation of the solid boundary condition straightforward.

Readers can find more information on the alternative formulations of NSEs and their numerical solutions from Quartapelle (1993).

2.2 Potential flow theory

2.2.1 *Basic assumptions*

Ideal fluid: All real fluids are viscous just as they are compressible. However, the viscous effect can be significant or insignificant on different occasions when compared with the rest of the force contributions. If the viscous effect is negligible in a fluid flow, the fluid is referred to as an ideal fluid. Among the three types of external forces in (2.2), only viscous forces have shear components that act on the tangential direction of fluid particles. It is these shear forces that change the rotational status of fluid particles. Therefore, in an ideal fluid the vorticity will be neither created nor destroyed.

Irrotational flows: It is well known that a steady fluid flow around a solid body will induce a boundary layer near the body surface, within which the viscous effect is important. Outside the boundary layer, the viscous effect diminishes toward the far field. This implies that in a location away from the solid body, the fluid gradually loses the driving mechanism that changes its vorticity status. If the fluid flow is initially irrotational, it will remain so, i.e., $\omega = 0$. This type of flow is called irrotational flow.

Potential function: If the flow is irrotational, there exists a scalar velocity potential function ϕ that can be expressed as follows:

$$\mathbf{u} = -\nabla\phi = -\frac{\partial\phi}{\partial x}\mathbf{i} - \frac{\partial\phi}{\partial y}\mathbf{j} - \frac{\partial\phi}{\partial z}\mathbf{k} \tag{2.21}$$

By substituting (2.21) into (2.13), it is not difficult to prove that the existence of the potential function is the necessary and sufficient condition for the irrotationality of the fluid flow.

2.2.2 *Potential flow and the Laplace equation*

The Laplace equation: If the potential function exists, the continuity equation for an incompressible fluid can be rewritten in a new form by substituting (2.21) into (2.11):

$$\nabla \cdot \mathbf{u} = \nabla(-\nabla\phi) = -\nabla^2\phi = 0 \tag{2.22}$$

This is the Laplace equation. Although the Laplace equation is the result of the continuity equation and it is for the purpose of ensuring mass conservation, additional enforcements (i.e., fluid incompressibility and flow irrotationality) have been added along the line of derivation. The flow that can be described by the above Laplace equation is called potential flow. Compared with equation (2.11), equation (2.22) is simplified in the sense that only one scalar ϕ needs to be dealt with, rather than a vector $\mathbf{u}$ that has three components.

Bernoulli equation: At this point, we still have the remaining problem of how to relate ϕ to pressure p. By substituting (2.21) into the momentum equation (2.12) and integrating the resulting equation in space after neglecting the viscous term (see Appendix III), we have:

$$-\frac{\partial \phi}{\partial t} + \frac{1}{2}\left[\left(\frac{\partial \phi}{\partial x}\right)^2 + \left(\frac{\partial \phi}{\partial y}\right)^2 + \left(\frac{\partial \phi}{\partial z}\right)^2\right] + \frac{p}{\rho} + gz = C(t) \qquad (2.23)$$

where $C(t)$ is the integration constant that is uniform in space and changes with time. The value of it can be determined from boundary conditions. To obtain the above equation, we also assume that the gravitational acceleration is in the vertical direction, i.e., $g_x = g_y = 0$ and $g_z = -g$. This equation is called the Bernoulli equation, and it relates fluid pressure with fluid velocity potential between any two positions in space. Equation (2.23) is the simplified momentum equation under the potential flow assumption. By solving (2.22) and (2.23) together, one is able to find the velocity and pressure variation in time and space. It is noted that there is another version of the Bernoulli equation along the streamline (not the entire space) for viscous fluid, which has been elaborated in many fluid mechanics books and will not be further discussed here.

2.2.3   Formulation based on the stream function for 2D problems

For a 2D irrotational flow, the stream function also satisfies the Laplace equation because the vorticity is zero throughout [see (2.18)]:

$$\nabla^2 \psi = -\omega = 0 \qquad (2.24)$$

With the definition of ψ and ϕ in terms of u and v by (2.20) and (2.21), it is ready to prove:

$$\nabla \psi \cdot \nabla \phi = 0 \qquad (2.25)$$

This implies that the streamline is always perpendicular to the contour line of velocity potential function.

It is also ready to have the following relationship:

$$\frac{\partial \phi}{\partial x} = \frac{\partial \psi}{\partial z}, \quad \frac{\partial \phi}{\partial z} = -\frac{\partial \psi}{\partial x} \qquad (2.26)$$

This pair forms the Cauchy–Riemann conditions and the problem can thus be solved by complex variable analysis.

2.3 Turbulent flows and turbulence modeling

2.3.1 Definition of turbulent flows

For a real fluid, viscous effect plays an important role in balancing the fluid inertia and dissipating the fluid energy. When the viscous effect is relatively important, the flow tends to be *laminar*. It would become *turbulent* as fluid inertia increases. The dimensionless parameter commonly used to characterize the tendency of flow transition from laminar flow to turbulent flow is known as the Reynolds number (Re):

$$\mathrm{Re} = \frac{UL}{\nu} \tag{2.27}$$

where U and L are the characteristic velocity scale and the length scale of the flow, respectively.

When a flow becomes turbulent, chaos will develop inside the flow. The resulting flow contains both organized components and random fluctuations. An instantaneous velocity in a turbulent flow can thus be expressed as the sum of two contributions:

$$\mathbf{u}(\mathbf{x}, t) = \langle \mathbf{u}(\mathbf{x}, t) \rangle + \mathbf{u}'(\mathbf{x}, t) \tag{2.28}$$

where $\langle \mathbf{u}(\mathbf{x}, t) \rangle$ is the organized velocity field that can be resolved by the repetition of flow sampling and it is also called ensemble average velocity; $\mathbf{u}'(\mathbf{x}, t)$ is the velocity fluctuation that is random for a particular sample but has certain statistical characteristics from a large number of samplings for the same problem.

Usually, $\langle \mathbf{u}(\mathbf{x}, t) \rangle$ covers the large-scale mean flow motion, whereas $\mathbf{u}'(\mathbf{x}, t)$ spans a wide spectrum of length scales from those close to the mean motion down to those at the Kolmogorov scale, in which the turbulence energy is dissipated (Tennekes and Lumley, 1972). Between the large-scale mean flow motion and the smallest turbulence scale are the energy cascades that are in the form of eddies in various sizes. This range of flow is classified as the inertial subrange, which serves as a bridge to pass the mean flow energy to the smallest turbulence eddies on which the viscous effect acts to dissipate the kinetic energy to heat. With the continuous reduction of the size of eddies in the inertial subrange, the turbulence becomes increasingly isotropic and similar to each other among various turbulent flows. Therefore, a turbulent flow is always 3D even though the mean flow can be 2D.

Compared to a laminar flow, a turbulent flow has stronger flow randomness, wider flow motion scales, larger capacity of momentum transfer, and greater energy dissipation rate. Despite the above facts, the original

NSEs are still valid to describe various turbulent flow motions. This consensus is the basis for different turbulence models to be discussed below.

2.3.2   *Direct numerical simulation*

The logic of direct numerical simulation (DNS) is simple and straightforward: since all types of fluid flows, laminar and turbulent, can be described by the NSEs, the numerical solution to a turbulent flow requires no special treatment except to solve the original NSEs accurately. To adequately capture a turbulent flow, however, the following requirements are needed:

1.   All turbulence structures, including the smallest Kolmogorov turbulence scale, must be adequately resolved by the numerical scheme.
2.   The numerical solution must be accurate enough to simulate the energy dissipation rate correctly (e.g., the numerical errors/dissipation should not overwhelm the actual energy dissipation).
3.   The proper statistical methods need to be used to analyze the numerical results in order to extract the turbulence information.

Based on Kolmogorov (1962), the smallest turbulence length scale (e.g., Kolmogorov η) can be estimated as:

$$\eta \sim \left(\frac{\nu^3}{\varepsilon}\right)^{1/4} \tag{2.29}$$

where ε is the dissipation rate of the turbulent flow that can be estimated by the dimensional analysis:

$$\varepsilon \sim \frac{U^3}{L} \tag{2.30}$$

This gives the ratio of the mean flow length scale L to the Kolmogorov scale η to be:

$$\frac{L}{\eta} \sim \left(\frac{L^3 U^3}{\nu^3}\right)^{1/4} = \mathrm{Re}^{3/4} \tag{2.31}$$

This ratio is also the minimum grid number required in one space to solve a turbulent flow. Therefore, the total number of grids in a 3D space is:

$$\left(\frac{L}{\eta}\right)^3 \sim \mathrm{Re}^{9/4} \tag{2.32}$$

For a problem with $\mathrm{Re} = 10^4$, the total number of grids is therefore 10^9, which is nearly the upper limit of the computer power at this stage.

This explains why DNS is applicable only to a limited number of practical problems with relatively small Re.

DNS offers an opportunity to a modeler to obtain an instantaneous flow field for a turbulent flow. The DNS results are analogous to high-precision experimental data, from which turbulence information can be extracted. Both ensemble average (for unsteady flows) and time average (for quasi-steady flows) can be employed to separate the mean quantities and random fluctuations. In order to obtain the statistical properties of flow turbulence, the Monte Carlo method is often used, which introduces random perturbation to the flow to trigger the turbulence generation in the simulation. Readers are referred to Pope (2000: 344) for more information on DNS.

2.3.3 Large eddy simulation

For most of the high Re turbulent flows, DNS is not a practical choice with the current computational power (probably it is also not an optimal choice for most engineering computations even when computer power catches up in the future). The natural thinking is that we only need to compute the resolvable large-scale turbulence and mean flow, while the unresolvable small-scale turbulence is modeled based on some closure models. This is the fundamental of large eddy simulation (LES), which can be regarded as a partial DNS because only part of the turbulence structure is directly calculated.

While a DNS provides information on the instantaneous flow field including both the mean flow and the full spectrum of turbulence motion, an LES truncates the computation of turbulence at some cutoff scale. In an LES calculation, the following division can be done:

$$\mathbf{u}(\mathbf{x}, t) = \langle \mathbf{u}(\mathbf{x}, t) \rangle + \mathbf{u}'(\mathbf{x}, t) = \langle \mathbf{u}(\mathbf{x}, t) \rangle + \mathbf{u}'_{\text{res}}(\mathbf{x}, t) + \mathbf{u}'_{\text{non-res}}(\mathbf{x}, t) \qquad (2.33)$$

and LES only captures $\langle \mathbf{u}(\mathbf{x}, t) \rangle + \mathbf{u}'_{\text{res}}(\mathbf{x}, t)$.

Obviously, both $\mathbf{u}'_{\text{res}}(\mathbf{x}, t)$ and $\mathbf{u}'_{\text{non-res}}(\mathbf{x}, t)$ vary with the change of the cut-off scale, which is often set to be the same as the mesh resolution. As the turbulence scale becomes smaller, the turbulence becomes more isotropic and self-similar in the inertial subrange. This implies that the universal form of a closure model for small-scale turbulence modeling may be possible, as long as the unresolvable turbulence structure is small enough. One of the earliest subgrid scale (SGS) models was proposed by Smagorinsky (1963). In his approach, the original NSEs are first transformed into the following form by applying the spatial average filter that has a size equivalent to the mesh size:

$$\frac{\partial \bar{u}_i}{\partial x_i} = 0 \qquad (2.34)$$

$$\frac{\partial \bar{u}_i}{\partial t} + \bar{u}_j \frac{\partial \bar{u}_i}{\partial x_j} = -\frac{1}{\rho} \frac{\partial \bar{p}}{\partial x_i} + \frac{1}{\rho} \frac{\partial}{\partial x_j}(\bar{\tau}_{ij} + R_{ij}) \tag{2.35}$$

where $\bar{u}_i = \langle u_i \rangle + u'_{i\text{res}}$, $\bar{p} = \langle p \rangle + p'_{\text{res}}$, $\bar{\tau}_{ij} = 2\mu S_{ij}$ with $S_{ij} = (\partial \bar{u}_i/\partial x_j + \partial \bar{u}_j/\partial x_i)/2$, and the overbar represents the spatial average operation. $R_{ij} = -\rho \overline{u'_{i\text{non-res}} u'_{j\text{non-res}}}$ is the additional stress induced by the unresolved turbulence fluctuation. This term can be modeled by the eddy viscosity concept, i.e., the additional effect of turbulence mixing and energy dissipation can be modeled as a process similar to molecular viscous effect, but with a different viscosity (eddy viscosity) that is flow- and filter-dependent. In the SGS model, the closure model for this stress is as follows:

$$R_{ij} = 2\rho \nu_t S_{ij} = 2\rho (L_S^2 \sqrt{2S_{ij}S_{ij}}) S_{ij} \tag{2.36}$$

where L_S is the characteristic length scale that equals $C_s \Delta$, with $C_s \approx 0.17$ (Lilly, 1967) being the Smagorinsky coefficient. The mean filter size Δ can be related to the mesh size in a 3D problem by $\Delta = (\Delta_1 \Delta_2 \Delta_3)^{1/3}$, where Δ_1, Δ_2, and Δ_3 are the mesh sizes in three directions, respectively.

As Δ approaches zero, the value of R_{ij} also approaches zero. This implies that the effect of small turbulence becomes negligible to the resolved flow computation and LES returns to DNS. In contrast, as Δ increases, more turbulence needs to be modeled. This causes the difficulty in having a simple yet universal closure model because larger size turbulence is more anisotropic and mean flow-dependent that require more sophisticated closure models. When the mesh size Δ is out of the turbulence inertial subrange, the simple closure model (2.36) is no longer accurate. Researchers have proposed the so-called dynamic SGS model (e.g., Lilly, 1992) that intends to incorporate the information on the local resolved flow into the determination of the coefficient in (2.36) so that the closure model is more adaptable to various turbulent flows. The dynamic SGS can generally provide more accurate simulation results compared with the simple SGS model.

Note that since $\bar{u}_i = \langle u_i \rangle + u'_{i\text{res}}$ contains the turbulence flow structure $u'_{i\text{res}}$ which is generally 3D, an LES model is normally required to run under a 3D framework. An exception occurs only when the mean flow $\langle u_i \rangle$ is 2D and the filter size Δ is so chosen that $u'_{i\text{res}}$ is also primarily 2D.

2.3.4   Reynolds stress model

It is quite natural to ask the following question: What happens when Δ in LES approaches the largest turbulence scale, i.e., the length scale of the mean flow? In this case, all turbulence effects will be lumped into $u'_{i\text{non-res}}$ and $u'_{i\text{res}} = 0$. We then have $\bar{u}_i = \langle u_i \rangle$ and $u'_{i\text{res}} = u'_{i\text{non-res}}$. The term R_{ij} would therefore include all of the turbulence effects on the mean flow. In this case, R_{ij} is called Reynolds stress and historically it is derived by performing the ensemble average (or Reynolds average, represented by the notation of $\langle\rangle$ in the following equations) on the original NSEs:

$$\frac{\partial\langle u_i\rangle}{\partial x_i} = 0 \tag{2.37}$$

$$\frac{\partial\langle u_i\rangle}{\partial t} + \langle u_j\rangle\frac{\partial\langle u_i\rangle}{\partial x_j} = -\frac{1}{\rho}\frac{\partial\langle p\rangle}{\partial x_i} + \frac{1}{\rho}\frac{\partial}{\partial x_j}(\langle\tau_{ij}\rangle + R_{ij}) \tag{2.38}$$

where $R_{ij} = -\rho\langle u_i'u_j'\rangle$. The above equations are called Reynolds-averaged Navier–Stokes (RANS) equations or simply Reynolds equations. There are many ways of modeling R_{ij} and they are summarized below.

2.3.4.1 Reynolds stress transport model

Starting from the original NSEs (2.8) and (2.9), one may derive the transport equations for the Reynolds stresses divided by $-\rho$ (Launder *et al.*, 1975) as follows:

$$\begin{aligned}
\frac{\partial\langle u_i'u_j'\rangle}{\partial t} + u_k\frac{\partial\langle u_i'u_j'\rangle}{\partial x_k} = &-\frac{1}{\rho}\frac{\partial}{\partial x_k}(\langle u_j'p'\rangle\delta_{jk} + \langle u_j'p'\rangle\delta_{ik}) \\
&- \frac{\partial}{\partial x_k}\left[\langle u_i'u_j'u_k'\rangle - \nu\frac{\partial\langle u_i'u_j'\rangle}{\partial x_k}\right] \\
&- \left(\langle u_i'u_k'\rangle\frac{\partial\langle u_j\rangle}{\partial x_k} + \langle u_j'u_k'\rangle\frac{\partial\langle u_i\rangle}{\partial x_k}\right) \\
&+ \left\langle\frac{p'}{\rho}\left(\frac{\partial u_i'}{\partial x_j} + \frac{\partial u_j'}{\partial x_i}\right)\right\rangle - 2\nu\left\langle\frac{\partial u_i'}{\partial x_k} + \frac{\partial u_j'}{\partial x_k}\right\rangle
\end{aligned} \tag{2.39}$$

where δ_{ij} is the Kronecker delta. The LHS of the equation calculates the rate of change of Reynolds stress following the mean flow field. The first two rows on the RHS represent the total diffusion of Reynolds stress through the turbulent pressure work, turbulent fluxes, and molecular viscous force. The third row on the RHS denotes the production of Reynolds stress due to the work done by Reynolds stresses against the mean flow gradients. The first term of the last row on the RHS represents the interaction between the pressure fluctuation and the rate of strain of turbulence, which does not contribute to the total change of turbulence energy but redistributes the turbulence energy in different directions. The second term of the RHS is the tensor of energy dissipation rate ε_{ij} caused by the viscous effect.

The above transport equations for Reynolds stresses contain a few higher order correlation terms, i.e., diffusion terms, pressure–strain rate correlation term, and dissipation term, which need to be closed by certain models. The turbulence diffusion terms can be modeled by the gradient diffusion models. However, depending on different levels of approximation, the diffusion coefficients can be either anisotropic (Daly and Harlow, 1970; Hanjalic and Launder, 1972) or isotropic (Mellor and Herring, 1973). The pressure–strain rate correlation term contributes to the redistribution of turbulence energy, and thus, it is important for the characteristics of turbulence anisotropy. This term is normally modeled by the combination of Rotta's linear model of return-to-isotropy (Rotta, 1951; Tennekes and Lumley, 1972) for the slow pressure effect and some nonlinear models for the rapid pressure effect. In view of the different assumptions, at least five different closure models have been proposed. Demuren and Sarkar (1993) conducted the numerical tests with the use of different diffusion models and pressure–strain rate correlation models and concluded that no model can predict the turbulence characteristics completely satisfactorily when compared with experimental data and DNS data. A further systematical study is still necessary to improve these closure models, especially for complex flows.

With regard to the dissipation term, it is obtained by solving the transport equation in most approaches. First, the isotropic dissipation is assumed that makes $\varepsilon_{ij} = \frac{2}{3}\varepsilon\delta_{ij}$, where $\varepsilon = \nu\langle(\partial u_i'/\partial x_k)^2\rangle$. The transport equation for ε can also be derived from the NSEs with higher order correlation terms as follows:

$$
\frac{\partial \varepsilon}{\partial t} + u_j \frac{\partial \varepsilon}{\partial x_j} = -2\nu \left\langle \frac{\partial u_i'}{\partial x_k} \frac{\partial u_i'}{\partial x_j} \frac{\partial u_k'}{\partial x_j} \right\rangle - 2 \left\langle \nu \left(\frac{\partial^2 u_i'}{\partial x_j \partial x_k} \right)^2 \right\rangle
$$

$$
- \frac{1}{\rho} \frac{\partial}{\partial x_k} \left[\rho\nu \left\langle u_k' \left(\frac{\partial u_i'}{\partial x_j} \right)^2 \right\rangle + 2\nu \left\langle \frac{\partial u_k'}{\partial x_i} \frac{\partial \rho'}{\partial x_i} \right\rangle - \rho\nu \frac{\partial \varepsilon}{\partial x_k} \right]
$$

$$
- 2\nu \left(\left\langle \frac{\partial u_i'}{\partial x_j} \frac{\partial u_k'}{\partial x_j} \right\rangle + \left\langle \frac{\partial u_j'}{\partial x_i} \frac{\partial u_j'}{\partial x_k} \right\rangle \right) \frac{\partial \langle u_i \rangle}{\partial x_k} - 2\nu \left\langle u_k' \frac{\partial u_i'}{\partial x_j} \right\rangle \frac{\partial^2 \langle u_i \rangle}{\partial x_j \partial x_k}
$$

$$
\tag{2.40}
$$

The physical meaning of each term on the RHS of the equation is as follows. The first term represents the production by vortex stretching due to the turbulence vorticity, while the second term represents the viscous dissipation due to the spatial gradients of turbulence vorticity. The third term, which is in the form of spatial divergence, represents the molecular and turbulence diffusion of ε. The last two terms represent the production due to the interaction between the turbulence correlations and the mean velocity gradients.

To close up the problem, the RHS of (2.40) will be modeled and the final transport equation reads as follows:

$$
\frac{\partial \varepsilon}{\partial t} + u_j \frac{\partial \varepsilon}{\partial x_j} = \frac{\partial}{\partial x_j} \left[\left(\nu + \frac{\nu_t}{\sigma_\varepsilon} \right) \frac{\partial \varepsilon}{\partial x_j} \right] + C_{1\varepsilon} \frac{\varepsilon}{k} 2\nu_t S_{ij} \frac{\partial \langle u_i \rangle}{\partial x_j} - C_{2\varepsilon} \frac{\varepsilon^2}{k} \tag{2.41}
$$

in which σ_ε, $C_{1\varepsilon}$ and $C_{2\varepsilon}$ are empirical coefficients and $k = \frac{1}{2}\langle u_i' u_i' \rangle$ is the turbulence kinetic energy.

The system of equations (2.39) and (2.41) is closed and can be solved numerically. However, since the numerical solution to the Reynolds stress transport model is computationally expensive and the closure model for the pressure–strain rate correlation may contain large uncertainties, simpler and more robust turbulence models are often used instead.

2.3.4.2   The algebraic stress model

One of the simplified approaches is to use an algebraic equation rather than the transport equation to express six Reynolds stresses in a 3D turbulent flow. The model based on this approach is called the algebraic stress model (ASM; e.g., Rodi, 1972). The model has the advantage of having a simple algebraic expression. However, the correct physics may be lost in the modeling of complex turbulent flows. For this reason, such a model was not popularly adopted in general engineering computation.

Another approach that is popularly used to model the Reynolds stresses is the combination of the algebraic model and reduced transport equations. The algebraic model makes use of the eddy viscosity concept, in which the Reynolds stresses are related to the local rate of strain of the mean flow and the effective (eddy) viscosity as follows:

$$R_{ij} = -\rho \langle u_i' u_j' \rangle = 2\rho \nu_t S_{ij} - \frac{2}{3}\rho k \delta_{ij} \tag{2.42}$$

ν_t in (2.42) includes all turbulence scale effects and it is always greater than that in the LES model (2.36). To determine ν_t, which is mean flow-dependent, there are a few so-called two-equation closure models, namely the $k-\epsilon$ model, $k-w$ model, and $k-kl$ model.

2.3.4.3   The $k-\varepsilon$ model

The $k-\varepsilon$ model is the best well-known turbulence model. In this model, ν_t is related to k and ε as follows:

$$\nu_t = C_d \frac{k^2}{\varepsilon} \tag{2.43}$$

where C_d is an empirical coefficient:

$$C_d = 0.09 \tag{2.44}$$

To obtain k, we need to solve the transport equation for k that can be obtained by summing up the three normal Reynolds stress transport equations together:

$$\frac{\partial k}{\partial t} + u_j \frac{\partial k}{\partial x_j} = -\frac{1}{\rho}\frac{\partial}{\partial x_j}\left(\langle u_j' p'\rangle + \rho \langle u_j' k\rangle - \mu\frac{\partial k}{\partial x_j}\right) - \langle u_i' u_j'\rangle\frac{\partial\langle u_i\rangle}{\partial x_j} - \varepsilon \quad (2.45)$$

The above equation is simpler than (2.39) not only because the number of the equation is reduced from 6 to 1 but also because the pressure–strain rate correlation term disappears and the dissipation term becomes scalar. The diffusion term, the only unknown correlation term, can be modeled by the gradient diffusion again (Rodi, 1980). The final equation reads:

$$\frac{\partial k}{\partial t} + u_j \frac{\partial k}{\partial x_j} = \frac{\partial}{\partial x_j}\left[\left(\frac{\nu_t}{\sigma_k}+\nu\right)\frac{\partial k}{\partial x_j}\right] - \langle u_i' u_j'\rangle\frac{\partial\langle u_i\rangle}{\partial x_j} - \varepsilon \qquad (2.46)$$

in which σ_k is an empirical coefficient.

The variable ε above has the same physical meaning as defined before, and therefore, the same transport equation (2.41) used in the Reynolds stress transport model can still be used here. The coefficients in the k and ε transport equations have been determined by performing many simple experiments; the recommended values for these coefficients are (Rodi, 1980):

$$C_{1\varepsilon} = 1.44, \quad C_{2\varepsilon} = 1.92, \quad \sigma_\varepsilon = 1.3, \quad \sigma_k = 1.0 \qquad (2.47)$$

In the above traditional $k-\varepsilon$ model, the Reynolds stress is linearly proportional to the rate of strain, similar to Newtonian fluid flows. The immediate effect of such treatment is that the turbulence production, which is the product of Reynolds stress and the rate of strain of the mean flow, is always positive, meaning turbulence always extracts energy from the mean flow, which may not be true in certain circumstances. Besides, this linear relationship may not adequately represent the physics for anisotropic turbulence in complex turbulent flows. To resolve this problem, Pope (1975) proposed a more general nonlinear Reynolds stress model. In this model, the Reynolds stresses are the function of not only the linear terms of the strain rate of the mean flow but also the higher order terms. Shih et al. (1996) proposed a set of coefficients for all quadratic terms for this type of model and calibrated the coefficients using the turbulent flow over a step as follows:

$$\langle u_i' u_j'\rangle = \frac{2}{3}k\delta_{ij} - C_d\frac{k^2}{\varepsilon}\left(\frac{\partial\langle u_i\rangle}{\partial x_j} + \frac{\partial\langle u_j\rangle}{\partial x_i}\right)$$
$$-\frac{k^3}{\varepsilon^2}\left[C_1\left(\frac{\partial\langle u_i\rangle}{\partial x_l}\frac{\partial\langle u_l\rangle}{\partial x_j} + \frac{\partial\langle u_j\rangle}{\partial x_l}\frac{\partial\langle u_l\rangle}{\partial x_i} - \frac{2}{3}\frac{\partial\langle u_l\rangle}{\partial x_k}\frac{\partial\langle u_k\rangle}{\partial x_l}\delta_{ij}\right)\right]$$
$$-\frac{k^3}{\varepsilon^2}\left[C_2\left(\frac{\partial\langle u_i\rangle}{\partial x_k}\frac{\partial\langle u_j\rangle}{\partial x_k} - \frac{1}{3}\frac{\partial\langle u_l\rangle}{\partial x_k}\frac{\partial\langle u_l\rangle}{\partial x_k}\delta_{ij}\right)\right]$$

$$-\frac{k^3}{\varepsilon^2}\left[C_3\left(\frac{\partial\langle u_k\rangle}{\partial x_i}\frac{\partial\langle u_k\rangle}{\partial x_j}-\frac{1}{3}\frac{\partial\langle u_l\rangle}{\partial x_k}\frac{\partial\langle u_l\rangle}{\partial x_k}\delta_{ij}\right)\right] \tag{2.48}$$

in which C_1, C_2, and C_3 are empirical coefficients. When $C_1 = C_2 = C_3 = 0$ the above model returns to the conventional linear eddy viscosity model. The values of these coefficients were further corrected by Lin and Liu (1998b) based on the Couette flow experiments and applied to the breaking wave studies:

$$C_1 = 0.0054, \quad C_2 = -0.0171, \quad C_3 = 0.0027 \tag{2.49}$$

Under the extreme flow conditions, model (2.48) may predict the unphysical situations, i.e., negative turbulence velocity in one direction or unbounded Reynolds stress component. To enforce the correct physics in complex flows, certain realizability requirements are necessary, which correct the coefficients in (2.49) to be (Lin and Liu, 1998b):

$$C_d = \frac{2}{3}\left(\frac{1}{7.4+S_{max}}\right), \quad C_1 = \frac{1}{185.2+D_{max}^2}, \quad C_2 = -\frac{1}{58.5+D_{max}^2},$$
$$C_3 = -\frac{1}{370.4+D_{max}^2} \tag{2.50}$$

where

$$S_{max} = \frac{k}{\varepsilon}\text{max}\left[\left|\frac{\partial\langle u_i\rangle}{\partial x_i}\right| \text{ (indices not summed)}\right]$$

and

$$D_{max} = \frac{k}{\varepsilon}\text{max}\left[\left|\frac{\partial\langle u_i\rangle}{\partial x_j}\right|\right].$$

The adoption of the above modifications will ensure the nonnegativity of turbulence velocity and bounded Reynolds stress. It is noted that all coefficients will return back to their originally proposed values as in (2.44) and (2.49) when S_{max} and D_{max} approach zero.

According to Apsley et al. (1998), the employment of the nonlinear Reynolds stress model can greatly improve the accuracy of numerical results because of the fulfillment of more physical constraints. The nonlinear model not only captures most of the physics described by the Reynolds stress transport model but also retains the simple form of the $k-\varepsilon$ model. For simplicity, many nonlinear models have included only the quadratic terms (Shih et al., 1996; Lin and Liu, 1998b), which represent the most important nonlinear anisotropy characteristics of turbulence.

2.3.4.4   The k–ω model

The k–ω model is a turbulence model, especially applicable to turbulent boundary layer flows. The variable ω can be related to k and ε by $\omega = \varepsilon/(C_d k)$. The ω equation was originally proposed by Kolmogorov (1942), who employed a physical reasoning and dimensional arguments similar to those involved in the derivation of the ε equation. The idea was further extended by Wilcox (1988), who provided the following k–ω model popularly used today. In the model, the eddy viscosity is determined as follows:

$$v_{\rm t} = \frac{k}{\omega} \tag{2.51}$$

The transport equation for turbulence kinetic energy is expressed as follows:

$$\frac{\partial k}{\partial t} + u_j \frac{\partial k}{\partial x_j} = \frac{\partial}{\partial x_j}\left[(\sigma^* v_{\rm t} + v)\frac{\partial k}{\partial x_j}\right] + v_{\rm t}\left(\frac{\partial\langle u_i\rangle}{\partial x_j} + \frac{\partial\langle u_j\rangle}{\partial x_i}\right)\frac{\partial\langle u_i\rangle}{\partial x_j} - \beta^* k\omega \tag{2.52}$$

The equation for ω is:

$$\frac{\partial \omega}{\partial t} + u_j \frac{\partial \omega}{\partial x_j} = \frac{\partial}{\partial x_j}\left[(\sigma v_{\rm t} + v)\frac{\partial \omega}{\partial x_j}\right] + \alpha\frac{\omega}{k}v_{\rm t}\left(\frac{\partial\langle u_i\rangle}{\partial x_j} + \frac{\partial\langle u_j\rangle}{\partial x_i}\right)\frac{\partial\langle u_i\rangle}{\partial x_j} - \beta\omega^2 \tag{2.53}$$

where $\alpha = 5/9$, $\beta = 3/40$, $\beta^* = C_d = 9/100$, $\sigma = 1/2$, $\sigma^* = 1/2$. Later, the k–ω model was further modified by Speziale *et al.* (1992) and Wilcox (2004).

2.3.4.5   The k–kl model

Besides the k–ε model, the k–kl model is another two-equation model used in some engineering computations. In this model, instead of solving the transport equation for ε, the transport equation for the product of turbulence kinetic energy k and turbulence scale l, kl, is proposed and solved (Rotta, 1951). In the k–kl model, the eddy viscosity is closed by:

$$v_{\rm t} = k^{1/2}l \tag{2.54}$$

While the k equation is the same as that in the k–ε model, the turbulence length scale is tracked by solving the following equation:

$$\frac{\partial(kl)}{\partial t} + u_j \frac{\partial(kl)}{\partial x_j} = \frac{\partial}{\partial x_j}\left[v\frac{\partial}{\partial x_j}(kl) + (v_{\rm t}/\sigma_{L1})l\frac{\partial k}{\partial x_j} + (v_{\rm t}/\sigma_{L2})k\frac{\partial l}{\partial x_j}\right]$$
$$- C_{L1}l < u_i'u_j' > \frac{\partial\langle u_i\rangle}{\partial x_j} - C_{L2}k^{3/2} \tag{2.55}$$

where $C_{L1} = 0.98$, $C_{L2} = 0.059 + 702(l/y)^6$, $C_D = 0.09$, and $\sigma_k = \sigma_{L1} = \sigma_{L2} = 1$.

Although the turbulence length scale has a clear physical meaning, the direct measurement of it is almost impossible. For this reason, the k–kl model has received less attention than the k–ε and k–ω models. Generally speaking, the predictions from this model are comparable to the k–ε and k–ω predictions for simple constant pressure flows, but the model has not been pursued to any greater extent.

Note that there are other types of two-equation models, e.g., the k–ω^2 model (Saffman, 1970) and the k–τ model (Speziale *et al.*, 1992). These models are less popular and thus they shall not be discussed in further detail in this book.

2.3.4.6 *One-equation models (the k-equation model)*

In the k–ε two-equation model, the k equation can be exactly derived except for the turbulence diffusion term, which can be modeled by a simple gradient diffusion model. Compared to the k equation, the ε equation contains many assumptions in the closure model. The same deficiency applied to the second equation of the other two-equation models. It is fair to say that the accuracy of the k–ε model is mainly affected by the imperfect modeling of ε. Because of this, researchers have proposed the one-equation model in which only k is solved with the use of the transport equation, whereas the other physical quantity, which can be ε, l, or ω, is obtained by some simpler (yet reliable) algebraic turbulence closure models (e.g. Spalart and Allmaras, 1992).

2.3.4.7 *Zero-equation models (mixing-length hypothesis)*

When the characteristic turbulence scales can be explicitly related to physical conditions, the solution of the transport equation(s) may be unnecessary. In this case, the so-called zero-equation model can be applied. One of the earliest zero-equation models is Prandtl's mixing-length hypothesis for solving the turbulent boundary layer flow. This model lays the basis for the derivation of the log-law velocity profile in the turbulent boundary layer above a flat plate. In this approach, ν_t is modeled by the product of the turbulence velocity U_t and the characteristic length scale L_t that are functions of the local mean flow gradient $d\langle u \rangle / dz$ and the normal distance to the wall z:

$$\nu_t = L_t \cdot U_t = L_t \left(L_t \frac{d\langle u \rangle}{dz} \right) = (\kappa z)\left(\kappa z \frac{d\langle u \rangle}{dz} \right) = \kappa^2 z^2 \frac{d\langle u \rangle}{dz} \tag{2.56}$$

where κ is called the von Karman constant.

Since a zero-equation model is unable to account for the history effects (e.g., diffusion and convection) on turbulence, this model is not applicable for general transient turbulent flows. It, however, serves as a good wall turbulence model that is popularly used together with the k–ε model when the spatial resolution is insufficient to fully resolve the turbulent boundary layer near a wall. With the assumption (Liu and Lin, 1997) of local equilibrium, the turbulence wall boundary conditions are as follows:

$$k = \frac{u_*^2}{\sqrt{C_d}}, \quad \varepsilon = \frac{u_*^3}{\kappa z}, \quad \nu_t = \kappa u_* z \tag{2.57}$$

where u_* is the friction velocity and it is related to wall shear stress by $\tau_w = \rho u_*^2$. The friction velocity can be obtained by solving the log-law equation, given a one-point velocity in the boundary layer:

$$\langle u \rangle = \frac{u_*}{\kappa} \ln \frac{z}{z_0} \tag{2.58}$$

where z_0 is the coefficient related to local wall roughness and Re. It will be discussed in more detail in Section 3.14.1 when the current-induced boundary layer is discussed.

2.3.5   *The renormalization group theory*

Renormalization group (RG or RNG) is a term in theoretical physics, and it refers to the concepts and techniques related to the change of physics with the observation scale. RNG theory studies the phenomena of scale invariance. One good example of RNG theory is a turbulence flow for which the effective viscosity seems enhanced in the scale of the mean flow although the actual molecular viscosity remains the same. When RNG theory is applied to turbulence modeling, a simple iteration will be used to eliminate the highest wave number modes (i.e., the smallest turbulence scale) and replace their effect on the remaining flow by a small increase of effective viscosity. The resulting equations are rescaled (renormalized) to be "equivalent" to the original equations. The iteration will continue until the rescaled equations are identical between two successive iterations.

This idea shares a number of similarities to LES, except that the iterative procedure is used in the RNG to achieve a better representation of small-scale turbulence effect on the remaining flow. The pioneering work was done by Yakhot and Orszag (1986) for the development of the dynamic RNG method for hydrodynamic turbulence. The method was applied to a SGS turbulence model for LES. One of the major advantages of the RNG theory is that by scale expansion, the important turbulence coefficients can be theoretically determined rather than being adjusted empirically. This, of course, is at the cost of increased computational time in the iterative procedure. Later, Yakhot *et al.* (1992) extended the RNG to the development of the generalized $k-\varepsilon$ turbulence model and Reynolds stress transport model. The RNG analysis results in the same k equation but a modified ε equation:

$$\frac{\partial \varepsilon}{\partial t} + u_j \frac{\partial \varepsilon}{\partial x_j} = \frac{\partial}{\partial x_j} \left[\left(\nu + \frac{\nu_t}{\sigma_\varepsilon} \right) \frac{\partial \varepsilon}{\partial x_j} \right] + C_{1\varepsilon} \frac{\varepsilon}{k} 2 \nu_t S_{ij} \frac{\partial \langle u_i \rangle}{\partial x_j} - R - C_{2\varepsilon} \frac{\varepsilon^2}{k} \tag{2.59}$$

where R is an *ad hoc* model not derived from RNG analysis and it plays an important role in the modeling of turbulent flows. Combining a few earlier studies, Orszag *et al.* (1996) suggested the following values of the turbulence transport coefficients:

$$C_d = 0.0845, \quad C_{1\varepsilon} = 1.4, \quad C_{2\varepsilon} = 1.68, \quad \sigma_\varepsilon = 0.72, \quad \sigma_k = 0.72$$

$$(2.60)$$

2.3.6 Detached eddy simulation

As flow approaches a solid boundary (e.g., a wall), the turbulence length scale is reduced. In this case, when LES is used, the grid size must be sufficiently small to resolve the representative turbulence length scale. The computational cost will consequently be increased. To reduce the computing effort, a wall model that is used in the near-wall region can be combined with the LES model so that LES is performed only in the interior region away from the wall. More generally, a RANS model can be used in the region where the turbulence length scale is smaller than the grid size, and it is switched smoothly to LES in the region where the turbulence length scale is adequately resolved by the LES model. Such a model is called detached eddy simulation (DES) model, in which a single velocity field is smooth across the solution regions of the RANS and LES models (e.g. Strelets, 2001).

3   Water wave theories and wave phenomena

In this chapter, various wave theories will first be introduced to describe linear and nonlinear waves under the idealized situations of a flat bottom and nonbreaking free surface. It is followed by the introduction of realistic wave phenomena in nature and their engineering applications. In connection to each wave phenomenon, we will also briefly introduce various techniques of analysis. Considering that the numerical techniques will be further elaborated in later chapters, we shall focus the discussion in this chapter on the nonnumerical methods such as theoretical approaches and empirical approaches.

3.1   Linear wave theory

The flow region under a water wave train can be decomposed into two parts, namely, the bottom boundary layer, within which the fluid viscous effect is significant, and the flow outside the boundary layer, where the viscous effect is negligible. In the region where the viscous effect can be neglected, the flow is essentially irrotational unless there are other mechanisms (e.g., wave breaking or wave–structure interaction) to transport vorticity from the boundaries of the bottom, the structure, or the free surface. In this case, it is justified to assume that the flow is the potential flow outside the boundary layer.

The thickness of the laminar boundary layer δ induced by a wave train can be estimated by the following simple formula derived from the dimensional analysis:

$$\delta \sim O\left(\sqrt{\nu T}\right) \tag{3.1}$$

where ν is the kinematic viscosity of the fluid and T is the wave period. Given the typical value of $\nu = 1.0 \times 10^{-6}$ m²/s for water and $T = 2 - 30$ s (from short wind waves to longer storm waves), the corresponding boundary layer thickness ranges from 1.4 to 5.5 mm. In the coastal region where the water depth is from a few meters to a few tens of meters, the boundary layer region is much smaller than the entire flow region. Therefore, it is justified to assume that the water waves can be governed by the Laplace equation based on the potential flow theory.

3.1.1 Small-amplitude waves

Without loss of generality, we will look into the case where a 2D wave train is propagating on a flat bottom (constant water depth h) toward the positive x-direction and the crest line is in the y-direction (see illustration in Figure 3.1).

The governing equation and the boundary conditions are summarized as follows:

Governing equation:

$$\nabla^2 \phi = 0 \tag{3.2}$$

Bottom boundary condition:

$$w = -u\frac{\partial h}{\partial x} - \frac{\partial h}{\partial t} = 0 \text{ at } z = -h \tag{3.3}$$

Kinematic free surface boundary condition:

$$w = \frac{\partial \eta}{\partial t} + u\frac{\partial \eta}{\partial x} \text{ at } z = \eta(x, t) \tag{3.4}$$

Dynamic free surface boundary condition:

$$-\frac{\partial t}{\partial t} + \frac{1}{2}\left(u^2 + w^2\right) + \frac{p_\eta}{\rho} + gz = 0 \text{ at } z = \eta(x, t) \tag{3.5}$$

Lateral periodic boundary condition:

$$\phi(x, t) = \phi(x + L, t) \tag{3.6}$$

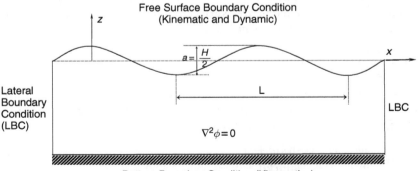

Figure 3.1 Illustration of a periodic wave train propagating on a flat bottom.

Temporal periodic condition:

$$\phi(x, t) = \phi(x, t + T) \tag{3.7}$$

where L is the wavelength, η is the free surface displacement measured from the still water depth where $z = 0$, and p_η is the atmospheric pressure on the free surface and is often set to be 0 (i.e., gauge pressure). It is noted that both the bottom and the free surface dynamic boundary conditions are derived from the general material surface preservation equation:

$$\frac{DF}{Dt} = 0 \Rightarrow \mathbf{u}.\mathbf{n} = -\frac{\partial F/\partial t}{|\nabla F|}$$

where $F(x,y,z,t) = 0$ represents the impermeable solid surface or the immiscible fluid surface.

The above governing equation (3.2) possesses the following analytical solution that satisfies all the constraints imposed by the conditions from (3.3) to (3.7) up to the first-order approximation of η, u, and w about $z = 0$:

$$\phi = -\frac{H}{2}\frac{g}{\sigma}\frac{\cosh k(h+z)}{\cosh kh}\sin(kx - \sigma t) \tag{3.8}$$

where $H = 2a$ is the wave height with a being the wave amplitude, σ is the wave angular frequency ($\sigma = 2\pi/T = 2\pi f$ with f being the wave frequency), and k is the wave number ($k = 2\pi/L$). By definition, the wave celerity (or wave phase velocity) can be calculated by $c = L/T = \sigma/k$. The validity of the solution relies on the assumptions of both $ka << 1$ (wave steepness; an important measure in deep water) and $a/h << 1$ (an important measure in shallow water; see proof in Appendix III) so that all the nonlinear terms (e.g., products of η, u, and w) can be neglected in the derivation. Therefore, besides the name of "linear wave theory," solution (3.8) also bears the name of "small-amplitude wave theory." The same theory sometimes is also referred to as the Airy (who made the earliest attempt to derive the theory of tides) wave theory or sinusoidal wave theory (the form of the expression).

From the linearized dynamic free surface boundary condition, the corresponding free surface displacement is found to be:

$$\eta = \frac{1}{g}\frac{\partial \phi}{\partial t} = \frac{H}{2}\cos(kx - \sigma t) \tag{3.9}$$

This represents a progressive wave (Figure 3.2), in which the wave form moves from the left to the right direction without change of wave shape.

The fluid particle velocities are found to be:

$$u = -\frac{\partial \phi}{\partial x} = \frac{H}{2}\sigma\frac{\cosh k(h+z)}{\sinh kh}\cos(kx - \sigma t) \tag{3.10}$$

$$w = -\frac{\partial \phi}{\partial z} = \frac{H}{2}\sigma\frac{\sinh k(h+z)}{\sinh kh}\sin(kx - \sigma t) \tag{3.11}$$

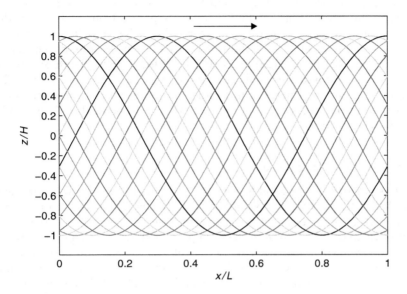

Figure 3.2 Illustration of a linear progressive wave train.

The local fluid particle accelerations are:

$$a_x = \frac{\partial u}{\partial t} = \frac{H}{2}\sigma^2 \frac{\cosh k(h+z)}{\sinh kh} \sin(kx - \sigma t) \tag{3.12}$$

$$a_z = \frac{\partial w}{\partial t} = -\frac{H}{2}\sigma^2 \frac{\sinh k(h+z)}{\sinh kh} \cos(kx - \sigma t) \tag{3.13}$$

The pressure under the wave can be obtained from the Bernoulli equation (2.23). It consists of two parts, namely the hydrostatic pressure that is independent of water waves and the dynamic pressure associated with the wave:

$$p = p_H + p_D = -\rho g z + \rho g \frac{H}{2} \frac{\cosh k(h+z)}{\cosh kh} \cos(kx - \sigma t) \tag{3.14}$$

With the use of kinematic free surface boundary condition, the angular wave frequency is related to the wave number and the local water depth by the wave dispersion equation:

$$\sigma^2 = gk \tanh kh \tag{3.15}$$

The above equation can be rewritten as:

$$c^2 = \frac{g}{k} \tanh kh \tag{3.16}$$

It implies that waves with different wave numbers (or wavelengths) will propagate at different speeds and therefore separate from each other (the effect of dispersion as light travels in a medium). By knowing any two of the three variables in (3.15), the third variable can be found. The most commonly encountered situation is that both local water depth and wave period are known and one needs to find the wave number. This needs to be done by iteration in general unless under some special circumstances such as short wave or long wave assumptions, which will be further discussed in the following sections.

3.1.2   Short and long waves

The wave theory presented in Section 3.1.1 represents a general linear wave train. When the wavelength becomes much shorter or longer than the local water depth, some special wave properties will be present and the wave theory can be simplified due to the asymptotic behavior of hyperbolic functions present in the linear wave theory. Table 3.1 gives a summary of linear wave theory in finite, deep, and shallow water depth. For both short and long waves, the dispersion equation can be simplified and therefore no iteration is needed when the equation is solved for k given σ and h.

Short waves: Short waves refer to waves with $kh >> 1$. Based on the asymptotic behavior of the hyperbolic tangential function in the dispersion equation (3.15), the most distinct feature of short waves is that the wave number is independent of the local water depth, i.e., $k = \sigma^2/g$. This is easy to understand because for a given wave period, the water depth will lose its effect on surface waves once it exceeds a certain threshold, beyond which further increase of water depth will not be "felt" by the wave train. This is a usual situation in deep oceans where the wind waves are generated. The effect of wind on the generation of waves will not reach the sea bottom. Another interesting observation for short waves is that all the wave-induced flow properties (i.e., velocity, dynamic pressure, and acceleration) decay exponentially toward the sea bottom where they all reduce to practically zero. The third feature of short waves is that the horizontal and vertical flow motions have the same magnitude but 90° of phase shift. This makes a circular orbit of the fluid particle under the wave.

Long waves: Long waves refer to waves whose wavelength is much larger than water depth (i.e., $kh << 1$). In contrast to short waves, the bottom effect now becomes a predominant factor for long waves. The dispersion equation can be reduced such that the wave phase velocity is dependent upon the local water depth only, i.e., $c = \sqrt{gh}$. This formula is the same as that used to calculate the propagation speed of disturbance in an open channel. The equation suggests that a group of waves with different wave periods will not "disperse" (separate) when they propagate in a shallow water region. Another observation for the long wave is that the horizontal velocity, acceleration, and dynamic pressure are uniform across the water

Table 3.1 Summary of wave parameters in finite, deep, and shallow water depths

Wave parameters	Finite water depth ($0 < kh < \infty$)	Deepwater depth ($kh \gg 1$)	Shallow water depth ($kh \ll 1$)
Velocity potential, ϕ	$\phi = -\dfrac{H}{2}\dfrac{g}{\sigma}\dfrac{\cosh k(h+z)}{\cosh kh}\sin(kx-\sigma t)$	$\phi = -\dfrac{H}{2}\dfrac{g}{\sigma}e^{kz}\sin(kx-\sigma t)$	$\phi = -\dfrac{H}{2}\dfrac{g}{\sigma}\sin(kx-\sigma t)$
Horizontal velocity, u	$u = \dfrac{H}{2}\sigma\dfrac{\cosh k(h+z)}{\sinh kh}\cos(kx-\sigma t)$	$u = \dfrac{H}{2}\sigma e^{kz}\cos(kx-\sigma t)$	$u = \dfrac{H}{2}\dfrac{\sigma}{kh}\cos(kx-\sigma t)$
Vertical velocity, w	$w = \dfrac{H}{2}\sigma\dfrac{\sinh k(h+z)}{\sinh kh}\sin(kx-\sigma t)$	$w = \dfrac{H}{2}\sigma e^{kz}\sin(kx-\sigma t)$	$w = \dfrac{H}{2}\sigma\left(1+\dfrac{z}{h}\right)\sin(kx-\sigma t)$
Local horizontal acceleration, a_x	$a_x = \dfrac{H}{2}\sigma^2\dfrac{\cosh k(h+z)}{\sinh kh}\sin(kx-\sigma t)$	$a_x = \dfrac{H}{2}\sigma^2 e^{kz}\sin(kx-\sigma t)$	$a_x = \dfrac{H}{2}\dfrac{\sigma^2}{kh}\sin(kx-\sigma t)$
Local vertical acceleration, a_z	$a_z = -\dfrac{H}{2}\sigma^2\dfrac{\sinh k(h+z)}{\sinh kh}\cos(kx-\sigma t)$	$a_z = -\dfrac{H}{2}\sigma^2 e^{kz}\cos(kx-\sigma t)$	$a_z = -\dfrac{H}{2}\sigma^2\left(1+\dfrac{z}{h}\right)\cos(kx-\sigma t)$
Dynamic pressure, p_D	$p_D = \rho g\dfrac{H}{2}\dfrac{\cosh k(h+z)}{\cosh kh}\cos(kx-\sigma t)$	$p_D = \rho g\dfrac{H}{2}e^{kz}\cos(kx-\sigma t)$	$p_D = \rho g\dfrac{H}{2}\cos(kx-\sigma t)$
Dispersion equation	$\sigma^2 = gk\tanh kh$	$\sigma^2 = gk$	$\sigma^2 = gk^2 h$
Wave celerity, c (phase velocity)	$c = \dfrac{\sigma}{k} = \dfrac{g}{\sigma}\tanh kh$	$c = \dfrac{\sigma}{k} = \dfrac{g}{\sigma} = \dfrac{gT}{2\pi}$	$c = \dfrac{\sigma}{k} = \sqrt{gh}$
Wave group velocity, c_g	$c_g = \left[\dfrac{1}{2}\left(1+\dfrac{2kh}{\sinh 2kh}\right)\right]c$	$c_g = \dfrac{c}{2}$	$c_g = c$
Wavelength, L	$L = cT = \dfrac{2\pi g}{\sigma^2}\tanh kh$	$L = cT = \dfrac{gT^2}{2\pi}$	$L = cT = T\sqrt{gh}$
Particle orbits	Elliptical orbit (horizontal > vertical)	Circular orbit	Linear orbit (horizontal only)

depth, whereas the vertical velocity and acceleration decay linearly from water surface to bottom, where both of them become zero. Because the vertical fluid motion under a long wave is much smaller than the horizontal fluid motion, the orbit of the fluid particle is reduced to a horizontal line.

3.1.3 Wave energy

The potential and kinetic energy contained in a wave train per unit horizontal area can be obtained by the integration of flow properties under a linear wave:

$$E_{\text{pot}} = \frac{1}{L}\int_0^L \rho g \frac{(h+\eta^2)}{2}\, dx - \rho g \frac{h^2}{2} = \frac{1}{16}\rho g H^2 \tag{3.17}$$

$$E_{\text{kin}} = \frac{1}{L}\int_0^L \int_{-h}^{\eta} \rho \frac{u^2+w^2}{2}\, dz\, dx = \frac{1}{16}\rho g H^2 \tag{3.18}$$

To calculate the wave-induced potential energy, the datum is selected on the bottom and the reference value of the potential energy of the fluid without the wave is subtracted from it. It is interesting to note that under a linear wave, the energy is equally distributed into potential and kinetic energy. The total energy is proportional to the square of the wave height, i.e.:

$$E = E_{\text{pot}} + E_{\text{kin}} = \frac{1}{8}\rho g H^2 = \frac{1}{2}\rho g a^2 \tag{3.19}$$

3.2 Nonlinear properties of linear waves

In linear wave theory, velocity, acceleration, and dynamic pressure are linearly proportional to the wave amplitude. The energy, however, does not follow this linear proportionality. Some interesting nonlinear wave properties can result from linear waves. In this section, the time-averaged nonlinear quantities, which are correct to the second order of wave steepness ka, will be derived from the linear wave theory.

3.2.1 Mean energy flux and group velocity

It is obvious that a wave train transmits energy in its propagation direction. Wave energy propagates at a speed that is different from the phase velocity. Based on the definition, the average energy flux, which is the product of the average wave energy defined above and the mean velocity at which the wave energy is transmitted, can be obtained as follows:

$$F = \frac{1}{T}\int_0^T \int_{-h}^{0} p_{\text{D}} u\, dz\, dt = E\left[\frac{1}{2}\left(1+\frac{2kh}{\sinh 2kh}\right)\right]c \tag{3.20}$$

Apparently, the speed at which the wave energy is transmitted is:

$$c_g = \left[\frac{1}{2}\left(1 + \frac{2kh}{\sinh 2kh}\right)\right] c \tag{3.21}$$

where c_g is the wave group velocity whose origin of the name will be discussed in Section 3.5.

It is not difficult to show that $c/2 \leq c_g \leq c$ in general; for long waves ($kh << 1$) we have $c_g = c$ and for short waves ($kh >> 1$) we have $c_g = c/2$. Since a wave train cannot propagate without the supply of wave energy, this implies that the wave front, behind which the wave energy is contained, will propagate at a generally slower speed than the following waves. Physically, this means that the following short waves that propagate at c will eventually catch up with the wave front that propagates at c_g and die out there. This is clearly shown in Figure 3.3 in which the wave propagates freely at c in the wave train before it reaches the wave front, where it disappears.

3.2.2 Mean mass flux and Stokes drift

A linear wave train not only transmits energy but also transports mass in its propagation direction. The trajectory of the particle motion under the wave can be found by taking the time integration of u and w. As a first-order approximation, with the use of the particle velocity at the center of the orbit

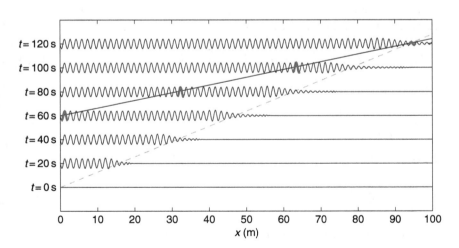

Figure 3.3 Propagation of a short wave train with $T = 1s$ in constant depth of $h = 1m$ ($kh = 4.03$); the solid and dashed lines represent the traces of wave profile and wave front following c and c_g; the highlighted wave profile with the thick line represents the same wave; computational results are from the time-dependent MSE model (Lin, 2004a).

motion, we are able to find the approximate particle displacement in x- and z-directions (ζ, ξ) as follows:

$$\zeta = -a\frac{\cosh k(h+z)}{\sinh kh}\sin(kx - \sigma t) = -\zeta_0 \sin(kx - \sigma t) \qquad (3.22)$$

$$\xi = a\frac{\sinh k(h+z)}{\sinh kh}\cos(kx - \sigma t) = \xi_0 \cos(kx - \sigma t) \qquad (3.23)$$

where ζ_0 and ξ_0 are particle displacement amplitudes. The fluid particle follows a closed-circle motion that starts from the forward motion under the crest, followed by a downward and then backward motion under the trough, and an upward and forward motion under the next crest.

Careful inspection of fluid particle motion reveals, however, that the magnitude of the forward velocity under the crest is always larger than the magnitude of the backward velocity under the trough, where the fluid particle is located at a lower elevation. This implies that after one wave period, the fluid particle will not come back exactly to its original location. Instead, it moves forward. This small forward movement is termed "Stokes drift" and it plays a pronounced role in long-term wave-induced mass transport. By taking the time average of the particle velocity following its near-orbit trajectory, we can find the expression for Stokes drift as:

$$u_d(z) = \frac{ga^2 k^2 \cosh 2k(h+z)}{\sigma \sinh 2kh} \qquad (3.24)$$

The mean mass flux induced by Stokes drift M_d can be found by the vertical integration of the above velocity:

$$M_d = \int_{-h}^{\eta} \rho u_d \, dz \simeq \int_{-h}^{0} \rho u_d \, dz = \frac{\rho g H^2 k}{8\sigma} = \frac{E}{c} \qquad (3.25)$$

The mean velocity of Stokes drift is then:

$$U_d = \frac{M_d}{\rho h} = \frac{g H^2}{8ch} \qquad (3.26)$$

In deep water, the equation for Stokes drift (3.24) can be reduced to:

$$u_d(z) = \frac{H^2}{4}\sigma k e^{2kz} \qquad (3.27)$$

3.2.3   Mean momentum flux and radiation stress

Consider a 2D free surface flow with negligible viscosity and turbulence. The momentum equation in the x-direction reads:

$$\frac{\partial u}{\partial t} + \frac{\partial (u^2)}{\partial x} + \frac{\partial (uw)}{\partial z} = -\frac{1}{\rho}\frac{\partial p}{\partial x} \qquad (3.28)$$

Integrating the equation in the vertical direction from $-h$ to η, we have:

$$\frac{\partial}{\partial t}\int_{-h}^{\eta} u\,dz + \frac{\partial}{\partial x}\int_{-h}^{\eta} u^2\,dz = -\frac{1}{\rho}\frac{\partial}{\partial x}\int_{-h}^{\eta} p\,dz + \frac{p_{-h}}{p}\frac{\partial h}{\partial x} \qquad (3.29)$$

In arriving at the above simple form, the Leibniz rule of integration is used, e.g.:

$$\int_{-h}^{\eta} \frac{\partial u}{\partial t}\,dz = \frac{\partial}{\partial t}\int_{-h}^{\eta} u\,dz - u(\eta)\frac{\partial \eta}{\partial t} + u(-h)\frac{\partial(-h)}{\partial t} \qquad (3.30)$$

In addition, the bottom boundary condition (3.3) and the kinematic free surface boundary condition (3.4) are also used to cancel out the resulting additional terms on the LHS. Furthermore, the pressure on the free surface is assumed to be zero (i.e., gauge pressure).

The pressure p in (3.29) can be obtained by taking the vertical integration of the momentum equation in the z-direction, i.e.:

$$\int_{-h}^{\eta} \frac{\partial w}{\partial t}\,dz + \int_{z}^{\eta} + \frac{\partial(uw)}{\partial x}\,dz + \int_{z}^{\eta} \frac{\partial(w^2)}{\partial z}\,dz = -\frac{1}{\rho}\int_{z}^{\eta} \frac{\partial p}{\partial z}\,dz - \int_{z}^{\eta} g\,dz$$

$$\Rightarrow \quad p = \rho g(\eta - z) - \rho w^2 + \rho\left(\frac{\partial}{\partial t}\int_{-h}^{\eta} w\,dz + \frac{\partial}{\partial x}\int_{-h}^{\eta} uw\,dz\right) \qquad (3.31)$$

Now let us consider a steady periodic wave train propagating over an uneven bottom in the x-direction (Figure 3.4). In this case, the flow motion contains wave motion only, i.e.:

$$u = \tilde{u}, \quad w = \tilde{w} \qquad (3.32)$$

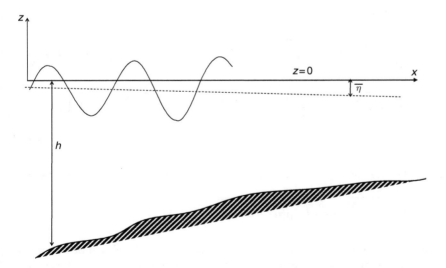

Figure 3.4 Illustration of a wave train propagating over an uneven bottom.

When the bottom slope is mild, the linear wave theory is still applicable as a first-order approximation. However, the wave height may vary slowly in space and the mean water level may deviate from the still water level $z = 0$. The free surface displacement can then be expressed as follows:

$$\eta = \bar{\eta} + a\cos(kx - \sigma t) \tag{3.33}$$

where the overbar denotes the time average over a wave period, i.e.:

$$\bar{\eta}(x) = \frac{1}{T}\int_0^T \eta(x, t)\,dt \tag{3.34}$$

By taking the time average of the depth-integrated momentum equation in the x-direction (3.29), we have:

$$\overline{\frac{\partial}{\partial t}\int_{-b}^{\eta} u\,dz} + \overline{\frac{\partial}{\partial x}\int_{-b}^{\eta} u^2\,dz} = -\frac{1}{\rho}\overline{\frac{\partial}{\partial x}\int_{-b}^{\eta} p\,dz} + \overline{\frac{p_{-b}}{\rho}\frac{\partial b}{\partial x}} \tag{3.35}$$

The periodicity of the dynamic pressure and velocity in time immediately reduces the above equation to:

$$\frac{\partial}{\partial x}\overline{\int_{-b}^{\eta} u^2\,dz} = -\frac{1}{\rho}\frac{\partial}{\partial x}\overline{\int_{-b}^{\eta} p\,dz} + g(b + \bar{\eta})\frac{\partial b}{\partial x} \tag{3.36}$$

The last term is the mean reaction force from the bottom. Now if we substitute the definition of p in (3.31) in (3.36), we have:

$$\frac{\partial}{\partial x}\overline{\int_{-b}^{\eta} u^2\,dz}$$

$$= -\frac{1}{\rho}\frac{\partial}{\partial x}\overline{\int_{-b}^{\eta}\left(\rho g(\eta - z) - \rho w^2 + \rho\left(\frac{\partial}{\partial t}\int_{-b}^{\eta} w\,dz + \frac{\partial}{\partial x}\int_{-b}^{\eta} uw\,dz\right)\right)dz} + g(b + \bar{\eta})\frac{\partial b}{\partial x}$$

$$= -g(b + \eta)\frac{\overline{\partial(b + \eta)}}{\partial x} + \frac{\partial}{\partial x}\overline{\int_{-b}^{\eta} w^2\,dz} - \frac{\partial}{\partial x}\overline{\int_{-b}^{\eta}\left(\frac{\partial}{\partial t}\int_{-b}^{\eta} w\,dz + \frac{\partial}{\partial x}\int_{-b}^{\eta} uw\,dz\right)dz}$$

$$+ g(b + \bar{\eta})\frac{\partial b}{\partial x} \tag{3.37}$$

By neglecting the third term on the RHS of the above equation, we have:

$$-\rho g(b + \bar{\eta})\frac{\overline{\partial\bar{\eta}}}{\partial x} = \frac{\partial}{\partial x}\overline{\int_{-b}^{\eta}\rho(u^2 - w^2)\,dz} \tag{3.38}$$

This equation implies that the change of mean water level will be balanced by the gradient of wave-induced time-averaged stresses on the RHS. To further simplify the expression, we shall make the following assumptions:

(a) For small amplitude waves, the term on the RHS in the order of $O(a^3)$ is negligible so that:

$$\frac{\partial}{\partial x}\overline{\int_{-h}^{\eta}\rho u^2 dz} = \frac{\partial}{\partial x}\overline{\int_{-h}^{0}\rho u^2 dz} + \frac{\partial}{\partial x}\overline{\int_{0}^{\eta}\rho u^2 dz} = \frac{\partial}{\partial x}\overline{\int_{-h}^{0}\rho u^2 dz}$$

$$+ \frac{\partial}{\partial x}O(a^3) \approx \frac{\partial}{\partial x}\overline{\int_{-h}^{0}\rho u^2 dz} \tag{3.39}$$

(b) The pressure distribution from the still water level to the free surface can be approximated by hydrostatic pressure, i.e., $p = \rho g(\eta - z)$ and thus $\rho w^2 \sim 0$.

This gives:

$$\frac{\partial}{\partial x}\overline{\int_{-h}^{\eta}\rho w^2 dz} = \frac{\partial}{\partial x}\overline{\int_{-h}^{0}\rho w^2 dz} + \frac{\partial}{\partial x}\overline{\int_{0}^{\eta}\rho w^2 dz} \approx \frac{\partial}{\partial x}\overline{\int_{-h}^{0}\rho w^2 dz} \tag{3.40}$$

This reduces (3.38) to:

$$-\rho g(h + \bar\eta)\frac{\partial \bar\eta}{\partial x} = \frac{\partial}{\partial x}\left[\overline{\int_{-h}^{0}\rho u^2\, dz} - \overline{\int_{-h}^{0}\rho w^2\, dz} + \left(\frac{1}{2}\rho g\overline{\eta^2}\right)\right] \tag{3.41}$$

The first term on the RHS is identified as the excess depth-integrated momentum flux in the wave propagation direction and the second and third terms are the contributions from the wave-induced dynamic pressure that acts in all directions. The sum of these three terms is called "radiation stress" (Longuet-Higgins and Stewart, 1964) that can be readily evaluated as follows:

$$S_{xx} = S_{xx}{}^{(1)} + S_{xx}{}^{(2)} + S_{xx}{}^{(3)} = \overline{\int_{-h}^{0}\rho u^2\, dz} + \left(-\overline{\int_{-h}^{0}\rho w^2\, dz}\right) + \left(\frac{1}{2}\rho g\overline{\eta^2}\right)$$

$$= \left(\frac{kh}{\sinh 2kh} + \frac{1}{2}\right)E + \left(\frac{kh}{\sinh 2kh} - \frac{1}{2}\right)E + \frac{1}{2}E = \left(\frac{2kh}{\sinh 2kh} + \frac{1}{2}\right)E \tag{3.42}$$

In the direction orthogonal to wave propagation, the radiation stress contains only pressure contribution:

$$S_{yy} = \left(\frac{kh}{\sinh 2kh} - \frac{1}{2}\right)E + \frac{1}{2}E = \left(\frac{kh}{\sinh 2kh}\right)E \tag{3.43}$$

Note that pressure makes no contribution to the depth-averaged shear radiation stress, which in this case is zero because the transverse velocity v equals zero:

$$S_{xy} = 0 \tag{3.44}$$

The above definition can be extended to the general case where the wave train propagates at an angle θ to the x-axis:

$$
\begin{aligned}
S_{xx} &= \left(\frac{kh}{\sinh 2kh} + \frac{1}{2}\right) E \cos^2\theta + \left(\frac{kh}{\sinh 2kh} - \frac{1}{2}\right) E + \frac{1}{2}E \\
&= \left[(1 + \cos^2\theta)\frac{kh}{\sinh 2kh} + \frac{1}{2}\cos^2\theta\right] E
\end{aligned}
\tag{3.45}
$$

$$
\begin{aligned}
S_{yy} &= \left(\frac{kh}{\sinh 2kh} + \frac{1}{2}\right) E \sin^2\theta + \left(\frac{kh}{\sinh 2kh} - \frac{1}{2}\right) E + \frac{1}{2}E \\
&= \left[(1 + \sin^2\theta)\frac{kh}{\sinh 2kh} + \frac{1}{2}\sin^2\theta\right] E
\end{aligned}
\tag{3.46}
$$

$$
S_{xy} = \left(\frac{kh}{\sinh 2kh} + \frac{1}{2}\right) E \sin\theta\cos\theta = \left(\frac{kh}{2\sinh 2kh} + \frac{1}{4}\right) \sin 2\theta E
\tag{3.47}
$$

In an even simpler notation, the radiation stresses are expressed in the tensor form as (Phillips, 1977):

$$
S_{ij} = E\frac{c_g}{c}\frac{k_i k_j}{|k|^2} + \frac{E}{2}\left(\frac{2c_g}{c} - 1\right)\delta_{ij}
\tag{3.48}
$$

where k_i is the wave number in the ith direction.

The radiation stress concept is important in understanding coastal processes, i.e., wave set-down and set-up (Section 3.2.4), generation of bound infragravity waves (Section 3.13.2), and generation of longshore current (Section 3.14.2.4). Recently, efforts have been made to derive 3D radiation stress (e.g., Lin and Zhang, 2004). In this section, a brief discussion of the general idea of the depth-dependent radiation stress is presented and further discussion will be addressed in Section 3.14.3.2 when wave–current interaction is considered. Considering again a wave train propagating over an uneven bottom in the x-direction and performing the time average directly (without depth averaging) on the same horizontal momentum equation (3.28), we have:

$$
\frac{\partial \bar{u}}{\partial t} + \frac{\overline{\partial(u^2)}}{\partial x} + \frac{\overline{\partial(uw)}}{\partial z} = -\frac{1}{\rho}\frac{\overline{\partial p}}{\partial x}
\tag{3.49}
$$

By substituting p defined in (3.31) into the above equation and realizing the periodicity of u under the wave, the above equation becomes:

$$\frac{\partial \overline{(u^2)}}{\partial x} + \frac{\partial \overline{(uw)}}{\partial z} = -\frac{1}{\rho}\frac{\partial}{\partial x}\left[\rho g(\eta - z) - \rho w^2 + \rho\left(\frac{\partial}{\partial t}\int_{-b}^{\eta} w\,dz + \frac{\partial}{\partial x}\int_{-b}^{\eta} uw\,dz\right)\right]$$

$$\approx -g\frac{\partial \bar{\eta}}{\partial x} + \frac{\partial \overline{w^2}}{\partial x} \Rightarrow -\rho g\frac{\partial \bar{\eta}}{\partial x} = \frac{\partial \overline{\rho\,(u^2 - w^2)}}{\partial x} + \frac{\partial \overline{\rho\,(uw)}}{\partial z}$$

$$(3.50)$$

Again the last term for pressure is neglected. The above equation states that the local wave-induced stress gradient must be balanced by the mean water level gradient. Based on the linear wave theory, we are ready to obtain the depth-dependent wave-induced normal stress as:

$$W_{xx}{}^{(1)} = -\rho\overline{(u^2 - w^2)} = -\left[\rho\overline{(u^2)} - \rho\overline{(w^2)}\right]$$

$$= -2kE\frac{\cosh^2 k(b+z)}{\sinh 2kb} + 2kE\frac{\sinh^2 k(b+z)}{\sinh 2kb}$$

$$= -\frac{2kE}{\sinh 2kb}$$

$$(3.51)$$

and the shear stress for irrotational wave (e.g. Rivero and Arcilla, 1995) as:

$$W_{xz} = -\rho\overline{(uw)} = \frac{2kE}{\sinh 2kb}\frac{\partial b}{\partial x} + (z+b)\frac{\partial}{\partial x}\left(\frac{kE}{\sinh 2kb}\right)$$

$$(3.52)$$

The wave-induced shear stress becomes zero when the bottom is flat and there is no change of wave amplitude in space. Such a conclusion can also be drawn from the fact that u and w have 90° phase shift for nondissipative waves on a flat bottom.

Interestingly, the wave-induced normal stress is constant across the depth due to the cancellation of $\overline{u^2}$ and $\overline{w^2}$. Integrating W_{xx} from the bottom to $z = 0$, where the time-averaged flow computation ends, the first two terms of the depth-averaged radiation stresses in (3.42), i.e., $S_{xx}{}^{(1)} + S_{xx}{}^{(2)}$, can be recovered. The term $S_{xx}{}^{(3)}$ that comes from the pressure contribution from $z = 0$ to free surface, however, cannot be resumed from the above expression. By realizing that $S_{xx}{}^{(3)}$ essentially takes into account the mean pressure effect of the orbit flow motion under the wave, Lin and Zhang (2004) proposed the depth-dependent correction term arising from the Lagrangian mean of the vertical fluid particle motion under the wave. The method is similar to the generalized Lagrangian mean (GLM) (Andrews and McIntyre, 1978) that can be used to obtain the depth-dependent Stokes drift except that this time it is applied to mean pressure contribution. The correction term takes the following form:

$$W_{xx}{}^{(2)} = -\frac{Ek\sinh 2k(z+b)}{2\sinh^2 kb}$$

$$(3.53)$$

The above additional normal stress reduces linearly to the bottom for small kh (shallow water) but exponentially for large kh (deep water). Under such a correction, the radiation stress appearing in (3.42) can be fully recovered by taking the depth integration of $-W_{xx}^{(1)} - W_{xx}^{(2)}$ from $-h$ to 0.

Thus, when applied to the region up to the mean water level with the presence of waves, equation (3.50) will be rewritten as follows (when the time-averaged flow computation goes only to the mean water level rather than the maximum height of the free surface):

$$-\rho g \frac{\partial \overline{\eta}}{\partial x} = -\frac{\partial}{\partial x}\left[W_{xx}^{(1)} + W_{xx}^{(2)}\right] - \frac{\partial W_{xz}}{\partial z}$$

$$= \frac{\partial}{\partial x}\left(\frac{kE}{\sinh 2kh} + \frac{Ek \sinh 2k(z+h)}{2 \sinh^2 kh}\right) \qquad (3.54)$$

3.2.4   Set-down and set-up of mean water level

Consider a wave train that propagates toward a plane beach with the normal incidence in the x-direction. As the local water depth reduces, both the wave number [based on the dispersion equation (3.15)] and the wave height (based on the shoaling formula to be explained in Section 3.6) will increase, causing an increase of radiation stress. The gradient of radiation stress must be balanced by the change of local mean water depth based on (3.41):

$$-\frac{1}{\rho g h}\frac{\mathrm{d}S_{xx}}{\mathrm{d}x} = \frac{\mathrm{d}\overline{\eta}}{\mathrm{d}x} \qquad (3.55)$$

The general conclusion we can draw here is that with the increase in wave height and thus S_{xx} toward the breaking point, the mean water level will be drawn down below the still water level $z = 0$. This process is called wave set-down. After the waves break, the wave height will decay as the result of energy dissipation and therefore the radiation stress will decrease. Consequently, the mean water level rises shoreward from the breaking point. This process is referred to as wave set-up.

3.3   Nonlinear wave theory

As indicated earlier, linear wave theory is constructed on the assumption of $ka << 1$ and $a/h << 1$. When wave amplitude increases beyond a certain range, the linear wave theory may become inadequate. The reason is that those higher order terms that have been neglected in the derivation become increasingly important as wave amplitude increases. Nonlinear wave theories are required for describing large-amplitude waves.

3.3.1 Stokes wave theory

The immediate extension of the linear wave theory is to retain all the second-order terms during the derivation. This will end up with the so-called second-order Stokes wave theory. The theory stipulates that for larger amplitude waves, the wave profile is the sum of two sinusoidal waves, one of which is the same as that obtained from the linear wave theory and the other from the second-order correction terms, i.e.:

$$\eta = \frac{H}{2}\cos(kx-\sigma t) + \frac{H^2 k}{16}\frac{\cosh kh}{\sinh^3 kh}(2+\cosh 2kh)\cos 2(kx-\sigma t) \quad (3.56)$$

The velocity, acceleration, and dynamic pressure are as follows:

$$u = \frac{Hgk}{2\sigma}\frac{\cosh k(h+z)}{\cosh kh}\cos(kx-\sigma t) + \frac{3H^2\sigma k}{16}\frac{\cosh 2k(h+z)}{\sinh^4 kh}\cos 2(kx-\sigma t)$$

$$(3.57)$$

$$w = \frac{Hgk}{2\sigma}\frac{\sinh k(h+z)}{\cosh kh}\sin(kx-\sigma t) + \frac{3H^2\sigma k}{16}\frac{\sinh 2k(h+z)}{\sinh^4 kh}\sin 2(kx-\sigma t)$$

$$(3.58)$$

$$a_x = \frac{\partial u}{\partial t} = \frac{Hgk}{2}\frac{\cosh k(h+z)}{\cosh kh}\sin(kx-\sigma t)$$

$$+ \frac{3H^2\sigma^2 k}{8}\frac{\cosh 2k(h+z)}{\sinh^4 kh}\sin 2(kx-\sigma t) \quad (3.59)$$

$$a_w = \frac{\partial w}{\partial t} = -\frac{Hgk}{2}\frac{\sinh k(h+z)}{\cosh kh}\cos(kx-\sigma t)$$

$$- \frac{3H^2\sigma^2 k}{8}\frac{\sinh 2k(h+z)}{\sinh^4 kh}\cos 2(kx-\sigma t) \quad (3.60)$$

$$p_D = \rho g\frac{H}{2}\frac{\cosh k(h+z)}{\cosh kh}\cos(kx-\sigma t)$$

$$+ \frac{3}{8}\rho gkH^2\frac{1}{\sinh 2kh}\left[\frac{\cosh 2k(h+z)}{\sinh^2 kh} - \frac{1}{3}\right]\cos 2(kx-\sigma t)$$

$$- \frac{1}{8}\rho gkH^2\frac{1}{\sinh 2kh}[\cosh 2k(h+z) - 1] \quad (3.61)$$

Note that for second-order Stokes waves, the linear dispersion equation (3.15) remains valid. Figure 3.5 shows the comparison of a linear wave and a second-order Stokes wave for the same given wave parameters.

The second-order Stokes wave theory may develop a secondary crest at the wave trough in shallow water. To prevent this from occurring, the following constraint is imposed:

$$\frac{L^2 H}{h^3} < \frac{8\pi^2}{3} \quad (3.62)$$

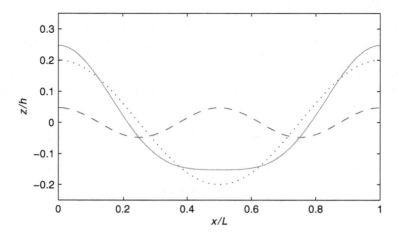

Figure 3.5 Comparison of a linear wave (dotted line) and second-order Stokes
wave (solid line) for the wave with $H = 0.4\,\text{m}$, $h = 1.0\,\text{m}$, and $T = 2.0\,\text{s}$; the difference is the second-order wave mode (dashed line).

The term on the LHS of the expression is called the Ursell (1953) parameter,
which is essentially the ratio of relative wave height ($\alpha = H/h$) and the square
of the relative water depth ($\beta = h^2/L^2$). While H/h represents nonlinear
effect, h/L represents dispersive effect.

As the wave height increases further, more nonlinear terms are needed to
be retained to represent the physics of nonlinear waves correctly. Besides the
second-order Stokes waves, there also exist third-, fourth-, fifth-, and higher-
order Stokes waves. In the third-order Stokes waves, the linear dispersion
equation becomes invalid and the correction must be included in the dispersion equation to reflect the wave height effect on wave dispersion. For most
of the engineering computations, the fifth-order Stokes waves are sufficient
(Skjelbreia and Hendrickson, 1961):

$$\eta = \sum_{i=1}^{5} a_i \cos(i\theta) = \frac{1}{k}[\lambda \cos\theta + (\lambda^2 B_{22} + \lambda^4 B_{24}) \cos 2\theta$$

$$+ (\lambda^3 B_{33} + \lambda^5 B_{35}) \cos 3\theta + \lambda^4 B_{44} \cos 4\theta + \lambda^5 B_{55} \cos 5\theta] \quad (3.63)$$

where k is the wave number and λ is the wave steepness coefficient ($\lambda = kH/2$ for a linear wave), the determination of which will be described later.
In the above equation, for simplicity, θ is introduced and represents the
phase angle ($= kx - \sigma t$). The coefficients B in the above equation can be
obtained from the following equations:

$$B_{22} = \frac{(2c_h^2 + 1)\,c_h}{4s^3} \quad (3.64)$$

$$B_{24} = \frac{(272c_h^8 - 504c_h^6 - 192c_h^4 + 322c_h^2 + 21)\,c_h}{384s_h^9} \quad (3.65)$$

$$B_{33} = \frac{3\left(8c_h{}^6 + 1\right)}{64 s_h{}^6} \tag{3.66}$$

$$B_{35} = \frac{88128 c_h{}^{14} - 208224 c_h{}^{12} + 70848 c_h{}^{10} + 54000 c_h{}^8 - 21816 c_h{}^6 + 6264 c_h{}^4 - 54 c_h{}^2 - 81}{12288 s_h{}^{12}\left(6 c_h{}^2 - 1\right)} \tag{3.67}$$

$$B_{44} = \frac{\left(768 c_h{}^{10} - 448 c_h{}^8 - 48 c_h{}^6 + 48 c_h{}^4 + 106 c_h{}^2 - 21\right) c_h}{384 s_h{}^9\left(6 c_h{}^2 - 1\right)} \tag{3.68}$$

$$B_{55} = \frac{192000 c_h{}^{16} - 262720 c_h{}^{14} + 83680 c_h{}^{12} + 20160 c_h{}^{10} - 7280 c_h{}^8 + 7160 c_h{}^6 - 1800 c_h{}^4 - 1050 c_h{}^2 + 225}{12288 s_h{}^{10}\left(6 c_h{}^2 - 1\right)\left(8 c_h{}^4 - 11 c_h{}^2 + 3\right)} \tag{3.69}$$

For simplicity, we denote $s_h = \sinh(kh)$ and $c_h = \cosh\,(kh)$. The nonlinear wave height and the dispersion equation are functions of k and λ:

$$H = \frac{2}{k}\left[\lambda + B_{33}\lambda^3 + (B_{35} + B_{55})\lambda^5\right] \tag{3.70}$$

$$\sigma^2 = gk \tanh kh\left(1 + C_1\lambda^2 + C_2\lambda^4\right) \tag{3.71}$$

where the expressions for the coefficients C_1 and C_2 are:

$$C_1 = \frac{8 c_h{}^4 - 8 c_h{}^2 + 9}{8 s_h{}^4} \tag{3.72}$$

$$C_2 = \frac{3840 c_h{}^{12} - 4096 c_h^{10} + 2592 c_h{}^8 - 1008 c_h{}^6 + 5944 c_h{}^4 - 1830 c_h{}^2 + 147}{512 s_h{}^{10}\left(6 c_h{}^2 - 1\right)} \tag{3.73}$$

The iteration scheme (e.g., Newton–Raphson method) can be used to solve the system of the nonlinear equations (3.70) and (3.71) to obtain the values of k and λ, given H, h, and T (see Appendix IV).

The velocities under the fifth-order Stokes waves are expressed as follows:

$$\begin{aligned}
u =\,& c\{\lambda(A_{11} + \lambda^3 A_{13} + \lambda^5 A_{15})\cosh\left[k(h+z)\right]\cos\theta \\
& + 2(\lambda^2 A_{22} + \lambda^4 A_{24})\cosh\left[2k(h+z)\right]\cos 2\theta \\
& + 3(\lambda^3 A_{33} + \lambda^5 A_{35})\cosh\left[3k(h+z)\right]\cos 3\theta \\
& + 4\lambda^4 A_{44}\cosh\left[4k(h+z)\right]\cos 4\theta + \lambda 5^5 A_{55}\cosh\left[5k(h+z)\right]\cos 5\theta\}
\end{aligned}$$
$$\tag{3.74}$$

$$\begin{aligned}
w =\,& c\{(\lambda A_{11} + \lambda^3 A_{13} + \lambda^5 A_{15})\sinh\left[k(h+z)\right]\sin\theta \\
& + 2(\lambda^2 A_{22} + \lambda^4 A_{24})\sinh\left[2k(h+z)\right]\sin 2\theta \\
& + 3(\lambda^3 A_{33} + \lambda^5 A_{35})\sinh\left[3k(h+z)\right]\sin 3\theta \\
& + 4\lambda^4 A_{44}\sinh\left[4k(h+z)\right]\sin 4\theta + \lambda 5^5 A_{55}\sinh\left[5k(h+z)\right]\sin 5\theta\}
\end{aligned}$$
$$\tag{3.75}$$

where the coefficients A are expressed as follows:

$$A_{11} = \frac{1}{s_h} \tag{3.76}$$

$$A_{13} = -\frac{c_h^2 \left(5c_h^2 + 1\right)}{8s_h^5} \tag{3.77}$$

$$A_{15} = -\frac{1184c_h^{10} - 1440c_h^8 - 1992c_h^6 + 2641c_h^4 - 249c_h^2 + 18}{1536s_h^{11}} \tag{3.78}$$

$$A_{22} = \frac{3}{8s_h^4} \tag{3.79}$$

$$A_{24} = \frac{192c_h^8 - 424c_h^6 - 312c_h^4 + 480c_h^2 - 17}{768s_h^{10}} \tag{3.80}$$

$$A_{33} = \frac{13 - 4c_h^2}{64s_h^7} \tag{3.81}$$

$$A_{35} = \frac{512c_h^{12} + 4224c_h^{10} - 6800c_h^8 - 12808c_h^6 + 16704c_h^4 - 3154c_h^2 + 107}{4096s_h^{13} \left(6c_h^2 - 1\right)} \tag{3.82}$$

$$A_{44} = \frac{80c_h^6 - 816c_h^4 + 1338c_h^2 - 197}{1536s_h^{10} \left(6c_h^2 - 1\right)} \tag{3.83}$$

$$A_{55} = -\frac{2880c_h^{10} - 72480c_h^8 + 324000c_h^6 - 432000c_h^4 + 163470c_h^2 - 16245}{61440c_h^{11} \left(6c_h^2 - 1\right) \left(8c_h^4 - 11c_h^2 + 3\right)} \tag{3.84}$$

and $c = \sigma/k$ is the nonlinear wave phase velocity.

Figure 3.6 shows the wave profile predicted by the fifth-order Stokes wave theory. It can be observed that the wave has a sharper and higher crest and

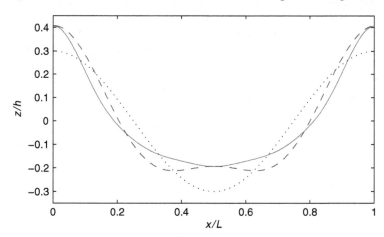

Figure 3.6 Comparison among a fifth-order Stokes wave (solid line), a second-order Stokes wave (dashed line), and a linear wave (dotted line) for a wave with $H = 0.6\,\mathrm{m}$, $h = 1.0\,\mathrm{m}$, and $T = 2.0\,\mathrm{s}$.

a flatter trough compared with the linear wave. The second-order Stokes wave theory fails here due to the presence of the secondary hump on the trough resulting from the strong wave nonlinearity.

3.3.2 Cnoidal wave theory

Stokes wave theory is basically for nonlinear waves in deep and intermediate water depth. For nonlinear waves in shallow water, Wiegel (1960) proposed the cnoidal wave theory:

$$\eta = \eta_t + H cn^2 \left[2K(k) \left(\frac{x}{L} - \frac{t}{T} \right), k \right] \tag{3.85}$$

where cn is the Jacobian elliptic function associated with the cosine, $K(k)$ is the complete elliptic integral of the first kind with modulus $k = \sqrt{m}$, and η_t is the displacement of the trough from the still water level (always negative) and is expressed as follows (Mei, 1989: 546):

$$\eta_t = \frac{H}{m} \left[1 - m - \frac{E(k)}{K(k)} \right] \tag{3.86}$$

Here $E(k)$ is the complete elliptic integral of the second kind with modulus $k = \sqrt{m}$.

Once the modulus k (or m) is known, the cnoidal wave profile can be determined completely. To solve m, the following two equations that relate m with H and L need to be solved simultaneously:

$$HL^2 = \frac{16}{3} h^3 m K(k)^2 \tag{3.87}$$

$$L^2 = ghT^2 \left\{ 1 + \frac{H}{h} \frac{1}{m} \left[2 - m - 3 \frac{E(k)}{K(k)} \right] \right\} \tag{3.88}$$

The role of these two equations is similar to the nonlinear wave height and dispersion equations for fifth-order Stokes waves (3.70) and (3.71). Substituting (3.88) into (3.87), the equation becomes:

$$T^2 = \frac{16 h^3 m^2 K(k)^2}{3gH \left[mh + H \left(2 - m - 3 \frac{E(k)}{K(k)} \right) \right]} \tag{3.89}$$

Given H, h, and T, the above equation can be solved iteratively to obtain m (Appendix V).

The corresponding horizontal and vertical velocities under a cnoidal wave are expressed as follows:

$$u = \sqrt{gh}\left\{ -\frac{5}{4} + \frac{3(\eta_t + h)}{2h} - \frac{(\eta_t + h)^2}{4h^2} + \left[\frac{3H}{2h} - \frac{(\eta_t + h)H}{2h^2}\right]cn^2() - \frac{H^2}{4h^2}cn^4() \right.$$

$$\left. -\frac{8HK^2(k)}{L^2}\left[\frac{h}{3} - \frac{(z+h)^2}{2h}\right]\left[-m\cdot sn^2()\cdot cn^2() + cn^2()\cdot dn^2() - sn^2()\cdot dn^2()\right]\right\}$$

$$(3.90)$$

$$w = \sqrt{gh}\,(z+h)\frac{2HK(k)}{Lh}\cdot sn()\cdot cn()\cdot dn()\left\{1 + \frac{(\eta_t + h)}{h} + \frac{H}{h}cn^2() + \frac{32K^2(k)}{3L^2}\right.$$

$$\left.\left[h^2 - \frac{(z+h)^2}{2}\right]\left[m\cdot sn^2() - m\cdot cn^2() - dn^2()\right]\right\}$$

$$(3.91)$$

where:

$$cn^2() = cn^2\left[2K(k)\left(\frac{x}{L} - \frac{t}{T}\right), k\right]$$

$$sn^2() = 1 - cn^2()$$

$$dn^2() = 1 - m\left[1 - cn^2()\right]$$

The Jacobian elliptic function *cn* is periodic with a period of $4K$ and cn^2 has a period of $2K$. Generally, cn^2 will give a function that has a sharp peak and a flat trough, as shown in Figure 3.7.

The cnoidal wave theory has two asymptotes: (1) $m = k = 0$ and (2) $m = k = 1$. When $m = k = 0$, the *cn* function is reduced to the cosine function and $\eta_t = -H/2$. The cnoidal wave becomes a simple sinusoidal wave, i.e.:

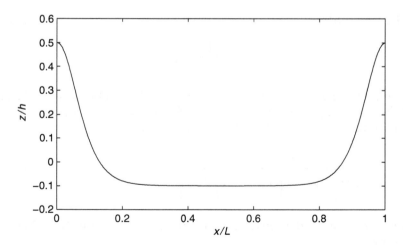

Figure 3.7 Illustration of a cnoidal wave with $H = 0.6\,\mathrm{m}$, $h = 1.0\,\mathrm{m}$, and $T = 5.0\,\mathrm{s}$.

$$\eta = -\eta_t + \lim_{k \to 0} H\mathrm{cn}^2\left[2K(k)\left(\frac{x}{L}-\frac{t}{T}\right),k\right] = -\frac{H}{2}+H\cos^2\left[\pi\left(\frac{x}{L}-\frac{t}{T}\right)\right]$$

$$= -\frac{H}{2}+\frac{H}{2}+\frac{H}{2}\cos\left[2\pi\left(\frac{x}{L}-\frac{t}{T}\right)\right] = \frac{H}{2}\cos(kx-\sigma t) \qquad (3.92)$$

3.3.3 Solitary wave theory and N-wave theory

When $m = k = 1$, the cn function is reduced to hyperbolic function and $\eta_t = 0$ (Lee et al., 1982):

$$\eta = \lim_{k \to 1} H\mathrm{cn}^2\left[2K(k)\left(\frac{x}{L}-\frac{t}{T}\right),k\right]$$

$$= H\mathrm{sech}^2\left[\sqrt{\frac{3\,H}{4\,h^3}}\,(x-ct)\right] = H\mathrm{sech}^2\left(\sqrt{\frac{3\,H}{4\,h^3}}X\right) \qquad (3.93)$$

where the phase velocity is expressed as follows:

$$c = \sqrt{gh\left(1+\frac{H}{h}\right)} \qquad (3.94)$$

and the water particle velocities are expressed as follows:

$$u = \sqrt{gh}\frac{\eta}{h}\left\{1-\frac{1}{4}\frac{\eta}{h}+\frac{h}{3}\frac{h}{\eta}\left[1-\frac{3}{2}\frac{(z+h)^2}{h^2}\right]\frac{d^2\eta}{dX^2}\right\} \qquad (3.95)$$

$$w = -\sqrt{gh}\frac{(z+h)}{h}\left\{\left(1-\frac{1}{2}\frac{\eta}{h}\right)\frac{d\eta}{dX}+\frac{1}{3}h^2\left[1-\frac{1}{2}\frac{(z+h)^2}{h^2}\right]\frac{d^3\eta}{dX^3}\right\} \qquad (3.96)$$

From (3.93):

$$\frac{d\eta}{dX} = -2H\sqrt{\frac{3\,H}{4\,h^3}}\,\frac{\sinh\left(\sqrt{\frac{3}{4}\frac{H}{h^3}}X\right)}{\cosh^3\left(\sqrt{\frac{3}{4}\frac{H}{h^3}}X\right)} \qquad (3.97)$$

$$\frac{d^2\eta}{dX^2} = 2H\left(\sqrt{\frac{3\,H}{4\,h^3}}\right)^2\frac{\left[2\cosh^2\left(\sqrt{\frac{3}{4}\frac{H}{h^3}}X\right)-3\right]}{\cosh^4\left(\sqrt{\frac{3}{4}\frac{H}{h^3}}X\right)} \qquad (3.98)$$

$$\frac{d^3\eta}{dX^3} = -8H\left(\sqrt{\frac{3\,H}{4\,h^3}}\right)^3\frac{\sinh\left(\sqrt{\frac{3}{4}\frac{H}{h^3}}X\right)\left[\cosh^2\left(\sqrt{\frac{3}{4}\frac{H}{h^3}}X\right)-3\right]}{\cosh^5\left(\sqrt{\frac{3}{4}\frac{H}{h^3}}X\right)} \qquad (3.99)$$

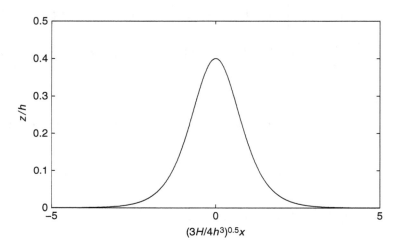

Figure 3.8 Illustration of a solitary wave with $H = 0.4$ m and $h = 1.0$ m.

Figure 3.8 shows the free surface profile for a solitary wave.

A solitary wave is regarded as a weakly nonlinear and dispersive wave, in which wave nonlinearity is perfectly balanced by wave dispersion. While wave nonlinearity tends to make the wave front steepened, wave dispersion will counterbalance it. For this reason, a solitary wave can travel a long distance without significant shape distortion and energy loss.

A solitary wave is often used to approximate the leading wave front of a tsunami. In reality, a tsunami may contain a wave packet with a few wave crests and troughs. Since the wave packet profile is similar to a series of "Ns" connected, the so-called N-wave theory has been proposed to describe the nonlinear wave packet (Tadepalli and Synolakis, 1995). Similar to a solitary wave, an N-wave is also weakly nonlinear and dispersive.

3.3.4   Validity of linear and nonlinear wave theories

All the wave theories presented so far are based on the assumptions that the flow is irrotational and the sea bottom is flat. While linear wave theory is valid for small-amplitude waves, Stokes wave theory and cnoidal wave theory are applicable for large-amplitude waves in deep and shallow waters, respectively. These wave theories describe waves with permanent shapes, but they fail under two circumstances. One is when bottom geometry changes so rapidly in space that it causes significant reflection and/or change of the wave form. The other is when the wave amplitude is so large that the wave front breaks. The former is related to wave shoaling (Section 3.6) and reflection (Section 3.9). The latter refers to wave breaking in deep and shallow waters with different breaking forms (Section 3.7), and it sets the outer limit of the validity range of the previously introduced wave theories. Figure 3.9 shows

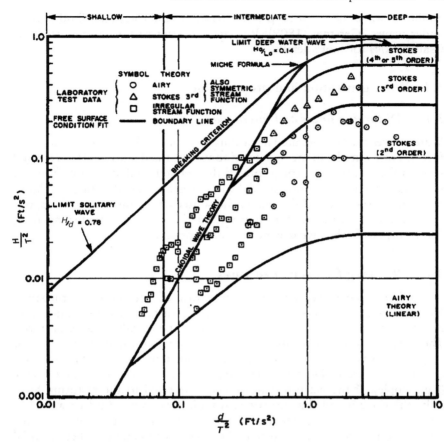

Figure 3.9 Validity range of various wave theories. (Courtesy of Prof. Subrata Chakrabarti at University of Illinois at Chicago)

the validity range of various wave theories. Readers are referred to Dean (1970) and Chakrabarti (1987:77) for more discussion on the analytical and physical validity of different wave theories.

3.3.5 Other wave theories

So far, all the wave theories have been based on the velocity potential function formulation. As shown in Section 2.2.3, both potential function and stream function satisfy the Laplace equation and they form a pair of Cauchy–Riemann equations. In principle, all orders of wave theories can be equally formulated in terms of stream function. However, so far, the formulation based on stream function was reported only for higher order nonlinear waves, attempting to provide the theoretical basis for simpler computation possible for any order of wave theory (Dean, 1965; Cokelet, 1977). There is also the nonlinear wave theory in deep water based on the rotational fluid particle under waves (Gerstner, 1809). Readers can refer to the review article by Craik (2004) for more details.

3.4   Wave generation and propagation

3.4.1   *Disturbing forces and forced waves*

Water waves can be generated in various ways under the action of disturbing forces. For example, in the case of wind waves, the disturbing force is wind stress; for tides, the disturbing force is the gravitational attraction from the sun and the moon. Because a tidal motion is associated with the relative positions of the sun and the moon, the source of the disturbing force, this type of wave is called a forced wave. In this case, the disturbing force contributes to both wave generation and propagation.

3.4.2   *Restoring forces and free waves*

All surface water waves start with some form of free surface disturbance. Once leaving its equilibrium position, water surface has the tendency to return back to its equilibrium position under the so-called restoring force. The restoring force for the majority of water waves on Earth is gravity. For very short waves (e.g., $T < 0.1$ s), the surface tension can dominate over gravity as the restoring force. This type of wave is called capillary wave (e.g., ripple). If the wave propagation is connected to the restoring force only, the wave is called a free wave. A typical example of a free wave is a wind wave outside of the fetch area. Another example is tsunami generated by seismic motion. It is noted that a tsunami is sometimes called a "tidal wave," although it has no physical connection at all to astronomical tides.

3.4.3   *Wind wave generation and swell formation*

Over 50 percent of wave energy in oceans is in the form of wind waves, most of which have a wave period ranging from 1 s to 30 s. These wind waves are often generated from a large fetch area in deep oceans where strong wind can continue to blow over the area for many hours. In the fetch region, water will receive energy from winds through wind-induced surface stress. The continuous feed of wind energy to water results in the growth of mean wave height and wave period over a distance (Garrison, 2002). However, due to the intermittent feature of wind, the generated waves often have a wide range of frequency and they propagate forward with a significant directional spreading. If the fetch area is large enough, a fully developed sea can be attained.

Wind will cease to blow toward the edge of the fetch outside of which gravity becomes the only predominant force. Under this circumstance, in the continued propagation of the wave packet, waves with similar frequency will group together and separate from other wave groups. This process of self-sorting is called wave dispersion. The longest waves that propagate at the fastest speed will arrive at the farther places earliest. After leaving behind the shorter waves, which cause the choppy sea state, the ocean surface becomes smoother and consists mainly of long and fast-moving smooth undulation, which is called swell. Compared to the shorter waves behind, the swell normally has large wave amplitude and it is of most critical consideration in coastal and offshore design.

3.4.4 Extreme waves

Storm wave height can reach up to 8–10 m in the deep sea under extreme wind conditions. If propagating against strong currents, the wave amplitude can be further enhanced. In a random sea, a very large wave can occasionally appear due to the superposition of a few short wave crests, creating a very large wave called a rogue wave or freak wave (Figure 3.10; see Section 5.3.4.5 for more discussion). The earlier observed maximum wave height was about 25 m estimated from the picture taken by Philippe Lijour on the coast of Durban in 1980. This massive ocean wave was suspected of being the possible cause of the sinking of supertankers and container ships exceeding 200 m in length, most of which, however, are only built to withstand maximum wave heights of 15 m. The recent satellite data collected by the European Space Agency revealed that more than 10 individual waves over 25 m in height were found around the globe within 3 weeks in 2001.

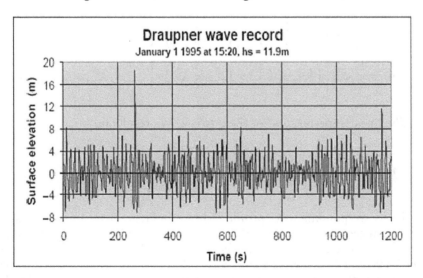

Figure 3.10 Draupner wave record in the North Sea, the first confirmed scientific evidence of a freak wave. (Courtesy of Dr. Sverre Haver at Marine Structures and Risers, Norway)

Tsunamis are infamous for their catastrophic power of sweeping through the coast. When it is in deep ocean, however, a tsunami's amplitude hardly exceeds a few meters. It is only when it approaches the coast that the wave amplitude is amplified significantly by the combined wave shoaling, refraction, and diffraction, during which wave energy can be focused and trapped near a small area. The maximum tsunami run-up can easily go beyond 30 m based on many post-tsunami surveys. The highest wave on record, however, was caused by a landslide in Lituya Bay, Alaska, in 1958 after a magnitude 8 earthquake. The wave was rather localized in the bay, but it had an astonishing height of 520 m. In land, the largest water wave is probably the dam break wave that is generated after the failure of a dam. A shock wave front with the wave height close to the reservoir height will be formed and flood the downstream area.

3.5   Wave superposition and wave group

Linear waves can be superposed together to create new waves. A random sea is regarded as the result of wave superposition of an infinite number of small linear waves with different wave amplitudes and frequencies traveling in different directions.

Now let us consider a special case of wave superposition to illustrate how wave group and wave group velocity are defined. Two linear wave trains propagate in the x-direction and have a small frequency difference $\Delta\sigma$ and the corresponding wave number difference Δk:

$$\eta_1 = \frac{H}{2}\cos\left[\left(k - \frac{1}{2}\Delta k\right)x - \left(\sigma - \frac{1}{2}\Delta\sigma\right)t\right] \tag{3.100}$$

$$\eta_2 = \frac{H}{2}\cos\left[\left(k + \frac{1}{2}\Delta k\right)x - \left(\sigma + \frac{1}{2}\Delta\sigma\right)t\right] \tag{3.101}$$

The linear superposition of these two waves results in:

$$\eta = \eta_1 + \eta_2 = H\cos(kx - \sigma t)\cos\left[\frac{1}{2}\Delta k\left(x - \frac{\Delta\sigma}{\Delta k}t\right)\right] \tag{3.102}$$

The corresponding wave profile is shown in Figure 3.11. It is seen that the wave packet (or "wave group") is bracketed within the "wave envelope" that also propagates forward. The envelope is defined by the last term on the RHS of (3.102). While waves within the envelope propagate at the speed of $c = \sigma/k$, the wave envelope propagates at a different speed of $\Delta\sigma/\Delta k$. The wave envelope has the nominal wavelength of $L_g = 2\pi/(\Delta k/2) = 4\pi/\Delta k$ and the length between any two peaks is half of L_g, i.e., $2\pi/\Delta k$.

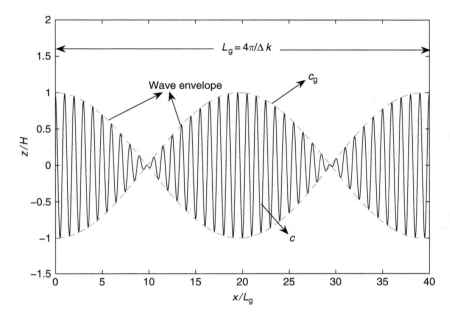

Figure 3.11 Wave envelope resulting from two linear wave trains propagating in constant water depth; $T = 4\,\text{s}$, $h = 100\,\text{m}$, $\Delta\sigma = 0.04\,(1/\text{s})$, and $\Delta k = 0.0128\,(1/\text{m})$.

The speed at which the wave envelope (or the group of the waves) propagates is called wave group velocity $c_\text{g} = \Delta\sigma/\Delta k$. In the limit of $\Delta\sigma, \Delta k \to 0$, c_g can be evaluated from the linear dispersion equation as follows:

$$c_\text{g} = \lim_{\Delta\sigma,\Delta k\to 0} \frac{\Delta\sigma}{\Delta k} = \frac{d\sigma}{dk} = \left[\frac{1}{2}\left(1 + \frac{2kh}{\sinh 2kh}\right)\right]c \qquad (3.103)$$

Note that the above expression is exactly the same as that for the speed at which the wave energy propagates [see equation (3.21)]. This suggests that the wave energy is transmitted together with the wave envelope rather than the individual wave form.

3.6 Wave shoaling

When a wave train propagates toward a gentle plane slope from a normal incidence, the train will gradually slow down in the shallow water region due to the reduction of water depth. If no major energy dissipation and wave reflection occur during the process, we would expect that the energy flux defined in (3.20) is a constant, i.e.:

$$F = Ec_\text{g} \qquad (3.104)$$

which will lead to the change of wave height as follows:

$$\frac{\rho g H_0^2}{8} c_{g0} = \frac{\rho g H^2}{8} c_g \Rightarrow H = H_0 \sqrt{\frac{c_{g0}}{c_g}} \tag{3.105}$$

where c_{g0} is the wave group velocity in deep water. This process is called wave shoaling, during which an approaching wave train will change its wave height based on its offshore condition and local water depth.

Figure 3.12 is the illustration of wave shoaling on a plane beach. Two observations can be made from the figure. One is that the wave crests are packed together in the nearshore region where the wave speed slows down and the wavelength is reduced. The other is that the wave height increases toward the shoreline due to the conservation of energy flux. In fact, based on the shoaling formula (3.105), the wave height will reach infinity at the shoreline where the still water depth is zero. This will of course never happen in reality due to the presence of wave nonlinearity, bottom friction, and wave breaking, which limit the growth of wave height.

For a long wave (e.g., tsunami), (3.105) can be reduced to:

$$H = H_0 \sqrt{\frac{c_{g0}}{c_g}} = H_0 \sqrt{\frac{\sqrt{gh_0}}{\sqrt{gh}}} = H_0 h_0^{1/4} h^{-1/4} \propto h^{-1/4} \tag{3.106}$$

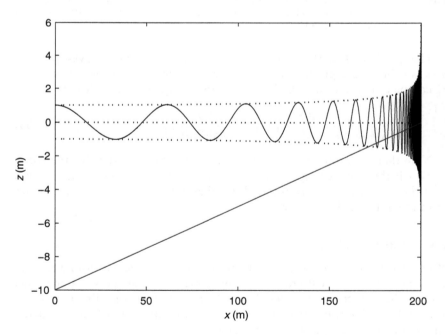

Figure 3.12 Linear wave shoaling on a plane beach with a slope of 1/20; the incident wave has a wave height of 1 m and wave period of 8 s.

It is seen that the wave height is reciprocally proportional to local depth to a power of 1/4. This is the so-called Green's law for long wave height (or amplitude) variation on a slope. The above equation is valid before wave nonlinearity becomes important. With the use of the theory of soliton, for which wave nonlinearity is balanced by wave dispersion, it is able to deduce the nonlinear Green's law (e.g., Shuto, 1973) that instead gives the change of wave height being reciprocally proportional to local water depth:

$$H = H_0 h_0 / h \propto h^{-1} \tag{3.107}$$

In reality, an actual wave amplitude evolution is often enveloped by the two lines from deeper water to shallow water before wave breaks.

3.7 Wave breaking

Wave height can be increased due to many reasons, e.g., wave shoaling, wave growth due to the continuous wind action, and wave energy focusing by either the superposition of various wave modes or the combined wave refraction and diffraction. When the wave height exceeds a certain threshold, the wave system will become unstable. Wave breaking is an important process for the unstable wave system to release the "excessive" wave energy into turbulence and to return back to the stable stage. The wave will break differently depending on whether the wave is in deep or shallow water.

3.7.1 *Wave breaking in deep and intermediate water*

In deep water, wave steepness is the only sensible index for judging whether a wave will break. Stokes (1880) found that the maximum crest angle for a nonbreaking wave is 120°, which leads to the following critical wave height in deep water:

$$H_b = \frac{1}{7} L_b \tag{3.108}$$

where L_b is the wavelength at the breaking location. The above equation can be modified to predict the breaking wave height in the intermediate water depth (e.g., Miche, 1944):

$$H_b = \tfrac{1}{7} L_b \tanh k_b h_b \tag{3.109}$$

where $k_b = 2\pi / L_b$ and h_b are the wave number and the water depth at the breaking point, respectively.

3.7.2 *Wave breaking in shallow water*

In shallow water, the ratio of wave height to local water depth is the natural index for judging whether a wave will break. The most commonly used formula to estimate the initial breaking wave height H_b on a mild-slope beach is the simple formula originally proposed by McCowan (1894):

$$H_b = K_b h_b \tag{3.110}$$

where $K_b = 0.78$. The wave-breaking criteria in (3.108)–(3.110) set the upper limits for the validity range of various nonlinear wave theories (refer to Section 3.3.4).

Generally speaking, there are three typical types of wave breakers in shallow water. When the beach slope is mild compared with the wave steepness, waves will break in the form of a spilling breaker, which forms a series of aerated wave crests on the beach. As the beach slope increases or the wave steepness decreases, the breaker type changes to a plunging breaker, during which the crest of the wave front overturns forward and impinges on the wave trough in front. As a result, an air tube is enclosed inside and the strong air entrainment will take place when the plunging jet collapses. As the beach slope increases further, waves will break in the form of a collapsing or surging breaker, which breaks on the dry land during the run-up process.

The classification of various breakers can be made based on the surf similarity parameter ξ_b, which is defined as follows:

$$\xi_b = \frac{\tan \beta}{\sqrt{H_b/L_0}} \tag{3.111}$$

where β is the beach slope and L_0 is the wavelength in deep water. When $\xi_b < 0.4$, waves break in the form of spilling breakers (Battjes, 1974). When $0.4 \leq \xi_b < 2.0$, plunging breakers occur. When $\xi_b \geq 2.0$, waves break as collapsing or surging breakers, during which most of the wave energy is reflected back from the beach with a small amount of wave energy lost in the breaking process. When the slope is very steep, the surging wave experiences near complete reflection with little breaking after it runs up to the maximum elevation.

Since (3.110) does not include the slope effect and breaker type, there have been a few modified breaking criteria developed in the past decades for coastal engineering design. Some commonly used formulas include the CERC formula (1984):

$$\frac{H_b}{h_b} = \left(c_1 - c_2 \frac{H_b}{gT^2}\right) \tag{3.112}$$

where

$$c_1 = 43.75 \left(1 - e^{-19 \tan \beta}\right), \quad c_2 = \frac{1.56}{1 + e^{-19.5 \tan \beta}}$$

Goda's formula (1970):

$$\frac{H_b}{h_b} = 0.17 \frac{L_0}{h_b} \left(1 - \exp\left(-\left(\frac{1.5\pi h_b}{L_0}\left(1 + 15 \left(\tan \beta\right)^{4/3}\right)\right)\right)\right) \tag{3.113}$$

and Kamphuis' formula (1991) for irregular waves:

$$\frac{H_{sb}}{h_b} = 0.56e^{3.5\tan\beta} \tag{3.114}$$

where H_{sb} is the significant wave height at the breaking point.

3.7.3   Bore

After waves break in shallow water and if the shallow water region is long, moving bores can be formed where broken wave fronts propagate forward at the celerity of local wave speed $\sqrt{gh}$. A moving bore can be regarded as a continuously breaking wave. Violent turbulence is generated inside the bore that causes continuous air entrainment. Similar to a hydraulic jump, a bore can be either uniform or undulatory. The inherent similarity among spilling breaker, bore, and hydraulic jump has been discussed by Peregrine and Svendsen (1978).

3.8   Wave run-up, run-down, and overtopping

3.8.1   Wave run-up

Waves on a sloping beach, breaking or nonbreaking, will eventually run up on the beach, causing inundation. While wave run-up is defined as the vertical distance between the mean water level and the wave front, inundation is defined as the horizontal distance between the mean shoreline and the wave front. The maximum run-up height R is used to measure the maximum level of the wet line caused by the wave, and it depends on both incident wave height and beach geometry and material type. Carrier and Greenspan (1958) proposed an analytical solution for nonlinear shallow water waves on a sloping beach. The solution, however, did not provide a simple and explicit relationship among R, incident wave height H_0, and beach slope β. Hunt (1959a), using a series of regular wave run-up experiments on smooth and impermeable slopes, summarized the empirical equation for R, which, after transforming from the original dimensional inhomogeneous equation to the dimensionless homogenous equation reads:

$$\xi = \frac{R}{H_0} = \frac{\tan\beta}{\sqrt{H_0/L_0}} \tag{3.115}$$

where H_0 is the wave height in deep water. This formula is valid for both nonbreaking and breaking waves. The ratio of R/H_0 in (3.115) has a maximum value of 3.0 for surging waves.

For nonbreaking solitary wave run-up on a smooth plane beach, Synolakis (1987) proposed an expression of the maximum wave run-up:

$$\frac{R}{h} = 2.831\sqrt{\cot\beta}\left(\frac{H_0}{h}\right)^{5/4} \Rightarrow \frac{R}{H_0} = 2.831\sqrt{\cot\beta}\left(\frac{H_0}{h}\right)^{1/4} \qquad (3.116)$$

The above formula is valid for $\sqrt{H_0/h} \gg 0.288\tan\beta$. The above theoretical result was extended by Li and Raichlen (2001) by including higher order nonlinear contribution:

$$\frac{R}{h} = 2.831\sqrt{\cot\beta}\left(\frac{H_0}{h}\right)^{5/4} + 0.293\sqrt{(\cot\beta)^3}\left(\frac{H_0}{h}\right)^{9/4} \qquad (3.117)$$

The theoretical work was also extended by Kanoglu and Synolakis (1998) for piecewise linear topographies and by Carrier *et al.* (2003) for various forms of tsunami wave fronts. With the introduction of the energy balance model (EBM), Li and Raichlen (2003) managed to predict breaking solitary wave run-up.

For irregular waves, the maximum run-up is often based on the run-up exceeded by 2 percent of the waves $R_{2\%}$. Due to the large scattering of experimental data, there were many empirical curves (e.g., CERC, 1984) and formulas (e.g., Ahrens, 1981). The recent *Coastal Engineering Manual* (Burcharth and Hughes, 2002) made the following recommendations for irregular wave run-up on a steep slope of $1/4 < \tan\beta < 1$:

$$\frac{R_{2\%}}{H_{m0}} = \begin{cases} 1.6\dfrac{\tan\beta}{\sqrt{H_{m0}/L_{0p}}}, & \dfrac{\tan\beta}{\sqrt{H_{m0}/L_{0p}}} \le 2.5 \\[3mm] 4.5 - 0.2\dfrac{\tan\beta}{\sqrt{H_{m0}/L_{0p}}}, & 2.5 < \dfrac{\tan\beta}{\sqrt{H_{m0}/L_{0p}}} \le 9.0 \end{cases} \qquad (3.118)$$

and on a mild slope $1/8 < \tan\beta < 1/3$ (De Waal and Van der Meer, 1992):

$$\frac{R_{2\%}}{H_{1/3}} = \begin{cases} 1.5\dfrac{\tan\beta}{\sqrt{H_s/L_{0p}}}, & 0.5 < \dfrac{\tan\beta}{\sqrt{H_s/L_{0p}}} \le 2.0 \\[3mm] 3.0, & 2.0 < \dfrac{\tan\beta}{\sqrt{H_s/L_{0p}}} \le 4.0 \end{cases} \qquad (3.119)$$

where H_{m0} is the energy-based zeroth-moment wave height, L_{0p} is the deep-water wavelength associated with the peak wave period T_p, and H_s is the significant wave height that approximately equals H_{m0} if the irregular wave spectrum follows the Rayleigh distribution. Recently, Hughes (2004) attempted to use depth-integrated momentum flux to unify various expressions.

Note that all the above formulas are based on smooth and impermeable slopes. For the rough and/or permeable slopes, the formulas need to be modified. Teng *et al.* (2000) found that for solitary wave run-up on a mild slope ($\beta < 10°$), the run-up height can be reduced by more than 50 percent. In most of the engineering applications, a surface roughness correction coefficient, which accounts for the roughness and permeability and ranges from 1.0 (smooth surface like asphalt or smooth concrete) to 0.3 (two-layer tetrapods), is multiplied by the original estimation from the above formulas. The correction coefficients for different material surfaces are summarized in Reeve *et al.* (2004: 362).

In the last two decades, the numerical modeling of wave run-up on sloping beaches has become popular. With the use of nonlinear SWEs, Hibberd and Peregrine (1979) investigated the behavior of bore run-up on a beach. Kobayashi *et al.* (1987) used the same approach to investigate wave run-up on rough slopes. Zelt (1991) used a Boussinesq model to investigate breaking solitary wave run-up on a beach. Lin *et al.* (1999) employed a RANS model to solve both nonbreaking and breaking wave run-up on a plane beach.

3.8.2 Wave run-down and undertow

A wave run-up is followed by wave run-down (or draw-down), during which water retreats seaward. The area of run-up and backwash of water is known as the swash zone (Figure 3.13). For illustration, a normal incident wave train on a uniform plane beach is considered. The retreating water from the beach will be partially blocked by the next incoming wave front, which most of the time is in the form of a moving bore. Since the breaking wave in the surf zone is highly nonlinear with the net mass flux near the free surface being much larger than that underneath, the blocking effect for the retreating water will be more significant near the free surface. The only possible channel for the receding water to return back to the sea on a

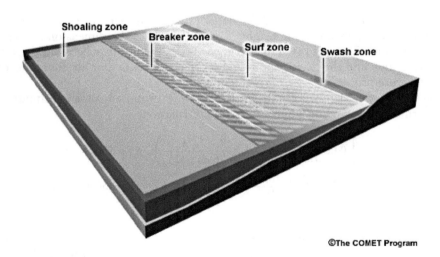

The Nearshore Environment

©The COMET Program

Figure 3.13 Illustration of a swash zone. (Courtesy of COMET program)

uniform beach is by means of the bottom current. This returning current underneath the breaking bore is called undertow. When it is strong enough, it can also push up a small wave that rolls back toward the incoming waves.

Depending on different types of breaking waves, undertow character-istics can be very different. This has a direct impact on beach morphology. Undertow flow has been extensively studied experimentally (e.g., Stive and Wind, 1986), numerically (e.g., Péchon and Teisson, 1994), and analytic-ally (e.g., Hansen and Svendsen, 1984). It is believed that the undertow profile is affected by many factors including radiation stresses, the mean wave set-up, and the turbulence shear stresses generated by breaking waves. The theoretical relationship between undertow and radiation stresses will be further elaborated in Section 3.14.2.3. Recently, Lin and Liu (2004) applied a RANS model that incorporates all the above mechanisms to simu-late the undertow velocity profiles at various locations in the surf zone under spilling and plunging breaking waves. Their results agree well with laboratory measurements by Ting and Kirby (1995, 1996).

3.8.3   Wave overtopping

When wave run-up takes place on a slope of finite length and height, there exists the possibility that the run-up will continue with wave overtopping, during which part of the fluid overtops the crown of the structure. This may happen in an emerging low-crested breakwater, a seawall, or a deck of an offshore plat-form under attack by large waves. One of the most popular empirical formulas to estimate the overtopping rate was proposed by Owen (1980):

$$\frac{q}{gH_S T} = a \exp\left(-bR_n\right) \tag{3.120}$$

where q is the time-averaged discharge rate per unit length, H_s is the post-breaking significant wave height, $R_n R_c/(T\sqrt{gH_s})$ is the dimensionless free-board with R_c being the actual freeboard (i.e., the distance between the still water line and the deck), and a and b are coefficients that are functions of structure profile.

Later, van der Meer and Janssen (1995) proposed another empirical formula by including slope and other information:

$$\frac{q}{\sqrt{gH_s^3}} = \begin{cases} 0.06\dfrac{\xi_p}{\sqrt{\tan\beta}}\exp\left(-5.2\dfrac{R_c}{H_s\xi_p}\dfrac{1}{\gamma_r\gamma_b\gamma_h\gamma_\beta}\right), & \xi_p < 2 \text{(breaking waves)} \\[4mm] 0.2\exp\left(-2.6\dfrac{R_c}{H_s}\dfrac{1}{\gamma_r\gamma_b\gamma_h\gamma_\beta}\right), & \xi_p \geq 2 \text{(nonbreaking waves)} \end{cases} \tag{3.121}$$

where the surf similarity parameter is $\xi_p = \tan\beta/\sqrt{H_s/L_{0p}}$ [see (3.119)] and the coefficients γ_b, γ_h, γ_r, and γ_β account for the influence of berm, shallow foreshore, roughness, and angle of wave attack, respectively.

Recently, Hedges and Reis (1998), based on Owen's experimental data, proposed another simple formula:

$$\frac{q}{\sqrt{g\,(C_{RH}H_s)^3}} = q_0 \left(1 - \frac{R_c}{C_{RH}H_s}\right)^b \tag{3.122}$$

where C_{RH} is the ratio of the maximum wave run-up on a smooth slope to the significant wave height (R/H_s), q_0 represents the dimensionless discharge when the dimensionless freeboard is zero, and b depends on the water surface profile on the seaward face of the structure. Other empirical formulas also exist (e.g., Franco and Franco, 1999), but they will not be discussed in this book. Readers can refer to Allsop *et al.* (2005) for a complete review of existing empirical formulas.

3.9 Wave reflection

Waves will be reflected back from an obstacle in the form of wave reflection. Depending on the material and shape of the obstacle, the ratio of the reflected wave energy and the incident wave energy can range from 0 (fully absorbed) to 1 (completely reflected).

3.9.1 *Standing waves*

Consider a smooth vertical wall that is fixed, rigid, impermeable, and infinitely long; a normal incident linear wave train will be completely reflected back and form the so-called standing waves, which is basically the superposition of two progressive waves propagating in opposite directions:

$$\eta = \eta_I + \eta_R = \frac{H}{2}\left[\cos(kx - \sigma t) + \cos(kx - \sigma t)\right] = H\cos kx \cos \sigma t \tag{3.123}$$

In a standing wave, there exist nodal points where the free surface displacement remains zero at all time and antinodal points where the wave amplitude is amplified two times (Figure 3.14). For a standing wave, the net energy flux is 0.

3.9.2 *Partially standing waves*

In cases when only part of the wave energy is reflected (and thus the reflected wave height H_R is smaller than the incident wave height H_I), the resulting wave profile can be obtained by the superposition of two wave trains:

$$\eta = \eta_I + \eta_R = \frac{H_I}{2}\cos(kx - \sigma t) + \frac{H_R}{2}\cos(kx + \sigma t)$$

$$= \left(\frac{H_I}{2} + \frac{H_R}{2}\right)\cos kx \cos \sigma t + \left(\frac{H_I}{2} - \frac{H_R}{2}\right)\sin kx \sin \sigma t \tag{3.124}$$

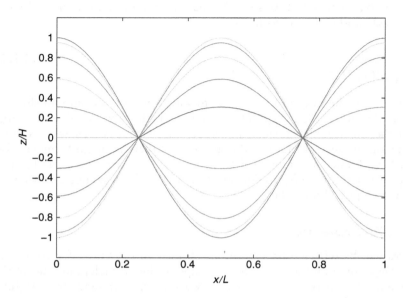

Figure 3.14 Wave envelope for a standing wave.

This is the so-called partially standing wave (Figure 3.15), in which the quasi-nodal points have the minimum wave height H_{min} and the quasi-antinodal points have the maximum wave height H_{max}. It is ready to find the following relation:

$$H_{max} = H_I + H_R, \quad H_{min} = H_I - H_R \tag{3.125}$$

A common practice to estimate the reflection coefficient $K_R = H_R/H_I$ in the laboratory is based on the measurement of the maximum and minimum wave heights at the quasi-nodal point and the quasi-antinodal point and the substitution into the following equation derived from (3.125):

$$K_R = \frac{H_R}{H_I} = \frac{H_{max} - H_{min}}{H_{max} + H_{min}} \tag{3.126}$$

3.9.3   Mach reflection and stem waves

For an obliquely incident linear wave train in front of a perfectly reflecting boundary, the interference between the incident waves and the reflected waves will form a honeycomb pattern as shown in Figure 3.16. The relative density of the nodal and antinodal points in two directions is dependent upon the wave incident angle. The larger the incident angle is, the denser the nodal and antinodal points will be along the wall. The pattern will propagate

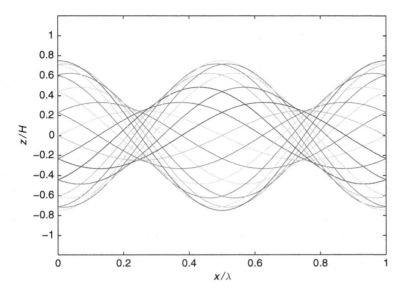

Figure 3.15 Wave envelope for a partial standing wave ($H_R/H_I = 0.5$).

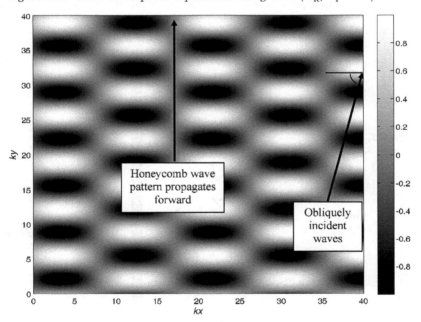

Figure 3.16 Instantaneous wave pattern in front of a perfectly reflecting vertical wall (at $kx = 40$ on the right) for an obliquely incident linear wave train with the incident angle $\theta = 70°$; the gray scale represents the free surface displacement normalized by the incident wave height.

forward along the wall direction, similar to a special type of progressive wave.

In the case of a large-amplitude wave approaching a vertical seawall obliquely, waves often break in front of the wall and there is a chance for the development of growing progressive waves in front of the wall. Such waves are termed stem waves and they are the result of the so-called Mach reflection, which was first discovered in aerodynamics. It is a well-known fact that when a shock wave (with its Mach number greater than 1) approaches obliquely a rigid surface, sometimes the reflection does not take place at the solid wall but a distance away from it. A so-called slipstream will be developed in the downstream. In the region between the down-stream wall and the slipstream line, a special reflection with a growing strength will occur with its Mach number being smaller than 1. This phenomenon is called Mach reflection or Mach stem (Figure 3.17). The Mach reflection in water waves was first found by Perroud (1957) when he studied the solitary wave reflection from a wall. It was further confirmed by Yue and Mei (1980) using numerical models and by Melville (1980) through laboratory experiments. Berger and Kohlhase (1976) conducted experiments to investigate periodic wave reflection from a wall. Yoon and Liu (1989a) employed a parabolic approximation to study the stem waves induced by an oblique cnoidal wave train in front of a vertical breakwater.

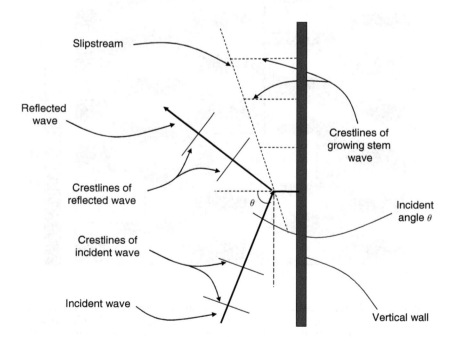

Figure 3.17 Sketch of the development of a stem wave due to Mach reflection.

Recently, Mase *et al.* (2002) extended the study of stem waves induced by random wave incidence.

3.9.4 Bragg reflection

Waves will experience reflection when they propagate over submerged structure and varying bottom topography. Further discussion will be made separately for wave interaction with submerged structures in Section 3.15.1.1. In this section, only the wave reflection from a periodic bathymetry, e.g., sandbar or sand ripple, will be discussed. It was found that strong reflection could occur when waves propagate through a long and periodically varying bottom, even though the magnitude of bottom variation is small. For normal incident waves, the condition for this strong reflection to take place is when the wavelength of the topography is equal to half of the incident wavelength.

The above strong reflection is termed Bragg reflection (or Bragg resonance, Bragg diffraction, Bragg scattering) as it was originally discovered by William Henry Bragg and William Lawrence Bragg (father and son) in 1913 when they performed the study of X-ray wave diffraction by crystalline solids. Strong reflection will be produced if the following Bragg's condition is satisfied:

$$2\lambda_m \cos \theta = nL \tag{3.127}$$

where θ is the incident angle, λ_m is the wavelength of periodic materials such as crystal matrix, and n is any integer. Their work won them the Nobel Prize in 1915; by then William Lawrence Bragg was only 25 years old, the youngest Nobel laureate to date.

The significance of Bragg reflection in water wave dynamics is that if a periodic bottom is formed specifically, the transmitted wave energy can be significantly reduced. On the basis of the Laplace equation, Mei (1985) proposed an asymptotic approximation for Bragg reflection of water waves from periodic sandbars. Considering the nonlinear wave interaction, Liu and Yue (1998) studied various classes of the Bragg scattering of surface waves by bottom ripples. Cho and Lee (2000) furthered the study by considering an arbitrarily varying topography and applied the theory to the study of singly and doubly sinusoidally varying topography. Enlarging on Floquet's theorem for electronics and optics, Chou (1998) derived the equation for small-amplitude water wave propagation in the presence of an infinite array of periodically arranged surface scatterers. Using Miles's theory (1981), Wang *et al.* (2006) studied Bragg reflection of water waves from double composite artificial bars.

Bragg reflection was also studied by using various simplified theories. For example, using shallow water approximation, Yoon and Liu (1987) studied resonant wave reflection by corrugated boundaries. The method was extended by Chen and Guza (1998) to study resonant scattering of a trapped edge wave by longshore periodic topography with the use of multiple-scale expansion. Starting from the Boussinesq equation, Liu and Cho (1993) showed that long waves associated with a wave group of long and short waves can be reflected resonantly by a field of periodic sandbars. Recently, extensive studies have been made by using the modified MSE, which has the capability of resolving rapidly changing topography, to the study of Bragg reflection over periodic bottom geometry (e.g., Miles and Chamberlain, 1998; Zhang *et al.*, 1999).

3.9.5   Wave reflection from a slope

When a wave train propagates normally to a plane slope, part of the wave energy will be reflected back from the slope. Most of the time, the reflected wave energy can be estimated only by empirical formulas. On a smooth impermeable slope, Battjes (1974) proposed the following formula for the estimation of reflection coefficient:

$$K_R = 0.1\xi^2 \tag{3.128}$$

where $\xi = \tan\beta/\sqrt{H_0/L_0}$ is the Iribarren number (Iribarren and Nogales, 1949) that appeared in (3.115). The only difference between the Iribarren number and the surf similarity parameter defined in (3.111) is that the breaking wave height H_b is used in (3.111).

On a rough permeable slope, Allsop and Channell (1988) proposed another formula:

$$K_R = 0.125\xi^{0.73} \tag{3.129}$$

Using the field data, Davidson *et al.* (1996) proposed an improved formula for porous breakwaters as follows:

$$K_R = 0.151R_c^{0.11} \tag{3.130}$$

where:

$$R_c = \frac{h_t L^2 \tan\beta}{H d_{50}^2}$$

is called the reflection parameter with h_t being the water depth at the toe of the structure and d_{50} the mean diameter of the porous material.

3.10   Wave refraction

The word "refraction" originated from optics that defines the process by which light changes its direction when entering obliquely from one medium to another. In water wave theory, it refers to a similar process except that the "medium" is replaced by "local water depth" that affects wave speed. For this argument, wave refraction will occur only for waves in shallow and intermediate water depths.

3.10.1 Conservation equation of waves

Consider a steady linear wave train propagating over a changing topography (see Figure 3.18 for definition). The free surface displacement of the waves can be described by the real part of the following expression:

$$\eta(x, y, t) = a(x, y)e^{iS(x,y,t)} = a(x, y)e^{i(k_x x + k_y y - \omega t)} \tag{3.131}$$

where $S(x, y, t)$ is the phase function and $k = \sqrt{k_x^2 + k_y^2}$ is the wave number with the local wave propagation angle $\theta = \tan^{-1}(k_y/k_x)$. The relationship among k_x and k_y and k is $k_x = k \cos \theta$ and $k_y = k \sin \theta$. The phase function is a scalar whose time derivative gives angular frequency, i.e., $\partial S/\partial t = -\omega$, and whose spatial gradient gives the wave number vector, i.e.:

$$\nabla S = \frac{\partial(k_x x)}{\partial x}\mathbf{i} + \frac{\partial(k_y x)}{\partial y}\mathbf{j} = k_x\mathbf{i} + k_y\mathbf{j} = \mathbf{k} \tag{3.132}$$

By substituting $S(x, y, t) = \mathbf{k} \cdot \mathbf{x} - \omega t$ into the trivial identity $\partial(\nabla S)/\partial t + \nabla(-\partial S/\partial t) = 0$, we have:

$$\frac{\partial \mathbf{k}}{\partial t} + \nabla\omega = 0 \Leftrightarrow \frac{\partial k_i}{\partial t} + \frac{\partial \omega}{\partial x_i} = 0 \tag{3.133}$$

The above equation is called the conservation equation of waves (or wave number). This equation describes the kinematics of wave propagation, during which the change rate of the wave number with time must be balanced

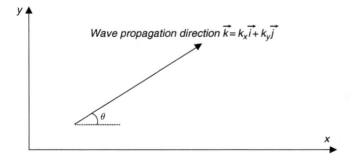

Figure 3.18 A steady linear wave train propagating over a changing topography.

by the spatial variation of wave angular frequency. For a steady wave field without current, the wave frequency is uniform in space and thus the wave number at a particular location is a constant.

3.10.2 Snell's law and wave refraction

From the identity of vector operation (see Appendix II), we have the irrotationality of the wave number vector as follows:

$$\nabla \times (\nabla S) = \nabla \times \mathbf{k} = 0 = \frac{\partial k_i}{\partial x_j} - \frac{\partial k_j}{\partial x_i} = \frac{\partial k_y}{\partial x_x} - \frac{\partial k_x}{\partial x_y} = \frac{\partial (k \sin \theta)}{\partial x_x} - \frac{\partial (k \cos \theta)}{\partial x_y}$$

(3.134)

Let us consider a simple case of a linear wave train propagating toward a plane beach from an angle θ (Figure 3.19). Since all variations in the y-direction along the plane beach are zero, equation (3.134) can be reduced to:

$$\frac{\mathrm{d}\,(k \sin \theta)}{\mathrm{d}x} = 0 \Rightarrow k \sin \theta = \text{constant}$$

(3.135)

This implies that the projection of the wave number on the longshore direction is a constant. Dividing by the constant angular frequency σ on both sides of the equation (3.135), we have:

$$\frac{\sin \theta}{c} = \text{constant} = \frac{\sin \theta_0}{c_0}$$

(3.136)

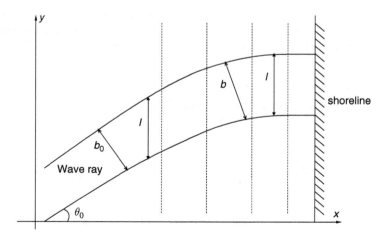

Figure 3.19 Illustration of a wave refraction on a plane beach for an obliquely incident wave train.

This is Snell's law originally found in geometry optics, and it can be equally applied to water waves. Based on the above equation, the trajectory of a particular wave ray, for which the incident wave angle in the offshore region is known, can be traced when the wave propagates toward the shoreline. This equation explains the common phenomenon that waves always turn to be normal to a shoreline when they are making the final approach to the beach, regardless of their incident angle in the offshore region (suppose near the shoreline $c \to 0$, which leads to $\theta \to 0$).

3.10.3 Combined wave shoaling and refraction

Because wave refraction always occurs on a changing topography, the change of wave direction is always accompanied by the change of wave height. When a wave train approaches a beach from an angle, the process of wave refraction is combined with wave shoaling. In this case, the shoaling formula derived earlier for a normally incident wave (3.105) must be modified. On the basis of the same assumptions of negligible wave energy dissipation and wave reflection, the conservation of energy flux within a band of wave rays ensures:

$$Ec_g b = E_0 c_{g0} b_0 \tag{3.137}$$

where b and b_0 are the widths of the band of wave rays (normal to its propagation direction) in nearshore and offshore regions. By realizing that the projected length (or wave number) of the wave rays in the longshore direction is a constant, i.e., $l = b/\cos\theta = b_0/\cos\theta_0$, we then have:

$$H = H_0 \sqrt{\frac{C_{g0}}{C_g}} \sqrt{\frac{b_0}{b}} = H_0 \sqrt{\frac{C_{g0}}{C_g}} \left(\frac{1 - \sin^2\theta_0}{1 - \sin^2\theta} \right)^{1/4} \tag{3.138}$$

Compared with the original shoaling formula (3.105), a correction term is added to account for the effect of wave refraction. The propagation angle θ at any location can be obtained with the use of Snell's law, given the incident wave angle and bottom geometry.

3.11 Wave diffraction

Diffraction refers to the process in which the wave bends around an obstruction and part of wave energy "diffuses" into the "shadow" area. Wave diffraction becomes a predominant process when waves pass a large object. The examples include wave diffraction behind a breakwater, wave diffraction around a large circular pile, and wave diffraction around a very large floating structure (VLFS). For structures with a simple shape, an analytical solution exists. For structures with a complex geometry, numerical modeling must be used in the analysis.

Two classical analytical solutions were developed by Sommerfeld (1896) for wave diffraction around a semi-infinite breakwater and by MacCamy and Fuchs (1954) for wave diffraction around a large fixed vertical circular cylinder. Both theories are based on potential flow and linear wave assumptions. Thus, the theories are applicable only to cases where flow separation around the structure is negligible.

3.11.1   Wave diffraction around a semi-infinite breakwater

Consider a linear wave approaching a semi-infinite breakwater from a normal direction (Figure 3.20). By reducing the 3D Laplace equation to 2D Helmholtz equation on a horizontal plane, Sommerfeld (1896) found the analytical expression for the complex velocity potential function $\theta(x, y, z, t) = Z(z)F(x, y)e^{i\sigma t}$, which contains both wave amplitude and phase information. The expression of $F(x, y)$ is:

$$F(x, y) = \frac{1+i}{2}\left[e^{-ikx}\int_{-\infty}^{\beta} e^{-i(\pi/2)u^2}\,du + e^{ikx}\int_{-\infty}^{\beta'} e^{-i(\pi/2)u^2}\,du\right] \qquad (3.139)$$

where:

$$\beta^2 = \frac{4}{L}(r-x), \quad \beta'^2 = \frac{4}{L}(r+x), \quad r = \sqrt{x^2+y^2} \qquad (3.140)$$

The sign of (β, β') is $(+,+)$ for $x < 0$ and $y < 0$ $(-,-)$ for $x > 0$ and $y < 0$, and $(+,-)$ for $y > 0$. Based on $F(x, y)$, the diffraction pattern around the breakwater can be determined (see Appendix VI).

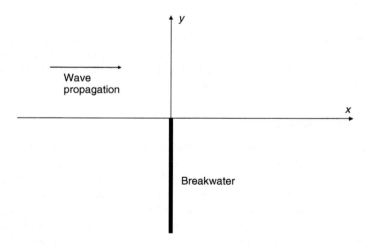

Figure 3.20 Sketch of a wave diffraction behind a semi-infinite breakwater.

The original Sommerfeld solution is exact but complicated in algebra. For large values of x (i.e., $x > 2L$), the approximate solution was proposed by Penney and Price (1952a):

$$D(x, y) = H(x, y)/H_0 = \left| \frac{1}{\sqrt{2}} \left\{ \left[\frac{1}{2} \pm C(\sigma_\pm) \right] + i \left[\frac{1}{2} \pm S(\sigma_\pm) \right] \right\} e^{-i\pi/4} \right|$$

(3.141)

where D is the diffraction coefficient that represents the ratio of local wave height to incident wave height H_0, and the Fresnel integrals $S(\sigma)$ and $C(\sigma)$ are:

$$S(\sigma) = \int_0^\sigma \sin\left(\frac{\pi u^2}{2} \right) du, \quad C(\sigma) = \int_0^\sigma \cos\left(\frac{\pi u^2}{2} \right) du \text{ with } \sigma_\pm = \frac{y_\pm}{\sqrt{\pi x / k}}.$$

Figure 3.21 gives the comparison of the wave diffraction coefficient behind the breakwater based on the exact and approximate theories. It is clear that

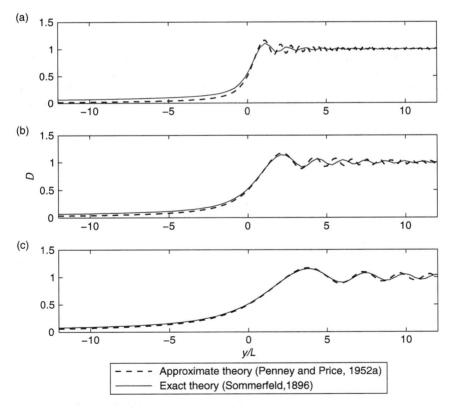

Figure 3.21 The diffraction coefficient behind a semi-infinite breakwater at (a) $x/L = 1.62$, (b) $x/L = 6.50$, and (c) $x/L = 19.50$.

the approximate theory underestimates the wave energy spreading into the shadow area near the breakwater. Away from the breakwater, the two theories are rather close.

3.11.2   Wave diffraction around a large vertical circular cylinder

Wave diffraction around a large body is often an important concern to offshore engineers. Based on the linear wave theory, MacCamy and Fuchs (1954) found the analytical solution of the velocity potential function for the wave field around a surface-piercing vertical circular cylinder. This leads to the closed-form solution for the diffraction coefficient around the cylinder:

$$D(r, \theta) = \frac{H(r, \theta)}{H_0}$$

$$= \left| \frac{1}{2} \sum_{m=0}^{\infty} \delta_m i^{m+1} \left[J_m(kr) - \frac{J'_m(ka)}{H_m^{(1)'}(ka)} H_m^{(1)}(kr) \right] \cos m\theta \right|$$

(3.142)

where $\delta_0 = 1$ and $\delta_m = 2$ for $m \geq 1$; $H_m^{(1)}$ is the Hankel function of the first kind of order m, and J_m is the Bessel function of the first kind of order m.

Figure 3.22 gives the analytical solution for the diffraction coefficient along the circular cylinder (see Appendix VII). It is seen that in the frontal part of the cylinder, the wave amplitude is about doubled due to the presence

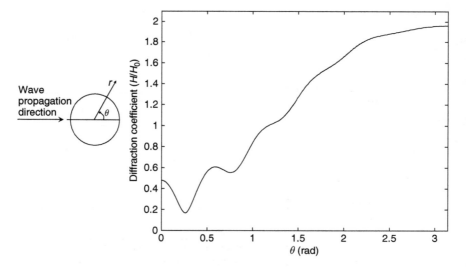

Figure 3.22 The closed-form solution of the wave diffraction coefficient around a circular vertical cylinder; wave period $T = 2$ s, water depth $h = 80$ m, and the diameter of cylinder $D = 10$ m.

of a partial standing wave by reflection. The wave amplitude reduces toward the rear part of the cylinder in an oscillatory manner due to wave diffraction in the sheltered region. MacCamy and Fuchs's theory was later extended to multiple circular structures (Spring and Monkmeyer, 1974; Linton and Evans, 1990) and porous cylinders (Wang and Ren, 1994).

3.11.3   Wave diffraction around a large floating circular thin plate

Recently, efforts have also been made to develop the exact solution for wave diffraction around a floating circular thin plate with zero drift. By using long-wave approximation, Zilman and Miloh (2000) proposed an exact solution in terms of angular eigenfunctions to express the wave velocity potential around a body. Later, Watanabe *et al.* (2003) extended the analytical solution to finite water depth. In their solution, the deflection of the plate due to wave–structure interaction is also formulated based on Mindlin's (1951) plate theory. As a result, the hydroelastic responses of the plate to the wave force and its influence on the wave diffraction field are coupled together. The solution is used to evaluate the behavior of a pontoon-type circular VLFS. The wave velocity potential around the plate can be written as the sum of diffraction and radiation potentials:

$$\phi(r, \theta, z) = \sum_{n=0}^{N} \phi_{Dn}(r, z)\cos(n\theta)$$

$$+ i\sigma \sum_{n=0}^{N} \sum_{\substack{m=0(n=0,1) \\ m=1(n\geq2)}}^{M} \zeta_{nm}\phi_{Snm}(r, z) \cos(n\theta) \tag{3.143}$$

where $\phi_{Dn}(r, z)$ is the potential function associated with wave diffraction, $\phi_{Snm}(r, z)$ is the potential function associated with wave radiation due to body deformation, and ζ_{ns} is the modal amplitude. The details of the evaluation of these variables can be found in Watanabe *et al.* (2003). When Young's modulus of elasticity approaches infinity, the solution represents the wave diffraction around a rigid and fixed floating thin circular plate.

3.11.4   Combined wave refraction and diffraction around an island

When a wave train approaches an island, it is subjected to combined wave refraction and diffraction due to the change of bottom topography and the obstruction from the island. Wave height can be amplified by many times in the rear part of the island provided there is favorable bottom geometry. Using the long-wave theory, Homma (1950) derived a closed-form solution of wave height distribution around a circular island mounted on a paraboloidal shoal. Recently, Liu *et al.* (2004) extended the problem to shorter waves and proposed an analytical solution in power series form based on the MSE.

3.11.5   Wave diffraction and scattering around a porous structure

When a linear wave train propagates over a porous structure, the permeability of the structure will affect the diffraction pattern around the structure. Theoretical expressions of wave reflection and transmission over a rectangular porous breakwater were derived for linear waves (e.g., Sollitt and Cross, 1972; Dalrymple *et al.*, 1991). Using potential flow theory, Yu and Chwang (1994) studied the general characteristics of wave transformation through porous structures of various shapes. Yu (1995) further extended the study to wave diffraction around porous breakwaters.

3.12   Wave damping

Wave damping refers to the continuous reduction of wave amplitude in the course of wave propagation. There are many causes for wave damping, such as bottom friction, bed percolation, wave breaking, and adverse surface wind effect. Besides, the surface tension and viscous effect will result in additional wave damping given sufficient time, which is especially important for ship hull design. Zilman and Miloh (2001) studied the effect of surface tension on surface wave behavior. Wu *et al.* (2001) presented a linearized analytical solution for viscous wave damping in a confined tank. Recently, Chen *et al.* (2006) proposed a study of gravity wave under the effects of both surface tension and fluid viscosity. In this section, only wave damping due to bottom effects, the major cause for most nonbreaking waves, will be elaborated.

3.12.1   Wave damping due to bottom friction

3.12.1.1   Laminar boundary layer

For a rigid and impermeable bottom, wave damping is mainly caused by bottom friction inside the wave-induced boundary layer, whose thickness can be estimated as follows:

$$\delta \approx O(\sqrt{\nu T}) \tag{3.144}$$

For most wind waves with wave periods less than 10 s, the boundary layer thickness is less than 1 cm, much smaller than the water depth being considered. The viscous effect within the boundary layer is significant. In this case, the wave-induced bottom shear stress can be derived analytically (Lamb, 1945: Art. 345; Dean and Dalrymple, 1991):

$$\tau_{b} = \rho\sqrt{\frac{\nu}{\sigma}}\frac{gak}{\cosh kh}\cos\left(kx - \sigma t\frac{\pi}{4}\right) \tag{3.145}$$

The bed shear stress has a phase delay of 45° from the free surface displacement. The average bed shear stress over a wave period is zero.

If the phase lag is neglected, we can express the bed shear stress in the following conventional form similar to the current-induced bottom shear stress (see Section 3.14.1):

$$\tau_b = \rho c_{fw} u_b |u_b| \tag{3.146}$$

where c_{fw} is the effective bottom friction coefficient under the wave and $u_b = \zeta_b \sigma \cos(kx - \sigma t)$ is the near-bottom velocity derived from the potential flow theory with:

$$\zeta_b = \frac{gak}{\sigma^2 \cosh kh}$$

being the maximum horizontal excursion of the fluid particle on the bottom.

Relating (3.145) and (3.146) for the maximum values by dropping the time-dependent terms, we have:

$$\rho \sqrt{\frac{\nu}{\sigma} \frac{gak}{\cosh kh}} = \rho c_{fw} \left(\frac{gak}{\sigma \cosh kh} \right)^2 \tag{3.147}$$

which gives:

$$c_{fw} = \sqrt{\frac{\nu}{\zeta_b u_{b\,max}}} = \frac{1}{Re^{1/2}} \tag{3.148}$$

where $u_{b\,max} = \zeta_b \sigma$. The above expression is valid for $Re < 10,000$ for a smooth bed.

The wave damping rate is determined by the rate of work done by bed shear stress. By assuming that the amplitude of a progressive wave train decays exponentially, i.e.:

$$a(t) = a_0 \exp(-\alpha t) \tag{3.149}$$

where a_0 is the initial wave amplitude and α is the damping coefficient, we can find the following expression for α (Dean and Dalrymple, 1991: 266):

$$\alpha = \frac{\nu k \sqrt{\sigma/2\nu}}{\sinh 2kh} \tag{3.150}$$

The wave damping in space can readily be obtained, given wave celerity.

The laminar boundary layer under other types of waves, e.g., Stokes waves, solitary waves, and cnoidal waves, has been studied theoretically, experimentally, and numerically. Readers are referred to Lin and Zhang (2008) for more information.

3.12.1.2   Turbulent boundary layer

As the boundary layer flow becomes turbulent due to increased bottom roughness and/or wave amplitude, two effects will result. One is the reduced phase lag between the bottom shear stress and the free surface displacement due to the stronger vertical mixing rate. The other is the increased bottom shear stress by turbulence. Both effects increase the energy dissipation rate within the boundary layer. Generally speaking, the friction coefficient is the function of Re and bottom roughness. For the laminar wave boundary layer discussed earlier, the bottom roughness is small and thus the friction coefficient depends on Re only. At the other extreme, when Re becomes large enough for a fully developed turbulent boundary layer, the friction coefficient tends to depend on bottom roughness only.

Putnam and Johnson (1949) performed an early theoretical and experimental study of wave damping on an impermeable bed due to bottom friction. Based on the available theory and experiments, Kamphuis (1975) generated the wave-induced friction coefficient diagram similar to the Moody diagram. With the use of the friction coefficient in the turbulent boundary layer, the wave damping in space is found to be (Dean and Dalrymple, 1991: 269):

$$a(x) = \frac{a_0}{1 + \dfrac{2c_{fw}}{3\pi} \dfrac{k^2 a_0 x}{(2kh + \sinh 2kh)\sinh kh}} \tag{3.151}$$

Due to the nonlinearity in the bottom friction, the decay rate no longer follows an exponential pattern. The larger the value of c_{fw} is, the faster the wave amplitude reduces.

3.12.2   Wave damping by other bottom mechanisms

Besides bottom friction, other bottom mechanisms can also cause wave damping. These mechanisms include a viscous bottom mud layer that moves with wave propagation, a rigid porous bed that allows flow percolation inside, and a sandy bottom whose sediment can be suspended. The theoretical study of wave damping over a viscous mud bottom was reviewed by Dean and Dalrymple (1991:273). Lian *et al.* (1999) proposed a nonlinear viscoelastic model for mud transport under waves and currents, which was later improved and validated against laboratory data by Zhao *et al.* (2006). The wave damping over a porous bed is mainly caused by friction inside the porous bed. The earliest study of wave damping due to percolation was performed by Putnam (1949) using Darcy's unsteady law. Hunt (1959b), Murray (1965), and Liu (1973) proposed analytical solutions of wave damping over the porous bed of infinite depth by considering the linear friction effect. Liu and Dalrymple (1984) extended the study to a finite depth of the porous bed. The study was further extended by Gu and Wang (1991) to include the linearized nonlinear effect. Recently, Karunarathna and Lin

(2006) developed a fully nonlinear numerical model to study wave damping over various porous beds. The wave damping over a sandy bottom with sediment suspension involves more complex mechanisms that include the effect of sediment suspension on bottom shear stress as well as the effect of changed bottom geometry on wave energy dissipation. At this moment, there is no complete analytical study for this problem.

3.13 Nonlinear wave interaction

Nonlinear wave interaction is the process during which waves exchange energy among different wave modes. Depending on the local wave and wind condition, wave energy can be transferred from high frequency to low frequency and vice versa. The most active wave energy exchange occurs between two wave modes having close frequencies, i.e., one has the frequency of $\sigma + \Delta\sigma/2$ and the other $\sigma - \Delta\sigma/2$. In this case, the nonlinear wave interaction transfers energy to waves whose frequencies equal the difference frequency $\Delta\sigma$ and sum frequency 2σ. The phenomenon can be explained by the secondary interaction theory proposed by Tick (1963). In addition, higher order energy exchanges for ocean waves also exist. Hasselman (1968) established the theory for calculating the net energy transfer among different wave modes in a wave frequency–direction spectrum. The theory is the basis for all wave energy spectral models to compute nonlinear wave energy transfer. We shall discuss this more in Section 5.3.4.2.

3.13.1 Harmonic generation

The most common nonlinear process for wave energy transfer from low frequency to high frequency is the process of frequency doubling, during which the input wave transfers energy to the wave that has the frequency twice that of the input wave. This can easily be demonstrated by a laboratory experiment, in which a large-amplitude sinusoidal wave generated from the wave maker will eventually evolve into a second-order (or higher order) Stokes wave, given sufficiently long propagation distance. The process is also called second harmonic generation (SHG), a famous nonlinear process also observed in optics. With the further nonlinear interaction of the input wave and the second harmonic wave, the third harmonic wave with the frequency triple of the input wave can also be generated. In principle, through a series of nonlinear wave interactions, wave energy can be redirected to all discrete wave modes that have the frequency of the integer times the fundamental frequency. Such a process is often termed harmonic generation. The process sometimes is also referred to as "nonlinear resonant interaction."

An informative but not rigorous way to view how a wave transfers energy to its lower and higher harmonics by nonlinear interaction is provided below. Consider a wave group containing two wave components with two close frequencies $\sigma + \Delta\sigma/2$ and $\sigma - \Delta\sigma/2$. The free surface displacement at a particular location is then given by:

$$\eta(t) = a_1 \cos\left[(\sigma + \Delta\sigma/2)\,t + \varepsilon_1\right] + a_2 \cos\left[(\sigma - \Delta\sigma/2)\,t + \varepsilon_2\right] \tag{3.152}$$

where a_1 and a_2 are wave amplitudes associated with the two wave components. Based on linear wave theory, the local horizontal velocity under the two waves is linearly proportional to the free surface displacement and thus can be expressed similarly. A few quantities such as wave energy and wave-induced pressure (think about the Bernoulli equation) are associated with the quadratic term of the free surface displacement and velocity, i.e., η^2 or u^2. Let us now evaluate η^2 based on (3.152) as follows:

$$\eta^2 = \frac{a_1^{\,2}}{2} + \frac{a_2^{\,2}}{2} + \frac{a_1^{\,2}}{2}\cos\left[2\,(\sigma + \Delta\sigma/2)\,t + \varepsilon_1\right]$$

$$+ \frac{a_2^{\,2}}{2}\cos\left[2\,(\sigma - \Delta\sigma/2)\,t + \varepsilon_2\right] + a_1 a_2 \cos\left[2\sigma t + \varepsilon_1 + \varepsilon_2\right]$$

$$+ a_1 a_2 \cos\left[\Delta\sigma t + \varepsilon_1 - \varepsilon_2\right] \tag{3.153}$$

It is seen that the quadratic term η^2 contains two additional terms containing the second-order nonlinear interaction of the two wave modes (i.e., the terms with $a_1 a_2$). One term is related to the sum frequency (or frequency doubling when two frequencies are close) and the other is related to the difference frequency.

3.13.2   Bound and free infragravity waves

Another way to explain the transfer of the wave energy from the fundamental frequency to the lower frequency wave mode is the concept of radiation stress introduced in Sections 3.2.3 and 3.2.4. Consider again a group of two waves with the same wave amplitude and two close frequencies $\sigma + \Delta\sigma/2$ and $\sigma - \Delta\sigma/2$. In Section 3.5, we have shown that the wave group will be confined in a wave envelope that propagates at a speed of c_g (see Figure 3.11). It is easy to see that the wave height in the envelope varies in space and time with the wave number and frequency of Δk and $\Delta\sigma$, respectively. According to the concept of radiation stress, the mean water level will be drawn down in the region where the wave amplitude is large. This causes the formation of a forced long wave that will propagate together with the wave group. This wave has the angular wave frequency of $\Delta\sigma$ and is termed "bound long wave" or sometimes "bound infragravity wave." The bound long wave has 180° of phase shift to the wave envelope (Figure 3.23). The wave amplitude is generally rather small in the offshore region and the corresponding wave period ranges from 20 to 300 s. This bound long wave was first described by Munk (1949) as surf beat, which

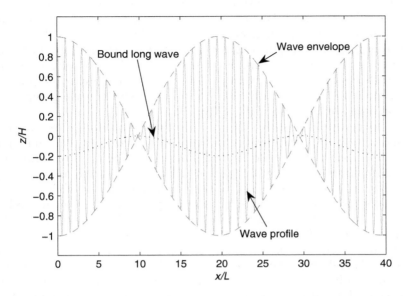

Figure 3.23 Illustration of wave group propagation and the associated bound long wave generated by a nonlinear wave interaction; $h = 100\,\text{m}$, $H = 1\,\text{m}$, $T = 4\,\text{s}$, $\Delta\sigma = 0.04\,\text{s}^{-1}$.

can also be generated by the time-varying wave set-down and set-up during the wave shoaling on a beach.

Once waves break on the beach, the mechanism that locks the bound infragravity wave to the carrier wave envelope disappears. The released long wave is called "free long wave" or "free infragravity wave" that has two fates. For a normal incident wave, it is reflected back to the offshore and is classified as the leaky mode of the free long wave. For an obliquely incident wave, most of the wave energy will be trapped in the nearshore region and form the edge wave that will propagate in the longshore direction.

3.13.3 *The Benjamin–Feir instability*

Nonlinear wave interaction also results in the so-called Benjamin–Feir instability, which was named after the theoretical work by Benjamin and Feir (1967) and sometimes is also called sideband instability (Mei, 1989: 620) or modulational instability (Dingemans, 1997: 929). In their works, the difficulty in maintaining a permanent shape of nonlinear wave trains (e.g., Stokes wave train) for long-distance propagation was explained theoretically. It was found that under certain circumstances, the wave mode of the fundamental frequency can become unstable and grow exponentially in time under the small disturbance of the wave modes with a close frequency, which is termed sideband disturbance. It is already known that with the introduction of a small sideband disturbance, a slowly modulated wave envelope will be formed. Due to nonlinear effects, the crest at the peak of the envelope will propagate faster than the nearby waves, thus shortening the waves ahead and lengthening the waves behind. The group velocity, at which the wave energy is transmitted, will increase behind but decrease in front with the change of wavelength. As a result, the crest at the peak of the envelope will continuously gain net energy from the wave groups and develop instability (see Lighthill, 1978: 462).

The phenomenon of Benjamin–Feir instability has been observed and measured for both gravity waves and capillary waves (e.g., Benjamin and Feir, 1967; Su, 1982a,b; Perlin and Hammack, 1991). The instability can be stabilized by energy dissipation, which was recently discussed by Segur *et al.* (2005) and Wu *et al.* (2006). Benjamin–Feir instability can be more rigorously explained with the use of nonlinear Schrödinger (NLS) equations, which will be detailed in Section 5.3.3.4. Using a cubic NLS equation, Hara and Mei (1991) included the wind effect to explain the downshift of frequency in Benjamin–Feir instability. In contrast, a proof within the Hamiltonian framework was given by Bridges and Mielke (1995) for Benjamin–Feir instability of the Stokes periodic wave train. Recently, Osborne *et al.* (2000) and Slunyaev *et al.* (2002) found that Benjamin–Feir instability may also contribute to the generation of rogue waves under certain circumstances.

3.14 Wave–current interaction

3.14.1 *Current-induced turbulent bottom boundary layer*

Consider a steady and uniform current above a flat impermeable bottom. The horizontal momentum equation can be reduced to:

$$\frac{\partial \cancel{u}}{\cancel{\partial} t} + \frac{\partial (\cancel{u^2})}{\cancel{\partial} x} + \frac{\partial (\cancel{u} w)}{\cancel{\partial} z} = -\frac{1}{\rho}\frac{\partial \cancel{p}}{\cancel{\partial} x} + \frac{1}{\rho}\frac{\partial R_{\cancel{xx}}}{\partial \cancel{x}} + \frac{1}{\rho}\frac{\partial R_{xz}}{\partial z} \qquad (3.154)$$

This implies that the turbulence-induced Reynolds stress in the boundary layer is a constant, i.e.:

$$R_{xz} = \rho v_{\mathrm{t}}\frac{\mathrm{d}u}{\mathrm{d}z} = \tau_{\mathrm{b}} = \rho u_*^2 \qquad (3.155)$$

where $u_* = \sqrt{\tau_{\mathrm{b}}/\rho}$ is the friction velocity with τ_{b} being the bottom shear stress. Using the mixing-length hypothesis, the eddy viscosity can be modeled by:

$$v_{\mathrm{t}} = L_{\mathrm{t}} U_{\mathrm{t}} = (\kappa z)\left(\kappa z \frac{\mathrm{d}u}{\mathrm{d}z}\right) \qquad (3.156)$$

where κ is the von Karman constant. By substituting the above definition into (3.155) and taking the vertical integration, we obtain the well-known log-law velocity profile in the turbulent boundary layer:

$$u = \frac{u_*}{\kappa} \ln \frac{z}{z_0} \tag{3.157}$$

where z_0 is the zero-crossing point of the log-law profile that is defined as:

$$z_0 = \begin{cases} \dfrac{k_n}{30} & \text{for} \quad \text{Re}_* = \dfrac{k_n u_*}{\nu} \geq 3.3 \\[2mm] \dfrac{\nu}{9u_*} & \text{for} \quad \text{Re}_* = \dfrac{k_n u_*}{\nu} < 3.3 \end{cases} \tag{3.158}$$

where k_n is the physical bottom roughness.

The above log-law profile is often used to describe the velocity profile within the thin layer of turbulent boundary, in which the shear stress does not vary significantly in the vertical direction. For free surface flows on a flat bottom, this is not true because the current is driven by a pressure gradient and thus the bed shear varies linearly in the vertical direction. In this case, we have:

$$\frac{\cancel{\partial u}}{\cancel{\partial t}} + \frac{\partial \left(u^2\right)}{\cancel{\partial x}} + \frac{\partial \left(uw\right)}{\cancel{\partial z}} = -\frac{1}{\rho}\frac{\partial p}{\partial x} + \frac{\partial R_{\cancel{xx}}}{\cancel{\partial x}} + \frac{\partial R_{xz}}{\cancel{\partial} \partial z} \Rightarrow \frac{\partial R_{xz}}{\partial z} = -\rho g \frac{\partial \eta}{\partial x} = -\rho g S \tag{3.159}$$

where S is the surface slope. Integrating the above equation in the vertical direction and applying the boundary conditions on the bottom $(\tau_{xz} = \tau_b)$ and free surface $(\tau_{xz} = 0)$, we have:

$$R_{xz} = \rho \nu_t \frac{du}{dz} = \tau_b \left(1 - \frac{z}{h}\right) = -\rho g S h \left(1 - \frac{z}{h}\right) = \rho u_*^2 \left(1 - \frac{z}{h}\right) \tag{3.160}$$

This time the eddy viscosity needs to be modeled differently by considering the fact that it reduces to zero on both the bottom and the free surface. It is found that when $\nu_t = \kappa u_* z(1 - z/h)$, the same log-law profile (3.157) can result. The same derivation is also applicable for open channel flows on a slope, where S represents the net slope of the free surface and bottom.

With the use of equation (3.157), any single velocity at an arbitrary depth can be used to determine the friction velocity and bottom shear stress. Besides, for the given water depth h, the mean velocity and the friction velocity have the following relation:

$$U = \frac{1}{h} \int_0^h u\,dz \approx \frac{1}{h} \int_{z_0}^h u\,dz = \frac{1}{h} \int_{z_0}^h \ln \frac{z}{z_0} dz$$

$$= \frac{1}{h}\frac{u_*}{\kappa}\left(h \ln \frac{h}{z_0} - h + z_0\right) \approx \frac{u_*}{\kappa}\left(\ln \frac{h}{z_0} - 1\right) \tag{3.161}$$

The bed shear stress can then be related to the mean flow velocity in the following way:

$$\tau_b = \rho u_*^2 = \rho \left(\frac{\kappa}{\ln(h/z_0) - 1} \right)^2 U^2 = \rho c_f U^2 \tag{3.162}$$

where c_f is the friction coefficient. The above equation has a form similar to the Chezy formula or Manning's equation, and they can be interchanged with each other when the coefficients associated with each equation are properly converted.

3.14.2   Wave-induced current

Wave motion can induce current in various ways and they are summarized below.

3.14.2.1   Stokes drift

Stokes drift is one of the most common wave-induced currents (see Section 3.2.2). The mechanism of generating Stokes drift has been discussed before. In open sea, Stokes drift creates a net mass flux in the wave propagation direction. In a close basin (e.g., wave tank, wave basin, continental shelf), because the net mass transport is zero, a returned flow will be formed near the sea bottom (see Figure 3.24 for illustration).

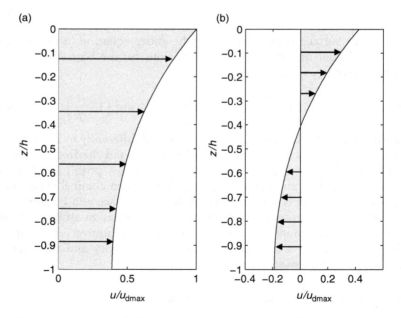

Figure 3.24 Stokes drift in (a) open sea and (b) closed basin; water depth $h = 10\,\text{m}$, wave height $H = 2\,\text{m}$, and wave period $T = 8.71\,\text{s}$; $u_{\text{dmax}} = 0.0943\,\text{m/s}$ is the maximum Stokes drift at $z = 0$.

3.14.2.2 Boundary layer drift

Due to the viscous effect, the horizontal and vertical velocities within the boundary layer do not have 90° of phase shift as predicted by linear wave theory. As a result, the time average over a wave period for the product of u and w does not reduce to zero. This will cause an additional stress similar to Reynolds stress. The vertical gradient of such stress will induce a flow in the same direction as wave propagation. Longuet-Higgins (1953) gave the explanation of this flow and termed it the boundary layer drift.

3.14.2.3 Undertow

While Stokes drift arises purely from the net mass flux of a linear wave based on its nonlinear property, the excess momentum flux can induce additional current under certain circumstances. In Sections 3.2.3 and 3.2.4, we introduced the concept of radiation stress and stated that the imbalance of radiation stress is the cause of wave set-down and set-up, which create mean surface slope to balance the radiation stress gradient. The mean wave set-down and set-up can be calculated by considering the depth-integral quantities of radiation stress. After waves break, the magnitude of the radiation stress gradient is quite large near the free surface and reduces quickly toward the bottom. The excess pressure gradient, however, is nearly uniform across the water depth. As a result, the net mass flux will be created near the free surface, accompanied by the returning flow near the bed. This flow is called undertow and it plays a profound role in sediment transport in surf zones. Nielsen (1992: 60) has discussed the quantification of undertow in a surf zone.

3.14.2.4 Longshore current

Besides the undertow that is in the cross-shore direction, there is another type of current in the longshore direction called longshore current. A longshore current is normally induced by an obliquely incident wave train that bends toward the shoreline due to refraction, during which the wave height changes locally and causes the change of local radiation stress S_{xx}, S_{yy}, and S_{xy} (see Section 3.2.3). The gradient of S_{xy} in the onshore–offshore direction will create a current in the longshore direction. However, such a current is generally quite weak before the wave breaks because the variation of wave height is small. After the wave breaks, the wave height decays rapidly. A strong longshore current can be generated that can reach up to 0.5 m s^{-1} in certain areas (Figure 3.25).

3.14.2.5 Rip current

Rip currents will form when waves approach a curved beach from a nearly normal incidence. Due to the nonuniformity of the beach profile and/or the

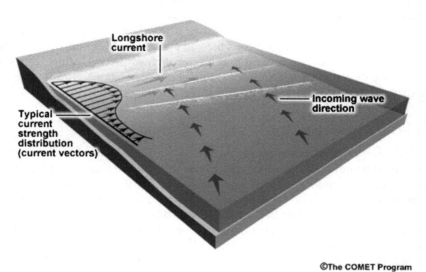

Figure 3.25 Illustration of the generation of longshore current by obliquely incident waves (Courtesy of COMET program).

wave field, waves can break strongly in one place and weakly in another place. A horizontal flow circulation will be formed within which a narrow channel of water flows fast in the offshore direction (Figure 3.26). This flow is called rip current, which has a width ranging from 15 m to 50 m and a speed from 0.3 m/s to 2.0 m/s. The generation of rip current is basically the result of the mass balance of the excess water driven by wind and waves from offshore. The effect of the rip current generally extends to deep water beyond the sandbars that separate the onshore and offshore regions.

3.14.3 Wave effects on current

3.14.3.1 Apparent bed roughness

Besides the wave-induced currents, there are other types of currents in oceans. Examples include large-scale ocean circulations driven by wind or the difference of density due to the change of salinity and/or temperature, tidal currents, and river flows in estuaries. One of the main wave effects on current is to increase the apparent bed roughness felt by the current. In other words, when a current enters into a wave field, it flows over a rougher bed and thus the effective bottom friction increases.

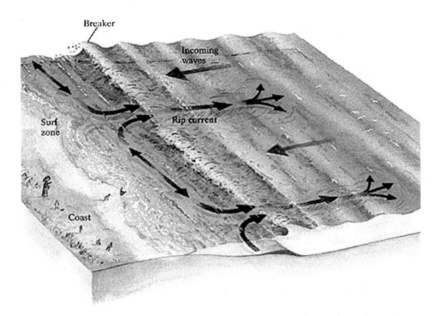

Figure 3.26 Illustration of the generation of rip currents on a nonuniform beach.

The increased bed shear stress enhances the current capacity of initiating bottom sediments and suspends them. This explains why large-amplitude waves riding on a current can significantly increase the turbidity level in coastal regions. In this case, the bed shear stress is the function of the cross flow (current), the physical roughness of the bed, and the wave condition. It has been found that the wave–current interaction angle does not have significant influence on bottom friction (e.g., Arnskov *et al.*, 1993). Grant and Madsen (1986) gave the formula to estimate the time-varying bed shear stress under the combined wave–current action.

3.14.3.2 Current profile under a wave

It is generally believed that under the influence of a wave, the near-bed current profile can still be described by the classical log-law, though the bed friction velocity and the apparent roughness will be modified. The current profile near the mean water level, however, can be significantly modified by waves. Such a change in current profile has important applications in the study of the fate of substances in large-scale ocean circulations. It was reported by Kemp and Simons (1982, 1983) from laboratory measurements that the current profile will be tilted back when waves propagate with the current and

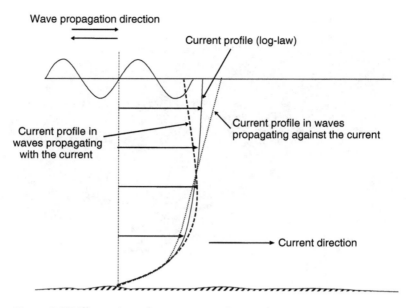

Figure 3.27 Illustration of wave–current interaction.

extended outward when waves propagate against the current (Figure 3.27). This is somehow against the intuition and knowledge from Stokes drift, but it was confirmed by Klopman's experiments later (1994).

To explain this phenomenon, let us start with the time-averaged equation (3.54) we derived earlier for 3D wave-induced radiation stress:

$$-\rho g \frac{\partial \overline{\eta}}{\partial x} = \frac{\overline{\partial \rho (u^2 - w_{\mathrm{D}}^2)} + W_{xx}^{(2)}}{\partial x} + \frac{\overline{\partial \rho (uw)}}{\partial z} \tag{3.163}$$

where $W_{xx}^{(2)}$ is the additional normal radiation stress caused by nonzero mean pressure of the Lagrangian orbit motion of fluid particles under waves. Also note that we add a subscript D for w^2 to indicate that the contribution of this term is caused by the dynamic pressure associated with waves only. Although the above equation was derived for waves only in Section 3.2.3, it can be extended to include current, waves (variables with tilde sign), and turbulence:

$$u = \bar{u} + \tilde{u} + u', \quad w = \bar{w} + \tilde{w} + w' \tag{3.164}$$

Substitute (3.164) into (3.163), we have:

$$-\rho g \frac{\partial \bar{\eta}}{\partial x} - \rho \frac{\partial \left(\bar{u}^2 + \overline{u'^2} \right)}{\partial x} = -\frac{\partial}{\partial x} \left[-\rho \left(\bar{u}^2 - \bar{w}^2 \right) + W_{xx}^{(2)} \right]$$

$$+ \frac{\partial \rho \left(\overline{uw} + \overline{\tilde{u}\tilde{w}} + \overline{u'w'} \right)}{\partial z} \qquad (3.165)$$

To arrive at the above equation, the correlations between mean flow and wave, between mean flow and turbulence, and between wave and turbulence are assumed to be zero under the time average. The time averages of $\bar{w}$ and w' are also set to zero. By realizing that in general $\overline{u'^2} \ll \overline{u}^2 = \bar{u}^2$ and $W_{xx}^{(1)} = -\rho \left(\overline{\tilde{u}^2} - \overline{\tilde{w}^2} \right)$ based on the previous definition, we then have:

$$-\rho g \frac{\partial \bar{\eta}}{\partial x} - \rho \frac{\partial \bar{u}^2}{\partial x} = -\frac{\partial}{\partial x} \left[W_{xx}^{(1)} + W_{xx}^{(2)} \right] + \rho \frac{\partial \left(\overline{uw} + \overline{\tilde{u}\tilde{w}} + \overline{u'w'} \right)}{\partial z} \qquad (3.166)$$

The above equation suggests that the change in the mean velocity and the mean water level in the x-direction is balanced by the gradient of wave-induced radiation stresses $\rho \overline{\tilde{u}\tilde{w}}$, $W_{xx}^{(1)}$, and $W_{xx}^{(2)}$, turbulence-induced Reynolds shear stress $\rho \overline{u'w'}$, and additional shear stress caused by nonzero mean flows $\rho \overline{uw}$.

The above equation can be rearranged into:

$$\frac{\partial R_{xz}}{\partial z} - \rho g \frac{\partial \bar{\eta}}{\partial x} = \left\{ -\frac{\partial}{\partial x} \left[W_{xx}^{(1)} + W_{xx}^{(2)} \right] + \rho \frac{\partial \overline{\tilde{u}\tilde{w}}}{\partial z} \right\} + \left[\rho \frac{\partial \bar{u}^2}{\partial x} + \rho \frac{\partial \left(\overline{uw} \right)}{\partial z} \right]$$

$$(3.167)$$

where $R_{xz} = -\rho \overline{u'w'}$ is the Reynolds shear stress and the RHS represents the mean stress contribution from waves and mean flow. When the RHS is neglected, this returns back to the open channel flow whose mean shear stress can readily be derived by taking the vertical integration as follows:

$$\rho \nu_t \frac{\partial u}{\partial z} = R_{xz} = \tau_b \left(1 - \frac{z+h}{h} \right) \qquad (3.168)$$

where $\tau_b = -\rho g h \left(\partial \bar{\eta} / \partial x \right)$.

Now consider the waves propagating with the current. Based on Klopman's experiments (1994), the mean current velocity near the free surface curls back, giving negative $\partial \bar{u} / \partial z$ and implying that the Reynolds shear stress becomes negative near the free surface. One possible explanation is that the presence of the waves introduces additional shear near the free surface. By neglecting the mean flow stress contributions in (3.167), the equation can be solved again to obtain with the use of (3.51), (3.52), and (3.53):

$$R_{xz} = \tau_b \left(1 - \frac{z+h}{h}\right) + \int_{-h}^{z} \left\{-\frac{\partial}{\partial x}\left[W_{xx}^{(1)} + W_{xx}^{(2)}\right] + \rho \frac{\partial \overline{\overline{uw}}}{\partial z}\right\} dz$$

$$= \tau_b \left(1 - \frac{z+h}{h}\right) + \frac{\partial}{\partial x}\int_{-h}^{z}\left[-W_{xx}^{(1)} - W_{xx}^{(2)}\right] dz + \left[-W_{xx}^{(1)} - W_{xx}^{(2)}\right]\frac{\partial h}{\partial x} - W_{xz}$$

$$= \tau_b \left(1 - \frac{z+h}{h}\right) + \frac{\partial}{\partial x}\left(\frac{2kE(z+h)}{\sinh 2kh} + \frac{E\sinh^2 k(z+h)}{2\sinh^2 kh}\right) + \left[\frac{2kE}{\sinh 2kh}\right.$$

$$\left. - \frac{E\sinh 2k(z+h)}{2\sinh^2 kh}\right]\frac{\partial h}{\partial x} - \frac{2kE}{\sinh 2kh}\frac{\partial h}{\partial x} - (z+h)\frac{\partial}{\partial x}\left(\frac{kE}{\sinh 2kh}\right)$$

$$= \tau_b \left(1 - \frac{z+h}{h}\right) + (z+h)\frac{\partial}{\partial x}\left(\frac{kE}{\sinh 2kh}\right) + \frac{\partial}{\partial x}\left(\frac{E\sinh^2 k(z+h)}{2\sinh^2 kh}\right)$$

$$+ kE\left(\frac{2}{\sinh 2kh} - \frac{\sinh 2k(z+h)}{2k\sinh^2 kh}\right)\frac{\partial h}{\partial x} \tag{3.169}$$

To make the Reynolds shear stress negative near the free surface, the RHS must be negative near $z = 0$. On a flat bottom when h is a constant, this can happen only when wave amplitude decays in the wave propagation direction (i.e., $\partial E/\partial x < 0$). With an uneven bottom where $\partial h/\partial x \neq 0$, the situation becomes more complex. Note that the above derivation is based on the pure linear wave theory. In the presence of both wave and current, nonlinear interaction among waves, current, and turbulence can be rather complicated and the above equation is inadequate. You (1996) made an attempt to model the wave-induced stresses under the influence of current as a lumped term by an empirical formula. But the empirical formula does not include important wave parameters such as wave height and wave number and thus it is not applicable for general wave–current interaction problems.

On a flat bottom, the change of mean water level in the current direction will inevitably cause the flow to be nonuniform. In this case, the additional contribution caused by $\rho \overline{u}^2$ and $\rho \overline{uw}$ on the RHS of equation (3.167) is not negligible. If the flow accelerates or decelerates in the x-direction, a nonzero $\overline{w}$ will result. The additional stress contribution $\rho \overline{uw}$ may become relatively significant near the free surface where the turbulent Reynolds stress approaches zero, even though $\overline{w}$ can be a few orders of magnitude smaller than $\overline{u}$.

In recent years, intensive studies have been performed in surface wave interaction with current, driven by the need to better understand the sediment transport and contaminant transport in coastal water. Huang and Mei (2003) showed theoretically the boundary layer effect on wave–current interaction. Yang *et al.* (2005) discussed the role of mean vertical velocity $\overline{w}$ in the study of wave–current interaction. Umeyama (2005) measured Reynolds stresses and velocity distribution in the combined wave and current flow. Shi and Lu (2007) simulated the turbulent flow structure under the combined wave and current. However, the explanations of the changed current profiles due to the presence of waves are still controversial and it requires more experimental evidence and theoretical exploration to arrive at consensus on this subject.

3.14.3.3   *Langmuir circulation*

During wave–current interaction, the mean current profile can also be affected by the transverse flow that is identified as the secondary flow (circulation) in open channels (e.g., Dingemans *et al.*, 1996). In open seas or lakes, both waves and current will be generated in the direction of the wind. In addition, the so-called Langmuir circulation may also be formed, similar to the secondary flow in wide open channels (Figure 3.28). As a result, the water particles will move forward in a spiral fashion (readers should not confuse this horizontal spiral motion with the wind-induced vertical Ekman spiral in oceans due to the Coriolis effect; see Figure 3.28 for illustration). Substances with smaller density, for example, air bubbles or flotsam, will tend to converge into lines parallel to the wind. These lines are called windrows, below which there exist downwelling water motions. In oceans with strong winds, the downwelling motion can be so strong that it will pull down air bubbles and other planktons even hundreds of meters below the water surface. The substances will gradually return to the surface, following the spiral path together with nutrients and cool water. This is an important mechanism in oceans for vertical mixing and substance exchange. The review article for Langmuir circulation can be found in Leibovich (1983).

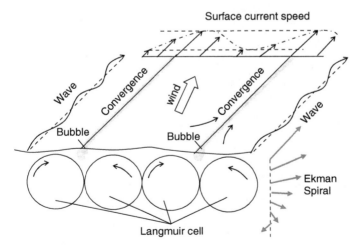

Figure 3.28 Sketch of Langmuir circulation and Ekman spiral.

3.14.4　*Current effects on wave*

Current can also affect waves in a few different ways, which are summarized as follows.

3.14.4.1　*Change of wavelength*

One of the most pronounced effects of current on a wave is the change of wavelength, which depends on the current speed and current angle to the wave propagation direction. When a wave train meets a following current, the wavelength will be elongated. In contrast, when a wave train propagates against a current, the wavelength will be shortened. A simple equation to relate a current to waves is as follows:

$$\omega = \omega(\mathbf{k}, x, y) = \sigma + \mathbf{k} \cdot \mathbf{U} \tag{3.170}$$

where ω is the absolute (or apparent) angular frequency and σ is the relative (or intrinsic) angular frequency that is related to wave number k by the linear dispersion equation. The variables ω and σ represent the observed wave frequencies from a fixed frame and a frame moving with the current, respectively. For a steady-state current and wave field, the absolute angular frequency ω is a constant over the field and the relative angular frequency σ is the function of $\mathbf{U}(x,y)$.

Generally speaking, this change of wave property perceived by an observer moving relative to the source of the waves is called *Doppler effect*, named after Christian Andreas Doppler. Figure 3.29 shows an example of Doppler

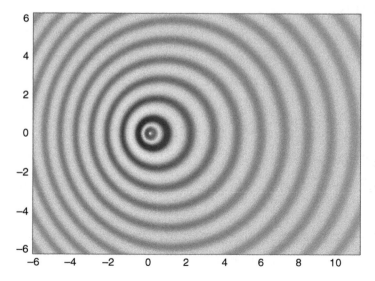

Figure 3.29 Illustration of Doppler effect when a point-source of wave is exposed to a current moving to the right.

effect for a stationary wave source exposed to a uniform current moving to the right. The wave propagates against the current on the left side where the observed wavelength is shortened and it propagates following the current on the right side where the observed wavelength is elongated.

To better explain this phenomenon mathematically, let us first consider a current that follows a linear wave train. Equation (3.170) can be rewritten as follows:

$$\omega - k|U| = \pm\sigma = \pm\sqrt{gk\tanh kh} \qquad (3.171)$$

For a given absolute frequency and current speed, there are two possible mathematical solutions for the intrinsic frequency. One solution has smaller magnitudes of k and σ and thus a longer wavelength than that without current. This corresponds to a slow current that waves always catch up with when the observer is riding on the current from behind. The other possible solution is associated with the negative values of σ and k. This solution corresponds to a fast current that overtakes the wave.

For an opposing current, the equation takes the following form:

$$\omega + k|U| = \sigma = \sqrt{gk\tanh kh} \qquad (3.172)$$

Mathematically, this will always result in the increase of k and thus a shorter wave. Physically, since wave energy is transmitted at the speed of group velocity, when the opposing current speed exceeds wave group velocity, the waves will be essentially blocked and the wave energy will dissipate into turbulence by wave-breaking processes. This is the typical situation in a river mouth where the incoming waves feel the increasing current when they are making the way against the river flow. Shorter waves are blocked first due to their smaller group velocities. At the location of wave blocking, strong air entrainments can be induced by wave breaking, forming a visible breaker line.

Longer waves, however, can penetrate further into the river. Some strong high tides can propagate against the river flow over 100 km further into the river delta region if the river mouth has a slowly expanding width. During the propagation of the tidal flow, the so-called tidal bore will be formed that looks like a breaking wave front. There are only a few places in the world where tidal bores form and propagate for very long distances. One of the well-known tidal bores is the Qiantang river tidal bore, which is formed due to the high tidal level and special topography in Hangzhou Bay that is in the shape of a trumpet (Figure 3.30). In a full-moon high tide, the tidal bore can propagate all the way to Hangzhou city, which is about 130 km from the Qiantang river mouth (see Figure 3.30).

Note that a similar equation with the change of sign is also applied to a ship moving in a wave field, i.e.:

$$\omega = \omega(\vec{k}, x, y) = \sigma - \vec{k} \cdot \vec{U} \qquad (3.173)$$

Figure 3.30 Map of Hangzhou Bay (top); and the Qiantang Tidal Bore at Haining (bottom). (Permission from Google Earth™ Mapping Service)

where U is the ship velocity. In this case, ω is called the encounter angular frequency and σ the intrinsic angular frequency, representing the observed frequency from the moving ship and a fixed frame, respectively. Different from the case of wave–current interaction, the intrinsic angular frequency σ is a constant whereas the encounter frequency ω is the function of $U(x,y)$.

3.14.4.2 Wave refraction by current

When waves propagate through a nonuniform current, they will experience a similar refraction effect as they propagate over a changing topography. The theoretical framework was developed by Longuet-Higgins and Stewart (1961) and extended by many researchers later (e.g., Dingemans, 1997: 336). For the complete review of the theory, readers are referred to Mei (1989: 89).

3.15 Wave–structure interaction

Wave–structure interaction is a branch of more general fluid–structure interaction (FSI). In the study of wave–structure interaction, focus is often on one or all the following aspects: (a) the change of wave motion due to the presence of a structure; (b) the wave loads on the structure; and (c) the dynamic response (motion) of the structure due to the wave loads and its effect from the further change of wave motion. We will now discuss these issues respectively.

3.15.1 Wave scattering by structure

The presence of a structure changes flow motion locally, which in turn causes wave transformations including wave diffraction, wave refraction, wave reflection, and wave breaking. The characteristics of wave transformation and diffraction have been discussed in earlier sections. In this section, we will further discuss wave transmission and reflection from bottom geometry (e.g., breakwater, sills, and trench) and porous structures that have important applications in coastal engineering.

3.15.1.1 Impermeable structure

For a submerged impermeable structure, part of the wave energy will transmit over the structure and form the transmitted wave that has smaller wave amplitude. Part of the wave energy will be reflected back from the structure and form the reflected waves. Near the structure, local vortex and turbulence may be generated that cause energy dissipation. The effectiveness of a structure in blocking wave energy is measured by the reduction of wave transmission. While the use of coastal protection is to minimize the wave transmission, a strong reflection is normally not preferred because of navigation concerns. In this case, a design can be made to dissipate wave energy locally near the structure through wave breaking or vortex shedding and turbulence dissipation. This, however, may increase the potential damage and failure of a structure by direct wave impact and/or foundation erosion. A practical design needs to be a compromise to balance wave energy transmission, reflection, and dissipation.

Long-wave approximation: Lamb (1945: Art. 176) gave analytical solutions of reflection and transmission coefficients for wave propagation over a step by using the linear long-wave approximation:

$$\frac{H_R}{H_I} = \frac{1 - \sqrt{1 - h_1/h_0}}{1 + \sqrt{1 - h_1/h_0}}, \quad \frac{H_T}{H_I} = \frac{2}{1 + \sqrt{1 - h_1/h_0}} \tag{3.174}$$

where H_I is the incident wave height, H_T is the transmitted wave height, H_R is the reflected wave height, h_0 is the still water depth for the incident wave, and h_1 is the still water depth on the step. In the above expressions, we have $1 \geq H_R/H_I \geq 0$ and $2 \geq H_T/H_I \geq 1$. It is easy to prove that the above equations satisfy the total energy conservation among the incident, transmitted, and reflected waves for a linear wave train, i.e.:

$$c_{gd} H_R^2 + c_{gs} H_T^2 = c_{gd} H_I^2 \tag{3.175}$$

where $c_{gd} = c_d = \sqrt{gh_0}$ is the group velocity in the deep water while $c_{gs} = c_s = \sqrt{g(h - b)}$ is the group velocity on the step. Equation (3.175) can be applied to all long-wave transformations over a changing topography with negligible wave energy dissipation.

Mei (1989: 130) extended the analysis to more general case of long waves passing through a step with a finite length of $2a$. The still water depths in front of, on, and behind the step are h_0, h_1, and h_2, respectively. The reflection coefficient was found to be:

$$\frac{H_R}{H_I} = -\frac{(1 - s_{01})(1 + s_{21})e^{-i\phi} - (1 + s_{01})(1 - s_{21})e^{i\phi}}{(1 + s_{01})(1 + s_{21})e^{-i\phi} - (1 - s_{01})(1 - s_{21})e^{i\phi}} \tag{3.176}$$

where $\phi = k_1(2a)$, $s_{01} = (k_0 h_0)/(k_1 h_1)$, and $s_{21} = (k_2 h_2)/(k_1 h_1)$ with k_0, k_1, and k_2 representing wave numbers in h_0, h_1, and h_2. The corresponding transmission and reflection coefficient can then be obtained by:

$$K_R = \left| \frac{H_R}{H_I} \right| = \sqrt{\frac{(s_{21} - s_{01})^2 + (s_{01}^2 - 1)(s_{21}^2 - 1)\sin^2\phi}{(s_{21} + s_{01})^2 + (s_{01}^2 - 1)(s_{21}^2 - 1)\sin^2\phi}}, \quad K_T = \sqrt{1 - K_R^2} \tag{3.177}$$

It is interesting to note that full transmission will be obtained if $h_0 = h_2$ and $\phi = k_1(2a) = n\pi$, during which the structure looks transparent to the incident waves. In contrast, the maximum reflection will be attained when $\phi = k_1(2a) = (n - 1/2)\pi$.

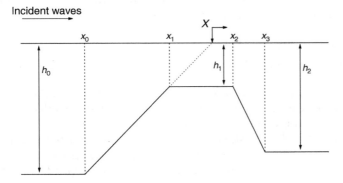

Figure 3.31 Sketch of wave transmission over a submerged obstacle of trapezoidal shape; the origin is defined at the intersection of the frontal slope extrapolation and the still water level.

When a slope is presented in front of a step, analytical solution is available only for some special slope profiles (e.g., power law function). Kajiura (1961) obtained an analytical solution for a step behind a parabolic slope and Dean (1964) obtained an analytical solution for a step behind a linear slope.

To assess the performance of a submerged breakwater of trapezoidal shape (Figure 3.31; note the origin is defined at the intersection between the extension of the frontal slope and the still water depth), Lin and Liu (2005) unified the earlier study by giving the more general form of analytical solution as follows:

$$\frac{H_R}{H_I} = -\frac{z_1 - iz_2}{z_1 + iz_2} e^{-i\alpha_0} \tag{3.178}$$

where:

$$z_1 = C_{11}(\alpha_0, \alpha_1)(P\cos\phi + Q\sin\phi) + C_{10}(\alpha_0, \alpha_1)(Q\cos\phi - P\sin\phi) \tag{3.179}$$

$$z_2 = C_{01}(\alpha_0, \alpha_1)(P\cos\phi + Q\sin\phi) + C_{00}(\alpha_0, \alpha_1)(Q\cos\phi - P\sin\phi) \tag{3.180}$$

with:

$$\phi = k_1(x_2 - x_1), \quad P = C_{01}(\beta_1, \beta_2) + iC_{00}(\beta_1, \beta_2),$$
$$Q = C_{11}(\beta_1, \beta_2) + iC_{10}(\beta_1, \beta_2) \tag{3.181}$$

and:

$$C_{ij}(\mu, \nu) = \begin{vmatrix} J_i(\mu) & J_j(\nu) \\ Y_i(\mu) & Y_j(\nu) \end{vmatrix}, \quad i, j = 1, 2 \tag{3.182}$$

where J_i and Y_i are the ith order Bessel function of the first and second kind. The definitions of α and β are given below:

$$\alpha_0 = -2k_0 x_0, \qquad \alpha_1 = -2k_1 x_1$$
$$\beta_1 = 2k_1 h_1 (x_3 - x_2)/(h_2 - h_1), \qquad \beta_2 = 2k_2 h_2 (x_3 - x_2)/(h_2 - h_1) \quad (3.183)$$

Using the above closed-form solution, one can easily predict the reflected and transmitted wave heights, given the information of water depths in front of, above, and behind the breakwater, the incident wave heights and wave periods, the frontal and back slopes of the breakwater, and the length of the crown of the breakwater. Liu and Lin (2005) extended the study to the trench case where $h_1 > h_2$. A Matlab script is provided in Appendix VIII for this purpose.

The above equation can be reduced to the simpler solutions by Lamb (1945: Art. 176) and Mei (1989: 130) for a step with infinite and finite length as well as the solution by Dean (1964) for wave transformation above a shelf behind a linear slope, i.e.:

$$K_R = \sqrt{\frac{\Delta_0 \mu_1 + \Delta_1 \mu_0 - 2\nu_0 \nu_1 - \dfrac{8}{\pi^2} \dfrac{s_{01}}{\alpha_0^2}}{\Delta_0 \mu_1 + \Delta_1 \mu_0 - 2\nu_0 \nu_1 + \dfrac{8}{\pi^2} \dfrac{s_{01}}{\alpha_0^2}}} \qquad (3.184)$$

where:

$$\Delta_0 = J_0^2(\alpha_0) + J_1^2(\alpha_0), \quad \Delta_1 = J_0^2(\alpha_1) + J_1^2(\alpha_1)$$
$$\mu_0 = Y_0^2(\alpha_0) + Y_1^2(\alpha_0), \quad \mu_1 = Y_0^2(\alpha_1) + Y_1^2(\alpha_1)$$
$$\nu_0 = J_0(\alpha_0) Y_0(\alpha_0) + J_1(\alpha_0) Y_1(\alpha_0), \quad \nu_1 = J_0(\alpha_1) Y_0(\alpha_1) + J_1(\alpha_1) Y_1(\alpha_1)$$

$$(3.185)$$

Potential flow theory: To study shorter wave transmission and reflection from a submerged obstacle, potential flow theory needs to be used. Bartholomeusz (1958) was probably the first to study wave reflection from an infinite step with the use of full potential flow theory. He obtained an integral equation that governs the horizontal component of velocity at the shelf discontinuity but solved the equation only in the limiting cases of long waves and recovered the results by Lamb (1945: Art. 176). Newman (1965a, b) considered both a long obstacle and a step in finite and infinite water depths using the method of matching eigenfunction expansions. In his approach, there are two propagative modes in any domain of constant water depth and an infinite number of evanescent modes at the step discontinuity. It was found by Newman (1965a) that the reflection coefficient could become zero for some lengths of the obstacle, as confirmed later by Mei (1989) using linear long-wave approximation.

In contrast, Miles (1967) applied the variational principle to a continental shelf. He introduced the so-called scattering matrix related to the coefficients of two propagative modes on each side of the obstacle, in which the elements were determined by means of variational integrals. This variational method was later employed by Mei and Black (1969) for a rectangular obstacle in a channel of finite water depth and extended by Devillard *et al.* (1988) by using the renormalized transfer matrix to study wave propagation over a set of successive steps. In contrast, Massel (1983) employed the Galerkin method to determine the wave reflection coefficient from a shelf. A new approach was introduced by Evans and Linton (1994) who transformed the problem into a uniform strip before the approximate analytical solution was sought. Recently, Rhee (1997) proposed a second-order solution for wave transmission over a shelf and compared his results to the earlier results by Newman (1965b), Miles (1967), and Massel (1983).

For more general problems of varying bathymetry, Roseau (1952) proposed an analytical solution for a special bottom profile based on the original Laplace equation. By employing the eigenfunction expansion method, Takano (1960) developed an approximate theoretical model to simulate the propagation of a monochromatic wave over an arbitrary bathymetry, where the topography is approximated by a series of small steps. This approximate theoretical model was later adopted by Kirby and Dalrymple (1983a), Liu *et al.* (1992), and Cho and Lee (2000).

Solitary wave: Besides periodic waves, special attention has been paid to the study of solitary wave transmission and reflection from a shelf, in an attempt to study the behavior of tsunami transformation over a continental shelf. Examples of these studies include the earlier attempts by Tappert and Zabusky (1971) and Johnson (1972). Miles (1979) considered nonlinear effects during wave transformation. Mei (1986) further explored the effect of sudden change of channel geometry on solitary wave transformation on a shelf. For solitary wave interaction with a rectangular obstacle of a finite length, so far there is no reported analytical study.

Roles of vortex generation and wave breaking: All the above analyses were based on the assumption that there is no energy dissipation in the process of wave interaction with obstacles. In reality, however, a vortex will be generated at the edge of the structure, which traps wave energy that will eventually be dissipated. Besides vortex generation, wave breaking may also take place when the wave amplitude is large on the structure. The energy dissipation will affect wave transmission and reflection. The way of quantifying the energy dissipation by vortex generation and wave breaking is through laboratory experiments or numerical modeling, the former of which will be discussed below and the latter will be detailed in Section 6.4.1.

Laboratory studies can be performed by taking both free surface displacement and velocity measurements. For example, based on wave gauge data, the nonlinear wave transformation over a submerged bar was analyzed by Beji and Battjes (1993). Rey *et al.* (1992) studied the vortex generation around the edge of the structure during the propagation of linear and weakly nonlinear waves. The more detailed point measurements were taken by using laser Doppler anemometer (LDA) for velocities around the structure under wave propagation by Ting and Kim (1994), who reconstructed spatial vortex structure around the obstacle and concluded that a viscous flow model is needed to resolve the complex eddies in the flow separation domain. Recently, Jung *et al.* (2004) used particle image velocimetry (PIV) to directly resolve the spatial structure of a vortex near the edge of the rectangular structure.

The laboratory study of solitary wave transformation over a shelf or an obstacle was performed by Seabra-Santos *et al.* (1987). Losada *et al.* (1989) examined the effect of the depth discontinuity on the solitary wave transformation over a shelf. Later, Zhuang and Lee (1996) and Tang and Chang (1998) studied the vortex motion behind the obstacle during the passage of a solitary wave. Chang *et al.* (2001) employed PIV to study detailed vortex structures near two edges of rectangular obstacles. The work was furthered by Chang *et al.* (2005) for the study of cnoidal waves and by Lin *et al.* (2005) for the study of solitary wave passage over a submerged vertical plate.

3.15.1.2   Porous structure

In the process of wave interaction with a porous structure, additional wave energy dissipation will be induced when the fluid flow passes through the pores between porous particles. Most of the existing theoretical studies of wave interaction with porous breakwaters have assumed linear periodic waves and a rectangular breakwater. In these studies, the energy dissipation inside the structure was taken into account through the linearized friction coefficient, which is evaluated by applying Lorentz's principle of equivalent work. By incorporating the linearized friction coefficient, several theoretical expressions for reflection and transmission coefficients have been derived for linear waves (e.g., Sollitt and Cross, 1972), linear long waves (e.g., Kondo and Toma, 1972; Madsen, 1974), and obliquely incident linear waves (e.g., Dalrymple *et al.*, 1991; Yu, 1995) interactions with rectangular porous breakwaters. Attempts have also been made to extend these models for breakwaters of trapezoidal shape (Madsen and White, 1975) and arbitrary shape (Sulisz, 1985). Wave reflection and transmission by other nonconventional porous breakwaters such as Jarlan-type structures (Isaacson *et al.*, 2000), multislice structures (e.g., Twu and Chieu, 2000; Twu *et al.*, 2002), semicircular breakwaters (Xie, 1999), and partially perforated walls (Li *et al.*, 2002) have also been studied theoretically and/or experimentally.

For nonlinear waves, the existing theoretical studies were mainly made for solitary waves, special weak nonlinear and dispersive shallow water waves. Due to the nonlinear characteristics of solitary waves and nonlinear flow motion inside the porous media, the available theoretical models have to make various assumptions for the linearization of resistance forces inside the porous media. Theoretical expressions for reflection and transmission coefficients for a solitary wave interaction with a rectangular porous break-water were derived by Vidal *et al.* (1988) by assuming that the length of the breakwater is small compared with wavelength. In their study, flow motion within the fluid domain and inside the porous breakwater was described by linear Boussinesq equations. The friction loss within the porous media was represented by an equivalent linearized term. Silva *et al.* (2000) derived another theory to evaluate reflection and transmission coefficients for solitary wave interaction with porous breakwaters. In the derivation, the incident solitary wave was decomposed into its harmonic series through Fourier analysis. The reflected and the transmitted waves were obtained by multiplying the harmonic series for the incident wave with transfer functions, which were obtained by using plane-wave approximation based on the linear wave theory.

To consider stronger nonlinear waves and/or breaking wave interaction with porous structure, experimental studies or numerical studies must be employed. This will be elaborated in Sections 5.2.1.7 and 6.4.1 when the numerical modeling of porous flows and wave–structure interaction is discussed.

3.15.2   Wave load on rigid and fixed structure

When flow passes around a structure, it will exert forces on the structure. These forces are induced by pressure difference around the body and the viscous stresses on the body surface. In principle, by taking the surface integration of pressure and viscous stresses around the object, the total fluid flow forces can be calculated. This, however, requires the detailed inform-ation of pressure and stress distribution on the surface of the body, which is generally unavailable unless the direct numerical computation is made.

3.15.2.1   The Morison equation

In-line forces: In many engineering computations, the simpler alternative based on semitheoretical and semiempirical approaches is adopted. Let us take a closer look at the contribution of pressure and viscous stress to the total fluid force on a body. For a simple case of a steady-state flow, the total force is called total drag that can be further divided into the form drag and the skin friction. The form drag is caused by the pressure difference in front of and behind the body due to flow separation, which creates a wake region behind the body where the pressure drops due to the existence of local vortices. The skin friction is caused by the viscous force within the boundary layer around the body. Generally speaking, the form drag force predominates for large Re and bluff body whereas the skin friction predominates for small Re and long body. In engineering applications, a popular way of estimating the total flow force on a body is by the use of the following empirical formula that merges the form drag and the skin friction:

$$F_D = \frac{1}{2} C_D \rho A U |U| \tag{3.186}$$

where A is the maximum cross-section area perpendicular to the flow, U is the far-field undisturbed velocity, and C_D is the drag coefficient that is the function of Re, body surface roughness, and body shape.

When an unsteady flow is considered, additional inertial force will result, in addition to the total drag force. This inertial force is basically the result of the additional pressure gradient across the body. One contribution to this additional pressure gradient comes from the pressure gradient that exists in an unsteady flow to accelerate or decelerate the flow itself, even without the presence of the body. The other contribution comes from the added mass effect, and it is equivalent to the case when the body is accelerating or decelerating in the opposite direction in a quiescent fluid. All together, the inertial force can be expressed as follows:

$$F_I = \rho V \frac{dU}{dt} + C_A \rho V \frac{dU}{dt} = (1 + C_A) \rho V \frac{dU}{dt} = C_M \rho V \frac{dU}{dt} \tag{3.187}$$

where V is the volume of the body, $C_M = 1 + C_A$ is the inertial coefficient, and C_A is the added mass coefficient that approaches zero for a long and thin body and approaches infinity for a short and thick bluff body.

When (3.186) and (3.188) are combined, we have the following formula that can be used to calculate the total force exerted by an unsteady flow on a body:

$$F_T = F_D + F_I = \frac{1}{2} C_D \rho A U |U| + C_M \rho V \frac{dU}{dt} \tag{3.188}$$

This equation was originally proposed by Morison *et al.* (1950) and it is called the Morison equation. It is widely used to calculate wave (a particular type of unsteady flow) loads on a body with arbitrary shape.

For a bottom-mounted vertical slender circular cylinder, the total wave force on it (where $x = 0$) can be obtained by using (3.188) and the formulas in Table 3.1:

$$F = \int_{-b}^{\eta} dF \approx \int_{-b}^{0} \frac{1}{2} C_D \rho D u |u| dz + \int_{-b}^{\eta} C_M \rho \frac{\pi D^2}{4} a_x dz$$

$$= C_D D \frac{C_g}{C} E \cos \sigma t |\cos \sigma t| - C_M \pi D E \frac{D}{H} \tanh kh \sin \sigma t \tag{3.189}$$

The overturning moment about the seafloor is as follows:

$$M = \int_{-h}^{\eta} dM \approx \int_{-h}^{0} (h+z)\,dF$$

$$= C_D D \frac{c_g}{c} E \cos \sigma t |\cos \sigma t| \left\{ h \left[1 - \frac{c}{2c_g} \left(\frac{\cosh 2kh - 1 + 2\,(kh)^2}{2kh \sinh 2kh} \right) \right] \right\}$$

$$- C_M \pi DE \frac{D}{H} \tanh kh \sin \sigma t \left\{ h \left(1 - \frac{\cosh kh - 1}{kh \sinh kh} \right) \right\} \qquad (3.190)$$

The terms within the braces on the RHS simply represent the lever arms, which are $h/2$ and h for long waves and short waves, respectively.

The relative importance of the inertial and drag forces can be estimated by evaluating the Keulegan–Carpenter (KC) number, which is named after the pioneer experimental work by Keulegan and Carpenter (1958) and is defined as $KC = u_m T / f D$, where u_m is the maximum fluid particle velocity in the oscillatory flow with the period of T and D is the characteristic length of the body in the flow direction. For a linear wave load on a vertical cylinder, when C_D and C_M are of the same order of magnitude, the drag force will predominate when $KC << 10$ and the inertial force will predominate when $KC << 10$.

Lift force: When the drag force predominates over the inertial force (i.e., $KC >> 1$), vortices will be formed in the lee side of the body. For a smooth circular cylinder, when $KC >> 1$ and $Re > 30$, these vortices will shed from the cylinder alternately, causing the time-varying asymmetric flow pattern in the transverse direction (Figure 3.32). The imbalance of pressure

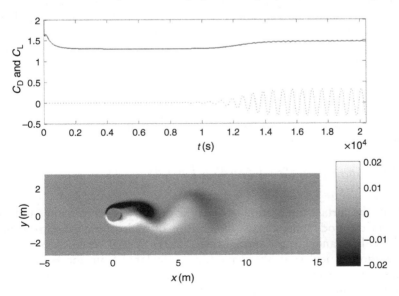

Figure 3.32 Steady current-induced vortex shedding behind a circular cylinder (bottom) and the associated time histories of drag (solid line) and lift (dotted line) coefficients (top).

on the two lateral sides of the cylinder results in the transverse force that is often termed the "lift force." This force has a zero mean but can oscillate strongly in time. The frequency of the oscillation of the lift force f is the same as the shedding frequency of vortices from the cylinder. The dimensionless frequency is represented by the so-called Strouhal number and is defined as:

$$St = f\frac{U}{D} \tag{3.191}$$

For a smooth circular cylinder, the Strouhal number is the function of Re and *KC*. For $KC \gg 1$ and $\mathrm{Re} < 10^5$, the Strouhal number is nearly constant $St \approx 0.2$. The lift force can be calculated in the same way as the drag force, i.e.:

$$F_{\mathrm{L}} = \frac{1}{2}C_{\mathrm{L}}\rho A U^2 \tag{3.192}$$

The only difference is that the lift coefficient has a different dependence on *KC* and Re and it is time dependent. Figure 3.32 shows the corresponding drag and lift coefficients as the function of time. It is seen that although the drag coefficient is also time dependent and has a frequency twice that of the lift coefficient, the fluctuation is much smaller and thus it can practically be taken as a constant in engineering computations.

3.15.2.2 *Froude–Krylov method*

When the inertial force predominates over the drag force for a body that is comparably smaller than wavelength, the Froude–Krylov (F–K) method may be used for force calculation. In this circumstance, the pressure distribution around the body can be obtained by using the linear wave theory as if there is no presence of the body:

$$p = p_{\mathrm{H}} + p_{\mathrm{D}} = -\rho gz + \rho g\frac{H}{2}\frac{\cosh k(h+z)}{\cosh kh}\cos(kx - \sigma t) \tag{3.193}$$

The force can then be obtained by taking the surface integration of the pressure around the body:

$$\mathbf{F} = \int_S (p_{\mathrm{H}} + p_{\mathrm{D}})\,\mathbf{n}\,\mathrm{d}S = \int_S p_{\mathrm{H}}\mathbf{n}\,\mathrm{d}S + \int_S p_{\mathrm{D}}\mathbf{n}\,\mathrm{d}S \tag{3.194}$$

While the surface integration of dynamic pressure generates wave-induced force on the body, the surface integration of hydrostatic pressure simply results in buoyancy. In the special case that the spatial variation of velocity acceleration is neglected, the wave force calculated from the F–K method is equivalent to the inertial force calculated from the Morison equation without considering the added mass effect, i.e.:

$$\int_S p_{\mathrm{D}}\,\vec{\mathbf{n}}\,\mathrm{d}S = \rho \mathbb{V}\,(\mathrm{d}\mathbf{u}/\mathrm{d}t)$$

Chakrabarti (1987: 234) gave the closed-form expression of the F–K method for a few bodies with common shapes of circular cylinder, circular plate, hemisphere, etc. In principle, the wave force on structures with a complex shape can also be evaluated with the use of the F–K method, as long as its overall size falls into the applicable range (i.e., $D << L$).

3.15.2.3 Diffraction theory

As the body size increases to be comparable to the wavelength, the interference of the body to the wave field becomes inevitable. Neither the Morison equation nor the F–K method can be used. In this case, the wave diffraction theory becomes a powerful tool to determine the disturbed wave field around the body, from which the wave force can be determined. Since the flow separation is no longer an important mechanism to be considered, it is justified to use potential flow theory to describe the wave field, from which various analytical solutions have been developed to describe wave diffraction and wave loads on bodies with different shape.

Diffraction theory finds its main application in offshore engineering and naval architecture, both of which involve large bodies, floating or bottom-mounted, exposed to the wave field. The classical work by MacCamy and Fuchs (1954) gave the closed-form solution of linear wave diffraction around a large vertical circular cylinder, from which the wave force is derived. Later, Spring and Monkmeyer (1974) obtained an analytical solution for wave diffraction around two unequal cylinders at an arbitrary wave incident angle. The solution of wave diffraction for multiple cylinders was given by Chakrabarti (1978). Kagemoto and Yue (1986) proposed an algebraic method to study wave diffraction around multiple 3D bodies. The method was extended by Linton and Evans (1990) to study wave interaction with arrays of vertical circular cylinders and by Yilmaz and Incecik (1998) to study wave diffraction around a group of truncated cylinders. Chakrabarti and Naftzger (1974) developed the analytical solution for wave forces on a horizontal half-circular cylinder and a bottom-seated hemisphere. For a floating circular plate (e.g., dock, pontoon), the analytical solution of wave forces on the body was proposed by Garret (1971).

People also attempted to develop nonlinear wave diffraction theory by including the higher order terms in the analysis. Using Stokes wave theory, Chakrabarti (1971), Raman *et al.* (1975), Isaacson (1977), and Molin (1979) proposed various treatments to develop the approximate analytical solution of second-order wave diffraction around a vertical circular cylinder. The second-order diffraction force on a vertical cylinder was calculated by Chau and Taylor (1992). For very large amplitude waves whose amplitude is on the order of magnitude as the cylinder diameter, Faltinsen *et al.* (1995) included third-order terms in the analysis of wave forces. McIver and McIver (1990) developed an analytical solution for wave diffraction around a submerged circular cylinder. Kim and Yue (1989) extended the analytical solution to a more general axisymmetrical body. Huang and Taylor (1996) developed a semianalytical solution for second-order wave diffraction by a truncated circular cylinder. Ghalayini and Williams (1991) presented an analytical solution for nonlinear wave forces on arrays of vertical cylinders. Sulisz (2002) presented the analytical solution for nonlinear wave forces on a horizontal rectangular cylinder.

Although the analytical studies of wave diffraction and wave forces on a large body offer good insight into physics and provide benchmark results, they are limited to relatively simple body shape. For composite structures (e.g., an offshore platform) or bodies with irregular shapes (e.g., a ship hull), the numerical modeling, which is based on the Laplace equation for large bodies or NSEs for small bodies, must be employed. More discussions on numerical modeling of wave loads on structures will be given in Sections 6.4.3 and 6.4.4.

3.15.2.4 Second-order wave drift force

Based on the linear wave theory, the mean wave force acting on a fixed body is zero for a regular wave. In reality, however, waves will exert a mean force on the body in the direction of wave propagation. This mean wave force is proportional to the square of the wave amplitude and is classified as the second-order wave force. Because this force is responsible for the slow drift motion of a floating body, it is also termed the wave drift force. A simple way of explaining the drift force is to look at a linear wave train propagating through a long floating body (e.g., the beam sea condition for a ship and thus the problem can be approximated by a 2D case near the center of the ship). Due to the presence of the body, part of the wave energy will be reflected back and the transmitted wave will have smaller wave amplitude. By examining the momentum balance in the control volume that encloses the ship, it is ready to derive the mean wave force (per unit length) by retaining the second-order terms (e.g. Longuet-Higgins, 1977):

$$\bar{f} = \frac{\rho g}{4}\left(1 + \frac{2kh}{\sinh 2kh}\right)\left(a_{\mathrm{I}}^2 + a_{\mathrm{R}}^2 - a_{\mathrm{T}}^2\right) \qquad (3.195)$$

When no energy dissipation is considered, we have $a_{\mathrm{I}}^2 = a_{\mathrm{R}}^2 + a_{\mathrm{T}}^2$. For the extreme case of full wave transmission, the wave drift force reduces to zero. In contrast, when waves are completely reflected back, the maximum drift force will be attained that equals E in deep water and $2E$ in shallow water. For a general 3D body, diffraction theory can be applied to obtain the wave drift force with the use of pressure and velocity under the diffracted wave field (Faltinsen, 1990: 136), i.e.:

$$\bar{F}_i = -\overline{\int_{S_\infty} [pn_i + \rho U_i U_n]\,\mathrm{d}S}, \quad i = 1, 2 \qquad (3.196)$$

where S_∞ can be any fixed vertical circular cylindrical surface away from the body.

When a random sea is considered, there exist additional second-order difference-frequency wave forces and sum-frequency wave forces. These forces are the result of the combined actions from different wave modes. The mechanism is similar to nonlinear wave–wave interaction that generates new waves with both difference frequency and sum frequency (see Section 3.13.1). The second-order difference-frequency wave force is responsible for the slowly varying drift motion of a floating body that will be discussed in Section 3.15.3.1, while the second-order sum-frequency wave force may induce high-frequency structure vibration.

3.15.3 Structure response to wave load

3.15.3.1 One-dimensional structure response to waves and currents and the response amplitude operator

A structure will react to wave loads by body deformation, body motion, or both of them. Let us first consider a one-DOF problem in the x-direction. The motion of a rigid body can be described by the dynamic equation of motion as follows:

$$I\ddot{x} = F_T + F_M + F_R \tag{3.197}$$

where x is the displacement of the body from its equilibrium position which is often defined as the origin (and thus $\ddot{x}$ is the acceleration of the body), I is the mass of the structure, F_T represents the force induced by fluid flow around the body, F_M represents the additional fluid force caused by the relative motion of the structure, and F_R is the restoration force (e.g., a mooring system for a floating body).

In reality, F_T and F_M are not separable. The sum of them can be obtained by the surface integration of actual pressure and stress around the body, i.e.:

$$F_T + F_M = \int_S pn_x \, dS \tag{3.198}$$

In engineering analysis, the Morison equation can be employed to approximate F_T and F_M separately, i.e.:

$$F_T = F_D + F_I = \frac{1}{2} C_D \rho A U |U| + C_M \rho V \frac{dU}{dt} \tag{3.199}$$

and

$$F_M = F_{DM} + F_{IM} = -\frac{1}{2} C_D \rho A x |\dot{x}| - C_A \rho V \ddot{x} \tag{3.200}$$

Note that the motion-induced force is also represented by the Morison equation with the change of signs and the replacement of the inertial coefficient by the added mass coefficient. Substituting the above definitions and Hook's law of linear restoration force $F_R = -Kx$ (where K is the restoring factor) into the dynamic equation, we have:

$$(I + C_A \rho V)\ddot{x} + \left(\frac{1}{2}C_D \rho A|\dot{x}|\right)\dot{x} + Kx = \frac{1}{2}C_D \rho A U|U| + C_M \rho V \frac{dU}{dt} \quad (3.201)$$

For a steady current, all the time-dependent terms will vanish. The above equation reduces to a simple algebraic equation describing the equilibrium stage where the restoring force is balanced by the flow-induced drag, i.e.:

$$x = \frac{1}{2K}C_D \rho A U|U| \quad (3.202)$$

For an unsteady flow, the above equation is a second-order nonlinear ODE. For a periodic forcing term (e.g. by waves), it is justified to assume that the motion of the structure is also periodic having the same frequency. Therefore, we have:

$$U = U_0 e^{i\sigma t} \quad (3.203)$$

$$x = x_0 e^{i(\sigma t + \delta)} \quad (3.204)$$

where U_0 is the maximum fluid particle velocity, x_0 is the maximum displacement of the structure to be determined, and δ is the phase shift. For wave-induced motion, U_0 is linearly proportional to wave amplitude, i.e., $U_0 = Ga$, where G is a coefficient. Substituting the above definitions into the equation, the ODE now becomes a nonlinear algebraic equation, from which x_0 can be solved. Since the equation is nonlinear with complex variables, only the numerical solution is available in general. However, when all nonlinear terms are neglected, the resulting algebraic equation can be simplified as follows:

$$\left[-(I + C_A \rho V)\sigma^2 + K\right]x_0 e^{i\delta} = iC_M \rho V \sigma U_0 = iC_M G \rho V \sigma a \quad (3.205)$$

This gives the closed-form expression for x_0, from which the so-called response amplitude operator (RAO) (sometimes called response transfer function) can be easily derived:

$$RAO = \frac{x_0}{a} = \frac{iC_M G \rho V \sigma}{\left[-(I + C_A \rho V)\sigma^2 + K\right]e^{i\delta}} = \frac{C_M G \rho V \sigma e^{i\left(\frac{\pi}{2} - \delta\right)}}{K\left[1 - \dfrac{\sigma^2}{K/(I + C_A \rho V)}\right]}$$

$$(3.206)$$

where $\omega_N = \sqrt{K/(I + C_A \rho V)}$ is the natural frequency of the structure that depends on the dynamic system (e.g., restoring force, body mass, and body

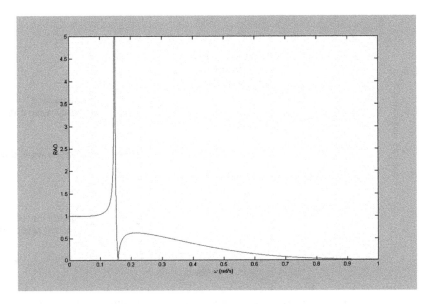

Figure 3.33 Illustration of the RAO curve for a floating body subject to wave action.

shape that affects added body mass). In general, RAO is a complex number whose amplitude is |RAO| and whose phase can be determined once δ is known; δ is $-\pi/2$ (i.e., the structure moves with the waves) when there is no damping and deviates from $-\pi/2$ when damping is present. Apparently, when $\sigma = \omega_N$, |RAO| tends to infinity, the indication of structural resonance (Figure 3.33). For a specific structure system, RAO is the function of incident wave frequency only. By plotting |RAO| against σ, one may locate the frequency range of the incident waves that may possibly induce structural resonance, i.e., |RAO| ≈ 1.

Under the combined wave and current, the motion of the body comprises both the current-induced response $(1/2K)C_D\rho A U_c|U_c|$ and the wave-induced response that is periodic and whose amplitude can be calculated as $a \times$ RAO. In addition, there also exists the steady drift response that can be estimated by $\overline{F}/K$, where $\overline{F}$ is the second-order wave drift force. For a random sea, due to the mean actions from wave components with different frequency, there is also a slow varying drift response of the body that can be calculated from difference-frequency wave drift force using wave spectrum information. The above four body motions represent the typical structure responses under the action of waves and currents in random seas (Figure 3.34), and it can be extended to the analysis of body motion in other directions and other modes (e.g., rotations).

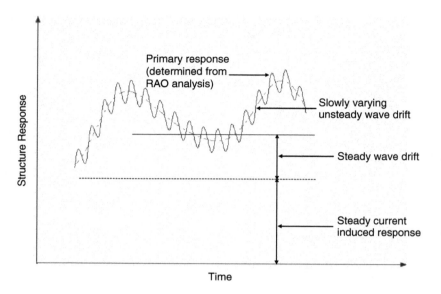

Figure 3.34 Typical structure responses under the combined action of random waves and a steady current.

3.15.3.2   *Two-dimensional structure responses to flow and vortex-induced vibration*

The above analysis can be extended to the analysis of 2D body motion under external forces. For a 2D body, three DOF motions (i.e., two translations and one rotation) can describe the body responses. When the rotation is negligible (e.g., a circular cylinder exposed to the fluid with negligible fluid shear on the body surface), we need to consider two translations in two orthogonal directions, and therefore, the similar equations and methodology introduced in Section 3.15.3.1 can be employed in the analysis of structure responses. A typical example is vortex-induced vibration (VIV) of a cylinder (e.g., a mooring line, a riser, a pipeline). When a steady flow passes over the body, both drag and lift forces will be induced. Because both forces have time-varying components, the body will oscillate about its equilibrium position in both in-line and transverse directions. Similar to RAO, the amplitude of the body motion depends on the dynamic system of the structure and the frequency of oscillating forces. The only difference is that in this case, this frequency is essentially the vortex-shedding frequency that depends on the incoming flow velocity and the body size.

From the definition of the Strouhal number, we can easily estimate the shedding period as follows:

$$T_s = \frac{1}{f} = \frac{D}{StU} \tag{3.207}$$

In contrast, the natural period of the dynamic system can be defined as T_N (for a simple spring system, we would have $T_N = 2\pi\sqrt{I/K}$). Obviously, the most severe vibration of the cylinder will occur when $T_s \approx T_N$, which gives:

$$\frac{D}{StU} \approx T_N \Rightarrow St\frac{UT_N}{D} = StU_R \approx 1 \qquad (3.208)$$

where U_R is called the reduced velocity and is dimensionless. Considering that $St \approx 0.2$, this gives the value of U_R around 5 for the occurrence of VIV. Practically, U_R varies from 2.5 to 8.3 (Chakrabarti, 2002: 328).

Since the lift force usually has a much stronger oscillation than the in-line drag force, the motion of a 2D cylinder under the influence of a steady flow often displaces a "figure-eight" (8) response during VIV. When wave loads are considered for VIV analysis, both wave frequency and vortex-shedding frequency need to be considered and this will complicate the analysis. For this reason, a numerical computation is often preferred. The VIV will have important effects on the fatigue life of a structure. When multiple structures in vicinity (e.g., risers for an offshore platform) experience VIV, possible collision may take place that will worsen the situation. The study of reducing VIV by employing various types of "VIV spoilers" is still being active researched in offshore engineering, and readers are referred to Chakrabarti (2002: 362) for more information.

3.15.3.3   Three-dimensional structure responses and global response analysis

For a 3D rigid body under the action of waves and current, there are six DOF motions. Referring to a ship, these six motions include three translations of surge, sway, and heave and three rotations of roll, pitch, and yaw (Figure 3.35).

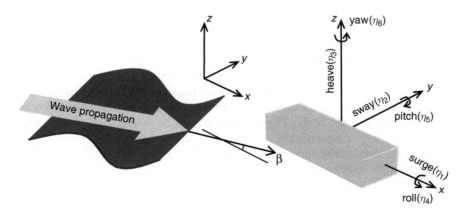

Figure 3.35 Rigid body motions under the action of waves and current.

The equations of motion can be expressed as follows:

$$\sum_{k=1}^{6} \left[\left(I_{jk} + A_{jk} \right) \ddot{x}_k + D_{0jk} \dot{x}_k / |\dot{x}_k| + D_{1ij} \dot{x}_k + D_{2jk} \dot{x}_k |\dot{x}_k| + K_{jk} x_k \right] = F_j,$$

$$j = 1, \ldots, 6$$

$$(3.209)$$

where I_{jk} is the generalized mass (for $j = 1,2,3$) and moment of inertia (for $j = 4,5,6$) matrix and A_{jk} is the added mass (for $j = 1,2,3$) and moment of inertia (for $j = 4,5,6$) matrix, D_{0jk}, D_{0jk}, and D_{0jk} are zeroth-, first-, and second-order damping coefficients, K_{jk} is the restoring coefficient matrix, and F_j is the excitation forces (for translations) and moments (for rotations) vector. The equations for $j = 1,2,3$ describe the body translation based on Newton's Second law and the equations for $j = 4,5,6$ describe body rotations based on the equations of balance of angular momentums.

In the above equation, the zeroth-order damping corresponds to the Coulomb friction, which is a constant independent of body velocity but dependent upon the mechanical system only. The first-order (or linear) damping is related to linear viscous damping that is significant only for small Re flow, the radiation damping, and the wave drift damping. Physically, the radiation damping can be explained by the simple fact that a moving body in a free surface flow will impart energy to the fluid system and part of the energy will radiate to infinity (and thus is not reversible) in an open space. As a result, the body will generally come to stop even if the fluid is inviscid. The wave drift damping is to realize the physical fact that a ship feels increased resistance in waves compared with calm water. This increased resistance is related to the mean wave drift force that is linearly proportional to ship speed (Faltinsen, 1990: 162). The second-order damping is related to damping due to flow separation and turbulence effect. This term is essentially represented by the drag force in the Morison equation for high Re flows.

For a particular dynamic system of structure exposed to known environmental conditions (e.g., wind, wave, and current), both K_{jk} and F_j can be directly determined, i.e., K_{jk} is known from the mechanical system and F_j is the sum of wind, current, and wave (first- and second-order) forces. The main difficulties lie on the determination of various damping coefficients and added mass and moment of inertia coefficients. These coefficients are primarily dependent upon the shape of the structure and vary with different heading angles of the body. The coefficients can be determined empirically, experimentally, or numerically.

With all the coefficients being known, the system of nonlinear ODEs (3.209) for six motions can be solved. Such a process of dynamic analysis for determining 3D structure motions is called global response analysis (GRA) in offshore engineering. The general solution to (3.209) will provide the time-varying information for the six motions, and the procedure of solving the six ODEs directly is called time-domain analysis. Although the theory ensures the unique solution for the motions, the way of seeking these solutions is not trivial due to the nonlinear terms appearing in both damping and forcing terms. The only feasible way is to solve the equation numerically, which can be computationally expensive for a random sea condition.

Alternatively, these nonlinear terms can be linearized or even neglected if the nonlinear effect is weak. With the equations being linearized, the immediate benefit is that the ODEs can be converted to algebraic equations by assuming harmonic motions. The total responses can be obtained by summing up the solutions of motions induced by each wave mode (see Section 3.15.3.1 for the 1D example). Since the analysis is based on the summation of structure responses to each incident wave component with different frequency, it is often referred to as frequency-domain analysis.

Although the above technique is mainly used in offshore engineering for the analysis of a floating body, it can be extended to the analysis of a bottom-seated structure as well. For example, Li *et al.* (1999) employed this analysis technique to develop the optimal design of tuned mass damper (TMD) for reducing the vibration of a jacket platform; Oumeraci and Kortenhaus (1994) conducted the analysis of dynamic responses of caisson breakwaters. Wang and Zheng (2001) analyzed the vibration of a caisson breakwater under breaking wave actions.

Further discussions of wave loads on ships and offshore structures and structure responses to wave action can be found in Sarpkaya and Isaacson (1981), Chakrabarti (1987), Faltinsen (1990), and Hudspeth (2006).

3.15.4 *Coupled fluid–structure system*

In the analyses discussed in Section 3.15.3, the external forcing term is independent of the structure motion, i.e., the wave and current conditions will not be changed by structure responses though the motion-induced additional forces have been considered. In reality, the motion of a structure may possibly change the flow condition around it and thus the fluid force on the structure is modified. To fully consider such effect, the coupled FSI must be considered. In this section, we discuss a few simple examples of such analyses.

3.15.4.1 *One-dimensional liquid sloshing in a U-tube with a moving object*

The simplest example that requires coupled FSI analysis is the liquid sloshing in a large U-tube, within which a moving body is deployed. The body will be moved by the fluid due to the flow loads on it and in turn it changes the fluid motion. Denote the length and the radius of the circular cross section of the U-tube to be L and R and the still water depth to be h (Figure 3.36).

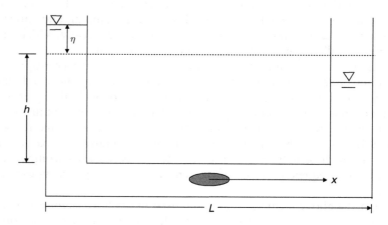

Figure 3.36 Sketch of a moving object in a U-tube.

The complete dynamic equation that governs the fluid motion inside the tube can be written as follows:

$$\rho(L+2h)\pi R^2 \ddot{\eta} = -C_D \rho A_1 u_0(\dot{\eta}+\dot{x}) - \rho V(C_M \ddot{\eta} + C_A \ddot{x}) - 2\rho g \eta \pi R^2$$

(3.210)

where u_0 is a linearized mean velocity for the damping, A_1 is the cross-sectional area of the body perpendicular to the flow, and V is the volume of the body. The three terms on the RHS of the equation represent the linearized drag force, the inertial and added mass force, and the restoring force by fluid weight. In contrast, the governing equation for the body motion is as follows:

$$M\ddot{x} = -C_M \rho V \ddot{\eta} - C_A \rho V \ddot{x} - C_D \rho A u_0(\dot{\eta}+\dot{x})$$

(3.211)

For the case that the inertial force is predominant, the drag term can be neglected and (3.211) can be simplified as follows:

$$\ddot{x} = -\frac{C_M \rho V}{M + C_A \rho V} \ddot{\eta}$$

(3.212)

Substituting (3.212) into (3.210) and neglecting the drag force we have:

$$\left[\rho(L+2h)\pi R^2 + C_M \rho V - C_A \rho V\left(\frac{C_M \rho V}{M + C_A \rho V}\right)\right]\ddot{\eta} + 2\rho g \pi R^2 \eta = 0$$

$$\text{or} \quad C\ddot{\eta} + E\eta = 0$$

(3.213)

If we define:

$$\eta = A e^{i\sigma t}$$

(3.214)

where A is the initial free surface displacement and σ is the complex angular frequency, the real part of σ then represents the oscillation frequency and the imaginary part represents the decaying rate. Substituting (3.214) into (3.213), we have:

$$-\sigma^2 C A e^{i\sigma t} + E A e^{i\sigma t} = 0 \tag{3.215}$$

This gives:

$$-\sigma^2 + E = 0 \Rightarrow \sigma = \sqrt{E/A} \tag{3.216}$$

Since σ is always a real number, both the fluid and the sphere will oscillate inside the U-tube forever. The oscillation frequency of the fluid system, however, is related to the body mass of the solid as well as the inertial coefficient.

In contrast, if the drag force predominates and the body motion is negligible, i.e., $\dot{x} = \ddot{x} = 0$, equation (3.210) can be simplified as follows:

$$\rho(L + 2h)\pi R^2 \ddot{\eta} + C_D \rho A u_0 \dot{\eta} + 2\rho g \eta \pi R^2 = 0$$

$$\text{or} \quad C\ddot{\eta} + D\dot{\eta} + E\eta = 0 \tag{3.217}$$

Substituting the definition $\eta = A e^{i\sigma t}$ into the dynamic equation, we have:

$$-\sigma^2 A e^{i\sigma t} + D i\sigma A e^{i\sigma t} + E A e^{i\sigma t} = 0 \tag{3.218}$$

This gives:

$$-\sigma^2 + Di\sigma + E = 0 \Rightarrow \sigma = \frac{Di \pm \sqrt{-D^2 + 4E}}{2} \tag{3.219}$$

If $X = -D^2 + 4E > 0$, then $\sigma = \left(Di \pm \sqrt{X}\right)/2$, the real component represents the oscillatory mode and the imaginary part the decay mode. If $X = -D^2 + 4E < 0$, there will be only decaying mode.

3.15.4.2 Two-dimensional vortex-induced vibration under wave and current loads

In Section 3.15.3.2, we have discussed the analysis of VIV by solving the dynamic equation of motion in two orthogonal directions. It must be realized, however, that the method is a simplified approach by neglecting the change of flow field around the cylinder in vibration, which may have profound impact on the forcing on the cylinder and the resulting VIV pattern. So far, there is no analytical technique to fully account for the effect of body motion on the changed vortex shedding. One possible way to address this issue is through a DNS. In the simulation, the viscous fluid flow is first solved around the structure, from which the flow force is obtained to calculate the body motion. The updated body position and velocity will subsequently feed back into the next flow computation. With this treatment, the fully coupled structure response can be recovered with the consideration of the body motion effect on the changed flow and flow force. One example of this approach was by Schultz and Kallinderis (1998), who developed a 2D FSI model to model the VIV and recover the well-known "figure-eight" motion of a vibrating cable.

3.15.4.3   Coupled response analysis of a large three-dimensional body using diffraction theory

For a large rigid body exposed to wave actions, the body responses include six DOF motions. These motions will affect the wave field that can be analyzed through diffraction theory. For a large flexible body, so-called hydroelastic analysis is required to account for the coupling effect between wave diffraction and wave radiation induced by structure motion and deformation. The hydroelastic analysis has also been used in the analysis of a flexible structure in response to fluid flow action, e.g., liquid sloshing in a container with flexible walls. Note, however, that the closed-form solution is available only for a limited number of problems with idealized wave condition and body shape.

Rigid body: The theoretical study of coupled structure response to the wave motion using diffraction theory started quite early. For example, Ijima *et al.* (1972) presented an analytical solution for body motion under wave action. Yeung (1981) presented an eigenfunction approach to study heave, sway, and roll motions of a floating vertical circular cylinder.

Deformable body (hydroelastic analysis): The application of hydroelastic analysis to a large deformable floating body was initiated in the 1990s. For example, Hamamoto and Tanaka (1992) developed an analytical method for the hydroelastic analysis of a circular floating island. Ertekin *et al.* (1993) developed an efficient method for the hydroelastic analysis of a VLFS. Using long-wave approximation, Zilman and Miloh (2000) obtained a closed-form solution of the hydroelastic response of a circular floating plate in shallow waters. For a circular thin plate, Watanabe *et al.* (2003) extended the analysis to deep waters. Yan *et al.* (2003) proposed a new plate Green function for the hydroelastic analysis of a VLFS. Ohkusu and Namba (2004) proposed an analytical solution for the hydroelastic analysis of a large rectangular plate. A combined analytical and numerical study was performed by Andrianov and Hermans (2005) for a circular plate in various water depths. The review articles about hydroelastic analysis of a VLFS were written by Kashiwagi (2000) and Watanabe *et al.* (2004).

3.15.4.4   Other coupled fluid–structure interactions

There are many other problems that require the coupled analysis of FSI. Examples include the water entry and slamming of a cylinder (Greenhow and Li, 1987), a ship approaching a dock or another ship. For a complex loading condition (e.g., combined wave and shear current) or a body with a complex shape, various numerical approaches (e.g., panel or strip method for irrotational flows or RANS model for turbulent flows) may have to be used. It will be further pursued in Section 6.4.5.

3.15.5   Waves generated by a moving body: ship waves

There is another category of wave–structure interaction, which is essentially the simplified version of coupled wave–structure interaction, by assuming that the body motion is prescribed and not affected by ambient flow. In short, this is the problem of wave generation by a moving body. In principle, any movement of a solid body in a fluid will produce a local disturbance of fluid pressure that will be the source of surface wave generation. There are a few cases commonly encountered, e.g., waves generated by wave makers, waves generated by ship motion, waves generated by seafloor movement (e.g., tsunami), and standing waves generated in a confined tank with external excitations (e.g., liquid sloshing). In the case of wave generation by wave maker, the paddle movement is driven by a mechanical device and will not be affected by the generated waves. The complete theory behind various wave makers can be found in Dean and Dalrymple (1991: 170) and will not be reiterated here. In this section, only ship waves and liquid sloshing will be discussed.

Ship waves are of great importance in naval architecture. As a ship advances steadily in water, surface waves will be generated. The wave originating from the bow of the ship is called a bow wave. As the bow wave spreads out, it defines the outer limits of the ship's wake within a "V"-shaped wedge, which is similar to the shock waves in aerodynamics. In general, the size of the bow wave is a function of the shape of the bow and the speed of the ship, ocean wind waves, and water depth. The bow wave will induce the so-called wave drag. The simple logic behind it is that the ship needs to impart energy to create the bow wave and such energy can be converted to the equivalent wave drag for a particular ship. In contrast, the wave originating from the ship stern is called a stern wave, which is present behind the ship and mixed up with the shed vortices from the ship in the vortex wake region (Figure 3.37). Since the pressure in the vortex region is low, this is another source of ship resistance.

The early study of ship waves can be traced back to Lord Kelvin, who discovered that in deep water, stationary ship waves are always left behind a moving ship and the angle of the wake formed by the bow wave is an invariant of roughly 39°, provided that the ambient wind waves and

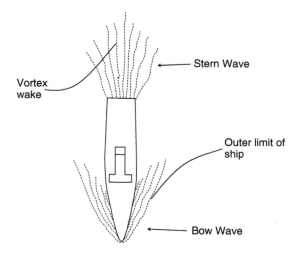

Figure 3.37 Sketch of bow wave and stern wave.

current effects are negligible. This phenomenon is evident from the photo in Figure 3.38, from which one can readily see that the angle of the wake (the wedge bracketed by the thin and dark traces) is independent of ship size, shape, and speed. Further inspection reveals that within the ship wake, which was later named Kelvin wake, there exist two types of waves, namely the divergent wave whose crest line is nearly parallel to the outlines of the wake on two sides of the ship and the transverse wave whose crest line is perpendicular to the track of the ship (see Figure 3.38). These two patterns of waves are called Kelvin wave system. Readers should not confuse this with the Kelvin wave resulting from the geostrophic effects (i.e., Coriolis forces) on long waves (e.g., Dean and Dalrymple, 1991: 155), which will not be discussed in this book.

Inspired by the observation of the stationary wave pattern around ships, the theoretical derivation was made by Lord Kelvin (Thomson, 1887) based on the small-amplitude wave theory. A moving ship can generate waves in all possible frequencies. However, only the wave that propagates at the same speed of the ship V is stationary; other waves will either overtake the ship or be left further behind and thus are unsteady and nonsustainable. This determines the phase velocity of the stationary ship wave to be $c = V$, from which the wavelength and wave frequency can be determined from the dispersion relationship. The spread half-angle α of the wedge of the Kelvin wake is the function of wave phase velocity and group velocity:

$$\sin \alpha = \frac{1}{\frac{2V}{c_g} - 1} \tag{3.220}$$

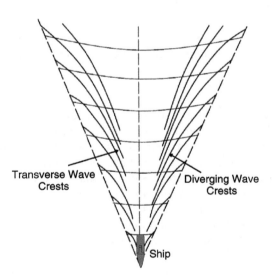

Transverse Wave
Crests

Diverging Wave
Crests

Ship

Figure 3.38 Photo of the aircraft carrier USS *John F. Kennedy* and its escorts in the
Arabian Gulf, from which the V-shape wave wakes are evident sketch
of transverse wave and diverging wave within Kelvin wake (bottom)
(Top photo take by Christian Eskelund; U.S. Navy imagery used in
illustration without endorsement expressed or implied).

In deep water, we have:

$$c_g = \frac{1}{2}c = \frac{1}{2}V$$

which leads to the following:

$$\alpha = \sin^{-1}\left(\frac{1}{3}\right) \approx 19.5° \tag{3.221}$$

The details of the above theoretical derivation can be found in Lamb (1945: 395), Stoker (1957: 181), Newman (1977: 270), and Lighthill (1978: 3.10).

In reality, the variation of the angle of Kelvin wake does exist. This is attributed to the fact that in the theoretical derivation, the ship motion was treated as a moving point source without consideration of the detailed shape of the hull, which can generate different shedding vortices and induce different wave breaking affecting the ship wake angle. Besides, the ambient current and ocean wind, waves will also have influence on the wake angle. Recently, Shugan *et al.* (2006) extended the analysis to include the wind wave and current effect on Kelvin wake.

When a ship moves in shallow water, both ship speed and depth effect have influence on Kelvin wake. Havelock (1908) presented an analysis with the inclusion of water depth effect. Under this circumstance, the half-angle α is dependent upon the depth Froude number (Sørensen, 1973):

$$Fr = V/\sqrt{gh}\left(= c/c_g\right).$$

When $Fr = 1$, the angle $\alpha = 90°$ and the transverse and divergent waves form a single large wave, which often breaks, with its crest normal to the sailing line. For $Fr < 1$, the ship speed is in the subcritical regime and for $Fr > 1$ it is in the supercritical regime. Lanzano (1991) proposed a theory to describe the linear ship waves from deep to shallow waters.

For fast ships in shallow water, the ship waves often possess the weak nonlinear and dispersive properties. Under this circumstance, the ship waves will have the form of Stokes waves or cnoidal waves. When F is around 1.0, a succession of solitons that propagate ahead of the ship may also be generated. Akylas (1984) proposed a theoretical study for the mechanism of the generation of these solitons, which were named by Wu (1987) as precursor solitons. Cole (1985) studied an equivalent problem of a flow past a bump in a channel. Mei (1986) proposed a study of soliton generation in shallow water by a slender body.

By treating the ship as the moving free surface pressure disturbance, Ertekin *et al.* (1986) used the depth-averaged long-wave Green–Naghdi (GN) equation to describe the ship-generated waves. In contrast, the forced KdV equation was proposed by Wu (1987) with a singular forcing function as a point source of disturbance. In all these approaches, the hull shape can be partially treated but the solution can be obtained only by numerical computation. Some other numerical computations include the use of BEM based on the potential flow theory (Cao *et al.*, 1993), the use of NSE solvers (e.g. Zhang and Chwang, 1999), and the use of modified Boussinesq equations (e.g. Liu and Wu, 2004). A review article for nonlinear ship wake and ship waves was written by Soomere (2006).

The total ship resistance includes wave drag, skin friction, and form drag. Thus, the boundary layer development around the ship hull, the vortex shedding behind the ship, the turbulence dissipation around the ship, and the wave breaking in the wake region will all have influence on the ship resistance. Besides, it was found that the capillary waves generated by ship motion will also contribute to wave drag (e.g., Raphael and deGennes, 1996). The short waves radiated from the turbulent boundary layer near the ship's stern were studied by Gu and Phillips (1994). The interaction among surfactant, vortices, and free surface is an important mechanism to affect ship waves (Tsai and Yue, 1995). In a review article, Sarpkaya (1996) reviewed the studies on the effects of surfactants and vortices on free surface waves.

3.15.6 *Liquid sloshing in tanks*

Liquid sloshing in confined tanks with prescribed excitations is a special type of standing wave generated by a moving boundary. The main difference between sloshed liquid in a tank and ship waves is that the wave motion during liquid sloshing is confined in a finite domain without energy radiation to a far field. Liquid sloshing is an important consideration in sea transport. In analysis, the sloshing can be decomposed into many standing wave modes, which can be expressed by the closed-form analytical solution for small-amplitude oscillation and particular shape of container. Below is the summary of existing analytical solutions for various liquid sloshing.

3.15.6.1 *Two-dimensional liquid sloshing in a stationary tank*

In the case that the free surface is initially displaced to some position and released, the liquid sloshing will continue without external forcing. This type of sloshing is classified as the free sloshing. For a 2D case, the available analytical solution exists when the rectangular tank with a length of W has a still water depth of h and a linear slope s of the initial free surface profile (Lin and Li, 2002):

$$\eta(x, t) = \sum_{n=1}^{\infty} A_n \sin(k_n x) \cos(\omega_n t) \tag{3.222}$$

where the origin is set at the center of the still water level, $k_n = (n\pi)/W$, $\omega_n = \sqrt{gk_n \tanh(k_n h)}$, and $A_n = 4sW/(n^2 \pi^2) \sin(n\pi/2)$.

3.15.6.2   Three-dimensional liquid sloshing in a stationary tank

For a 3D case, the analytical solution exists when the tank is of rectangular shape $L_x \times L_y$ and the initial free surface displacement is of Gaussian distribution about the center of the basin, i.e.:

$$\eta_0(x, y) = H_0 \exp\left\{-\beta\left[(x - L_x/2)^2 + (y - L_y/2)^2\right]\right\} \tag{3.223}$$

where H_0 is the initial height of the hump, β is the peak enhancement factor, and the origin is defined at the left-bottom corner of the basin. Wei and Kirby (1995) proposed a linear analytical solution for the free surface deformation within the basin:

$$\eta(x, y, t) = \sum_{n=0}^{\infty} \sum_{m=0}^{\infty} \bar{\eta}_{nm} e^{-i\omega_{nm}t} \cos(n\lambda x) \cos(m\lambda y) \tag{3.224}$$

in which:

$$\bar{\eta}_{mn} = \frac{4}{(1 + \delta_{n0})(1 + \delta_{m0})L_x L_y} \int_0^{L_x} \int_0^{L_y} \eta_0(x, y) \cos(n\lambda x) \cos(m\lambda y)\, dx\, dy \tag{3.225}$$

where δ_{nm} is the Kronecker delta function and:

$$\lambda = \frac{\pi}{L_x} = \frac{\pi}{L_y} \tag{3.226}$$

The (n,m) wave mode has the corresponding natural frequency, which is determined by the linear dispersion equation:

$$\omega_{nm}^2 = g k_{nm} \tanh(k_{nm} h_0) \tag{3.227}$$

where h_0 is the still water depth and:

$$k_{nm}^2 = (n\lambda)^2 + (m\lambda)^2 = \left(\frac{\pi}{L_x}\right)^2 (n^2 + m^2) \tag{3.228}$$

3.15.6.3   Two-dimensional sloshing of viscous fluid in a stationary tank

When viscous effect is considered, a linearized analytical solution for 2D small-amplitude viscous waves in a rectangular tank has been derived by Wu *et al.* (2001) where the initial interface elevation profile is specified as follows:

$$\eta_0 = a \cos k\left(x + \frac{W}{2}\right) \tag{3.229}$$

where the origin is set at the left edge of the still water level, a is the initial wave amplitude, and $k = 2\pi/W$ with W being the tank length. In the case when $\kappa = g/(v^2 k^3) > 0.5814122$, the analytical solution for free surface deformation within the basin can be expressed in terms of the following closed form:

$$\frac{\eta(t)}{\eta_0} = 1 - 2\kappa \text{Re} \sum_{m=1}^{2} \frac{A_m \left\{ -\gamma_m e^{(-1+\gamma_m^2)vk^2 t} \left[1 + \text{erf}\left(\gamma_m k\sqrt{vt}\right) \right] + \gamma_m + \text{erf}\left(k\sqrt{vt}\right) \right\}}{1 - \gamma_m^2}$$

(3.230)

where $\text{Re} = h\sqrt{gh}/v$ and γ_1 and γ_2 are any two nonconjugate roots of the four possible roots of the equation:

$$\left(x^2 + 1\right)^2 - 4x + \kappa = 0$$

(3.231)

Writing $\gamma_1 = \gamma_{1R} + i\gamma_{1I}$ and $\gamma_2 = \gamma_{2R} + i\gamma_{2I}$, we then have:

$$A_1 = -\frac{2}{\Delta}\gamma_{1R} - i\frac{2\gamma_{1R}^3 + 1}{\gamma_{1R}\gamma_{1I}\Delta} \quad \text{and} \quad A_2 = \frac{2}{\Delta}\gamma_{1R} - i\frac{2\gamma_{1R}^3 - 1}{\gamma_{1R}\gamma_{2I}\Delta}$$

(3.232)

where $\Delta = 4\left(6\gamma_{1R}^2 + 1\right)\left(2\gamma_{1R}^2 + 1\right) - 4\left(\kappa + 1\right)$.

3.15.6.4 *Two-dimensional and three-dimensional sloshing in a tank with periodic horizontal excitation*

A 2D rectangular tank has still water depth h and tank length $W = 2a$. For the periodic horizontal excitation of $u_e = -A\cos\omega t$, where u_e is the tank excitation velocity, $A = b\omega$ is the velocity amplitude with b being the displacement amplitude, and ω is the angular frequency of the excitation, Faltinsen (1978) gave the linear analytical solution for the velocity potential function, from which the free surface displacement η can be obtained as follows:

$$\eta = \frac{1}{g} \sum_{n=0}^{\infty} \sin\left\{\frac{(2n+1)\pi}{W}x\right\} \cosh\left\{\frac{(2n+1)\pi}{W}h\right\}$$
$$\left[-A_n\omega_n \sin\omega_n t - C_n\omega \sin\omega t \right] - \frac{1}{g}A\omega x \sin\omega t$$

(3.233)

where:

$$\omega_n^2 = g\frac{(2n+1)\pi}{W}\tanh\left\{\frac{(2n+1)\pi}{W}h\right\}, \quad C_n = \frac{\omega K_n}{\omega_n^2 - \omega^2}, \quad A_n = -C_n - \frac{K_n}{\omega}$$

$$K_n = \frac{\omega A}{\cosh\left\{\frac{(2n+1)\pi}{W}h\right\}}\frac{4}{W}\left[\frac{W}{(2n+1)\pi}\right]^2(-1)^n$$

The origin is set at the center of the still water level.

Note that for a 3D rectangular tank with tank lengths of W_x and W_y in x- and y-directions, the liquid sloshing under the horizontal excitations of combined surge and sway can be obtained by summing up the surge and sway contributions together (Liu and Lin, 2008), i.e.:

$$\eta = \eta_{\text{surge}}(x, y, t) + \eta_{\text{sway}}(x, y, t) \tag{3.234}$$

where $\eta_{\text{surge}}(x, y, t)$ and $\eta_{\text{sway}}(x, y, t)$ can be obtained with the use of (3.233) for 2D tank.

3.15.6.5　Liquid sloshing in a tank under vertical excitation or rotational excitation

It has been long since the first discovery of fluid instability in a tank under vertical oscillation (Faraday, 1831). Small waves will develop in a tank under vertical oscillation and some modes of these waves will be amplified under certain conditions of vertical oscillation. The Mathieu equation, a second-order ODE, can be used to describe the wave motions under some special cases when the separation of variables is allowed. With the introduction of coordinate transformation, Mathieu's equation can be transferred into algebraic form and thus allows the analytical solution in terms of special functions for some special cases. The theoretical study was performed by Benjamin and Ursell (1954). Penney and Price (1952b) proposed a nonlinear analytical solution to describe wave behavior in finite amplitude.

The liquid sloshing under the excitations of pitch and roll is also an important mechanism to consider. So far, there are few reports on the closed-form solution that describes the liquid sloshing under the rotational excitation. Numerical simulation (e.g., Kim *et al.*, 2004) or semianalytical solution based on the multimodal approach (e.g., Faltinsen *et al.*, 2000), however, has been used to investigate this problem.

3.15.6.6　Liquid sloshing in a cylindrical tank or a tank of arbitrary shape

Most of the theoretical studies were made for rectangular tanks. For cylindrical tanks that are often used as containers of liquids, it is difficult to separate sloshing modes in radial and angular directions. Thus, the complete theoretical solution was rarely reported except for some semianalytical ones (e.g., Papaspyrou *et al.*, 2004). Various types of numerical models were also developed for the analysis of liquid sloshing in tanks of arbitrary shapes.

3.15.6.7 Nonlinear liquid sloshing

When the sloshing amplitude becomes comparable to water depth, the linear solution will fail to provide accurate prediction of liquid sloshing. Under this circumstance, the nonlinear terms should be included in the solution to account for nonlinear contributions. Faltinsen *et al.* (2000) proposed a multidimensional modal analysis for liquid sloshing in a rectangular tank by considering the nonlinear modal contributions. The study was revised to match both shallow liquid and secondary (internal) resonance asymptote later (Faltinsen and Timokha, 2002). Recently, Frandsen (2004) and Hill and Frandsen (2005) presented third-order analytical solutions for liquid sloshing in a horizontally oscillating basin.

3.15.6.8 Liquid sloshing in a tank with flexible walls

Recently, using containers with flexible walls has become an attractive idea to reduce liquid sloshing. In this case, both liquid sloshing and tank response are of important consideration. So far there are few analytical solutions for the analysis of liquid sloshing in such a tank. Alternatively, numerical modeling of this type of problem is promising and under active research.

4 Numerical methods

4.1 Introduction

4.1.1 Background of numerical methods

Most of the traditional numerical methods are equation-based, i.e., the methods are developed to solve existing governing equations for particular physical problems numerically. Recently, there is a new methodology of problem-based numerical methods, i.e., the numerical methods are developed to solve physical problems directly.

4.1.1.1 Equation-based numerical methods

Strong and weak formulation: Most of the governing equations for fluid flows or water waves are in the form of PDEs. Examples include the momentum equation for general fluids (2.1) and the Laplace equation for potential flows (2.7). Various numerical algorithms have been developed to solve PDEs. Based on how an approximation is made, there are two types of numerical methods. One approximates various orders of the derivatives in the original PDE and is called strong formulation, and the other approximates the unknown solution function and is called weak formulation. Finite difference method (FDM) is a representative example for the former, finite element method (FEM) is for the latter.

Mesh-based and meshless methods: There is another way of categorizing numerical methods, i.e., whether the algorithm is constructed on meshes (or elements) or on particles (or points). The former type covers all the traditional numerical methods including FDM, finite volume method (FVM), FEM, and BEM, in which computational nodes are connected by a mesh or a grid to form elements or cells. The latter type is called a meshless (or gridless, mesh-free, particle Lagrangian) method that is constructed on particles without explicit connectivity. The popular meshless methods include generalized finite difference method (GFDM; also called finite point method, FPM), smoothed particle hydrodynamics (SPH), reproducing kernel particle method (RKPM), and element-free Galerkin method (EFGM). While GFDM is based on the strong formulation that solves the original PDE by approximating derivatives, SPH, the PKPM, and EFGM as well as other methods in the same family are based on various kernel or Galerkin weak formulation by approximating solutions.

The mesh-based methods generally suffer difficulties in resolving arbitrary boundary geometry. For example, FDM is not flexible near a complex boundary, and thus, numerical accuracy can be reduced due to the poor representation of boundary curvature. In contrast, FEM or FVM is able to resolve irregular boundary shape with unstructured mesh, but the mesh generation can be tedious and time-consuming. Furthermore, the accuracy of the result is highly dependent upon the quality of mesh generation. On the other hand, the meshless methods can overcome these difficulties because the methods operate on the distribution of scattered particles and their weak local connectivity information. Unlike the mesh-based methods where strong connectivity information is required, i.e., specific lines must be used to connect neighboring nodes, the weak connectivity information in the meshless method is loose, flexible, and implicit and can be easily changed with time.

4.1.1.2 *Problem-based numerical methods*

A numerical method can also be established on the basis of solving an intended problem directly. By doing so, the problem is formulated in discrete form based on fundamental physical laws and thus the governing equation in the form of PDE is no longer needed. The representative example of this method is the lattice Boltzmann method (LBM), which is constructed on a stochastic basis for microscopic scale particle motion to simulate the macroscopic behavior of fluids. For solids, the equivalent method is called discrete element method (DEM).

4.1.1.3 *Numerical discretization*

All numerical methods, meshless or mesh-based, require the approximation of the continuous time and space in a physical problem by the discrete time and space in numerical computations. While meshless methods tend to be less cumbersome in the arrangement of particles that are used to discretize the space and tracked by the Lagrangian approach, mesh-based methods must have a predetermined grid or mesh system on which the computation is performed. The quality of grid or mesh configuration is important to the numerical accuracy and stability, especially when the flow is violent and near an irregular boundary. Various ways of mesh generation will be introduced in Section 4.9 after numerical methods are discussed.

4.1.1.4　Outline of contents in this chapter

Since most of the numerical methods are equation-based (more specifically PDE-based), we shall first have a brief discussion of the mathematical characteristics of various types of PDEs that may be encountered in water wave modeling. After the introduction of PDEs, we will start with FDM, one of the most commonly used numerical methods in computational fluid dynamics (CFD). A detailed introduction will be provided for the construction of the finite difference (FD) scheme as well as the concept of numerical consistency, numerical accuracy, numerical stability, etc. The other traditional numerical methods such as FVM, FEM, and BEM will then be introduced and compared to each other. Next, different meshless methods and problem-based methods will be introduced. Finally, the mesh generation methods and matrix solvers will be discussed briefly.

4.1.2　Basic equations for fluid flows and waves

Let us start from the following 1D PDE with a first-order time derivative:

$$\frac{\partial \phi}{\partial t} + \sum_{n=1}^{i} C_n \frac{\partial^n \phi}{\partial x^n} = C_0 \tag{4.1}$$

where $\phi(x, t)$ is a general variable, C_n is the coefficient associated with the nth spatial derivative, and C_0 is the source/sink term. The first term represents the rate of change of the variable in time, which is balanced by the sum of the rate of change of the variable in 1D space x and the contributions from source and sink. For a well-posed problem with proper initial and boundary conditions, a unique solution of $\phi(x, t)$ exists.

When i and C_n in (4.1) are equal to different numbers and values, the equation represents different physical phenomena. For example, when $i = 1$, this is a convection equation representing that the variable is transported from one place to another. In case when C_1 is a constant and $C_0 = 0$, this becomes a 1D wave equation, the solution of which gives a wave form translating at a constant speed without change of shape. When C_1 is the function of x or $\phi(x)$, the term $C_1(\partial \phi / \partial x)$ becomes a nonlinear convective term. In this case, the wave changes its shape during propagation.

When $i = 2$, the second-order spatial derivative represents the diffusion process, and the equation is called convection–diffusion equation. In the case of $C_1 = 0$, the equation reduces to the transient diffusion equation. Analogous to the classification of polynomials, the diffusion equation is referred to as a parabolic type of PDE.

With the extension of i to 3, one more term of the third-order derivative is included. This term represents the "dispersion" process, during which wave modes with different wavelengths will propagate at different speeds and thus will separate (disperse) away from each other. Generally speaking, the odd number of n represents the process of wave propagation ($n = 1$)

and dispersion ($n = 3, 5, \ldots, 2m + 1$, where m is a positive integer), during which the total "energy" (ϕ^2) is conserved. In contrast, the even number of n represents the process of diffusion, during which the total "energy" can decrease [$(-1)^{n/2}C_n > 0$] or increase [negative diffusion; $(-1)^{n/2}C_n < 0$].

When a second-order time derivative is considered, the PDE will have different characteristics. Let us now use the following 2D second-order PDE as an example:

$$a\frac{\partial^2 \phi}{\partial t^2} - b\frac{\partial^2 \phi}{\partial x^2} - c\frac{\partial^2 \phi}{\partial y^2} = d \tag{4.2}$$

When $a, b, c > 0$, the equation is classified as the hyperbolic type of PDE that can be split into two first-order convection equations, representing wave propagation in two opposite directions. When $a = 0$ and $b \times c > 0$, the equation is classified as the elliptic type of PDE. The equation is the 2D Laplace equation if $d = 0$ or the 2D Poisson equation if $d \neq 0$. Including the parabolic type of PDE (diffusion equation) discussed earlier, these equations form the three basic types of PDEs. The majority of PDEs encountered in water wave modeling are in the form of one of these three basic PDEs or the combination of them. Below we shall review some common PDEs in wave mechanics.

Nonlinear wave equation: This equation describes nonlinear wave motion, during which the shock front will be developed during propagation due to wave nonlinearity:

$$\frac{\partial \phi}{\partial t} + \phi\frac{\partial \phi}{\partial x} = 0 \tag{4.3}$$

Burgers's equation: The equation is named after Johannes Martinus Burgers (1895–1981). It is a nonlinear convection and diffusion equation. The equation has been applied in many areas like wave propagation, gas dynamics, and traffic flow:

$$\frac{\partial \phi}{\partial t} + \phi\frac{\partial \phi}{\partial x} = \nu\frac{\partial^2 \phi}{\partial x^2} \tag{4.4}$$

When $\nu = 0$, the equation is reduced to a nonlinear wave equation, which sometimes is also called inviscid Burgers's equation.

Korteweg–de Vries (KdV) *equation*: The equation is used to describe nonlinear dispersive shallow water wave traveling in one direction. It was named after Diederik Korteweg and Gustav de Vries and has the following form:

$$\frac{\partial \phi}{\partial t} + 6\phi\frac{\partial \phi}{\partial x} + \frac{\partial^3 \phi}{\partial x^3} = 0 \tag{4.5}$$

The KdV equation allows a permanent positive wave form moving to the positive x-direction. Such a solution describes the so-called soliton, in which wave nonlinearity is balanced by dispersion.

Boussinesq equation: Boussinesq equation is the extension of the KdV equation, and it is used to describe nonlinear dispersive wave propagation in two opposite directions (Zwillinger, 1997):

$$\frac{\partial^2 \phi}{\partial t^2} - \frac{\partial^2 \phi}{\partial x^2} + 3\frac{\partial^2 \left(\phi^2\right)}{\partial x^2} - \frac{\partial^4 \phi}{\partial x^4} = 0 \tag{4.6}$$

For the 2D case, the Boussinesq equation is often expressed as the system of two PDEs in terms of first-order time derivatives.

Kadomtsev–Petviashvili (KP) equation: It is also called Kadomtsev–Petviashvili–Boussinesq equation, which is the generalization of the KdV equation in the 2D case (Kadomtsev and Petviashvili, 1970) for wave propagation primarily in one direction with a small lateral spreading angle. It has the following form:

$$\frac{\partial}{\partial x}\left(\frac{\partial \phi}{\partial t} + \phi\frac{\partial \phi}{\partial x} + \varepsilon^2\frac{\partial \phi}{\partial x^3}\right) \pm \frac{\partial^2 \phi}{\partial y^2} = 0 \tag{4.7}$$

The KP equation has been used to describe waves in porous media, ferromagnetic media, and the matter-wave pulses in Bose–Einstein condensates.

Nonlinear Schrödinger equation: This equation is used in fiber optics, water wave theory, and theoretical physics (e.g., second quantized bosonic theory to describe diffracted waves). It has the following form:

$$i\frac{\partial \phi}{\partial t} + \frac{1}{2}\frac{\partial^2 \phi}{\partial x^2} - \kappa|\phi|^2\phi = 0 \tag{4.8}$$

where ϕ represents a complex field.

4.1.3   Initial and boundary conditions

For a transient problem in a finite domain, both initial and boundary conditions must be specified to make the problem well posed and to have a unique solution. Such a problem is called an initial–boundary value problem. In the case where the transient problem is solved in an infinite domain, the problem can be simplified to initial value problem because no boundary condition is needed. In contrast, if a steady problem is solved in a finite domain, the problem is called boundary value problem and only boundary conditions are needed.

The number of boundary conditions required is equal to the order of the highest spatial derivatives in the governing equation for each coordinate space. For example, for a 1D diffusion problem, the highest order of the spatial derivative is two and thus two spatial boundary conditions are needed. Typically, three types of boundary conditions can be specified, namely the Dirichlet type where the variable value is given, the Neumann type where the normal gradient of the variable is provided, and the mixed type where the relationship of the variable and its normal gradient is defined, i.e.:

$$\phi = f_1 \quad \text{Direchlet}$$

$$\frac{\partial \phi}{\partial n} = f_2 \quad \text{Neumann} \tag{4.9}$$

$$a\phi + b\frac{\partial \phi}{\partial n} = f_3 \quad \text{Mixed}$$

Note that for higher order derivatives, there exist other combinations of mixed types of boundary conditions.

4.2 Finite difference method

4.2.1 Finite difference construction

The FDM is the most natural way of solving a PDE directly in an approximate manner. The idea behind FDM is to discretize the continuous time and space into a finite number of discrete grid points and then to approximate the local derivatives at these grid points with FD schemes. Let us first use the function $\phi(x) = x^2$ as an example to illustrate how an FD scheme is derived for $\partial \phi / \partial x$ and $\partial^2 \phi / \partial x^2$. To represent this curve with discrete points, we can either use fine and uniform grid distribution (e.g., square on the left curve in Figure 4.1) or coarse and nonuniform grid distribution (e.g., circle

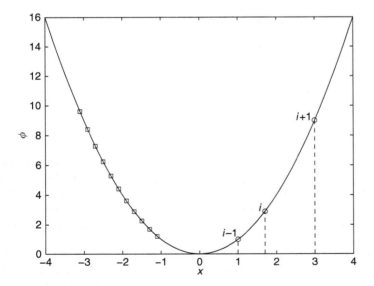

Figure 4.1 FDM representation of a continuous function by discrete points.

on the right curve in Figure 4.1). This implies that the discretization process is not unique. However, a "sufficient" discretization should be such that the main features of the variable function (e.g., maxima and minima) are captured.

Once the discretization is determined, an FD can then be developed. Without loss of generality, we shall use three circular points on the right side of the curve to illustrate how we can develop an FD scheme to approximate $\partial\phi/\partial x$ and $\partial^2\phi/\partial x^2$ numerically at node i. We thereby define the following notations: ϕ_{i-1}, ϕ_i, and ϕ_{i+1} represent the value of ϕ at the nodes $i-1$, i, and $i+1$, respectively. The distance between node $i-1$ and i is defined as $\Delta x_{i-1/2}$ and between i and $i+1$ as $\Delta x_{i+1/2}$. Although there are other ways of developing an FD scheme, e.g., polynomial representation (Jaluria and Torrance, 2003), the most common approach is to start from Taylor expansion, from which we have:

$$
\begin{aligned}
\phi_{i-1} = \phi_i &+ \left(-\Delta x_{i-1/2}\right)\left(\frac{\partial\phi}{\partial x}\right)_i + \frac{\left(-\Delta x_{i-1/2}\right)^2}{2!}\left(\frac{\partial^2\phi}{\partial x^2}\right)_i \\
&+ \frac{\left(-\Delta x_{i-1/2}\right)^3}{3!}\left(\frac{\partial^3\phi}{\partial x^3}\right)_i + \frac{\left(-\Delta x_{i-1/2}\right)^4}{4!}\left(\frac{\partial^4\phi}{\partial x^4}\right)_i \\
&+ \cdots + \frac{\left(-\Delta x_{i-1/2}\right)^m}{m!}\left(\frac{\partial^m\phi}{\partial x^m}\right)_i + \cdots
\end{aligned}
\tag{4.10}
$$

$$
\begin{aligned}
\phi_{i+1} = \phi_i &+ \left(\Delta x_{i+1/2}\right)\left(\frac{\partial\phi}{\partial x}\right)_i + \frac{\left(\Delta x_{i+1/2}\right)^2}{2!}\left(\frac{\partial^2\phi}{\partial x^2}\right)_i \\
&+ \frac{\left(\Delta x_{i+1/2}\right)^3}{3!}\left(\frac{\partial^3\phi}{\partial x^3}\right)_i + \frac{\left(\Delta x_{i+1/2}\right)^4}{4!}\left(\frac{\partial^4\phi}{\partial x^4}\right)_i \\
&+ \cdots + \frac{\left(\Delta x_{i+1/2}\right)^m}{m!}\left(\frac{\partial^m\phi}{\partial x^m}\right)_i + \cdots
\end{aligned}
\tag{4.11}
$$

The above expansion is valid as long as the function is smooth, and it is exact when $m \to \infty$. In practical computation, truncation needs to be made to include a finite number of terms in the above equation. The symbol $O\left[(\Delta x)^m\right]$ is often used to represent the order of the error of the approximation if the truncation is made up to $m-1$. Obviously, the larger the m is, the more accurate the approximation will be.

Now let us see how a particular FD scheme can be derived from the above Taylor series expansion. Assume that we want to use two points only, e.g., ϕ_{i-1} and ϕ_i, to approximate $(\partial\phi/\partial x)_i$. A linear combination of ϕ_{i-1} and ϕ_i will be used here, i.e.:

$$(\partial\phi/\partial x)_i = a\phi_{i-1} + b\phi_i$$

$$= a\left[\phi_i + (-\Delta x_{i-1/2})\left(\frac{\partial\phi}{\partial x}\right)_i + \frac{(-\Delta x_{i-1/2})^2}{2!}\left(\frac{\partial^2\phi}{\partial x^2}\right)_i\right.$$

$$\left. +\cdots+ \frac{(-\Delta x_{i-1/2})^m}{m!}\left(\frac{\partial^m\phi}{\partial x^m}\right)_i\right] + b\phi_i$$

$$= (a+b)\phi_i + (-a\Delta x_{i-1/2})\left(\frac{\partial\phi}{\partial x}\right)_i + \frac{a(-\Delta x_{i-1/2})^2}{2!}\left(\frac{\partial^2\phi}{\partial x^3}\right)_i$$

$$+\cdots+ \frac{a(-\Delta x_{i-1/2})^m}{m!}\left(\frac{\partial^m\phi}{\partial x^m}\right)_i \tag{4.12}$$

We need to have the coefficients in front of each term equal on both sides of the equation, which gives a system of equations as follows:

$$\begin{cases} a+b = 0 \\ -a\Delta x_{i-1/2} = 1 \\ \dfrac{a(-\Delta x_{i-1/2})^2}{2!} = 0 \\ \vdots \\ \dfrac{a(-\Delta x_{i-1/2})^m}{m!} = 0 \end{cases} \tag{4.13}$$

In the above system of equations, we have an infinite number of equations but only two unknowns. Therefore, the system of equations is overdetermined with more constraints than necessary. We can relax the system by removing the less important constraints (e.g., higher order terms) until the system is balanced. In this case, we will retain only the first two equations to solve the coefficients a and b that leads to the following:

$$a = -\frac{1}{\Delta x_{i-1/2}}, \quad b = \frac{1}{\Delta x_{i-1/2}} \tag{4.14}$$

This gives the final form of the FD scheme:

$$(\partial\phi/\partial x)_i = \frac{\phi_i - \phi_{i-1}}{\Delta x_{i-1/2}} + O(\Delta x_{i-1/2}) \tag{4.15}$$

The last term will not be evaluated in the computation and is the leading order of the error for this particular FD scheme. This FD is therefore *first-order accurate*. Because the scheme is based on the local node i and its backward node $i-1$, the scheme is called *backward difference*. The above form can also be deduced based on simple intuition. However, the use of Taylor expansion not only provides the theoretical explanation for the order of the error but also serves as the basis for the systematic development of other more complex FD schemes

4.2.2 Truncation error and order of accuracy

The exact expression of $O\left(\Delta x_{i-1/2}\right)$ in (4.15) can be obtained by substituting the Taylor series for each term in the FD equation. This term is called *truncation error* (TE), which is used to quantify the order of errors as well as the characteristics of the leading error terms for the FD scheme. For example, for the backward FD, the TE is found by the following operation:

$$
\begin{aligned}
TE &= \left(\frac{\partial\phi}{\partial x}\right)_i - \frac{\phi_i - \phi_{i-1}}{\Delta x_{i-1/2}} \\
&= \left(\frac{\partial\phi}{\partial x}\right)_i - \frac{1}{\Delta x_{i-1/2}} \\
&\quad \left[\phi_i - \left[\begin{array}{l} \phi_i + \left(-\Delta x_{i-1/2}\right)\left(\dfrac{\partial\phi}{\partial x}\right)_i + \dfrac{\left(-\Delta x_{i-1/2}\right)^2}{2!}\left(\dfrac{\partial^2\phi}{\partial x^2}\right)_i \\ + \dfrac{\left(-\Delta x_{i-1/2}\right)^3}{3!}\left(\dfrac{\partial^3\phi}{\partial x^3}\right)_i + \dfrac{\left(-\Delta x_{i-1/2}\right)^4}{4!}\left(\dfrac{\partial^4\phi}{\partial x^4}\right)_i + \cdots \\ + \dfrac{\left(-\Delta x_{i-1/2}\right)^m}{m!}\left(\dfrac{\partial^m\phi}{\partial x^m}\right)_i + \cdots \end{array}\right]\right] \\
&= \frac{\Delta x_{i-1/2}}{2!}\left(\frac{\partial^2\phi}{\partial x^2}\right)_i - \frac{\left(\Delta x_{i-1/2}\right)^2}{3!}\left(\frac{\partial^3\phi}{\partial x^3}\right)_i + \frac{\left(\Delta x_{i-1/2}\right)^3}{4!}\left(\frac{\partial^4\phi}{\partial x^4}\right)_i + \cdots
\end{aligned}
$$

$$(4.16)$$

Apparently, TE is the difference between the FD numerical approximation [e.g., $\left(\phi_i - \phi_{i-1}\right)/\Delta x_{i-1/2}$] and the exact derivative [e.g., $\left(\partial\phi/\partial x\right)_i$]. For this particular FD scheme, the leading TE is proportional to $\Delta x_{i-1/2}$.

Following the same procedure, we can find the FD form and the corresponding TE when nodes i and $i+1$ are used to approximate $\left(\partial\phi/\partial x\right)_i$:

$$\left(\partial\phi/\partial x\right)_i = \frac{\phi_{i+1} - \phi_i}{\Delta x_{i+1/2}} + TE \tag{4.17}$$

$$
\begin{aligned}
TE &= \left(\frac{\partial\phi}{\partial x}\right)_i - \frac{\phi_{i+1} - \phi_i}{\Delta x_{i-1/2}} \\
&= -\frac{\Delta x_{i+1/2}}{2!}\left(\frac{\partial^2\phi}{\partial x^2}\right)_i - \frac{\left(\Delta x_{i+1/2}\right)^2}{3!}\left(\frac{\partial^3\phi}{\partial x^3}\right)_i - \frac{\left(\Delta x_{i+1/2}\right)^3}{4!}\left(\frac{\partial^4\phi}{\partial x^4}\right)_i - \cdots
\end{aligned}
$$

$$(4.18)$$

Because the FD scheme uses local node i and its forward node $i+1$, the scheme is called *forward difference*. Compared with the backward difference, the TE of forward difference has a different sign in front of all even-order derivatives (diffusion). This difference of sign changes the characteristics of the two FD schemes, although the *order of accuracy* is the same for both schemes.

When we use nodes $i-1$ and $i+1$ to approximate the first-order derivative, the scheme is called *central difference* because the derivative is evaluated at the location between two nodal points. The FD scheme and the TE are found to be:

$$(\partial\phi/\partial x)_i = \frac{\phi_{i+1}-\phi_{i-1}}{\Delta x_{i-1/2}+\Delta x_{i+1/2}} + \text{TE} \tag{4.19}$$

$$\begin{aligned}
\text{TE} &= \left(\frac{\partial\phi}{\partial x}\right)_i - \frac{\phi_{i+1}-\phi_{i-1}}{\Delta x_{i-1/2}+\Delta x_{i+1/2}} \\
&= -\frac{\Delta x_{i+1/2}-\Delta x_{i-1/2}}{2!}\left(\frac{\partial^2\phi}{\partial x^2}\right)_i - \left[\frac{\left(\Delta x_{i+1/2}\right)^3+\left(\Delta x_{i-1/2}\right)^3}{3!\left(\Delta x_{i-1/2}+\Delta x_{i+1/2}\right)}\right] \\
&\quad \left(\frac{\partial^3\phi}{\partial x^3}\right)_i - \left[\frac{\left(\Delta x_{i+1/2}\right)^4-\left(\Delta x_{i-1/2}\right)^4}{4!\left(\Delta x_{i-1/2}+\Delta x_{i+1/2}\right)}\right]\left(\frac{\partial^4\phi}{\partial x^4}\right)_i -\cdots
\end{aligned} \tag{4.20}$$

An interesting finding here is that the scheme remains first-order accurate when $\Delta x_{i+1/2} \neq \Delta x_{i-1/2}$ (i.e., nonuniform grid spacing). However, the scheme becomes second-order accurate when uniform grids are applied in which $\Delta x_{i+1/2} = \Delta x_{i-1/2}$.

To have a second-order FD for $(\partial\phi/\partial x)_i$ in a nonuniform grid system, at least three grid points are needed. If we use nodes $i-1$, i, and $i+1$, we would have the following FD scheme:

$$\begin{aligned}
(\partial\phi/\partial x)_i &= a\phi_{i-1}+b\phi_i+c\phi_{i+1} \\
&= a\left[\phi_i+\left(-\Delta x_{i-1/2}\right)\left(\frac{\partial\phi}{\partial x}\right)_i \right. \\
&\quad \left. +\frac{\left(-\Delta x_{i-1/2}\right)^2}{2!}\left(\frac{\partial^2\phi}{\partial x^2}\right)_i+\frac{\left(-\Delta x_{i+1/2}\right)^3}{3!}\left(\frac{\partial^3\phi}{\partial x^3}\right)_i\cdots\right]+b\phi_i \\
&\quad +c\left[\phi_i+\left(\Delta x_{i+1/2}\right)\left(\frac{\partial\phi}{\partial x}\right)_i \right. \\
&\quad \left. +\frac{\left(\Delta x_{i+1/2}\right)^2}{2!}\left(\frac{\partial^2\phi}{\partial x^2}\right)_i+\frac{\left(\Delta x_{i+1/2}\right)^3}{3!}\left(\frac{\partial^3\phi}{\partial x^3}\right)_i+\cdots\right] \\
&= (a+b+c)\phi_i+\left(-a\Delta x_{i-1/2}+c\Delta x_{i+1/2}\right)\left(\frac{\partial\phi}{\partial x}\right)_i \\
&\quad +\frac{a\left(-\Delta x_{i-1/2}\right)^2+c\left(\Delta x_{i+1/2}\right)^2}{2!}\left(\frac{\partial^2\phi}{\partial x^2}\right)_i+O(\Delta x^3)
\end{aligned} \tag{4.21}$$

To have a second-order accurate scheme, we need to impose the following:

$$a + b + c = 0$$

$$-a\Delta x_{i-1/2} + c\Delta x_{i+1/2} = 1 \tag{4.22}$$

$$\frac{a\left(-\Delta x_{i-1/2}\right)^2 + c\left(\Delta x_{i+1/2}\right)^2}{2!} = 0$$

This renders:

$$\begin{cases} a = -\dfrac{\Delta x_{i+1/2}}{\Delta x_{i-1/2}\left(\Delta x_{i-1/2} + \Delta x_{i+1/2}\right)} \\[4mm] b = \dfrac{\Delta x_{i+1/2} - \Delta x_{i-1/2}}{\Delta x_{i-1/2}\Delta x_{i+1/2}} \\[4mm] c = \dfrac{\Delta x_{i-1/2}}{\Delta x_{i+1/2}\left(\Delta x_{i-1/2} + \Delta x_{i+1/2}\right)} \end{cases} \tag{4.23}$$

Substituting the above expressions of a, b, and c into (4.21), we can find that the FD is as follows:

$$(\partial\phi/\partial x)_i \approx \frac{\Delta x_{i+1/2}}{\Delta x_{i-1/2} + \Delta x_{i+1/2}}\left(\frac{\phi_i - \phi_{i-1}}{\Delta x_{i-1/2}}\right)$$

$$+ \frac{\Delta x_{i-1/2}}{\Delta x_{i-1/2} + \Delta x_{i+1/2}}\left(\frac{\phi_{i+1} - \phi_i}{\Delta x_{i+1/2}}\right) \tag{4.24}$$

Another way of understanding the above FD scheme is that it is the linear combination of the FDs for $(\partial\phi/\partial x)_{i-1/2}$ and $(\partial\phi/\partial x)_{i+1/2}$ with the weighting coefficients being reciprocally proportional to the lengths between two grid points. The above scheme is reduced to the conventional central difference scheme when $\Delta x_{i+1/2} = \Delta x_{i-1/2}$, in which node i is no longer needed.

Using the methods discussed in (4.21), we are in principle able to establish an FD scheme in any order of accuracy, given sufficient number of grid points. A few facts can be proven for the general case with a nonuniform grid:

1   To derive an FD scheme for an nth-order derivative, at least $m = n + 1$ grid points are needed.
2   The *maximum* order of accuracy of an FD scheme is equal to the number of grid points m minus the order of derivative n for a nonuniform grid.
3   When the required order of accuracy of an FD scheme is equal to $m - n$, the FD scheme is unique.
4   When the required order of accuracy of an FD scheme is smaller than $m - n$, there will be an infinite number of possible FDs. For example, if

we use nodes $i-1$, i, and $i+1$ to develop a first-order accurate scheme for $(\partial\phi/\partial x)$, based on (4.21), we only need to satisfy (4.22a) and (4.22b) whereas the RHS of (4.22c) can be any arbitrary value. This makes an infinite number of combinations of a, b, and c, within which the backward difference, forward difference, or central difference is only one of the many possible solutions.

5   For an FD equation that consists of a few PDE approximations, the overall order of accuracy is the lowest order of accuracy in all the FD approximations.

4.2.3 *Consistency and convergence*

In Section 4.2.2, we introduced the concept of FDs and the systematical way of establishing a particular FD for a derivative based on the order of accuracy requirement and the given grid points. Now let us consider a specific PDE with the combination of a few derivative terms. The simplest yet very useful PDE is the 1D convection equation:

$$\frac{\partial\phi}{\partial t} + C\frac{\partial\phi}{\partial x} = 0 \tag{4.25}$$

Consistency: The definition of consistency is related to whether an FD equation authentically represents a PDE. Based on the discussion in Section 4.2.2, we are able to approximate any spatial derivative term by an FD with the controlled order of accuracy. The error terms can be lumped into a TE whose leading term is proportional to certain power of mesh size. The same methodology can be applied to the temporal derivative. An FD representation of a PDE is *consistent* if the total TE approaches zero when Δx and Δt approach zero. It is not difficult to view the following fact: if the FD for each derivative in a PDE is at least first-order accurate, the finite different equation will be consistent with the PDE.

Convergence: While consistency ensures an FDM gives an approximation close enough to a PDE, convergence is related to whether the numerical solution from an FD equation is close enough to the true solution of a PDE. Mathematically, with the well-posed initial and boundary conditions, a unique solution $\phi(t, x)$ always exists, although most of the time it cannot be expressed as a closed-form analytical solution. By solving the FD equation, a numerical solution can be obtained. An FD scheme is called convergent if its numerical solution approaches the true solution as Δx and Δt approach zero. Generally, convergence is a stronger condition for an FD than is consistency. A convergent scheme is always consistent, whereas a consistent scheme may not be convergent due to inbuilt instability. However, although the above statement is true and mathematically it is possible to "create" such a scheme that is consistent but not convergent, in most of the practical applications, the convergence and the consistency of a numerical scheme imply the same; with the continuous refinement of mesh and time step, the numerical solution will converge to the true solution within a bounded finite TE.

4.2.4   Stability

4.2.4.1   *Heuristic analysis of numerical stability*

A consistent and convergent FD scheme may not necessarily lead to an acceptable numerical solution for a PDE. The classical example is the explicit "forward-time and central-space (FTCS)" FD for the simple convection equation. Under a uniform grid system, the FD is written as follows:

$$\frac{\phi_i^{n+1} - \phi_i^n}{\Delta t} + C \frac{\phi_{i+1}^n - \phi_{i-1}^n}{2\Delta x} = 0 \Rightarrow \phi_i^{n+1} = \phi_i^n - \frac{C\Delta t}{2\Delta x}\left(\phi_{i+1}^n - \phi_{i-1}^n\right) \quad (4.26)$$

Based on the above scheme, the variable value at any nodal point can be advanced from the current time step n to the next time step $n+1$ (time matching) with the known information at time step n. The "explicit" reflects the fact that the nodal information can be updated individually based on the previous time-step information without referring to its neighborhood at the current time step. The forward time method is also called "Euler method".

This is a simple scheme but will unfortunately lead to an unstable solution no matter how small Δx and Δt are. Figure 4.2 shows an example of the numerical result from the above scheme for calculating a step of discontinuity advancing to the right with a constant speed C. Without loss of

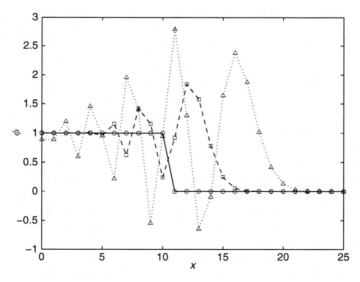

Figure 4.2 Numerical results for pure advection based on FTCS scheme at $t = 0$ (solid line), 5 (dashed line), and 10 (dotted line).

generality, $C = 1$, $\Delta x = 1$, and $\Delta t = 0.5$. In this case, the analytical solution will lead to the pure advection of the step without a change of shape. The numerical simulation, however, results in increasing size of wiggles, which will eventually blow up with the further progress of time. The above wiggles will not disappear even when Δx and Δt are reduced. This feature is called instability of the numerical scheme, and it prompts us to explore the stability property of an FD scheme.

First, let us try to find out what causes the instability of an FD. Based on the earlier introduction, an FD is an approximation of the original PDE. The difference is actually the TE that varies for different FD approximations. For the FTCS scheme, it is not difficult to perform the analysis to reveal the following fact:

$$\frac{\phi_i^{n+1} - \phi_i^n}{\Delta t} + C\frac{\phi_{i+1}^n - \phi_{i-1}^n}{2\Delta x} = 0$$

$$\overset{\text{equivalent}}{\leftrightarrow} \left(\frac{\partial \phi}{\partial t}\right)_i^n + C\left(\frac{\partial \phi}{\partial x}\right)_i^n = \text{TE} = -\frac{\Delta t C^2}{2}\left(\frac{\partial^2 \phi}{\partial x^2}\right)_i^n$$

$$-\frac{\Delta x^2 C}{3}\left(\frac{\partial^3 \phi}{\partial x^3}\right)_i^n + O(\Delta x^3) \quad (4.27)$$

In the above analysis, the approximation of $(\partial^2 \phi/\partial t^2)_i^n \approx C^2 (\partial^2 \phi/\partial x^2)_i^n$ is used to convert the second-order time derivative into the equivalent spatial derivative. It is seen that the FD is equivalent to solving the original PDE at the nodal point i with the additional TEs that are expressed on the RHS of the equation. The leading orders of the TE in (4.28) are the second-order spatial derivative (diffusion) and third-order spatial derivative (dispersion). It is noted that the "diffusion coefficient" in front of the second-order derivative is always negative, implying that in this case negative diffusion prevails. In contrast to the positive diffusion, a negative diffusion enhances the wiggles generated by the TE and round-off error. This makes the numerical scheme unstable, i.e., a small error can have unlimited amplification as the scheme moves in time. For this reason, numerical *stability* refers only to *transient problems*, whereas *convergence* applies to both *steady-state and transient problems*.

The above analytical technique can be applied to essentially all FD schemes. By inspecting the characteristics of the TE for a particular FD, we are able to not only examine the accuracy of a scheme but also judge whether a scheme can possibly become unstable. This type of stability analysis is called heuristic analysis of numerical stability.

4.2.4.2 *von Neumann analysis of numerical stability*

A more rigorous way of performing stability analysis is by using the well-known von Neumann stability analysis method, which is based on Fourier series expansion, and it is applicable to linear FD schemes only. In this method, an error is introduced at time step n for node i, and this error takes the form of one component of the Fourier series, i.e.:

$$E_i^n = A^n e^{Ik_x(i\Delta x)} \tag{4.28}$$

where $I = \sqrt{-1}$, k_x is the wave number in the x-direction, and A^n is the amplitude associated with this component. It is noted that here we consider only one space but it can be easily extended to multiple dimensions.

To examine whether a particular FD scheme is stable or not, we would assume that the error propagation follows the same FD and check whether the error will be bounded, i.e., $|G| = |A^{n+1}/A^n| \leq 1$, where G is the amplification factor. To illustrate how this method works, we shall still use the FTCS scheme discussed earlier as the example:

$$\frac{\phi_i^{n+1} - \phi_i^n}{\Delta t} + C\frac{\phi_{i+1}^n - \phi_{i-1}^n}{2\Delta x} = 0 \Rightarrow \phi_i^{n+1} = \phi_i^n - \frac{C\Delta t}{2\Delta x}\left(\phi_{i+1}^n - \phi_{i-1}^n\right) \tag{4.29}$$

Replacing ϕ by E and substituting (4.28) into the above FD, we obtain:

$$A^{n+1}e^{Ik_x(i\Delta x)} = A^n e^{Ik_x(i\Delta x)} - \frac{C\Delta t}{2\Delta x}\left[A^n e^{Ik_x[(i+1)\Delta x]} - A^n e^{Ik_x[(i-1)\Delta x]}\right]$$

$$= A^n e^{Iik_x x}\left[1 - \frac{C\Delta t}{2\Delta x}e^{Ik_x\Delta x} + \frac{C\Delta t}{2\Delta x}e^{-Ik_x\Delta x}\right] \tag{4.30}$$

From this equation, we are able to find the expression for the amplification factor:

$$G = \frac{A^{n+1}}{A^n} = 1 - \frac{C\Delta t}{2\Delta x}e^{Ik_x\Delta x} + \frac{C\Delta t}{2\Delta x}e^{-Ik_x\Delta x} = 1 - I\frac{C\Delta t}{\Delta x}\sin(k_x\Delta x) \tag{4.31}$$

from which we have:

$$|G| = \sqrt{1^2 + \left(\frac{C\Delta t}{\Delta x}\right)^2 \sin^2(k_x\Delta x)} > 1 \tag{4.32}$$

The magnitude of the amplification factor is always greater than 1.0, implying that the proposed scheme is unconditionally unstable.

4.2.5   *Finite difference schemes for convection equation*

A few well-known FD schemes have been proposed in the past decades to solve the convection equation. Their forms and properties are summarized in this section. We will use uniform grid and linear convection in all examples.

4.2.5.1 Implicit methods (Laasonen method, Crank–Nicolson method, and alternating direction implicit method)

Realizing that FTCS is an unconditionally unstable scheme for a pure convection problem, the immediate alternative is to weight the convection FD between the time step n and the time step $n+1$ as follows:

$$\frac{\phi_i^{n+1} - \phi_i^n}{\Delta t} + C\left[(1-\gamma)\left(\frac{\phi_{i+1}^n - \phi_{i-1}^n}{2\Delta x}\right) + \gamma\left(\frac{\phi_{i+1}^{n+1} - \phi_{i-1}^{n+1}}{2\Delta x}\right)\right] = 0 \quad (4.33)$$

where γ is the weighting factor that varies from 0 to 1. When $\gamma = 0$, this returns to FTCS, whereas when $\gamma \neq 0$, the $n+1$ time-step information is required in the solution procedure. The method is thus called "implicit method" because a matrix needs to be solved to update all variables simultaneously.

Laasonen method and Crank–Nicolson method: When $\gamma = 1$, the scheme is equivalent to the backward time difference and is referred to as "Laasonen method." When $\gamma = 1/2$, the FD scheme is second-order accurate in both time and space, and it is unconditionally stable based on von Neumann stability analysis. This scheme is called "Crank–Nicolson" method and is primarily used when a higher order accurate scheme is needed. Although the Crank–Nicolson method has the advantage in both accuracy and stability properties, it possesses two shortcomings. One is that when the scheme is used to solve a convection problem with a sharp front, spurious oscillation will be present near the front. This is caused by the leading TEs in dispersion form. Another is that since the scheme is implicit, a matrix must be solved to obtain the solution. The matrix can become very large when 2D and 3D problems are considered. In that case, either a sparse matrix solver must be introduced or some other alternatives [e.g. alternating direction implicit (ADI) method; see below] need to be used.

ADI method: ADI is the application of the time-splitting implicit method for multiple dimensional problems. It is known that when the implicit method is used, a matrix needs to be solved. For 1D problems, the matrix has a simple tridiagonal format that can be efficiently solved by the Thomas algorithm. For 2D or 3D problems, the matrix being formed, however, will be large and sparse. To avoid the numerical solution to a large sparse matrix, one can split the solution into a few substeps. In each substep, only 1D convection is solved. The efficient Thomas algorithm can then be applied to the numerical solution in each direction, and the final solution will be obtained after the ADI is applied to all directions alternately within one time step.

4.2.5.2 Explicit central-space methods (the Lax method and the Lax–Wendroff method)

Compared to implicit methods, explicit methods are attractive because they are easier to construct and compute. However, the explicit FTCS has proven to be an unconditionally unstable scheme. Alternatively, people have developed various stable explicit central-space methods to solve the convection equation.

Lax method: The simplest way of constructing an explicit central-space method is to revise the FTCS by replacing the previous time-step central point variable with its spatial average from its left and right neighborhood nodes:

$$\phi_i^{n+1} = \frac{1}{2}\left(\phi_{i+1}^n + \phi_{i-1}^n\right) - C\frac{\Delta t}{2\Delta x}\left(\phi_{i+1}^n - \phi_{i-1}^n\right) \tag{4.34}$$

This scheme has a TE of $O\left(\Delta t, (\Delta x)^2\right)$ and is stable if $C_r = C\Delta t/\Delta x \leq 1$, where C_r is the Courant number and the above condition is called Courant–Friedrichs–Lewy (CFL) condition, which is in fact applicable to most of the explicit FD schemes. The mathematical implication of the CFL condition is straightforward: Δt must be so small that the information cannot propagate over two grid points within one time step. One of the drawbacks of the Lax method is that the method is too dissipative due to the first-order TE in time.

Lax–Wendroff method: To construct a higher order accurate explicit scheme for the convection equation, the Lax–Wendroff scheme is used. Realizing that FTCS has the leading order TE of negative diffusion, to make it stable one can introduce an artificial positive diffusion to balance this term:

$$\frac{\phi_i^{n+1} - \phi_i^n}{\Delta t} + C\frac{\phi_{i+1}^n - \phi_{i-1}^n}{2\Delta x} = \frac{C^2\Delta t}{2\Delta x^2}\left(\phi_{i+1}^n - 2\phi_i^n + \phi_{i-1}^n\right) \tag{4.35}$$

This scheme is second-order accurate in both time and space and is conditionally stable when $C_r \leq 1$. The leading order error term for the Lax–Wendroff method has the form of third-order dispersion, which implies that wiggles may be developed near the shock front in the computation, similar to the Crank–Nicolson method.

4.2.5.3   Upwind schemes (first-order upwind, QUICK, QUICKEST, and method of characteristics)

First-order upwind scheme: To suppress numerical oscillation, some kind of numerical damping is needed. One of the most commonly used schemes for this purpose is the first-order upwind scheme, e.g.:

$$\frac{\phi_i^{n+1} - \phi_i^n}{\Delta t} + C\left[(1-\gamma)\left(\frac{\phi_{i+1}^n - \phi_i^n}{\Delta x}\right) + \gamma\left(\frac{\phi_i^n - \phi_{i-1}^n}{\Delta x}\right)\right] = 0$$

$$\text{where } \gamma = \begin{cases} 0, & C \geq 0 \\ 1, & C < 0 \end{cases} \tag{4.36}$$

In this scheme, the backward difference is used when C is positive, and the forward difference is used when C is negative. This scheme is explicit and only first-order accurate in time and space but can successfully suppress the oscillation near the sharp front. The scheme is conditionally stable when $C_r \leq 1$.

QUICK (quadratic upwind interpolation of convective kinematics) scheme: Since the above upwind scheme is only first-order accurate in time and space, researchers have proposed higher order upwind schemes to improve the accuracy of the scheme and to preserve the advantage of the scheme in suppressing numerical oscillations. QUICK is one such scheme (e.g., Leonard, 1979; Hayse *et al.*, 1992). The scheme is based on the local upwind-weighted quadratic interpolation for the convection term. The scheme is often used to solve the convection–diffusion equation with the employment of second-order central difference for the diffusion term. For the convection term, the scheme is third-order accurate in space and first-order accurate in time, and thus, it is sometimes called third-order upwinding. Due to the first-order TE in time, QUICK is most appropriate for steady or quasi-steady highly convective elliptic flows.

QUICKEST (quadratic upstream interpolation for convective kinematics with estimated streaming terms) scheme: To improve the numerical accuracy in time, Leonard (1988) proposed the QUICKEST scheme. By estimating the temporal behavior of the convection term (and diffusion term if any), more accurate averages can be obtained in the solution procedure. Formally, QUICKEST is third-order accurate in both time and space.

Method of characteristics: This method is a special FDM that is similar to the upwind scheme in concept but different in numerical treatment. In this method, numerical solution follows the characteristic lines describing wave propagation routes. Numerical interpolations are used to obtain the variable information between two grid points at the previous time step. The method is popularly used to solve a system of two convection equations describing wave propagation in two opposite directions. The examples include water wave equations and pressure wave equations (e.g., water hammer problems in pipe flows). Depending on how many grid points are used in the interpolation procedure, various orders of accuracy of the scheme can be achieved.

4.2.5.4   Oscillation control methods (flux limiter methods, essentially nonoscillatory methods, and other interface-tracking methods)

All the higher order (second-order and above) numerical schemes we have introduced so far will inevitably develop spurious oscillations when they are used to solve the convection process of a shock front. This problem can be alleviated by using lower order schemes (e.g., the first-order upwind or Lax method) or introducing artificial viscosity near the front. While the former practice reduces overall numerical accuracy, the latter practice has inherent difficulty in the determination of the artificial viscosity that is case-dependent. Alternatively, some more robust and accurate methods have been developed to solve the convection equation with a sharp front.

TDV methods using flux limiters: The introduction of various flux limiters is meant to eliminate the oscillation to achieve the criterion of total variation diminishing (TDV) (Harten, 1984). By using the carefully designed limiters, the oscillation near the discontinuity can be eliminated. The limiters will be active only near the sharp front, where the numerical accuracy can possibly degenerate to first order only, and will not affect the numerical accuracy in the rest of the computational domain. Some common limiters include van Leer limiter, ULTIMATE (universal limiter for transport interpolation modeling of the advective transport equation), and Chakravarthy–Osher limiter. These limiters are usually combined with higher order accurate numerical schemes to have a numerical scheme that is both accurate and oscillation-free, e.g., the ULTIMATE–QUICKEST scheme (Leonard, 1991).

ENO schemes: As shown earlier in (4.21), FDM is essentially an interpolation method to approximate the local derivative for a smooth function. The order of accuracy can be precisely analyzed with the use of Taylor expansion. For a specific derivative, higher order accurate schemes can be derived by using more nodal points. In the shock front where discontinuity of the function exists, the above conclusion will not hold. As a matter of fact, the more nodal points included, the more spurious oscillations will be developed. The ENO scheme (Harten *et al.*, 1987) is basically a self-similar and uniformly higher order scheme for piecewise smooth functions. A hierarchy is established in ENO to automatically determine the number of nodal points to be included in the interpolation process based on the local smoothness of the function. As a result, the scheme is especially suitable for the problem that both shock fronts and smooth flow regions are presented. The numerical solution will resolve the shock front with essentially no spurious oscillation but maintain the higher order numerical accuracy for the rest of the regions. The scheme was later extended by many other people to further enhance the stability and improve the accuracy (e.g., weighted ENO (WENO) scheme by Liu *et al.*, 1994).

Other interface-capturing methods: A shock front is essentially an interface where the function has a discontinuity. Sometimes, the motion of the interface is the only interest to a modeler when the convection equation is solved, e.g., the density equation is solved to capture the interface between two immiscible fluids. In this case, some special treatments can be developed near the interface to keep it sharp in the process of solving the convection equation. In many cases, the information of interface orientation is helpful to develop a higher order nonoscillation scheme for interface capturing. This requires that the interface reconstruction before the convection equation be solved. The typical examples include volume-of-fluid (VOF) method to track the free surface and level set method to track the interface. More details of these methods will be provided in Sections 5.2.1.3 and 5.5.1.1.

4.2.5.5 Time-splitting methods (MacCormack method and other predictor–corrector methods)

MacCormack method: MacCormack method is one of the simplest time-splitting methods that breaks down the solution procedure into two substeps in time:

$$\bar{\phi}_i^{n+1} = \phi_i^n - C\frac{\Delta t}{\Delta x}\left(\phi_{i+1}^n - \phi_i^n\right)$$

$$\phi_i^{n+1} = \phi_i^n - C\frac{\Delta t}{2\Delta x}\left(\phi_{i+1}^n - \phi_i^n + \bar{\phi}_i^{n+1} - \bar{\phi}_{i-1}^{n+1}\right)$$

(4.37)

In the predictor step, the tentative value of the variable $\bar{\phi}_i^{n+1}$ is obtained. This tentative value is corrected in the corrector step, during which the space derivative is approximated by the arithmetic average of the forward difference at time t^n and the backward difference of the tentative value at time t^{n+1}. The main merits of the scheme include simplicity and high accuracy (second-order in both space and time). However, similar to other second-order accurate central-space schemes (e.g., Lax–Wendroff and Crank–Nicolson methods), it has a drawback that in the computation of shock waves and spurious oscillations may appear.

Adams–Bashforth–Moulton method: This method was originally developed based on the integration of the polynomials that interpolate the known nodal information to approximate the nearby derivative. The method can be extended to any order of accuracy, given sufficient number of grid points. This method was often used to solve ODEs, but it can also be extended to solve PDEs. More commonly, this method is used to develop a higher order accurate FD scheme in time by using two-level or multiple-level time-splitting predictor–corrector methods.

Time-splitting methods for multiple-dimension problems: For multiple-dimension problems, the convection process takes place in all directions. In the numerical solution, the FD from time step n to $n+1$ can be split into m sublevels where m is the dimension of the equation. As a result, in each substep, only 1D convection is solved. This can simplify the coding effort considering that the convection in various directions can be solved by the same method. In connection to the implicit method, it can also improve the numerical efficiency (e.g., ADI) by reducing the resulting matrix size.

Time-splitting methods for multiple-variable problems (leapfrog method): The time-splitting method can also be applied to multiple-variable problems. For example, the so-called leapfrog method is a time-splitting method that is employed to solve a system of two (for 1D problems) or three (for 2D problems) PDEs, one of which describes the time derivative of a scalar and the other describes a vector (e.g., SWEs). By defining the scalar and vector on a staggered mesh system (i.e., the scalar defined at the center of a cell whereas the vector defined at cell boundaries), the scalar and vector will be solved at n and $n+1/2$

time steps, respectively, with the central-space method for the convection term. This method is second-order accurate in both time and space.

4.2.6   *Finite difference schemes for diffusion and dispersion equations*

So far, we have not discussed the FD approximation for other terms like diffusion and dispersion. These terms are normally discretized by various orders of central differences and the corresponding stability conditions can be obtained by von Neumann analysis if the problem is linear. Because the resulting leading order TEs from the approximation of these terms are associated with higher order derivative terms, the addition of diffusion and dispersion generally only modifies the stability criterion but not the characteristics of the leading TE. For the diffusion term, the typical stability criterion for an explicit central difference method is $\nu\Delta t/\Delta x^2 < O(1)$, similar to the CFL condition imposed by the explicit solution for convection.

When an FDM is used to solve a PDE with both lower and higher order derivatives, caution must be taken that the TE resulting from the lower order derivatives does not overwhelm the physical higher order derivatives. Consider a 1D convection–diffusion equation. If we use the first-order upwind scheme to discretize the convection term and the central difference to discretize the diffusion term, with the consideration of the leading order TE, we are equivalently solving the following equation:

$$\frac{\partial\phi}{\partial t} + C\frac{\partial\phi}{\partial x} - D\frac{\partial^2\phi}{\partial x^2} = TE = D_N(\Delta t, \Delta x)\frac{\partial^2\phi}{\partial x^2} + O(\Delta t^2, \Delta x^2) \qquad (4.38)$$

where D_N is the numerical diffusivity that results from the upwind scheme and has been used to stabilize the numerical scheme. If the diffusion process is important, we must have $D_N << D$ to ensure that the correct physics is represented by the FDM; otherwise we are actually solving a different problem. If the convection is dominant over diffusion (i.e., potential flow), the numerical results will then become less sensitive to D_N and it may be tolerable to have $D_N > D$. A similar argument applies to other equations, e.g., the convection–dispersion equation (i.e., Boussinesq equation). In this case, the numerical dispersion resulting from the FD approximation of the convective term must be kept small enough when compared to the physical dispersion terms.

4.3   Finite volume method

The FVM can be regarded as the integral formulation of FDM over a small space around a grid point. In contrast to the FDM, which is established on a nodal point based on Taylor expansion to approximate local derivatives, FVM is established over a small control volume, within which the conservation of the variable is ensured not only mathematically but also numerically. For this reason, FVM is also called the control the volume method.

Consider a convection equation as follows:

$$\frac{\partial \phi}{\partial t} + \frac{\partial (F_X)}{\partial x} + \frac{\partial (F_Y)}{\partial y} = 0 \Rightarrow \frac{\partial \phi}{\partial t} + \nabla \cdot \mathbf{F} = 0 \tag{4.39}$$

where $\mathbf{F} = (F_X, F_Y) = (u\phi, v\phi)$. Compared with the convection equation we studied earlier [e.g. (4.25)], the difference in the above equation is that the convection term is now expressed as the form of flux derivatives. The physical meaning of the above equation is clear: the time rate of change of variable ϕ is balanced by the variable flux in space. This is clearer when we integrate the equation over a control volume:

$$\frac{\partial}{\partial t} \int\limits_V \phi + \int\limits_S \mathbf{F} \bullet \mathbf{n} dS = 0 \tag{4.40}$$

where V is the control volume and S is the control surface with $\mathbf{n}$ being denoted as a unit outward normal vector on the surface. The conversion of the volume integration to surface integration is accomplished by using Gauss's divergence theorem.

To approximate the above equation with the use of an FVM, we shall first divide the physical domain into a group of small control volumes that are not overlapping. There are many ways of discretizing the physical

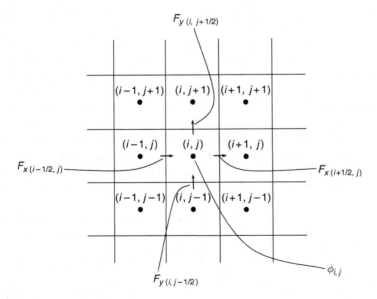

Figure 4.3 Sketch of a structured staggered mesh system for an FVM.

domain, but one of the simplest ways is to use a staggered mesh as shown in Figure 4.3. In this figure, a rectangular mesh is used for control volumes. In each control volume, the scalar variable ϕ is defined at the cell center [i.e., (i, j)], whereas the vectors $\mathbf{F}_X$ and $\mathbf{F}_Y$ are defined at the cell boundaries. The straightforward finite volume approximation for the integral equation (4.40) can be readily derived:

$$\frac{\partial}{\partial t} \int_{\Delta x \Delta y} \phi \, dx \, dy - \int_{i-1/2} F_X \, dy + \int_{i+1/2} F_X \, dy - \int_{j-1/2} F_Y \, dx + \int_{j+1/2} F_Y \, dx = 0 \quad (4.41)$$

It is not difficult to prove that with the above scheme, the conservation law will strictly satisfy for a local control volume. If we further assume that ϕ is uniformly distributed within the control volume and F is uniformly distributed along all boundaries, the above equation can be reduced to a form similar to an FD:

$$\frac{\partial \phi}{\partial t} \Delta x \Delta y - F_{X_{i-1/2,j}} \Delta y + F_{X_{i+1/2,j}} \Delta y - F_{Y_{i,j-1/2}} \Delta x + F_{Y_{i,j+1/2}} \Delta x = 0$$

$$\Leftrightarrow \frac{\phi_{i,j}^{n+1} - \phi_{i,j}^{n}}{\Delta t} + \frac{-F_{X_{i-1/2,j}} + F_{X_{i+1/2,j}}}{\Delta x} + \frac{-F_{Y_{i,j-1/2}} + F_{Y_{i,j+1/2}}}{\Delta y} = 0$$

$$(4.42)$$

Since the fluxes are defined at cell boundaries, their information will be shared by any two adjacent cells. Therefore, the conservation law is automatically satisfied not only locally for a particular cell but also globally for the entire physical domain. This global conservation is the result of the conservation of transport between control volumes, which ensures that the conservation law is satisfied for any arbitrary assembly of connected control volumes.

An FVM can be regarded as the equivalent FD expression of the integral (not differential) form of the governing equation. From the above analysis, it can be seen that the FVM in (4.42) is similar to the FTCS FD scheme. As a matter of fact, nearly all FD schemes (e.g., Lax–Wendroff method, QUICK, and QUICKEST) can find their counterpart expressions in FVM in a structured mesh system. Under such circumstances, the numerical accuracy and numerical stability for an FVM can be established by examining the leading TE using Taylor series expansion or by performing von Neumann stability analysis using Fourier analysis, similar to those used in the FDM. The major reason of constructing a finite volume scheme is often based on the motivation of having a fully conserved scheme, which is not explicitly enforced by an FD method based on the differential form of the governing equation.

Another major advantage of a finite volume approach is that it allows the use of unstructured meshes that can be conformal to irregular boundary geometry. This is especially useful when we are dealing with a complex problem with an arbitrary domain shape. In this case, a finite volume constructed on an unstructured mesh system (e.g., triangle mesh) can be used to follow closely the actual domain boundaries. The accuracy and stability analysis, however, will not be straightforward.

4.4 Finite element method

4.4.1 Background of finite element method

While both FDM and FVM approximate a governing PDE, in differential and integral form, respectively, FEM approximates the solution of a PDE. FEM provides regional approximations to the solution of a PDE by using piecewise functional approximations, usually polynomials. Similar to FVM, FEM has the flexibility of using various geometrical shapes in element approximations and therefore can be adapted to a complex boundary. Traditionally, FEM has been popularly used in structural and solid mechanics where the material is approximated by an elastic or an elastoplastic continuum. In recent years, its applications in fluid computation have grown.

4.4.2 Solution procedure in finite element method

The FEM has a well-established standard procedure to solve various types of problems. In the following section, we shall use the convection problem again as an example to illustrate how the FEM works. Consider the following equation:

$$\frac{\partial \phi}{\partial t} + C\frac{\partial \phi}{\partial x} = 0 \tag{4.43}$$

The integration of the above equation in the entire domain V or any subdomain V' will still be valid, e.g.:

$$\int_{V'} \left[\frac{\partial \phi}{\partial t} + C\frac{\partial \phi}{\partial x} \right] dV = 0 \tag{4.44}$$

4.4.2.1 Domain discretization

The first step of the finite element (FE) approach is the discretization of the domain into a finite number N of elements, each of which has N_e nodes. The total nodal number is M that is always less than the sum of N_e because a node can be shared by multiple elements. For the 1D case, line segments are the natural choice (Figure 4.4). In this example, we shall use N uniform linear elements with the element length $\Delta x = L/N$, where L is the total length of the computational domain and $0 \leq x \leq L$. As a result, we have $M = N + 1$ nodes. For 2D and 3D problems, various types of geometrical shapes are possible, e.g., triangles and tetrahedra.

4.4.2.2 Introduction of shape function

By dividing the entire domain into N elements, we are now able to approximate the true solution of $\Phi(x, t)$ in the entire domain by the approximate

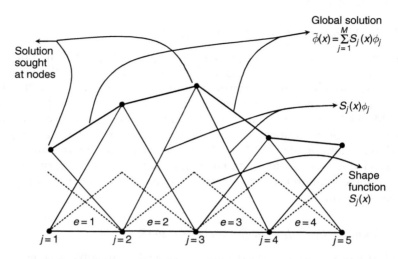

Figure 4.4 Illustration of a solution function approximation in the FEM for 1D problem; *e* represents the element number and *j* the nodal number.

function $\tilde{\Phi}(x, t)$ that can be expressed as the linear summation of M nodal values multiplied by the corresponding shape function (or interpolation function):

$$\Phi(x, t) \approx \tilde{\Phi}(x, t) = \sum_{j=1}^{M} S_j(x)\Phi_j(t) \tag{4.45}$$

where $S_j(x)$ is the linearly independent shape function of x associated with the nodal point j and Φ_j is the approximated value of Φ at the nodal point j and time t, the latter of which is the numerical solution to be sought that should satisfy all boundary conditions. Here Φ_j is in close proximity to the true solution of Φ at the nodal point j, and therefore, one shall expect that the shape function $S_j(x)$ is unity at this node j but zero far away. In practical computation, the shape function must be a known simple function. Polynomials are the most commonly used shape functions, within which a linear polynomial is the simplest. In this case, $S(x)$ is 1.0 at node j and linearly reduces to zero to its nearest neighborhood nodes:

$$S_i(x) = \begin{cases} \dfrac{x - (j-2)\Delta x}{\Delta x} & \text{for } (j-2)\Delta x \le x \le (j-2)\Delta x \\[2mm] 1 - \dfrac{x - (j-2)\Delta x}{\Delta x} & \text{for } (j-1)\Delta x \le x \le j\Delta x, \quad j = 1, 2, \ldots, M \\[2mm] 0 & \text{elsewhere} \end{cases}$$

$$\tag{4.46}$$

4.4.2.3 Error minimization by the method of weighted residuals (collocation method, Galerkin method, and least-squares method)

Since $\tilde{\Phi}(x, t)$ is the approximate solution of $\Phi(x, t)$, it is expected that by substituting it into the original PDE, a small error (or residual) will result:

$$\frac{\partial \tilde{\Phi}}{\partial t} + C \frac{\partial \tilde{\Phi}}{\partial x} = \varepsilon \tag{4.47}$$

Although we are unable to force ε to be zero everywhere in the domain [otherwise $\tilde{\Phi}(x, t)$ would have been the true solution already], we can minimize it globally by driving the weighted integral of the residual to zero or to its minimum for M linearly independent weighting functions W_k:

$$\int_V W_k \varepsilon dV = \int_V W_k \left(\frac{\partial \tilde{\Phi}}{\partial t} + C \frac{\partial \tilde{\Phi}}{\partial x} \right) dV = 0, \quad k = 1, 2, \ldots, M \tag{4.48}$$

The above method is called the method of weighted residuals and different weighting functions can be introduced in the computation.

Collocation method: When the weighting functions are chosen to be the delta function, i.e.:

$$W_k = \delta(\mathbf{x} - \mathbf{x}_k) \tag{4.49}$$

for all nodes, equation (4.48) becomes:

$$\int_V W_k \varepsilon dV = \int_V \delta(\mathbf{x} - \mathbf{x}_k) \varepsilon dV = \varepsilon_k = 0, \quad k = 1, 2, \ldots, M \tag{4.50}$$

The method is called collocation method. In this method, the residuals are driven to zero at all nodes but not necessarily on the elements. Possible instability may develop in practical computations with the use of this method.

Galerkin method: When the weighting functions are chosen to be the same as the shape function, i.e., $W_k = S_k$, the method is called Galerkin's method. This method requires the residual to be orthogonal to all the linearly independent shape functions W_k, i.e.:

$$\int_V W_k \varepsilon dV = \int_V S_k(x) \left(\frac{\partial \sum_{j=1}^{M} S_j(x) \Phi_j(t)}{\partial t} + C \frac{\partial \sum_{j=1}^{M} S_j(x) \Phi_j(t)}{\partial x} \right) dV = 0,$$

$$k = 1, 2, \ldots, M \tag{4.51}$$

It can be proven that the norm of the residual in the entire domain approaches zero as M approaches infinity. This is the most popular method in the FEM. Compared to the collocation method, the Galerkin method is more stable because the residual has been minimized in the entire domain. The price it pays is the increased computational effort in evaluating the integrals in (4.51).

Least-squares (LS) method: In formulating the weighted integral of the residual, the weighting function can also be chosen as the residual itself. In this case, the integral corresponds to the inner product of the residual, and it will always be positive. We are only able to minimize it by forcing its derivative with respect to $\Phi_k(L)$ to be zero:

$$\frac{\partial}{\partial \Phi_k} \int_V \varepsilon^2 dV = 2 \int_V \frac{\partial \varepsilon}{\partial \Phi_k} \varepsilon \, dV = 0 \tag{4.52}$$

This method is called the LS method. It is equivalent to choosing the weighting function as $\partial \varepsilon / \partial \Phi_k$ in (4.48).

In the following example, we shall focus on the Galerkin method.

4.4.2.4   Matrix assembly

When the shape function $S_j(x)$ is a simple polynomial, the closed-form expression of (4.51) is possible, provided that the time derivative $\partial \Phi_j(t)/\partial t$ can be expressed by the FD approximation $(\Phi_j^{n+1} - \Phi_j^n)/\Delta t$. At the end, a system of linear equations for all nodes can be formed:

$$A_{jk}\Phi_j^{n+1} = f_k(\Phi^n), \quad j, k = 1, 2, \ldots, M \tag{4.53}$$

where A_{jk} is the matrix of $M \times M$, Φ_j is the unknown solution to be solved at all M nodes, and f_k is the forcing vector of size M that incorporates all boundary conditions and previous time-step information. Both A_{jk} and f_k are determined by the property of the PDE to be solved and the choice of FE.

In practical computation, by realizing that the shape function S_k is nonzero only in the elements adjacent to node k, the global integration can be reduced to the sum of local element integration. In this problem with the specific numbering system for nodes and elements in Figure 4.4, by realizing that node k is shared by elements $k - 1$ and k, equation (4.51) can be recast into the following form:

$$\int_V S_k(x) \left(\frac{\partial \sum_{j=1}^M S_j(x)\Phi_j(t)}{\partial t} + C \frac{\partial \sum_{j=1}^M S_j(x)\Phi_j(t)}{\partial x} \right) dV$$

$$= \sum_{e=k-1}^{k} \int_{V^{(e)}} S_k(x) \left(\frac{\partial \sum_{m=1}^{N_e} s_m^{(e)}(x)\phi_m^{(e)}(t)}{\partial t} + C \frac{\partial \sum_{m=1}^{N_e} s_m^{(e)}(x)\phi_m^{(e)}(t)}{\partial x} \right) dV = 0$$

$$k = 1, 2, \ldots, M \tag{4.54}$$

where $V^{(e)}$ is the integration range for the eth element and $s_m^{(e)}(x)\phi_m^{(e)}(t)$ is equivalent to $S_{e+m-1}(x)\Phi_{e+m-1}(t)$, the former of which is defined based on element information and the latter on global nodal information. For a particular k, the above equation can be rewritten as follows:

$$
\int_{(k-2)\Delta x}^{(k-1)\Delta x} s_2^{(k-1)}(x) \left(\frac{\partial \sum_{m=1}^2 s_m^{(k-1)}(x)\phi_m^{(k-1)}(t)}{\partial t} \right.
$$

$$
\left. + C\frac{\partial \sum_{m=1}^2 s_m^{(k-1)}(x)\phi_m^{(k-1)}(t)}{\partial x} \right) dx
$$

$$
+ \int_{(k-1)\Delta x}^{k\Delta x} s_1^{(k)}(x) \left(\frac{\partial \sum_{m=1}^2 s_m^{(k)}(x)\phi_m^{(k)}(t)}{\partial t} \right.
$$

$$
\left. + C\frac{\partial \sum_{m=1}^2 s_m^{(k)}(x)\phi_m^{(k)}(t)}{\partial x} \right) dx = 0
$$

(4.55)

By expressing the integration in terms of each element, the integration can be performed on each element with reference to the local coordinate (i.e., from 0 to Δx) as shown above. When all elements share the same geometric property, the evaluation of integration can be greatly simplified because only a limited number of combinations of shape function products exist. In this case, by using (4.46) and the fact $s_m^{(e)}(x) = S_{m+e-1}$, the above equation can be further reduced to:

$$
\int_0^{\Delta x} \frac{x}{\Delta x} \left(\frac{\partial \left[\left(1 - \frac{x}{\Delta x}\right) \phi_1^{(k-1)}(t) + \frac{x}{\Delta x}\phi_2^{(k-1)} \right]}{\partial t} \right.
$$

$$
\left. + C\frac{\partial \left[\left(1 - \frac{x}{\Delta x}\right) \phi_1^{(k-1)}(t) + \frac{x}{\Delta x}\phi_2^{(k-1)} \right]}{\partial x} \right) dx
$$

$$
+ \int_0^{\Delta x} \left(1 - \frac{x}{\Delta x}\right) \left(\frac{\partial \left[\left(1 - \frac{x}{\Delta x}\right) \phi_1^{(k)}(t) + \frac{x}{\Delta x}\phi_2^{(k)} \right]}{\partial t} \right.
$$

$$
\left. + C\frac{\partial \sum_{m=1}^{N_e} \left[\left(1 - \frac{x}{\Delta x}\right) \phi_1^{(k)}(t) + \frac{x}{\Delta x}\phi_2^{(k)} \right]}{\partial x} \right) dx = 0
$$

(4.56)

The above localization facilitates the evaluation of integration of shape function. Once the integration for each element is obtained, it can be assembled to obtain the coefficients A_{jk} associated with each $\phi_m^{(e)}$ and thus $\Phi_{j=m+e-1}$ for the kth linear equation. When the operation is applied to all k from 1 to M, the entire matrix A_{jk} can be determined. The boundary condition and the time derivative term will be reflected in $f_k(\Phi^n)$ in (4.53).

4.4.2.5 Solution of linear system of equations

Equation (4.53) can be solved by any appropriate matrix solver to obtain the approximate solution of Φ_j^{n+1} at all nodes.

4.4.3 Comments on finite element method

Although being approached very differently, there actually exists an inherent relationship between FEM and FDM. With the time derivative being approximated by the forward time difference (and thus the spatial derivative term being expressed at the time step n) and replacing $\phi_m^{(e)}$ by $\Phi_{j=m+e-1}$, equation (4.56) actually gives the following:

$$\frac{1}{\Delta t}\left[\left(\frac{\Phi_{j-1}^{n+1}}{6} + \frac{2\Phi_j^{n+1}}{3} + \frac{\Phi_{j+1}^{n+1}}{6}\right) - \left(\frac{\Phi_{j-1}^n}{6} + \frac{2\Phi_j^n}{3} + \frac{\Phi_{j+1}^n}{6}\right)\right]$$
$$+ C\frac{\Phi_{j+1}^n - \Phi_{j-1}^n}{2\Delta x} = 0 \qquad (4.57)$$

A very interesting finding here is that (4.57) is similar to the FTCS FD scheme except that the time derivative is weighted among three neighborhood nodes. Presumably, the scheme will be more stable than the FTCS method, and in fact standard von Neumann analysis can be used to analyze its stability property.

The clear relationship between FEM and FDM, however, does not exist generally for 2D and 3D problems with complex shape functions. In contrast to FDM, there exists no well-established technique for the analysis of numerical accuracy and numerical stability in FEM. This may pose a great difficulty to CFD computation when the flow is complex and varies strongly in both time and space. For this reason, although FEM has the advantage of representing arbitrary surface geometry on the boundary, its application in fluid computation for long-term time integration is still less popular compared with FDM or BEM, the latter of which will be discussed later.

It is also noted that there also exist other types of FEM based on variational principles, and they are mainly applied to structural analysis. Readers shall refer to Zienkiewicz and Taylor (1989) for more details and discussions about FEM.

4.5 Spectral method

Similar to FEM, spectral method (SM) also approximates the solution of a governing equation, but via the Fourier series or other valid series. In SM, the true solution of a problem is approximated by a series of functions, i.e.:

$$\phi(x, t) \approx \tilde{\phi}(x, t) = \sum_{K=0}^{M} a_K(t)\Phi_K(x) \qquad (4.58)$$

where $\Phi_K(x) = e^{iKx}$ in Fourier spectral method (FSM), $\Phi_K(x) = T_K(x)$ in Chebyshev spectral method (CSM), and $\Phi_K(x) = L_K(x)$ in Legendre spectral method (LSM). The main advantage of SM is the unparallel accuracy of the solution with a relatively small amount of computational expense. However, the SM is often limited to relatively simple domains (e.g., with the shape of cube, cylinder, rectangle).

It is interesting to compare the functional approximation between SM and FEM, which approximates the solution as:

$$\phi(x, t) \approx \tilde{\phi}(x, t) = \sum_{j=1}^{M} S_j(x)\Phi_j(t) \tag{4.59}$$

Although (4.58) and (4.59) look similar, the interpretation of the two approximations is quite different. First, the number k in SM represents the modal number in wave number space, whereas j in FEM represents the nodal number of the discretization in physical space. Second, the function $\Phi_K(x)$ in SM is continuous in space, and it represents the wave modes in different wave number space for FSM. The unknown coefficient $a_K(t)$ is associated with mode K. In FEM, the interpolation function $S_j(x)$ is 0.0 at node j and 1.0 at other nodes. The value $\Phi_j(t)$ is the approximate solution to be sought at node j. If an analog is made for the variables in these two methods, $\Phi_K(x)$ in SM is similar to the interpolation function $S_j(x)$ in FEM and $a_K(t)$ in SM is similar to $\Phi_j(t)$ in FEM.

To illustrate how an SM works, we shall use FSM to solve a 1D diffusion equation:

$$\frac{\partial \phi}{\partial t} + \frac{\partial^2 \phi}{\partial x^2} = 0 \tag{4.60}$$

With the initial and periodic boundary condition:

$$\phi(x, 0) = \phi_0(x) \quad \text{and} \quad \phi_0(x, t) = \phi_0(x + 2\pi, t) \tag{4.61}$$

The FSM approximates the solution directly and globally:

$$\phi(x, t) = \sum_{K=0}^{\infty} a_K(t)e^{iKx} \approx \tilde{\phi} = \phi^M(x, t) = \sum_{K=0}^{M} a_K(t)e^{iKx}, \quad x \in [0, 2\pi) \tag{4.62}$$

where $a_K(t)$ is to be determined. Substituting the above approximation into (4.60), we have:

$$\sum_{K=0}^{M} \frac{\partial a_K(t)}{\partial t} e^{iKx} = \sum_{K=0}^{M} a_K(t)(iK)^2 e^{iKx} \tag{4.63}$$

This gives us the following:

$$\frac{\partial a_K(t)}{\partial t} = a_K(t)(iK)^2 \Rightarrow a_K(t) = a_K(0)e^{-K^2 t} \tag{4.64}$$

where $a_K(0)$ can be determined by the initial condition from the Fourier transformation:

$$a_K(0) = \frac{1}{2\pi} \int_0^{2\pi} \phi_0(x) e^{-iKx} dx \qquad (4.65)$$

If $\phi_0(x)$ is a simple continuous function, the coefficients can be evaluated analytically and therefore there is no need to use grids at all. The accuracy of the numerical results depends only on the total number of wave modes M included in the summation. In cases when the initial condition is provided at discrete grid points or the analytical integration of (4.65) is impossible, numerical integration is needed by using discrete grid points. Consider N collocation points at:

$$x_j = \frac{2\pi j}{N}, \quad j = 0, 2, \ldots, N-1 \qquad (4.66)$$

Then:

$$a_K(0) = \frac{1}{N} \sum_{j=0}^{N-1} \phi_0(x_j) e^{-ikx_j} \qquad (4.67)$$

The above method is called pseudospectral (collocation) method, which is more popularly used in practical computation.

A classical study was performed by Orszag and Patterson (1972) for homogenous turbulence with DNS with the use of pseudospectral method. Readers should refer to Gottlieb and Orszag (1977) and Canuto et al. (1987) for a more detailed description and application of this method.

4.6   Boundary element method

The BEM is based on the mathematical identity between volume integration and surface integration for certain types of PDEs. Let us use an example to illustrate how this method works, and along the solution procedure, we shall also discuss the advantages and disadvantages of the method.

Consider a 2D Laplace equation for velocity potential ϕ. The governing equation reads:

$$\frac{\partial}{\partial x}\left(\frac{\partial \phi}{\partial x}\right) + \frac{\partial}{\partial y}\left(\frac{\partial \phi}{\partial y}\right) = 0 \qquad (4.68)$$

This is a boundary value problem, in which ϕ, $\partial\phi/\partial n$, or their relationship (e.g., $a\phi + b\partial\phi/\partial n$ for linear problems) must be specified in order to have a well-posed problem. Based on Green's second identity, we have:

$$\int_V (Q\nabla^2 P - P\nabla^2 Q)\, dV = \int_S \left(Q\frac{\partial P}{\partial n} - P\frac{\partial Q}{\partial n}\right) dS \qquad (4.69)$$

where P and Q are arbitrary functions that are continuous in the domain except at finite points. In connection with the numerical solution to a 2D Laplace equation, we often make the following definition:

$$P = \ln r = \ln |x - x_0| \text{ and } Q = \phi \tag{4.70}$$

where P is called the free space Green's function between the boundary point x and any known point x_0 on the boundary or the interior. The expression of P will change accordingly if a 3D Laplace equation is considered.

Substituting (4.70) into (4.69), we have:

$$\alpha \phi (x_0) = \int_S \left[\phi (x) \frac{\partial P}{\partial n} - \ln r \frac{\partial \phi}{\partial n} (x) \right] dS \tag{4.71}$$

where

$$\alpha = \begin{cases} 2\pi & \text{interior point} \\ \pi & \text{boundary point} \end{cases}$$

In the numerical solution procedure, the boundary is first discretized by a series of N linear boundary elements with N nodes. Equation (4.71) is then applied at each node, for which a linear equation will be formed with $2N$ unknowns of ϕ and $\partial \phi / \partial n$ at all boundary nodes that have resulted from the surface integration on the RHS of the equation. With the application of N boundary conditions, which specify ϕ, $\partial \phi / \partial n$, or their relationship at each of N nodes, the number of unknowns can be reduced to N. This ensures the unique solutions of unknowns ϕ and $\partial \phi / \partial n$ on the boundary; together with the boundary conditions one can predict the value of ϕ in the interior domain using (4.71).

The major advantage of BEM is the possible numerical efficiency embedded in the solution procedure. After the application of the boundary integral equation, the problem will be converted into its equivalency with the reduction of dimension by 1. This means that the original 2D and 3D problems are reduced to 1D and 2D boundary problems. Such a reduction of problem dimension, however, does not necessarily ensure a reduction of CPU time in computer modeling, since the resulting matrix is dense and it can be computationally expensive for a large problem. We shall come back to the assessment of computational efficiency between BEM and FDM for a large 3D problem in Section 6.2.2.3 when the modeling of wave–structure interaction is discussed.

Although BEM was originally developed for the efficient solution to steady boundary value problems that can be described by the Laplace equation, it was later extended to the solution of the Helmholtz equation and the Poisson equation. BEM finds many applications in potential flows, groundwater flows, fracture mechanics, acoustics, electromagnetics, etc. In fluid mechanics, the application of BEM was also explored to solve incompressible viscous fluid flows (Wu, 1982).

Regardless of the success of BEM in many scientific and engineering computations, it must be realized that BEM is not a general solution procedure of PDEs. It can be applied only to a limited number of PDEs that allow for the transformation of the original equations to boundary integral equations. This is different from FDM, FVM, and FEM, which are more versatile and can be applied to any PDE.

4.7   Meshless particle method

Meshless methods have been developed to resolve the complex surface geometry that cannot be readily handled by mesh-based methods. SPH is the earliest meshless method developed in the 1970s for the study of astrophysics using Lagrangian particles. The GFDM was proposed in early 1980 based on particle information. Substantial development and improvement of meshless methods started from the mid-1990s, after which there was a booming of meshless methods in computational physics and mechanics. At this moment, SPH is the main meshless method that has many reported successful applications in water wave modeling and interfacial fluid flow computation.

In meshless methods, particles that can be either regularly or randomly distributed, moving or stationary are used to discretize the domain. The loose local connectivity of these particles (e.g., relative locations) can be established once the domain of influence for a particle of interest is determined. The collective behavior of many neighboring particles will be used to approximate the function and/or its derivatives. To adequately represent a problem, a certain amount of particles needs to be used. This is similar to the mesh-based methods where a certain mesh resolution is needed. Therefore, a meshless method is not meant to reduce computational effort. In fact, it can be more computationally expensive than the conventional method, but with the possibly improved flexibility and accuracy near irregular boundaries.

4.7.1   Meshless derivative approximation method

The GFDM is the meshless method that approximates the functional derivative in a PDE (e.g., Liszka and Orkisz, 1980). Let us use a 2D diffusion equation as an example:

$$\frac{\partial \phi}{\partial t} = \frac{\partial^2 \phi}{\partial x^2} + \frac{\partial^2 \phi}{\partial y^2} \tag{4.72}$$

Writing it in FD form, we have:

$$\frac{\phi_i^{n+1} - \phi_i^n}{\Delta t} = \left(\frac{\partial^2 \phi}{\partial x^2}\right)_i^n + \left(\frac{\partial^2 \phi}{\partial y^2}\right)_i^n, \quad i = 0, 1, 2, \ldots, M \tag{4.73}$$

where M is the total number of particles in the computational domain. At this juncture, GFDM is still the same as FDM.

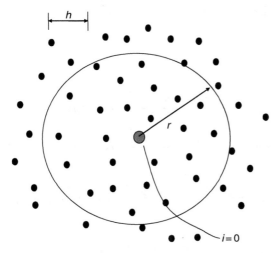

Figure 4.5 Illustration of GFDM; the circle represents the domain of influence for the central particle.

The key issue at this moment is how to evaluate $\partial^2 \phi/\partial x^2$ and $\partial^2 \phi/\partial y^2$ at all particle locations based on the neighborhood particle information. To achieve this objective, the domain of influence must be defined first. Generally, this domain must be large enough to include sufficient number of neighborhood particles for the accurate evaluation of local derivatives. However, the domain also needs to be small enough to be locally representative. Although there is no hard rule here, usually the radius of the search r is about twice the average distance between two particles h, i.e., $r \sim 2h$ (Figure 4.5).

Once the radius of influence for a particular node at (x_0, y_0) $(i = 0)$ is determined, the other nodes $(i = 1$ to $N)$ within the radius can be determined by searching the domain. Based on Taylor series expansion, we are able to write the expression to relate the functional values at these nodes ϕ_i to that at the interest node ϕ_0 as:

$$
\phi_i = \phi_0 + (x_i - x_0) \left(\frac{\partial \phi}{\partial x} \right)_0 + (y_i - y_0) \left(\frac{\partial \phi}{\partial y} \right)_0 + \frac{(x_i - x_0)^2}{2} \left(\frac{\partial^2 \phi}{\partial x^2} \right)_0
$$
$$
+ \frac{(y_i - y_0)^2}{2} \left(\frac{\partial^2 \phi}{\partial y^2} \right)_0 + (x_i - x_0)(y_i - y_0) \left(\frac{\partial^2 \phi}{\partial x \partial y} \right)_0 + O(\Delta^3),
$$
$$
i = 1, 2, \ldots, N \tag{4.74}
$$

We are now able to form a linear system of equations with the equation number of N and the number of unknowns of 5 for this case:

$$Ax = b \qquad (4.75)$$

where:

$$A = \begin{bmatrix} (x_1 - x_0) & (y_1 - y_0) & \dfrac{(x_1 - x_0)^2}{2} & \dfrac{(y_1 - y_0)^2}{2} & (x_1 - x_0)(y_1 - y_0) \\[2ex] (x_2 - x_0) & (y_2 - y_0) & \dfrac{(x_2 - x_0)^2}{2} & \dfrac{(y_2 - y_0)^2}{2} & (x_2 - x_0)(y_2 - y_0) \\[1ex] & & \vdots & & \\[1ex] (x_M - x_0) & (y_M - y_0) & \dfrac{(x_M - x_0)^2}{2} & \dfrac{(y_M - y_0)^2}{2} & (x_M - x_0)(y_M - y_0) \end{bmatrix}$$

$$\qquad (4.76)$$

$$x^T = \left[\left(\frac{\partial \phi}{\partial x} \right)_0 \left(\frac{\partial \phi}{\partial y} \right)_0 \left(\frac{\partial^2 \phi}{\partial x^2} \right)_0 \left(\frac{\partial^2 \phi}{\partial y^2} \right)_0 \left(\frac{\partial^2 \phi}{\partial x \partial y} \right)_0 \right]$$

and

$$b^T = \begin{bmatrix} \phi_1 - \phi_0 & \phi_2 - \phi_0 & \cdots & \phi_M - \phi_0 \end{bmatrix} \qquad (4.77)$$

Note that as long as there are at least five other nodes within the radius of influence, the five derivatives can be determined by error minimization of the system of linear equations, i.e., by LS method. The GFDM does not require the symmetric set of particles. It can be proven that under a special case of symmetry particles and for the particular choice of domain of influence, the GFDM will produce the results with the same accuracy as does the conventional FDM. The GFDM, however, excels in being able to make use of an irregular set of nodes, which are advantageous in simulating a problem with complex boundary configuration.

4.7.2   *Meshless function approximation method*

Compared to GFDM that approximates derivatives using the collective information of particles, the meshless function approximation method makes use of the particles to approximate solution function directly, analogous to FEM. The development of the general meshless function approximation method is rather recent. Similar to the FEM, there are two key issues related to this type of meshless method, namely the function approximation and the solution procedure for the function. The former is about how the function and its derivative are approximated, and the latter is

related to how the problem is formulated (i.e., Galerkin method, collocation method, or LS method). In the following discussion, various types of
meshless methods will be discussed based on the classification in terms of
the way of function approximation.

4.7.2.1 Moving least-squares method

The moving least-squares (MLS) method can be used in function approximation. A few methods with the use of MLS were proposed in the past
decade.

Diffuse element method: The diffuse element method is a meshless collocation method first proposed by Nayroles *et al.* (1992). In this method, a
"diffuse approximation method" was used to approximate functions from
a given set of particles. The so-called MLS method, which reconstructs the
continuous function from an arbitrary set of particles via the calculation of a
weighted LS around the evaluation point, is used to obtain the local approximate derivatives. The MLS is generally more accurate than the grid-based
numerical methods near irregular surfaces. This method can solve general
PDEs and it has been applied to both solid mechanics and fluid dynamic
problems.

Element-free Galerkin method: EFGM is a meshless Lagrangian method
developed for continuum problems (Belytschko *et al.*, 1994). The method
improved the diffuse element method by providing more stable and accurate
results in the numerical computation. The method shares some similarity to
SPH, but it is much more computationally expensive. Although the proposal
of EFGM is very recent, it has been quickly and widely applied to many
engineering analyses. Currently, EFGM is one of the most matured meshless
methods in engineering computation. EFGM method is most frequently
applied in solid mechanics that involves strain localization and fracture,
where there exist a large deformation and spatial gradient of material
properties.

4.7.2.2 Kernel-based method

Smoothed particle hydrodynamics: SPH is the first meshless method in
computational physics. The method was proposed in the 1970s (Gingold
and Monaghan, 1977; Lucy, 1977) for the study of astrophysics. SPH
is a Lagrangian meshless method where the solution is approximated by
a kernel function. An important interpretation of the method was made
by Monaghan (1992). The method has been successfully applied to many
fluid flow problems, including breaking waves. In Japan, a similar method
is named moving particle semi-implicit (MPS) method (Koshizuka *et al.*,
1998). The conventional SPH may be subjected to numerical instability near
boundaries. Various types of numerical damping have been introduced to
ensure stable numerical results.

Reproducing kernel particle method: SPH tends to produce unstable and/or less accurate results near boundaries, where the consistency conditions are not imposed. To overcome this difficulty, W.K. Liu *et al.* (1995) developed RKPM. Being still a kernel-based method, RKPM improves the numerical accuracy by constructing the shape function and its derivatives on the boundary in such a way that the consistency conditions are enforced explicitly. Thus, the method is especially suitable for problems with large deformation, around which the tensile instability can be easily developed in conventional SPH.

4.7.2.3   *Partition of unity methods and other meshless methods*

Belytschko *et al.* (1996) have shown that any kernel method whose parent kernel is identical to the weighting function of an MLS approximation is identical to the MLS method. Babuska and Melenk (1997) proposed the partition of unity method and argued that both MLS method and kernel-based method can be regarded as the special cases of the partition of unity method. Some other popular meshless methods under the categories of partition of unity method include h–p Clouds method (Duarte and Oden, 1996) and generalized finite element method (GFEM) (e.g., Strouboulis *et al.*, 2000).

There are also many other function approximation meshless methods such as free mesh method (Yagawa and Yamada, 1996), natural element method (NEM) (Sukumar *et al.*, 1998), and the meshless local Petrov–Galerkin (MLPG) method (Atluri and Zhu, 1998). The characteristics shared by all the methods discussed in Section 4.7.2 are that they are all based on various Galerkin or collocation weak formulations for the problem, against the strong formulation by GFDM.

Meshless methods have advanced very rapidly in the last decade, and new methods are still emerging for better function and/or its derivative approximation. Interested readers should refer to some dedicated books for more discussion on meshless methods (e.g., Liu, 2002; Li and Liu, 2004; Belytschko and Chen, 2007).

4.8   Problem-based discrete formulation methods

All the numerical methods introduced earlier are numerical techniques developed to approximately solve a PDE. Alternatively, there exists another type of numerical method that solves the problem directly with discrete formulation, usually based on particles, using fundamental physical principles such as mass and momentum conservation.

4.8.1   *Lattice Boltzmann method*

The LBM has a very different way of solving a fluid problem compared to traditional numerical methods. It is well known that fluid motion can be described on three levels: the molecular (microscopic) level at which the motion is reversible; the kinetic (mesoscopic) level at which the motion is irreversible and can be represented by Boltzmann approximation; and the macroscopic level at which the continuum approximation is applicable. It has been proven that the macroscopic features of fluid dynamics can be recovered by an appropriate Boltzmann model using a finite number of velocity vectors. Most of the Boltzmann models are based on probabilistic assumptions and they are continuous in time and space, but discrete in velocity.

In the 1980s, lattice gas (LG) automata became an attractive alternative to solve fluid problems by using discrete lattices (Frisch *et al.*, 1986). LG can be considered a simplified fictitious molecular dynamic in which space, time, and particle velocities are all discrete. LG employs the simplified kinetic equation to describe the microscopic particles, the collective behavior of which results in the desired macroscopic fluid dynamics. In LG automation, there can be either 1 or 0 particle at a lattice node that moves in a lattice direction. At the next time step, each particle will move to its neighboring node in its direction, and this process is called propagation. When there is more than one particle arriving at the same node from different directions, they collide and change their directions according to the collision laws. Suitable collision rules should conserve the particle number (mass), momentum, and energy before and after the collision. Because LG is constructed on simple linear kinetic equations for discrete particles, it has advantages in the implementation of complex boundary conditions and parallel computations. Because of the use of particle occupation variables that are defined by Boolean variables, statistic noise will result even for smooth flows. Besides, Galilean invariance (Galilean relativity), that the fundamental laws of physics are the same in all inertial frames, may not be strictly satisfied in LG formulation.

The LBM was designed to overcome the shortcoming of the original LG by replacing the Boolean particle number in a lattice direction with its ensemble average, the so-called density distribution function. In this way, individual particle motion is neglected and only the average of many particles is considered. The LBE that is in the form of a discrete kinetic equation and that governs the particle distribution function reads:

$$f_i\left(\mathbf{x} + c_i\Delta\mathbf{x}, t + \Delta t\right) = f_i(\mathbf{x}, t) + \Omega_i\left(f_i(\mathbf{x}, t)\right), \qquad i = 1, 2, \ldots, M \qquad (4.78)$$

where f_i is the particle distribution function in ith direction, $\Omega_i\left(f_i(\mathbf{x}, t)\right)$ is the collision function to be detailed later, Δx and Δt are space and time intervals, c_i is the local velocity, and M is the number of discrete directions of the particle velocity that will be determined by the modelers based on accuracy requirements (Figure 4.6).

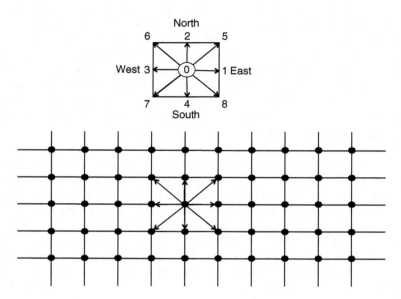

Figure 4.6 Illustration of LBM for a 2D problem with nine velocity components in square lattice elements.

The macroscopic fluid flow properties, i.e., density ρ and momentum $\rho\mathbf{u}$, are defined as the first and second moments of the particle distribution function as follows:

$$\rho = \sum_{i=0}^{M} f_i, \quad \rho\mathbf{u} = \sum_{i=0}^{M} f_i \mathbf{c}_i \tag{4.79}$$

The discrete collision rules in LG automation are modified to be a continuous collision operator. One of the most common methods is to use the Bhatnagar–Gross–Krook term (Bhatnagar *et al.*, 1954) to approximate the collision operator:

$$\Omega_i \left(f_i(\mathbf{x}, t) \right) = \frac{1}{\tau} \left[f_i(\mathbf{x}, t) - f_i^{\text{eq}}(\mathbf{x}, t) \right] \tag{4.80}$$

where τ is the relaxation time at which the local particle distribution is relaxed to its equilibrium state with the equilibrium particle distribution function as $f_i^{\text{eq}}(\mathbf{x}, t)$. The above collision term is the simplified linear operator to approximate the nonlinear collision process.

If we carefully examine the LBE, we would find that this simple kinetic equation that advances the distribution function in time and space looks very similar to the FD approximation discussed before. The difference is that in the FDM, the PDE is prescribed, whereas in LBM we do not know in advance which PDE the kinetic equation represents. Curiosity leads us to ask the following question: if Δx and Δt are small enough, can LBE be reduced to the PDE we are familiar with? The answer is yes! In fact, by using multiscale Chapman–Enskog expansion (Frisch *et al.*, 1986), it is possible to find that the LBE can be converted to the following system of PDEs up to the second order of small perturbation $\varepsilon \sim O(\Delta x) \sim O(\Delta t)$ (Chen and Doolen, 1998):

$$\frac{\partial \rho}{\partial t} + \nabla \cdot (\rho \mathbf{u}) = 0 \tag{4.81}$$

$$\frac{\partial \rho \mathbf{u}}{\partial t} + \nabla \cdot \Pi = 0 \tag{4.82}$$

where Π is the momentum tensor that can be expressed as:

$$\Pi_{jk} = \sum_{i=0}^{M} (\mathbf{c}_i)_j (\mathbf{c}_i)_k \left[f_i^{eq} + \left(1 - \frac{1}{2\tau} \right) f_i^{(1)} \right] \tag{4.83}$$

and $f_i^{(1)}$ is the first-order nonequilibrium distribution function, which satisfies the following:

$$f_i = f_i^{eq} + \varepsilon f_i^{(1)} + O(\varepsilon^2) \tag{4.84}$$

Equations (4.81) and (4.82) look similar to the mass and momentum conservation equations in NSEs. As the matter of fact, it has been proven that the incompressible NSEs can be recovered in the limiting case of small variation of density if (1) sufficient lattice symmetry is provided (so is the symmetry of discrete velocity direction) and (2) a proper local equilibrium distribution function $f_i^{eq}(\mathbf{x}, t)$ is chosen. The general form of $f_i^{eq}(\mathbf{x}, t)$ for a small Mach number is written as follows:

$$f_i^{eq} = \rho \left[A + B \mathbf{c}_i \cdot \mathbf{u} + C (\mathbf{c}_i \cdot \mathbf{u})^2 + D \mathbf{u} \cdot \mathbf{u} \right] \tag{4.85}$$

where A, B, C, and D are lattice constants, which can be obtained analytically based on the constraints of $\rho = \sum_{i=0}^{M} f_i^{eq}$ and $\rho \mathbf{u} = \sum_{i=0}^{M} f_i^{eq} \mathbf{c}_i$, given the specific lattice structure. Similarly, different state equations can be derived to relate the local pressure p to the local density and speed of sound and to relate the kinematic viscosity ν to the relaxation time τ.

Possessing both the microscopic statistical feature and the macroscopic dynamic feature, the LBM is especially promising in modeling some new physics where the existing knowledge is not adequate to explain the macroscopic phenomenon. Examples include air entrainment during wave breaking, sediment incipience and suspension, and other multiphase problems, which so far are still highly dependent upon very crude empirical formulas.

For more discussions on LBM, readers are referred to Benzi *et al.* (1992), Qian *et al.* (1995), and He and Luo (1997).

4.8.2   *The discrete element method*

The DEM was pioneered by Cundall and Strack (1979) in the study of rock mechanics. Readers should not confuse DEM with diffuse element method, a MLS meshless method we discussed before. DEM has another name: distinct element method. The method utilizes a large number of discrete particles (usually spheres or discs) to represent the bulk behavior of granular materials such as sands. The particle interaction laws are based on collision laws and the equations of motion are typically integrated explicitly in time. The resulting algorithms are simple and intuitive, but they are able to describe complex physical phenomena.

The DEM shares an important similarity to LBM, i.e., both of them start from the direct discrete formulation of the problem using particle representation, whereas the other numerical methods solve the existing governing equations in the form of PDEs on a discrete grid or mesh system. DEM can be regarded as the sister version of the LBM in the application in solid mechanics. Because DEM was primarily used in the simulation of soil, rock, and other granular solid particles, we will not discuss this method further in this book. Readers can refer to Bicanic (2004) for more details.

4.9   Grid and mesh generation

All mesh-based numerical methods require a grid or a mesh for discretizing the computational domain. Difficulty can easily arise when we apply a numerical model in a domain with complex boundary configuration. In this section, we discuss various ways of generating grids or meshes automatically, especially when an irregular boundary is present.

4.9.1   *Definition and classification of grid and mesh*

4.9.1.1   *Definitions of grid and mesh*

Let us use a 2D example to illustrate how grid and mesh are defined. Consider a rectangular domain of 12 m × 10 m. The simplest spatial discretization is to divide the domain into uniform cells of squares (Figure 4.7). The domain after spatial discretization is thus composed of a grid system where the numerical solution is sought at each grid point, the intersection of any two orthogonal lines. The line connecting two adjacent grid points shows the connectivity between grid points, on which no physical quantity is defined. This is a typical "grid system" when FDM is constructed. In contrast, the numerical solution can also be sought at the centroid of a cell in FVM. In this case, the line connecting grid points now becomes the boundary of the cells, on which the flux of the physical quantity is defined. This is a typical *mesh system*. Therefore, for a simple problem, a grid system

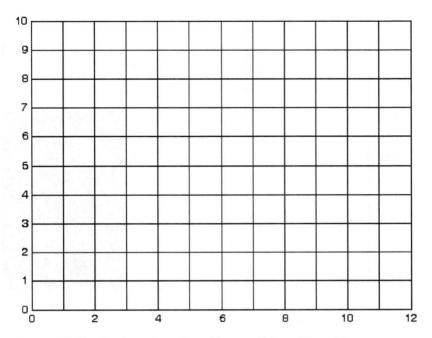

Figure 4.7 Sketch of a rectangular grid or mesh for a 2D problem.

can be identical to a mesh system. Whether it is regarded as a grid system or a mesh system depends on how it is used by a particular numerical scheme.

Sometimes, however, the difference between a mesh system and a grid system becomes blurred when a hybrid scheme is used. For example, a numerical model can be solved in the so-called staggered grid system, in which the scalar (e.g., pressure, free surface displacement) is defined and solved at the center of a rectangular cell and the vector (e.g., velocity, flux) is defined and solved at the cell boundary (Figure 4.8). For the scalar, the scheme is similar to the control volume method, whereas for the vector, the general FD scheme (e.g., upwind scheme) may be used that does not enforce the explicit conservation law within the control volume. In this case, both grid and mesh are used.

4.9.1.2 *Structured and unstructured grid and mesh*

A typical "structured grid" is characterized by regular (or structured) connectivity that can be expressed as a 2D or 3D array for 2D or 3D problems. The major feature of a structured grid is that all interior grid points (sometimes called nodes) in the grid system have an equal number of adjacent cells or elements. In contrast, an unstructured grid is characterized by

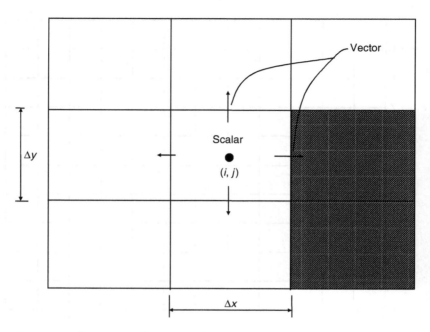

Figure 4.8 Illustration of a staggered grid system.

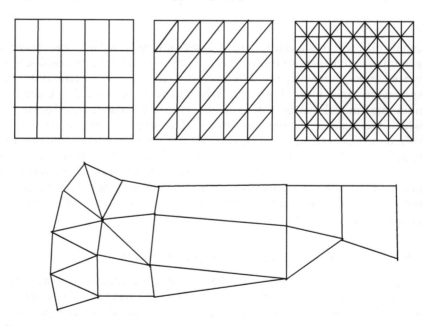

Figure 4.9 Examples of structured (top) and unstructured grids (bottom).

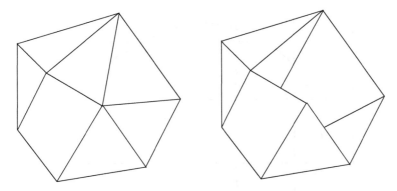

Figure 4.10 Conformal (left) and nonconformal (right) meshes.

irregular connectivity that allows for any possible element a solver might use (Figure 4.9). Compared to a structured grid, the storage requirements for an unstructured grid can be substantially larger since the neighborhood connectivity must be explicitly stored. Because an unstructured grid is often used in FEM or FVM, it is more commonly referred to as an unstructured mesh.

4.9.1.3 Conformal and nonconformal mesh

There is another classification of mesh based on whether the conformity condition is satisfied. A mesh is said to be conformal if the intersection of any two elements in the mesh is an empty set, a vertex, an edge, or a face of both elements. The examples of conformal and nonconformal meshes are shown in Figure 4.10. Although the conformal mesh has simpler connectivity information (and thus requires less memory for faster computation), the nonconformal mesh possesses better flexibility for resolving a computational domain.

4.9.2 Structured grid generation and coordinate transformation

4.9.2.1 Cartesian grid and mesh

When a structured grid is used, the grid shape is normally taken as either quadrangle for 2D problems or hexahedron for 3D problems. For a 2D problem, the simplest quadrangle is a rectangle and the grid system based on rectangles is also called Cartesian grid. With the use of the Cartesian grid, it is straightforward to construct FDM. However, it poses a problem for a modeler to resolve arbitrary geometry of the body surface that may not necessarily cross grid points. In this case, a grid point can be taken as either

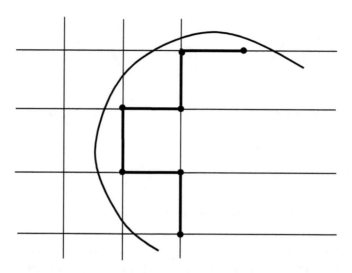

Figure 4.11 Sketch of stair-step of a structured grid in the vicinity of a curved body surface; the thick lines connecting dotted grid points represent the approximated body surface.

a fluid grid or a solid grid, depending on whether it is located outside or inside the body. As a result, an irregular surface is approximated by a series of stairs (Figure 4.11). The numerical accuracy may be greatly reduced in the vicinity of the solid surface, especially when the resolution is low.

4.9.2.2   General coordinate transformation

To resolve this problem, a popular method is to apply a coordinate transformation that recasts the irregular surface in the physical domain into a straight line in the computational domain. Thus, the body surface can be precisely represented in the computational domain by grids distributed exactly on the surface. Let us consider a 2D problem whose original governing equations in the form of PDEs are established in the physical domain (x,y,t). To map the physical problem, in which the solid surface is curved, into the computational domain where the curve surface is transformed into a straight line, we need to introduce the coordinate transformation that links the physical domain and the computational domain, i.e.:

$$\xi = \xi(x, y, t)$$
$$\eta = \eta(x, y, t) \tag{4.86}$$
$$\tau = \tau(t)$$

There are many ways of choosing ξ, η, and τ, which are to be discussed below. It is noted that since the problem will be solved in the transformed computational domain, the original governing equations also need to be transformed so that the derivatives with respect to (x, y, t) are converted into (ξ, η, τ) with the use of chain rules of differential calculus. For first-order derivatives, we have:

$$\left(\frac{\partial}{\partial x}\right)_{y,t} = \left(\frac{\partial}{\partial \xi}\right)_{\eta,\tau}\left(\frac{\partial \xi}{\partial x}\right)_{y,t} + \left(\frac{\partial}{\partial \eta}\right)_{\xi,\tau}\left(\frac{\partial \eta}{\partial x}\right)_{y,t} \tag{4.87}$$

$$\left(\frac{\partial}{\partial y}\right)_{x,t} = \left(\frac{\partial}{\partial \xi}\right)_{\eta,\tau}\left(\frac{\partial \xi}{\partial y}\right)_{x,t} + \left(\frac{\partial}{\partial \eta}\right)_{\xi,\tau}\left(\frac{\partial \eta}{\partial y}\right)_{x,t} \tag{4.88}$$

$$\left(\frac{\partial}{\partial t}\right)_{x,y} = \left(\frac{\partial}{\partial \xi}\right)_{\eta,\tau}\left(\frac{\partial \xi}{\partial t}\right)_{x,y} + \left(\frac{\partial}{\partial \eta}\right)_{\xi,\tau}\left(\frac{\partial \eta}{\partial t}\right)_{x,y} + \left(\frac{\partial}{\partial \tau}\right)_{\xi,\eta}\left(\frac{d\tau}{dt}\right)_{x,y}$$

$$\tag{4.89}$$

The subscripts above indicate that these variables will be held constant in the partial differentiation. Similarly, the second-order and higher order derivatives can also be obtained. With the use of these chain rules, the governing equations are re-expressed in the new coordinate ξ, η, τ. An alternative way of achieving the same objective is to use Jacobian matrix, which is defined as follows:

$$J = \frac{\partial(x, y, t)}{\partial(\xi, \eta, \tau)} = \begin{vmatrix} \partial x/\partial \xi & \partial y/\partial \xi & \partial t/\partial \xi \\ \partial x/\partial \eta & \partial y/\partial \eta & \partial t/\partial \eta \\ \partial x/\partial \tau & \partial y/\partial \tau & \partial t/\partial \tau \end{vmatrix} \tag{4.90}$$

Either way, there is a need to determine:

$$\left(\frac{\partial \xi}{\partial x}\right)_{y,t}, \left(\frac{\partial \eta}{\partial x}\right)_{y,t}, \left(\frac{\partial \xi}{\partial y}\right)_{x,t}, \left(\frac{\partial \eta}{\partial y}\right)_{x,t}, \left(\frac{\partial \xi}{\partial t}\right)_{x,y}, \left(\frac{\partial \eta}{\partial t}\right)_{x,y}, \cdots$$

from the functional dependency of (ξ, η, τ) on (x, y, t).

4.9.2.3 *Elliptic grid generation*

There are many ways to specify (ξ, η, τ). One of the most efficient ways to map a fluid domain around a body is to use elliptic grid generation (Thompson *et al.*, 1974). In this approach, the simple elliptic equations will be used to define ξ, η as follows:

$$\frac{\partial^2 \xi}{\partial x^2} + \frac{\partial^2 \xi}{\partial y^2} = 0$$

$$\frac{\partial^2 \eta}{\partial x^2} + \frac{\partial^2 \eta}{\partial y^2} = 0 \tag{4.91}$$

In principle, the solution of the above system of PDE gives ξ, η as the function of (x,y), from which:

$$\left(\frac{\partial\xi}{\partial x}\right)_{y,t}, \left(\frac{\partial\eta}{\partial x}\right)_{y,t}, \left(\frac{\partial\xi}{\partial y}\right)_{x,t}, \left(\frac{\partial\eta}{\partial y}\right)_{x,t}$$

can be evaluated, at least numerically, and adopted in the governing equation in the transformed plane. No transformation is used for time, i.e., $\tau = t$.

In actual computation, the information of (ξ, η) is easy to obtain at the interaction of any two straight lines, which prompts us to solve (x,y) with respect to (ξ, η) by using the inverse of the original elliptic equations:

$$\left[\left(\frac{\partial x}{\partial \eta}\right)^2 + \left(\frac{\partial y}{\partial \eta}\right)^2\right]\frac{\partial^2 x}{\partial^2 \xi} - 2\left[\left(\frac{\partial x}{\partial \xi}\right)\left(\frac{\partial x}{\partial \eta}\right) + \left(\frac{\partial y}{\partial \xi}\right)\left(\frac{\partial y}{\partial \eta}\right)\right]\frac{\partial^2 x}{\partial \xi \partial \eta}$$

$$+ \left[\left(\frac{\partial x}{\partial \xi}\right)^2 + \left(\frac{\partial y}{\partial \xi}\right)^2\right]\frac{\partial^2 x}{\partial \eta^2} = 0$$

$$\left[\left(\frac{\partial x}{\partial \eta}\right)^2 + \left(\frac{\partial y}{\partial \eta}\right)^2\right]\frac{\partial^2 y}{\partial^2 \xi} - 2\left[\left(\frac{\partial x}{\partial \xi}\right)\left(\frac{\partial x}{\partial \eta}\right) + \left(\frac{\partial y}{\partial \xi}\right)\left(\frac{\partial y}{\partial \eta}\right)\right]\frac{\partial^2 y}{\partial \xi \partial \eta}$$

$$+ \left[\left(\frac{\partial x}{\partial \xi}\right)^2 + \left(\frac{\partial y}{\partial \xi}\right)^2\right]\frac{\partial^2 y}{\partial \eta^2} = 0 \tag{4.92}$$

By specifying the proper boundary condition of (x,y) at the four boundaries of (ξ, η), the above equations can be solved numerically to map every discrete point (ξ, η) on the computational domain to (x,y) in the physical space. Figure 4.12 gives an example of using elliptic grid generation method to map the physical domain around a 2D body to the corresponding computational domain represented by a rectilinear orthogonal grid. The elliptic mesh generation method can be readily extended to 3D problems.

Instead of solving the problem in the transformed computational domain, another way of employing the method is to generate a curvilinear boundary-fitted mesh in the physical domain, on which the problem is solved directly with the use of FEM or FVM (e.g., the left side of Figure 4.12). In this case, the solution of the elliptic equations is simply generating a smoothly connected structured boundary-fitted mesh.

4.9.2.4   Structured grid generation using σ-coordinate transformation

When mapping between two wavy surfaces, σ-coordinate transformation is often used because of its simplicity. Consider a 2D physical domain (x,y)

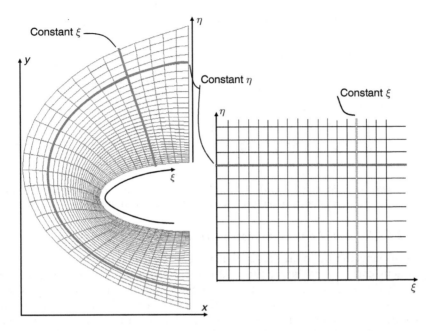

Figure 4.12 Sketch of elliptic grid generation that maps the irregular boundary conformal grid in the physical domain (left) to the Cartesian grid in the computational domain (right).

in which the bottom is at $y = -h(x)$ and the free surface at $y = \zeta(x, t)$; the σ-coordinate transformation is expressed as follows:

$$\xi = x, \quad \sigma = \frac{z + h(x)}{\zeta(x) + h(x)}, \quad \tau = t \tag{4.93}$$

where σ is similar to η in the elliptic grid generation. With the use of the above transformation, the irregular physical domain is transformed into a regular computational domain as shown in Figure 4.13. The great advantage of σ-coordinate transformation over elliptic coordinate transformation is that only the algebraic operation is involved in the transformation. Thus, the derivative terms of $\partial\sigma/\partial x$, $\partial\sigma/\partial y$, $\partial\sigma/\partial t$, etc., which will be used in the conversion of the governing equations to those in the computational domain, can be expressed analytically in terms of $h(x)$ and $\zeta(x, t)$.

Similar to elliptic grid generation, σ-coordinate transformation can also be simply viewed as a technique to generate structured boundary-fitted mesh if the problem is directly solved in the physical domain using FEM or FVM.

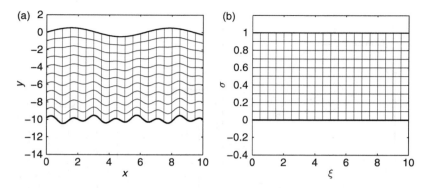

Figure 4.13 Sketch of grids in the physical (left) and computational (right) domains based on σ-coordinate transformation.

4.9.2.5   *Stretched and compressed structured grid*

In resolving the boundary layer that is very thin, within which the velocity gradient can be very large, the grid can be compressed in this region and stretched outside of the boundary layer for computational efficiency. There are many possible ways to stretch the grid in one direction using linear, exponential, or logarithmic functions. To illustrate how this works, we shall only use the following simple logarithmic function as an example:

$$\xi = x, \quad \eta = \ln(y + 1), \quad \tau = t \tag{4.94}$$

For a flat plate on the bottom, the grids in the physical domain and the computational domain are shown in Figure 4.14. Note that the grid-stretching technique above can be combined with the σ-coordinate transformation to resolve the boundary layer above an irregular bottom.

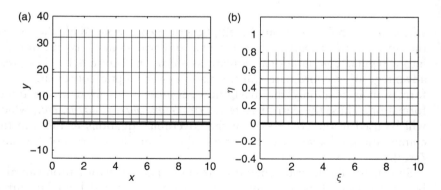

Figure 4.14 Sketch of grids in the physical (left) and computational (right) domains in a stretched grid system.

The stretched grid can be applied to 2D and 3D problems. This is necessary if FSI is to be modeled and the boundary layers in all directions are to be considered. A typical example would be wave interaction with a square cylinder. To model this process, finer grids will be deployed around the cylinder and stretched in all directions outward from the cylinder. In another example of fluid motion in a confined tank, the finer grids may be deployed around all walls and stretched toward the interior region.

The stretched grid technique is one of the simplest algebraic mesh generations. The generated mesh is also referred to as nonuniform mesh. Besides the logarithmic stretching, some other commonly used nonuniform mesh systems include linearly varying mesh (the difference of any two adjacent mesh sizes is a constant), constantly stretching mesh (the ratio of any two adjacent mesh sizes is a constant), and mesh stretching using hyperbolic sine function (Thompson *et al.*, 1985).

4.9.2.6 *Adaptive structured grid*

In some cases, whether the grids should be compressed or stretched at a particular location cannot be determined a priori. This can happen when the flow is transient and complex so that the high strain-rate flow region is unknown in advance. To resolve this problem, the so-called adaptive grids may be used, which automatically adjust the grid distribution to adapt to the local flow-field gradients. Automatic adjustment of local mesh size may be achieved by relating it to the local velocity gradient at every time step.

Besides resolving the unsteady flow, adaptive grids are also popularly used in the problems with moving boundaries when there is a moving body or a moving free surface. In treating the problems with free surface, the grid generated from σ-coordinate transformation is one typical example of adaptive grid. The purpose of using the adaptive grids in this case is to have a time-dependent boundary-fitted grid and the grid size in the vertical distance maintains certain ratio regardless of the free surface location.

4.9.2.7 *Comments on structured grid generation*

As mentioned earlier, all the coordinate transformations can also be employed as the tool for the generation of curvilinear boundary-fitted grids in the physical domain. If we use FVM or FEM, the computation can proceed on the mesh basis in the physical domain directly. In this case, the coordinate transformation becomes the tool for the generation of a structured mesh system in the physical domain.

There are essentially only two main branches of structured mesh generation, namely elliptic mesh generation and algebraic mesh generation. When the elliptic system of PDEs is used to generate regular quadrilateral and hexahedral meshes, the method is also referred to as *mesh extrusion technique* (e.g., Vassberg, 2000). In contrast, all algebraic mesh generation techniques are constructed on some kind of interpolation between the surface of the body and either another surface (e.g., free surface) or the computational boundary. The interpolation can be done independently in different directions or include the interdependence of the interpolation in three dimensions. The latter method is called *transfinite interpolation* and it is a simple yet reliable technique for algebraic generation of structured meshes around 3D bodies (e.g., Eriksson, 1982).

4.9.3   Special Cartesian grid and mesh

The Cartesian grid is the simplest structured grid. However, as mentioned earlier, it fails to represent an arbitrary shape of the body surface accurately. This prompts researchers to develop the modified Cartesian grid that retains the advantage of a rectilinear grid while overcoming its drawback in the vicinity of solid boundaries.

4.9.3.1   Cut-cell method and partial-cell method

The cut-cell method is used to reshape the mesh in the vicinity of an irregular solid boundary. Most of the time, the actual curved geometry of the solid surface is approximated by the linear (or higher order polynomial function) piecewise segments in the background of Cartesian cells, using the information of interactions between the solid surface and the grid lines. Away from the body, the Cartesian coordinate is resumed. Figure 4.15 shows how a curved body surface is represented by the cut-cell method.

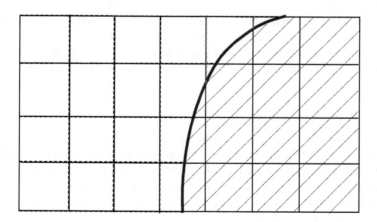

Figure 4.15 Sketch of cut-cell treatment; the shaded area on the right represents a solid body and the thick dashed line on the left represents the Cartesian grid whose length varies near the solid boundary.

Compared to the stair-step representation of a curved solid surface, the cut-cell method has a better representation of actual body geometry and it retains the simplicity of Cartesian grid in the far field. A special numerical scheme, however, has to be developed at these cut cells. If FVM is used, the cut cells can simply be regarded as some special unstructured meshes and the construction of a flux-conservation scheme does not require great additional effort. If FDM is employed, the FD scheme at the boundary cut cells needs to be written in a way that the distance to the solid surface (or the length of the cut-cell boundary) is formulated properly.

In the family of cut-cell method, there is another branch of the so-called partial-cell treatment, which is often used in the FD scheme constructed on a staggered grid system. In this method, the cut cells at the solid boundary will be treated as the interior cells, and thus, no special treatment will be made to change the FD scheme there. However, the effect of the solid surface will be accounted for by the modification of the mean flux around the cell boundary. This is equivalent to having these cells filled with a porous medium so that the mean velocity, which is continuous across every cell boundary, is locally equal to the pore velocity multiplied by the porosity. Apparently, the more the cell boundary is cut and occupied by the solid body, the less the space is left to the fluid and the smaller the "porosity" is, which slows down the local velocity. One example of partial-cell treatment is shown in Liu and Lin (1997), who used this method to approximate a sloping beach in their simulation of wave run-up on a slope.

4.9.3.2 Quadtree grids

Quadtree grids are composed of square cells with different sizes in a hierarchical tree structure. Starting from an initial unit square, recursive subdivision is applied within this cell according to prescribed criteria to accomplish the local refinement for either an irregular body surface or complex flow structure. Quadtree grids have some obvious advantages, i.e., fast and automatic grid generation, clear hierarchical data structures, and flexibility of local mesh refinement and adaptation. Figure 4.16 shows the quadtree grid representation of a circle.

Quadtree grids are mainly used in FDM, which can be easily constructed on a square grid system. In a 2D problem, the subdivision of any square cell will always generate four small squares of equal size. This facilitates the exchange of information between any two level cells. It is noted that quadtree can also be applied to FEM (e.g., Greaves and Borthwick, 1999). The early application of quadtree grids was made by Samet (1990) for processing digital image and it quickly spread into CFD computation (e.g., Yiu *et al.*, 1996). Recently, the method was also extended to water wave simulation (e.g., Park *et al.*, 2006). It is noted, however, that although a quadtree grid can improve accuracy by local refinement, the method has the inherent nature of stair-step approximation of curved surface.

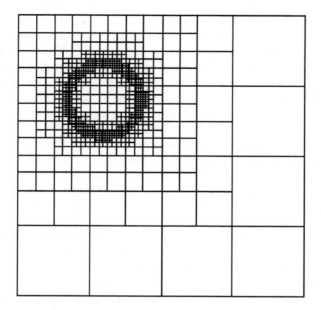

Figure 4.16 Quadtree grid for a circle.

4.9.4   Unstructured mesh generation

4.9.4.1   Unstructured mesh around a fixed and rigid body

Compared to a structured mesh, an unstructured mesh is more flexible to represent arbitrary surface geometry. Without the restriction of regular connectivity, unstructured meshes can be freely inserted into the area where higher resolution is needed. Because a modeler can shape the mesh cells based on his or her own preference, the generation of unstructured meshes in some sense is more like a work of art. The use of unstructured meshes simply increases the complex level of mesh connectivity but will not violate any principle in FVM and FEM.

Unlike the structured mesh generation, it is not easy to have a clear classification for the methods of unstructured mesh generation. Every method can be very different. Nevertheless, most of the unstructured mesh generation techniques make use of the principle of Delaunay triangulation (Delaunay, 1934) or Voronoi triangulation, which divides the 2D or 3D Euclidean place into triangles or tetrahedra. For a 2D case, given a set of points, a Delaunay triangulation of these points is the set of triangles such that no point is inside the circumcircle of a triangle. The triangulation is unique if no three points are on the same line and no four points are on the same circle. This process can be extended to a 3D problem with the triangle being replaced by a tetrahedra.

Triangle and tetrahedron are the most common shapes used for resolving the curved surface. Sometimes, a quadrangle and hexahedron can also be used in the generation of unstructured meshes in 2D and 3D cases. Other polygonal and polyhedral shapes may also be possible. The methods of quadtree/octree can be used for this purpose.

The generation of unstructured meshes is not a trivial task. The automatic generation of a mesh system with the given constraints is a branch in computation geometry and topology, and it is still an active research area. Nevertheless, many automatic (or semiautomatic) mesh generators, open source codes (e.g., GMSH; http://www.geuz.org/gmsh/), and commercial software (GAMBIT for FLUENT) are available.

4.9.4.2 Hybrid mesh (mixed mesh)

While the unstructured mesh is flexible to resolve the irregular and curved body surface, the structured mesh has the advantages of simple structure connectivity, easy generation, and higher accuracy with the same computational effort. It is natural to combine the two approaches to create a hybrid or mixed mesh, where the unstructured mesh is deployed near the body and the structured mesh is used in all other places.

4.9.4.3 Adaptive unstructured mesh for moving objects

In case the boundary (solid surface or free surface) is in motion, unless the coordinate is established on the moving body that is feasible only for the simple case of a single rigid body in consideration, the boundary-fitted mesh system needs to be adjusted at each time step to maintain the mesh boundary conformality. Such a mesh system is called adaptive mesh or moving mesh, which is constructed on either the global (one) mesh system or the overlapped mesh system.

Global mesh system: In the global mesh system, an unstructured mesh is used with the cell boundaries being conformal to solid surfaces at all times. When a body is in motion, either a mesh deformation (for small movement) or a remeshing process (for large movement) is required at every time step to maintain the conformity between cell boundary and solid boundary. For relatively simple and small movements such as body oscillation or liquid sloshing, the arbitrary Lagrangian–Eulerian (ALE) method can be used. In this method, the Lagrangian motion of fluid particles at all mesh nodal positions is computed at each time step, followed by the rezoning stage, in which the mesh nodes can move by following the Lagrangian velocity (Lagrangian description), the weighted Lagrangian velocity (combined Lagrangian–Eulerian description), or not move at all (Eulerian description). The method was introduced by Hirt *et al.* (1974) in fluid computation and was used widely in CFD because of its simple nature. For large movements, the automatic mesh generator needs to be used in the remeshing process (e.g., Hassan *et al.*, 2000 for two-object simulation and Lohner *et al.*, 1999 for multiple-object simulation), which can be computationally expensive.

Overlapped mesh system: Alternatively, the overlapped mesh system, which is also called multiblock mesh system or *Chimera grid system*, has been proposed (e.g., Benek *et al.*, 1985). In this approach, an unstructured mesh that conforms to the body surface and follows the body motion is arranged only in the vicinity of the moving body and the rest of the domain is discretized by fixed structured grids. Interpolation is used to exchange data between two mesh systems (or multiple mesh systems if multiple bodies are considered). The method is similar to the hybrid mesh system with the exception that the unstructured mesh moves with the body. Since no connectivity between the structured mesh and the unstructured mesh is pursued, the adaptive mesh around the body can be easily obtained by simple translation and rotation.

For more information on automatic mesh generation, readers are referred to Thompson *et al.* (1982, 1985) and Frey and George (2000).

4.10   Matrix solvers

The numerical solution of a PDE often involves the solution of a matrix (e.g., implicit method in FDM, all FEM, and BEM). The resulting system of linear algebraic equations takes the following general form:

$$Ax = b \tag{4.95}$$

The matrix A has a dimension of $n \times n$ where n is the total nodal number. The element in A is represented by a_{ij}. The solution of the above system of linear equations is found to be:

$$x = A^{-1}b \tag{4.96}$$

The numerical solution to A^{-1} (the inverse of A), directly or approximately, is the main challenge for a large matrix.

In the case when A is full and dense (e.g., BEM), Gauss elimination can be used to solve the system of equations. This, however, will take great computational effort. Although some decomposition techniques such as LU factorisation, QR factorisation, Cholesky decomposition, and singular value decomposition (SVD) can be used to enhance the solution efficiency (Press *et al.*, 2007), the computational cost is still rather high. Fortunately, in many other cases, the matrix can be sparse, meaning that there are only a few nonzeros in each row of the matrix. This is the situation in most of FDM and FEM schemes. In this case, various iteration techniques can be used to solve the matrix, taking advantage of the sparseness of the matrix by manipulating nonzero elements only. These iteration schemes include Gauss–Seidel method, successive over-relaxation (SOR) method, and conjugate gradient (CG) method.

4.10.1 Gauss–Seidel method and Jacobi method

The Gauss–Seidel method is a technique used to solve a linear system of equations iteratively. The solution procedure reads:

$$x_i^{(k+1)} = \frac{1}{a_{ii}} \left(b_i - \sum_{j=1}^{i-1} a_{ij} x_j^{(k+1)} - \sum_{j=i+1}^{n} a_{ij} x_j^{(k)} \right), \quad i = 1, 2, \ldots, n \qquad (4.97)$$

Note that the computation is performed only for nonzero a_{ij} on the RHS of (4.97) and thus the computation effort can be greatly reduced for a large sparse matrix. The calculation of $x_i^{(k+1)}$ makes the mixed use of the most recent value of $x_j^{(k+1)}$ that has already been computed (e.g., $j < i$) and the previous value of $x_j^{(k)}$ that has yet to be advanced to iteration $k+1$ (e.g., $j > i$). This means that no additional storage is required, and the computation can be done in place ($x^{(k+1)}$ replaces $x^{(k)}$). This is the only difference between the Gauss–Seidel method and the Jacobi method, the latter of which makes use of only previous step information $x^{(k)}$ in the iteration. The iteration continues until the changes between two iterations are below some tolerance. This process is called convergence of the iteration to the true solution of a linear equation system. The Gauss–Seidel method makes a faster convergence than the Jacobi method because of the use of updated solution information in the iteration.

4.10.2 Successive over-relaxation method

SOR is a numerical method used to further speed up the convergence of the Gauss–Seidel method. By realizing that the Gauss–Seidel method converges faster than the Jacobi method because of the use of the updated solution information, one may naturally expect that the convergence may be further enhanced by using the "future" solution information extrapolated from the previous and updated solution information. The algorithm then changes to:

$$x_i^{(k+1)} = (1 - \omega) x_i^{(k)} + \frac{\omega}{a_{ii}} \left(b_i - \sum_{j=1}^{i-1} a_{ij} x_j^{(k+1)} - \sum_{j=i+1}^{n} a_{ij} x_j^{(k)} \right), \quad i = 1, 2, \ldots, n$$

$$(4.98)$$

Compared to the Gauss–Seidel method, the difference is that the relaxation ω is multiplied by the original estimation of $x_i^{(k+1)}$, and it is counterbalanced by the previous solution $(1 - \omega) x_i^{(k)}$. For a symmetric (i.e., $A^T = A$) and positive-definite (i.e., $x^T A x > 0$ for all nonzero vectors x) matrix, it can be proven that $0 < \omega < 2$ will lead to convergence and Gauss–Seidel method corresponds to $\omega = 1$. When $\omega < 1$, the method is called SOR method. The optimal value of ω for the fastest convergence is case-dependent.

4.10.3   *Conjugate gradient method*

The CG method is an iterative algorithm for obtaining the numerical solution of a symmetric and positive-definite matrix. Two nonzero vectors $\mathbf{u}$ and $\mathbf{v}$ are said to be conjugate with respect to A if:

$$u^{\mathrm{T}} A v = 0 \tag{4.99}$$

If $\mathbf{u}$ is conjugate to $\mathbf{v}$, $\mathbf{v}$ is conjugate to $\mathbf{u}$ as well.

In CG method, the solution of the linear equation system is first expanded as follows (similar to Fourier expansion but the series number is finite here):

$$x = \alpha_1 p_1 + \cdots + \alpha_n p_n = \sum_{k=1}^{n} \alpha_k p_k \tag{4.100}$$

where $\mathbf{p}_k$ is a sequence of n mutually conjugate vectors representing n orthogonal directions. The coefficient associated with each $\mathbf{p}_k$ is as follows:

$$\alpha_k = \frac{p_k^{\mathrm{T}} b}{p_k^{\mathrm{T}} A p_k} \tag{4.101}$$

Now the key issue becomes how to find these n conjugate directions $\mathbf{p}_k$. If we can choose the conjugate vectors $\mathbf{p}_k$ properly, we will be able to obtain a good approximation for the true solution x with a small number of leading terms in (4.100). This idea is again similar to Fourier series expansion whose finite number of leading terms may provide a good approximation to the original function.

In the iteration, we shall start from any initial guesses of x as x_0 and $p_k = p_0 = b - Ax_0$. The corresponding first base vector $p_{k+1} = p_1$ is then calculated by:

$$p_{k+1} = r_k - \frac{p_k^{\mathrm{T}} A r_k}{p_k^{\mathrm{T}} A p_k} p_k \tag{4.102}$$

where $r_k = b - Ax_k$ is the residual vector.

In iteration, the calculation is terminated when the norm of the residual is smaller than certain small tolerance, i.e., $\|r_k\| < \varepsilon$. Theoretically, the solution to $Ax = b$ is equivalent to the minimization of $f(x) = x^T Ax/2 - b^T x$. This is achieved by setting the gradient of $f(x)$ to zero. The CG method bears its current name because the way to find p_k in (4.102) is equivalent to finding the gradient of $f(x)$ at $x = x_k$ and the following procedure corresponds to finding the conjugate vectors for this gradient.

5 Water wave models

5.1 Introduction

Water wave models are numerical tools for simulating ocean waves. In Chapter 1, we briefly introduced different types of wave models developed for different purposes. In this chapter, we will classify the wave models based on the level of their theoretical completeness. The major assumptions associated with each model are discussed. This allows readers to understand the limitations of each model and the theoretical linkages among different wave models. Examples of simulations are used to demonstrate the capability of each model in engineering applications.

Some case studies are presented that provide examples of the applications of wave models to realistic problems. There is special emphasis on the coupling between different wave models in order to advance the simulation from far field to near field, during which either the same wave model, but with different levels of grid resolution, or different wave models will be applied to resolve different physical processes.

Finally, we present a few examples of wave models and a series of benchmark tests for different wave problems. Adequate numerical details are provided so that advanced readers can build up their own wave models using the techniques introduced. In addition, readers can test the wave models with either the self-developed or existing ones against the benchmark tests where the comparisons of numerical results and available theories and experimental data have already been made. The main purpose of this section is to establish the validity range of different wave models.

5.2 Depth-resolved models

In this section, we will introduce wave models that are able to resolve depth-varying flow information in the simulations. We will start from the most generic wave models that are based on the NSEs. It is then followed by various simplified wave models.

5.2.1   *Numerical models based on full Navier–Stokes equations*

5.2.1.1   *Governing equations*

The NSEs are the governing equations for general fluid flows including water waves. For most of the water wave problems, we can assume that the fluid is incompressible, which leads to the incompressible NSEs, i.e.:

$$\frac{\partial u_i}{\partial x_i} = 0 \tag{5.1}$$

$$\frac{\partial u_i}{\partial t} + u_j \frac{\partial u_i}{\partial x_j} = -\frac{1}{\rho}\frac{\partial p}{\partial x_i} + g_i + \frac{1}{\rho}\frac{\partial}{\partial x_j}\left(\mu\frac{\partial u_i}{\partial x_j}\right) \tag{5.2}$$

where $i = 1, 2$ for 2D models and $i = 1, 2, 3$ for 3D models.

The main difficulty in solving the above equations is that the pressure is a passive variable that is not determined by a time-advancing transport equation. Instead, it depends on the instantaneous velocity field. The solution for pressure needs to be obtained by solving the PPE that incorporates the continuity equation in the momentum equation. Since the Poisson equation is an elliptic type of PDE, the numerical solution for pressure for 2D and 3D problems involves the solution of a large sparse matrix. Normally, an iterative algorithm such as the SOR method or CG method is required.

5.2.1.2   *NSE solvers*

Pressure–velocity iteration method and pressure correction technique: The first numerical model for solving incompressible NSEs was developed by Harlow and Welch (1965). In their model, the NSEs were discretised into forward-time FD form. By enforcing zero divergence of the velocity field at both previous and current time steps, the pressure was solved in an iterative way. With the employment of the updated pressure, the velocity information at the current time step can be obtained. The method was later referred to as pressure–velocity iteration method.

One branch of the later development of NSE solvers follows a similar idea. Examples include the more practical SIMPLE-like models based on a pressure correction technique. For example, SIMPLE (Semi-Implicit Method for Pressure-Linked Equations) algorithm was first proposed by Patankar and Spalding (1972) and SIMPLER (SIMPLE Revised) by Patankar (1981). By improving the model convergence, more members joined the SIMPLE family later, e.g., SIMPLEC (SIMPLE Consistent) algorithm by van Doormaal and Raithby (1984), SIMPLEST (SIMPLE ShorTened) algorithm by Spalding (1980), and SIMPLEM (SIMPLE Modified) algorithm by Acharya and Moukalled (1989). Besides, PISO (Pressure Implicit with Splitting of Operators) algorithm by Issa (1982) was another improvement to the original solution procedure, in which the pressure–velocity coupling scheme was used. Note that many of the above solvers are applicable for both incompressible and compressible flows. Recently, Moukalled and Darwish (2000) reviewed the performance of various NSE solvers and proposed a unified model for solving flow problems at all speeds.

Projection method: Another important branch stemmed from the projection method proposed by Chorin (1968, 1969). In the projection method, the calculation is split into two steps. At the first step, the tentative velocities are calculated with the absence of pressure and thus the velocity field carries the correct vorticity information. At the second step, the pressure is updated based on the PPE to obtain the final velocity field satisfying the continuity equation.

Based on the projection method, Kothe *et al.* (1991) and Kothe and Mjolsness (1991) developed an efficient and robust numerical scheme for free surface flows. The model, called RIPPLE, solves the PPE using the incomplete Cholesky conjugate gradient (ICCG) method. In RIPPLE, a new approach was also adopted to model the surface tension as a volume force. Because of its computational efficiency and numerical robustness, this model was selected as the base model by Liu and Lin (1997) for studying water wave problems. The model was further extended to COBRAS (Liu *et al.*, 1999b) and NEWFLUME (Lin and Xu, 2006) for more general turbulent free surface flow problems.

Artificial compressibility method: There exists another way of solving incompressible NSEs, with the use of artificial compressibility method (ACM). The method was originally proposed by Chorin (1967) in the attempt to resolve the difficulty in solving the elliptic PPE iteratively. By realizing that NSEs for compressible fluid flow are of a hyperbolic type that are easier to solve, one may introduce an artificial (or sometimes called pseudo-) compressibility into the flow. The pseudotime and subiteration, however, are needed for each physical time step to ensure numerical accuracy. The method has been successfully applied to many fluid flow problems, e.g., Rogers *et al.* (1991) for internal flows and Farmer *et al.* (1994) for free surface flows. Recently, by incorporating the $k-\varepsilon$ turbulence model and employing a boundary-conforming control volume mesh, Li (2003) developed an ACM model to study ship waves.

5.2.1.3 *Free surface tracking*

The accurate tracking of free surface is of paramount importance for water wave models solving NSEs. Traditionally, there have been two types of approaches for tracking free surfaces, namely the Lagrangian and the Eulerian methods. Although the Lagrangian method determines the exact free surface location within a computational cell, the Eulerian method provides the bulk fluid property (e.g., mean density) in a computational cell, from which the free surface can be reconstructed only approximately.

Sometimes, the Lagrangian method is classified as "interface tracking method," whereas the Eulerian method is classified as "interface capturing method."

Marker-and-cell (MAC) method: The Lagrangian approach follows the motion of particles on the free surface and in the interior domain based on the ambient flow velocity information. This kind of tracking philosophy forms the basis of the MAC method originally developed by Harlow and Welch (1965):

$$\frac{\mathrm{d}X_i}{\mathrm{d}t} = u_i \qquad (5.3)$$

where X_i is the fluid particle position (tracer) in the i-direction and u_i is the corresponding velocity at the particle location that can be obtained by interpolation. By following these particles originally deployed on the free surface, the updated free surface information can be obtained at each time step, assuming that particles on the free surface will remain so for nonbreaking waves. For breaking waves, the free surface particles need to be reconstructed because particles on the free surface may leave the surface during the breaking process. In fact, MAC can be regarded as the extension of the earlier particle-in-cell (PIC) method, which was originally developed for plasma simulation. With the use of MAC method, Hirt *et al.* (1975) developed the SOLA model to simulate free surface flows.

The VOF method: In the Eulerian approach, the normalized density at the fixed location (e.g., cell center) is tracked by solving the transport equation of the VOF function, i.e.:

$$\frac{\partial F}{\partial t} + u_i \frac{\partial F}{\partial x_i} = 0 \qquad (5.4)$$

where F is defined as the volume fraction of the fluid in the cell (control volume), which is equivalent to the normalized (by liquid density) mean density of the liquid–air mixture in the cell. Obviously, $F = 1$ in liquid, $F = 0$ in air, and $0 < F < 1$ on the free surface. Therefore, with the updated information of F in all computational cells, the free surface geometry can be approximated and reconstructed. This approach is the basis of the so-called VOF method originally developed by Nichols *et al.* (1980) and Hirt and Nichols (1981) in their code SOLA-VOF for the simulation of broken free surfaces.

Level set method: In recent years, the level set method developed by Osher and Sethian (1988) became popular to capture interface motion (e.g., Sussman *et al.*, 1994; Chang *et al.*, 1996). In this method, instead of tracking the density, which has a sharp front on the free surface or two-fluid interface, the so-called level set function ϕ, which is designed to be zero on the interface but smoothly changing across the interface, is tracked by solving the following equation:

$$\frac{\partial \phi}{\partial t} + u_i \frac{\partial \phi}{\partial x_i} = 0 \qquad (5.5)$$

There are many ways of designing ϕ and the most intuitive way is to define it as the normal distance function to the free surface, which leads to $\phi > 0$ in water, $\phi < 0$ in air, and $\phi = 0$ on the free surface.

After the level set function is known, it will be used to determine the density at each cell through the Heaviside function (Gu *et al.*, 2005) as follows:

$$
\gamma_\phi = \begin{cases} 0, & \text{if } \phi < 0, & f_p \notin \Gamma(p) \\ a_p, & \text{if} & f_p \in \Gamma(p) \\ 1, & \text{if } \phi > 0, & f_p \notin \Gamma(p) \end{cases} \tag{5.6}
$$

in which $f_p \notin \Gamma(p)$ represents the nonfree surface cell and $f_p \in \Gamma(p)$ the free surface cell. The coefficient a_p is defined as the volume ratio of liquid in the cell that can be calculated based on the tangential plane of the zero level set contour ($\phi = 0$) across the cell (Gueyffier *et al.*, 1999). The mean fluid density for a particular cell is then calculated as follows:

$$
\rho = \rho_a + (\rho_w - \rho_a)\gamma_\phi \tag{5.7}
$$

Kinematic free surface boundary condition: For a nonbreaking free surface that is the single-value function of the horizontal coordinate, one can solve the following equation to update the free surface location:

$$
\frac{\partial \eta}{\partial t} + u\frac{\partial \eta}{\partial x} + v\frac{\partial \eta}{\partial y} = w \text{ at } z = \eta(x, y, t) \tag{5.8}
$$

The above equation is essentially the extension of the kinematic free surface boundary condition (3.4) to 3D problems. To obtain the updated free surface location, the accurate velocity information on the free surface must be known.

Height function transport equation: More often, for a nonbreaking free surface, the so-called height function transport equation is solved instead as follows:

$$
\frac{\partial(\eta + b)}{\partial t} + \frac{\partial}{\partial x}\int_{-b}^{\eta} u \, dz + \frac{\partial}{\partial y}\int_{-b}^{\eta} v \, dz = 0 \tag{5.9}
$$

The equation is essentially the vertically integrated continuity equation. The equation is favored against (5.8) because it is less sensitive to the inaccuracy of velocity calculation near the free surface. The review of the advantages and shortcomings of various free-surface tracking methods can be found in Floryan and Rasmussen (1989) and Lin and Liu (1999a).

5.2.1.4 Turbulence modeling

To model general water wave problems, the inclusion of proper turbulence models is necessary. In recent years, in connection with the NSE solvers, there are two main approaches for turbulence modeling, namely $k-\varepsilon$ turbulence models linked with the RANS equations and LES models linked with spatially averaged Navier–Stokes (SANS) equations. The model based on the former approach is often referred to as RANS wave model (e.g., Lin and Liu, 1998a) and the latter as LES wave model (e.g., Li and Lin, 2001). Because both RANS equations and SANS equations have the same equation structure as NSEs, the NSE solvers discussed above can be equally applied to solving turbulent flows except that the additional turbulence-induced stresses must be properly modeled.

5.2.1.5   *Boundary conditions*

To model water waves, proper boundary conditions must be specified at all domain boundaries and physical boundaries. The domain boundaries include inflow and outflow boundaries, whereas physical boundaries include free surface, bottom boundaries, and boundaries on solid surfaces.

Inflow boundary: At the inflow boundary, normally both free surface displacement (or its derivative) and velocity distribution are given based on certain wave and current theories, i.e.:

$$\begin{cases} u_i = u_{i0}(t) \\ \eta = \eta_0(t) & \text{(wave or flow with } Fr > 1) \quad \text{or} \\ \dfrac{\partial \eta}{\partial n} = \alpha_0 & \text{(flow with } Fr < 1) \end{cases} \qquad (5.10)$$

where the variables on the RHS of the equation with the subscript 0 are the known values.

Radiation (open or nonreflecting) boundary: Generally, the computation can be performed only in a finite domain. This means that the domain must be truncated at some place where the artificial boundary condition is specified. For water waves, the Sommerfeld radiation boundary condition (1964) is applicable at the truncated boundary where outgoing waves are allowed to leave freely without reflection. This radiation condition was explained by Sommerfeld as "the sources must be sources, not sinks of energy. The energy which is radiated from the sources must scatter to infinity; no energy may be radiated from infinity into . . . the field." Mathematically, this ensures the following condition that is the solution of the Helmholtz equation in all directions of an m-dimensional "sphere":

$$\lim_{|\mathbf{x}| \to \infty} |\mathbf{x}|^{(m-1)/2} \left(\frac{\partial}{\partial |\mathbf{x}|} - ik \right) \phi(\mathbf{x}) = 0 \qquad (5.11)$$

where ϕ is any flow variable and $m = 1, 2, 3$ is the dimension of space. When $m = 1$, the above equation reduces to:

$$\lim_{|x| \to \infty} \phi(x) = \phi_0 e^{ikx} \qquad (5.12)$$

By including the time harmonic, the flow variable at infinity can be expressed as follows:

$$\lim_{|x|\to\infty} \phi(x) = \phi_0 e^{i(kx-\sigma t)} \tag{5.13}$$

When the truncated boundary is located at a place that is far from the source of disturbance and has a mildly varying bathymetry, it is justified to treat this boundary as the radiation boundary. For a right-propagating linear wave train, equation (5.13) can be reduced to:

$$\frac{\partial \phi}{\partial t} + \frac{\sigma}{k}\frac{\partial \phi}{\partial x} = \frac{\partial \phi}{\partial t} + c\frac{\partial \phi}{\partial x} = 0 \tag{5.14}$$

where c is the local wave phase velocity. For an oblique wave train, the radiation boundary condition should be revised to:

$$\frac{\partial \phi}{\partial t} + c\cos\theta\frac{\partial \phi}{\partial x} = 0 \tag{5.15}$$

where θ is the incident wave angle. In an actual simulation, the angle θ is often unknown a priori.

Engquist and Majda (1977) proposed a higher order 2D open boundary condition when the main propagation direction is in the x-direction:

$$\frac{\partial^2 \phi}{\partial t^2} + c\frac{\partial^2 \phi}{\partial t\partial x} - \frac{c^2}{2}\frac{\partial^2 \phi}{\partial y^2} = 0 \tag{5.16}$$

The condition allows the majority of wave trains leaving the domain with the reflected wave amplitude less than 3 percent of the incident wave amplitude for $\theta \leq 45°$. Extensive reviews of various nonreflecting open boundary conditions were made by Givoli (1991) and Tsynkov (1998), the latter of whom provided a comparative assessment of different existing methods in constructing the open boundary conditions for a truncated domain.

Sponge layer treatment: Note, however, that the open boundary conditions (5.15) and (5.16) are, in principle, applicable only to linear monochromatic waves. For nonlinear waves and waves with different frequencies, the definition of c is ambiguous. Significant errors may result when a wide-spectrum wave train is treated. To help resolve this problem, a sponge layer, which is effective to damp out short wave energy, can be added in front of the radiation boundary. Inside this sponge layer, a gradually increased artificial viscosity is introduced to damp out wave energy without causing significant wave reflection. If the artificial viscosity and the length of the sponge layer are well chosen, the majority of short waves will be damped out and the remaining waves will contain only long-wave components, for which the phase speed can be easily calculated by $c = \sqrt{gh}$. Lin and Liu

(2004) in their NSE solvers introduced an artificial damping term $-f(x)\langle u_i \rangle$ in the momentum equation of the RANS equations inside the sponge layer that has a length $x_s = 1.5\lambda$, with λ being the wavelength of the waves to be damped out. By measuring from the left boundary where $x = 0$, $f(x)$ takes the following form:

$$f(x) = \alpha \frac{\exp\left[\left(\frac{x_s - x}{x_s}\right)^n\right] - 1}{\exp(1) - 1} \tag{5.17}$$

where $n = 10$ and $\alpha = 200$ in their study but can be changed to other values when the length of the sponge layer is changed.

Infinite element method: In both the radiation the boundary condition and artificial sponge layer treatment, the domain is truncated into finite length and the treatment is meant to minimize the artificial wave reflecting from the truncated boundary. To have a real nonreflecting boundary, there is another treatment called infinite element method. In this method, the exterior domain is not truncated but is represented by elements of infinite extent. Some type of outward propagating wave-like function is then proposed to describe the general characteristics of waves in the element (Bettess and Zienkiewicz, 1977). Most of the time, the infinite element method is used with the FEM in solving wave problems.

Free surface boundary: On the free surface, the continuity of normal and shear stresses is imposed, i.e.:

$$-p + \tau_{nn} = -p_{0\_air} + \tau_{nn0\_air} \tag{5.18}$$

$$\tau_{tn} = \tau_{tn0\_air}$$

In most of the numerical computations, the air effect can be neglected and the RHS is set to zero.

Solid boundary: On the surface of a solid body, the fluid velocity will be the same as the solid velocity, i.e., the no-slip boundary condition:

$$u_i = u_{iS} \tag{5.19}$$

If the boundary is fixed and rigid (e.g., bottom), the above boundary condition becomes:

$$u_i = 0 \tag{5.20}$$

The pressure boundary condition is also needed when the PPE is solved. This boundary condition can be derived by using the momentum equation and requires the convection and diffusion terms to vanish on the solid boundary:

$$\frac{\partial p}{\partial x_i} = \rho \left(g_i - \frac{\partial u_i}{\partial t} \right) \tag{5.21}$$

For a stationary body, the above equation is reduced to:

$$\frac{\partial p}{\partial x_i} = \rho g_i \tag{5.22}$$

Boundary conditions for turbulence on solid surface and free surface: If turbulence modeling is included, proper boundary conditions for turbulence on the solid surface are also needed. In both the $k-\varepsilon$ model and LES model, if the viscous sublayer is not resolved by the numerical resolution, the log-law wall function is first used, i.e.:

$$u = \frac{u_*}{\kappa} \ln \frac{z}{z_0} \tag{5.23}$$

By employing the tangential velocity information u and its normal distance to the wall z at the fluid cell nearest to the wall, we are able to solve (5.23) to find the friction velocity, from which the wall shear stress can be obtained:

$$\tau_b = \rho u_*^2 \tag{5.24}$$

This wall shear stress will be used as the boundary condition when the momentum equation is solved.

For the $k-\varepsilon$ model, the boundary conditions for k and ε on a solid surface are needed, i.e:

$$k = \frac{u_*^2}{\sqrt{C_d}} \text{ and } \varepsilon = \frac{u_*^3}{\kappa y} \tag{5.25}$$

On free surface, the zero fluxes of k and ε are imposed, i.e.:

$$\frac{\partial k}{\partial n} = \frac{\partial \varepsilon}{\partial n} = 0 \tag{5.26}$$

5.2.1.6   Wave generation

Wave generation from inflow boundary: Various waves can be generated from the inflow boundary. The free surface displacement and velocities can be specified based on various wave theories (e.g., linear, irregular, Stokes, cnoidal, and solitary waves) at the inflow boundary as follows:

$$\phi = \phi_{int} \tag{5.27}$$

where ϕ represent both velocity and free surface displacement. Different waves can be generated from the inflow boundary and then propagate into the computational domain.

Wave generation with absorption of weak reflected waves: When there is a solid body inside the computational domain or the radiation boundary is not perfect so that wave reflection results, the reflected waves will propagate backward and interfere with the inflow boundary where waves are generated. In this event, the so-called secondary (unrealistic) wave reflection will be generated from the inflow boundary and contaminate the computation. To relieve this problem, the use of an absorbing wave maker has been adopted. Basically, the idea behind the absorbing wave maker is to detect the reflected waves and then to correct the inflow wave-making process accordingly to absorb the reflected waves during wave generation. Mathematically, the inflow boundary condition (on the left) is revised as follows:

$$\frac{\partial \phi_{\text{out}}}{\partial t} - c\frac{\partial \phi_{\text{out}}}{\partial n} = 0 \tag{5.28}$$

where $\phi_{\text{out}} = \phi - \phi_{\text{int}}$ is the detected wave reflection with ϕ_{int} being the known incident wave variable from wave theory and ϕ being the wave variable to be specified. This equation will be reduced to the conventional radiation boundary condition if there is no wave incidence. In the case when the wave reflection is zero, the above equation is reduced to $\phi = \phi_{\text{int}}$, the conventional wave generation boundary condition. The above treatment can be used to treat weak wave reflection that will not cause strong nonlinear wave interaction at the inflow boundary. The main difficulty in this method is the accurate detection of the reflected wave, which can be of multiple directions and multiple frequencies.

Wave generation by a moving paddle: Alternatively, if a moving boundary is coded in the computer program, a wave train can be generated by specifying the time history of the velocity and trajectory of the wave paddle, a mimic to the wake maker in a physical wave flume. Numerically, it requires the track of the Lagrangian motion of the paddle. Lin (2007) proposed the so-called locally relative stationary (LRS) method to model a moving object in the fixed-grid RANS model. The model was used to create a solitary wave by specifying the velocity and trajectory as follows:

$$u[x(t), t] = \frac{c\eta}{h + \eta} \tag{5.29}$$

where

$$\eta[x(t), t] = H\text{sech}^2\left\{\sqrt{\frac{3H}{4h^3}}[x(t) - ct + x_0]\right\} \text{ and } x(t) = \int_0^t u(\tau)\mathrm{d}\tau$$

x_0 is the distance between the origin and the crest of the wave train at $t = 0$ and $c = \sqrt{g(h + H)}$ is the phase velocity of the wave train.

Internal wave maker: All the above wave generation techniques may face problems when a strong reflection is present in the computation. To avoid the complication, we should try to avoid the simultaneous treatment of wave generation and wave absorption at the same place. This can be achieved by generating waves inside the computational domain with the use of artificial mass or momentum sources. If the source function is linear, it will not interfere with the background waves in the process of wave generation. Lin and Liu (1999b) were the first ones to implement mass source in the NSE solver to generate waves inside the computational domain. In their method, the continuity equation was modified in the selected source region inside the computational domain as follows:

$$\frac{\partial u_i}{\partial x_i} = s(\mathbf{x}, t) \tag{5.30}$$

where $s(\mathbf{x}, t)$ is the mass source function.

Depending on how the source function is specified, various types of linear and nonlinear waves can be generated inside the computational domain. For a 2D problem, by choosing a fully submerged rectangular domain and setting that using a uniform source function $s(t)$, Lin and Liu (1999b) suggested the following form of $s(t)$ for a linear wave train:

$$s(t) = \frac{cH}{A} \sin(\sigma t) \tag{5.31}$$

where A is the total area of the source region.

An irregular wave train can be generated as follows:

$$s(t) = \sum_{i=1}^{n} \frac{cH_i}{A} \sin(\sigma_i t + \delta_i) \tag{5.32}$$

where δ_i is the random phase angle.

A Stokes wave (second order or fifth order) can be generated as follows:

$$s(t) = \sum_{i=1}^{m} \frac{2c}{A} a_i \sin\left[i(2\pi - \sigma t - \delta_i)\right] \tag{5.33}$$

where the wave amplitude associated with each wave component a_i can be found in Section 3.3.1.

A cnoidal wave can be generated as follows:

$$s(t) = \frac{2cH}{A} \left\{ y_t/H + \operatorname{cn}^2\left[2K(m)\left(-\frac{t}{T} + \delta\right), m\right] \right\} \tag{5.34}$$

where y_t and m can be determined given H, h, and T (see Section 3.3.2).

A solitary wave can be generated as follows:

$$s(t) = \frac{cH}{A} \operatorname{sech}^2\left[\sqrt{\frac{3H}{4h^3}}(x_s - ct)\right] \tag{5.35}$$

where $x_s = 4h/\sqrt{H/h}$ in order to reduce the mass error below 1 percent.

The major advantage of this approach is that the inflow boundary condition is no longer needed and only the radiation boundaries are specified at all lateral boundaries. When combined with the use of a numerical sponge layer, wave reflection can be effectively controlled, which is especially important when long-term simulation of random waves is performed.

5.2.1.7   Modeling of porous flows and flows in vegetation

In the earlier numerical study of wave-induced porous flows, either potential flow theory (e.g., Sulisz, 1985) or SWEs (e.g., Kobayashi and Wurjanto, 1990) were used as the base model that incorporates the damping terms in porous media. Later, the more rigorous and accurate porous flow models in connection with NSE solvers were developed (e.g., van Gent, 1995; Liu *et al.*, 1999a). In their approach, the NSEs were averaged over space to have a new set of equations that have similar characteristics to NSEs but the additional friction in porous media is included.

To illustrate how the new equations are derived, we shall confine ourselves to saturated porous flows only. To simplify the derivation, we assume that the porous medium is of spherical shape with the mean diameter d_{50}. The original NSEs are still valid in the fluid domain, i.e.:

$$\frac{\partial u_i}{\partial x_i} = 0 \tag{5.36}$$

$$\frac{\partial u_i}{\partial t} + u_j \frac{\partial u_i}{\partial x_j} = -\frac{1}{\rho}\frac{\partial p}{\partial x_i} + g_i + \frac{1}{\rho}\frac{\partial}{\partial x_j}\left[\mu\left(\frac{\partial u_i}{\partial x_j} + \frac{\partial u_j}{\partial x_i}\right)\right] \tag{5.37}$$

Now consider a control volume (sometimes called filter space) as enclosed by the thick lines in Figure 5.1. In this control volume, the volume taken

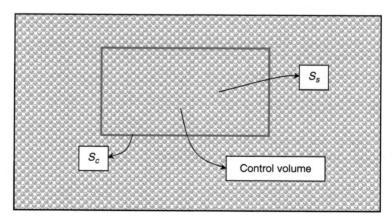

Figure 5.1 Sketch of porous skeleton and porous flow.

by the porous material and the pore space filled with fluid are V_s and V_f, respectively. The porosity is defined as $n = V_f/(V_f + V_s)$.

Let us now introduce the spatial average operation as:

$$\overline{\overline{\phi}} = (1/V_f) \int_{V_f} \phi \, dV.$$

By taking the spatial average of the NSEs within the control volume for the fluid only, we obtain the new set of equations by applying divergence theorem and assume that the operator is commutable for integration and differentiation of velocities:

$$\frac{\partial \overline{\overline{u_i}}}{\partial x_i} = 0 \tag{5.38}$$

$$\frac{\partial \overline{\overline{u_i}}}{\partial t} + \overline{\overline{u_j}}\frac{\partial \overline{\overline{u_i}}}{\partial x_j} + \frac{\partial \overline{\overline{u_i'' u_j''}}}{\partial x_j} = -\frac{1}{\rho V_f}\int_{S_c+S_p} pn_i \, dS + g_i + \frac{1}{\rho V_f}\int_{S_c+S_p} \tau_{ij}n_j \, dS \tag{5.39}$$

where S_c is the control surface of the fluid part for the chosen control volume, S_p is the total surface of the porous material inside the control volume, and **n** is the outward unit normal vector from the fluid to the solid surface. In the above equations, we have two terms requiring proper closures: (1) the additional stress term induced by the correlation of velocity variation in space $\overline{\overline{u_i'' u_j''}}$ (this is similar to Reynolds stress for turbulent flows) and (2) the additional forcing terms from the surface integration of the reaction force from the porous materials within the control volume. Physically, this is not difficult to understand because the surface of the porous material coincides with the surface of integrated fluid volume, through which the porous material acts as the source of external force.

We can now relate the control volume mean velocity $\bar{u}_i$ with the spatial-averaged velocity in fluid domain $\overline{\overline{u_i}}$ by $\bar{u}_i = n\overline{\overline{u_i}}$ (based on the mass conservation argument). By further assuming that the porous material shares the same pressure and stresses as those in fluid, we can apply the divergence theorem again in the entire control volume and obtain:

$$\frac{1}{\rho V_f}\int_{S_c} pn_i \, dS = -\frac{1}{\rho}\frac{\partial \bar{p}}{\partial x_i} \tag{5.40}$$

$$\frac{1}{\rho V_f}\int_{S_c} \tau_{ij}n_j \, dS = \frac{\nu}{n}\frac{\partial^2 \bar{u}_i}{\partial x_j \partial x_j} \tag{5.41}$$

If we define the friction term:

$$\overline{f}_i = \frac{1}{\rho V_f}\int_{S_p} pn_i \, dS - \frac{1}{\rho V_f}\int_{S_p} \tau_{ij}n_j \, dS$$

and substitute (5.40) and (5.41) into (5.39), we have the new set of governing equations for the mean flow averaged over the entire control volume:

$$\frac{\partial \bar{u}_i}{\partial x_i} = 0 \tag{5.42}$$

$$\frac{1}{n}\frac{\partial \bar{u}_i}{\partial t} + \frac{\bar{u}_j}{n^2}\frac{\partial \bar{u}_i}{\partial x_j} = -\frac{1}{\rho}\frac{\partial \bar{p}}{\partial x_i} + g_i + \frac{\nu}{n}\frac{\partial^2 \bar{u}_i}{\partial x_j \partial x_j} - \bar{f}_i - \frac{1}{n^2}\frac{\partial \overline{u_i'' u_j''}}{\partial x_j} \tag{5.43}$$

By using the control volume mean velocity, the flow inside and outside the porous media can be directly linked by the same velocity, which is continuous across the interface. As a result, there is no need to impose flow boundary conditions on the porous interface when the flow inside and outside of the porous domain is solved as one fluid flow system.

The main issue now becomes how to model the friction term $\bar{f}_i$ and the turbulence stress term $-\rho \overline{u_i'' u_j''}$. While the latter term can be modeled by the same kind of modified turbulence eddy viscosity model, i.e., $-\overline{u_i'' u_j''} = \nu_{tp}(\partial \bar{u}_i/\partial x_j + \partial \bar{u}_j/\partial x_i)$, the former can be modeled by the Morison equation (3.188) for general unsteady flow, i.e.:

$$\bar{f}_i = \overline{f_{Ii}} + \overline{f_{Di}} \tag{5.44}$$

where $\overline{f_{Ii}}$ and $\overline{f_{Di}}$ are the inertial and drag force contributions. The inertial force is caused by the resistance of the rigid and fixed porous particles to the ambient flow acceleration. Lin and Karunarathna (2007) derived the following expressions assuming that the control volume is filled with uniform spheres:

$$\overline{f_{Ii}} = \frac{1}{\rho V_f}\left(N\rho C_M V_p \frac{1}{n}\frac{\partial \overline{u_i}}{\partial t}\right) = \frac{N V_p}{V_f}\left(C_M \frac{1}{n}\frac{\partial \overline{u_i}}{\partial t}\right) = \frac{C_M(1-n)}{n^2}\frac{\partial \overline{u_i}}{\partial t} \tag{5.45}$$

where N is the total number of spheres in the control volume, V_p is the volume of one sphere, and C_M is the virtual mass coefficient. C_M is also called the inertial coefficient that is the sum of the Froude–Krylov force coefficient 1.0 and the added mass force coefficient C_a, i.e., $C_M = 1 + C_a$. The Froude–Krylov (F–K) force results from the pressure difference in an accelerating fluid flow and it behaves in a similar way to the buoyancy force. The added mass force is the additional force resulting from the presence of the body and it is a function of body shape. For a single sphere, $C_a = 0.5$ and $C_M = 1.5$.

However, in most of the existing literature, the value of C_M based on oscillatory flow experiments is generally smaller than 1.0. For example, the suggested values of C_M by van Gent (1995) [based on Smith's (1991) experiments] and Gu and Wang (1991) are $C_M = 0.34$ and $C_M = 0.46$, respectively. Lin and Karunarathna (2007) also adopted $C_M = 0.34$ in their study. One possible explanation for the small value of C_M in porous media could be that in a densely packed porous medium, the porous flow moves through the channels of pores similar to the flow in curved tubes with changing diameter, in which both F–K force coefficient and added mass coefficient are reduced. The reduction of the F–K force coefficient is due to the packing of porous materials that reduce the effective cross-sectional area exposed to the flow direction. Since the F–K force has the same expression and similar justification as buoyancy force, the reduced FK force coefficient in porous media can be explained by the reduced buoyancy for a composite cylinder sitting on the bottom (Figure 5.2). In such a case, the total lift force is reduced because part of the uplift force originally acting on the lower surface of the structure disappears. Besides the F–K force coefficient, the added mass coefficient is also reduced when we consider the porous materials as many variable-diameter pipelines rather than individual spheres.

Based on the Morison equation again, the drag force can be expressed as:

$$\overline{f_{Di}} = \frac{1}{\rho V_f}\left(N\frac{1}{2}\rho C_D A\frac{\overline{u}_c\,\overline{u}_i}{n\ n}\right) = \frac{N}{V_f}\left(\frac{1}{2}C_D\frac{\pi}{4}d_{50}{}^2\frac{\overline{u}_c\,\overline{u}_i}{n\ n}\right)$$

$$= \frac{NV_p}{V_f}\left(\frac{1}{V_p}\frac{1}{2}C_D\frac{\pi}{4}d_{50}{}^2\frac{\overline{u}_c\,\overline{u}_i}{n\ n}\right) = \frac{V_s}{V_f}\left(\frac{1}{\frac{\pi}{6}d_{50}{}^3}\frac{1}{2}C_D\frac{\pi}{4}d_{50}{}^2\frac{\overline{u}_c\,\overline{u}_i}{n\ n}\right)$$

$$= \frac{3}{4}C_D\frac{(1-n)}{n^3}\frac{1}{d_{50}}\overline{u}_c\overline{u}_i \tag{5.46}$$

where $\overline{u}_c$ is the characteristic velocity that can be estimated by $\overline{u}_c = \sqrt{\overline{u}_i\overline{u}_i}$, $A = \pi d_{50}/4$ is the maximum cross-sectional area of the sphere, and C_D is the drag force coefficient.

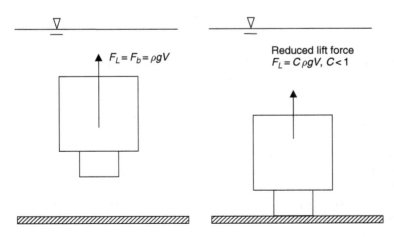

Figure 5.2 Illustration of the changed lift force on a composite cylinder due to the change of relative position of the structure.

For a single sphere in steady flow, the drag coefficient formula was proposed by Fair *et al.* (1968):

$$C_{DS} = \frac{24.0}{Re} + \frac{3.0}{\sqrt{Re}} + 0.34 \tag{5.47}$$

where $Re = \bar{u}_c d_{50}/\nu$. In reality, porous media have various shapes, surface roughness, and packing patterns. The wake flows behind particles also have influence on the drag force of the following particles. All these would have an impact on the value of C_D. By employing a large amount of experimental data, Lin and Karunarathna (2007) suggested the following modification for C_D in simulating wave-induced flows in porous materials:

$$C_D = c_1 \frac{24.0}{Re} + c_2 \left(\frac{3.0}{\sqrt{Re}} + 0.34 \right) \left(1 + \frac{7.5}{KC} \right) \tag{5.48}$$

where $c_1 = 7.0$ and $c_2 = 1.8 \sim 4.2$ are found in their study, with the smaller c_2 better describing the flow in a small porous medium. In this book, the value of $c_2 = 4.0$ is adopted. In the Keulegan–Carpenter number $KC = U_{max} T/(n d_{50})$, U_{max} is the maximum water particle velocity and T is the wave period. This term $(1 + 7.5/KC)$ is included to account for the additional influence of the flow unsteadiness on the turbulent boundary layer around the porous particles. For more general unsteady flows, T represents the characteristic timescale that has to be determined case by case.

Consider a 1D steady flow in the x-direction. Substituting (5.46) into (5.43), we have:

$$0 = -\frac{1}{\rho}\frac{\partial \bar{p}}{\partial x} - \frac{3}{4}C_D\frac{(1-n)}{n^3}\frac{1}{d_{50}}\bar{u}\bar{u} \Rightarrow -\frac{1}{\rho g}\frac{d\bar{p}}{dx} = I = \frac{3}{4}C_D\frac{(1-n)}{n^3}\frac{1}{gd_{50}}\bar{u}\bar{u} \tag{5.49}$$

For small Re (e.g., $Re < 10$), the porous flow is laminar and $C_D = 24c_1/Re$. Substituting it into (5.49), we have:

$$-\frac{1}{\rho g}\frac{d\bar{p}}{dx} = I = \frac{3}{4}\frac{24c_1\nu}{\bar{u}d_{50}}\frac{(1-n)}{n^3}\frac{1}{gd_{50}}\bar{u}\bar{u} = \frac{(1-n)}{n^3}\frac{18c_1\nu}{g(d_{50})^2}\bar{u} = \frac{\bar{u}}{K} \tag{5.50}$$

where K (m/s) is the permeability. This is the well-known Darcy's law for laminar flows in porous media where the mean velocity is linearly proportional to the mean pressure gradient.

When Re is large, $C_D = 0.34c_2$ becomes a constant and (5.49) is reduced to:

$$-\frac{1}{\rho g}\frac{d\bar{p}}{dx} = I = \frac{3}{4}0.34c_1\frac{(1-n)}{n^3}\frac{1}{gd_{50}}\bar{u}\bar{u} = 0.255c_1\frac{(1-n)}{n^3}\frac{1}{gd_{50}}\bar{u}\bar{u} \tag{5.51}$$

Now the drag is proportional to the mean velocity square and it is often called nonlinear friction that prevails when the porous flow is fully turbulent (e.g., Re > 1000).

In many earlier studies, only the linear and nonlinear terms are employed to describe flow resistance in porous media, neglecting the transitional flow range [i.e., the term $3.0c_2/\sqrt{Re}$ in (5.48)] that may be important for intermediate Re (e.g., $10 < Re < 1000$). For example, Forchheimer (1901) proposed the following formula:

$$-\frac{1}{\rho g}\frac{d\bar{p}}{dx} = I = a\bar{u} + b\bar{u}|\bar{u}| \qquad (5.52)$$

There have been many proposals for the expression of a and b based on different theoretical assumptions and experimental data (e.g., Ergun, 1952; Engelund, 1953). The comparison among different equations was made by Burcharth and Andersen (1995) and is extended here by using the following more general equation:

$$-\frac{1}{\rho g}\frac{d\bar{p}}{dx} = I = a\bar{u} + b\bar{u}|\bar{u}| + c\bar{u}\sqrt{|\bar{u}|} \qquad (5.53)$$

Authors	a	b	c
Carman (1937)	$180\dfrac{(1-n)^2}{n^3}\dfrac{\nu}{gd_{50}^2}$	0	0
Ergun (1952)	$150\dfrac{(1-n)^2}{n^3}\dfrac{\nu}{gd_{50}^2}$	$1.75\dfrac{(1-n)}{n^3}\dfrac{1}{gd_{50}}$	0
Engelund (1953)	$\alpha\dfrac{(1-n)^3}{n^2}\dfrac{\nu}{gd_{50}^2}$ $\alpha = 780\text{--}1500$	$\beta\dfrac{(1-n)}{n^3}\dfrac{1}{gd_{50}}$ $\beta = 1.8\text{--}3.6$	0
van Gent (1995)	$1000\dfrac{(1-n)^2}{n^3}\dfrac{\nu}{gd_{50}^2}$	$1.1\dfrac{(1-n)}{n^3}\dfrac{1}{gd_{50}}$	0
Liu *et al.* (1999a)	$200\dfrac{(1-n)^2}{n^3}\dfrac{\nu}{gd_{50}^2}$	$1.1\dfrac{(1-n)}{n^3}\dfrac{1}{gd_{50}}$	0
Lin and Karunarathna (2007) with $c_1 = 7.0$ and $c_2 = 4.0$	$126\dfrac{(1-n)}{n^3}\dfrac{\nu}{gd_{50}^2}$	$1.02\dfrac{(1-n)}{n^3}\dfrac{1}{gd_{50}}$	$4.5\dfrac{(1-n)}{n^3}\dfrac{\nu^{1/2}}{gd_{50}^{3/2}}$

The power dependency of a on $(1-n)$ varies from one another. Lin and Karunarathna's (2007) coefficient is smaller than all others because of the lower order dependency on $(1-n)$. The functional form for b is essentially the same among all models except that the coefficients vary. Lin and

Karunarathna's (2007) coefficient is again smaller than others, probably because of the additional term in their model for the transitional flow, where C_D is reciprocally proportional to $\sqrt{Re}$.

Comparisons can also be made by converting the earlier models in the form of (5.52) to the equivalent drag coefficient used in (5.46). The converted drag coefficient has similar expression to (5.48) except that there is no term associated with $1/\sqrt{Re}$. Since different dependencies on $(1-n)$ are used in different models, the comparisons need to be made for different values of n separately. Figure 5.3 shows the comparison of the effective C_D (Re) in different porous flow models. Most of these models give close prediction of C_D except for van Gent's model that has a much larger value of C_D for relatively small Re. Due to the inclusion of the transition flow correction, Lin and Karunarathna's (2007) model has larger C_D than Liu *et al.*'s (1999a) model for $10 < Re < 1000$ and $n = 0.3$ and 0.4.

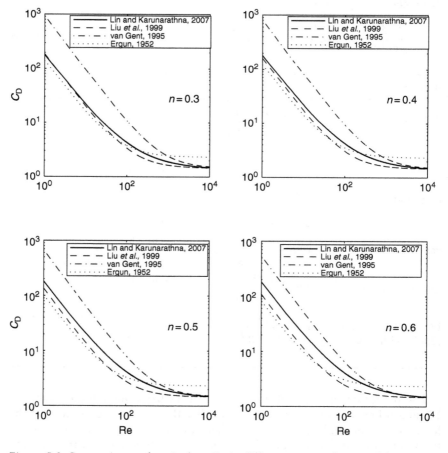

Figure 5.3 Comparisons of equivalent C_D in different porous flow models.

For the flows in vegetation, the spatially averaged governing equations can be similarly derived and they are essentially the same as (5.42) and (5.43) except that the friction term needs to be expressed differently. By assuming that the vegetation is of uniform rigid smooth cylinders with the diameter of D, Su and Lin (2005) proposed the following expression for the friction term:

$$\overline{f}_i = \overline{f}_{\mathrm{I}i} + \overline{f}_{\mathrm{D}i} = \frac{C_{\mathrm{M}}(1-n)}{n^2}\frac{\partial \bar{u}_i}{\partial t} + \frac{2}{\pi}C_{\mathrm{D}}\frac{(1-n)}{n^3}\frac{1}{D}\bar{u}_c\bar{u}_i \qquad (5.54)$$

where the added mass coefficient is defined as (Andersen, 1994):

$$C_{\mathrm{M}} = \begin{cases} 2 + 16(1-n) & \text{perpendicular to cylinder} \\[1mm] \begin{cases} 1 & \text{submerged} \\ 0 & \text{surface-piercing} \end{cases} & \text{parallel to cylinder} \end{cases} \qquad (5.55)$$

and the drag force coefficient is derived from that for smooth circular cylinders:

$$C_{\mathrm{D}} = \begin{cases} \dfrac{10^{[1.05+0.08(\log(\mathrm{Re}))^2]}}{\mathrm{Re}^{0.66}}, & 0.1 < \mathrm{Re} \leq 50; \\[3mm] \dfrac{10^{[0.75+0.06(\log(\mathrm{Re}))^2]}}{\mathrm{Re}^{0.44}}, & 50 < \mathrm{Re} \leq 4 \times 10^3; \text{ perpendicular to cylinder} \\[3mm] \dfrac{\mathrm{Re}^{0.77}}{10^{[1.72+0.08(\log(\mathrm{Re}))^2]}}, & 4 \times 10^3 < \mathrm{Re} \leq 7 \times 10^4 \\[3mm] 0.664/\sqrt{\mathrm{Re}}, & \mathrm{Re} \leq 5 \times 10^5 \quad \text{parallel to cylinder} \end{cases} \qquad (5.56)$$

For the drag force coefficients, a larger range of Re dependency needs to be developed if the prototype problems are to be modeled.

So far, little has been said about the modeling of turbulence inside porous media. The study of turbulence characteristics in porous media has been intensively made in meteorology (e.g., boundary layer air flow interaction with forest; see Raupach *et al.*, 1996) but relatively less in hydrodynamics. For a densely packed porous medium, the Reynolds stress is often relatively small compared to the friction term because the presence of porous particles restricts the growth of turbulence. For a loosely packed porous medium, e.g., vegetation, the turbulence can actually be enhanced due to the presence of porous materials. In this case, the realistic turbulence modeling inside the porous domain becomes crucial. In principle, the modified simple mixing-length eddy viscosity model (e.g., Nepf, 1999), $k-\varepsilon$ model (Shimizu and Tsujimoto, 1994), and the LES model (e.g., Su and Lin, 2005) can all be used but a robust and accurate turbulence model inside porous media for various types of porous flows is yet to be developed and tested.

5.2.1.8   Water wave models based on the Navier–Stokes equation solvers

Chan and Street (1970) were probably the first to apply NSE solvers with the MAC method to water wave modeling. The so-called irregular star method is used in the pressure solution on a free surface. The model is named SUMMAC. Along the line of using MAC to track free surface, SOLA was developed later for more general free surface flows (Hirt *et al.*, 1975). TUMMAC was developed in the 1980s to simulate ship waves (Miyata *et al.*, 1985; Miyata, 1986) and its improved version is still used nowadays for hull design (Park *et al.*, 1999) and breaking wave impact on ships (Yamasaki *et al.*, 2005). Gao and Zhao (1995) used the MAC model to study wave interaction with structures and sand beds.

Water wave models were also developed by using the VOF method to track free surface. For example, Austin and Schlueter (1982) extended SOLA-VOF model (Nichols *et al.*, 1980) to the study of breaking wave interaction with breakwaters. Lemos (1992) introduced the $k-\varepsilon$ turbulence model to SOLA-VOF to model turbulence under broken waves. The model SKYLLA developed by van der Meer *et al.* (1992), van Gent *et al.* (1994), and van Gent (1995) was applied to free surface flows on impermeable slopes and permeable structures (Doorn and van Gent, 2004). Independently, Iwata *et al.* (1996) developed a numerical code to investigate breaking waves and post-breaking wave deformation above submerged impermeable structures. Wang and Su (1993) simulated wave breaking on a sloping beach. Troch (1997) proposed a model called VOFbreak2 to study wave breaking on rubble mound breakwaters. Lin and Liu (1998a,b) presented a 2D RANS model named COBRAS that is coupled with a nonlinear Reynolds stress model and a $k-\varepsilon$ turbulence model to study breaking waves and their transformation over slopes and structures (Liu *et al.*, 1999a; Lin *et al.*, 1999). Lin and Xu (2006) proposed a more general model named NEWFLUME to study turbulent free surface flows in both open channels and oceans. In Figure 5.4, an example of numerical simulation of plunging breaking waves on a sloping beach by NEWFLUME is given. Recently, Liu and Lin (2008) developed a 3D NumErical Wave TANK (NEWTANK) to simulate more general free surface flow problems. The model can be used to simulate laminar and turbulent flows in porous media and vegetation. Figure 5.5 shows the simulation of wave run-up on a beach with the presence of coastal vegetation (e.g., mangrove), where the effect of vegetation on reflecting and attenuating the wave energy and reducing coastal inundation is clearly demonstrated.

Water wave models based on NSE solvers can employ other methods for free surface tracking. For example, using the level set method, Gu *et al.* (2005) proposed a 3D model to simulate liquid sloshing in a tank. By using the height function as the free surface tracking technique and the irregular star method for solving pressure on the free surface, Kim (2001) proposed a 3D model to simulate liquid sloshing in a tank under external excitation.

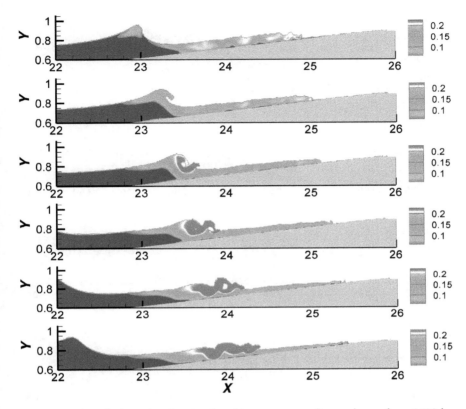

Figure 5.4 Simulation of a plunging breaking wave on a linear slope of $s = 1/10$ by using NEWFLUME; the periodic wave has $H = 0.24$ m and $T = 1.35$ s in constant water, and $h = 0.80$ m; the time interval between the two snapshots is 0.05 s; the uniform $\Delta x = \Delta z = 0.01$ cm is used; color represents turbulence intensity $\sqrt{2k}$ (m/s).

All the previous wave models are z-coordinate (z-level) models based on the FD formulation in a physical domain. Such models will face difficulty in resolving uneven bottom. Special treatment such as cut-cell method, partial-cell treatment, or immersed boundary (IB) method must be used, which increases the coding complexity and reduces the numerical accuracy near the bottom. Some models employ unstructured mesh that is boundary-fitted (e.g., Fringer *et al.*, 2006), but the mesh generation may become difficult and expensive.

An alternative way of handling the uneven bottom is the use of σ-coordinate transformation that maps the irregular physical domain between the wavy free surface and the uneven bottom into a regular computational domain, in which the solution is sought by solving the modified governing equations. For example, Stansby and Zhou (1998) and Li and

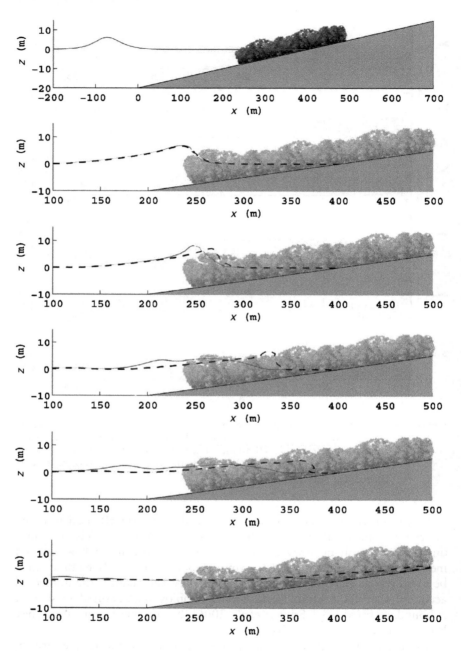

Figure 5.5 Comparison of a solitary wave run-up on a beach with (solid line) and without (dashed line) vegetation; the top figure is for the problem setup of $H = 6$ m, $h = 20$ m, and $s = 1/20$; the vegetation domain is from 250 m to 500 m and the vegetation has a mean stem diameter of 0.05 m and a volume density of 1%.

Johns (2001) proposed a vertical 2D σ-coordinate numerical model that considered nonhydrostatic pressure. Casulli (1999) proposed a semi-implicit model for 3D nonhydrostatic free surface flows. Lin and Li (2002) developed a 3D wave model and employed it to study wave interaction with current (Lin and Li, 2003), structures (Li and Lin, 2001), and vegetation (Su and Lin, 2005). Stelling and Zijlema (2003) also developed an efficient σ-coordinate for solving Reynolds equations. Recently, Lin (2006) developed a multiple-layer σ-coordinate to solve wave interaction with immersed and floating structures, which cannot be treated by the conventional σ-coordinate models.

The wave models based on meshless particle methods have advanced rapidly in the past few years. Most of these meshless methods are based on SPH method (e.g., Shao and Ji, 2006 for the study of 2D breaking waves; Dalrymple and Rogers, 2006 for the study of 3D breaking waves) or MPS method (e.g., Koshizuka *et al.*, 1998 for simulation of breaking waves) that solves NSEs. Currently, this type of water wave model is still under development and refinement. Figure 5.6 gives an example of the simulation of 2D plunging breaking waves on a plane beach using an SPH wave model (Shao and Ji, 2006).

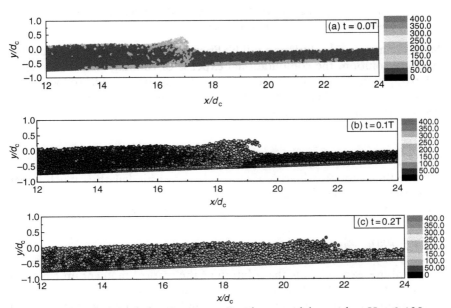

Figure 5.6 Simulation of plunging waves (incident cnoidal wave has $H = 0.128$ m, $h = 0.40$ m, and $T = 5.0$ s) on a plane beach ($s = 1/35$) from an SPH model; color represents eddy viscosity (mm^2/s). (Courtesy of Dr. Songdong Shao of the University of Bradford)

5.2.1.9   Commercial software for NSE solvers applicable for water wave modeling

So far, most of the available commercial software based on NSE solvers are not dedicated to water wave modeling. Instead, the software was designed to be a CFD tool for general fluid problems. There is a lot of CFD software available in the market, namely FLUENT (originally by FLUENT, Inc. and now acquired by ANSYS), FLOW3D (by Flow Science, Inc.), PHOENICS (by CHAM Ltd.), STAR-CD (by CD-adapco), and CFX (by ANSYS Ltd.) Most of the software packages are based on the FVM formulation, though some also have the version of, the FEM formulation. Freitas (1995) compared the numerical results from eight commercial codes (i.e., FLOW3D, FLOTRAN, STAR-CD, NS3, CFD-ACE, FLUENT, CFDS-FLOW3D, and NISA/3D-FLUID) with five benchmark experiments. He concluded that although these codes are in general very promising, they can be inaccurate under certain circumstances, even for laminar flow computation. Thus, these codes must be used with caution and proper validation. Almost all the software packages have the capability of treating turbulent flows with free surface. Their applications to water wave problems, especially breaking wave problems, however, require further validation.

5.2.2   Wave models based on the Navier–Stokes equations with hydrostatic pressure assumption

Since the numerical solution to full NSEs requires the iterative procedure for pressure solution, the computational expense of running an NSE model is relatively high. This limits the application of this type of model to local-scale computation. To efficiently model a large-scale wave and current problem, certain simplifications are necessary.

5.2.2.1   Hydrostatic pressure assumption

For the flow motion whose horizontal characteristic length is much larger than local water depth, e.g., long waves, storm surges, tides, bores, ocean circulation, ocean currents, the hydrostatic pressure assumption is valid, i.e.:

$$p = \rho g(\eta - z) \tag{5.57}$$

The above expression is the solution of the vertical momentum equation in the original NSEs after neglecting all inertial and viscous terms. The assumption of hydrostatic pressure also implies that the free surface has a mild slope and is the single function of the horizontal plane, i.e., $\eta = \eta(x, y)$.

Let us consider a 3D case where $g_1 = g_2 = 0$ and $g_3 = -g$. By substituting (5.57) into the horizontal momentum equations in NSEs, we have:

$$\frac{\partial u}{\partial t}+u\frac{\partial u}{\partial x}+v\frac{\partial u}{\partial y}+w\frac{\partial u}{\partial z}=-g\frac{\partial \eta}{\partial x}+\frac{1}{\rho}\frac{\partial \tau_{xx}}{\partial x}+\frac{1}{\rho}\frac{\partial \tau_{xy}}{\partial y}+\frac{1}{\rho}\frac{\partial \tau_{xz}}{\partial z} \qquad (5.58)$$

$$\frac{\partial v}{\partial t}+u\frac{\partial v}{\partial x}+v\frac{\partial v}{\partial y}+w\frac{\partial v}{\partial z}=-g\frac{\partial \eta}{\partial y}+\frac{1}{\rho}\frac{\partial \tau_{yx}}{\partial x}+\frac{1}{\rho}\frac{\partial \tau_{yy}}{\partial y}+\frac{1}{\rho}\frac{\partial \tau_{yz}}{\partial z} \qquad (5.59)$$

The continuity equation takes the same form as in the original NSEs:

$$\frac{\partial u}{\partial x}+\frac{\partial v}{\partial y}+\frac{\partial w}{\partial z}=0 \qquad (5.60)$$

This equation can be used to calculate w when u and v are solved from (5.48) and (5.49).

If we further integrate the continuity equation in the vertical direction from the bottom to free surface and apply the kinematic free surface boundary condition and the bottom boundary condition, we obtain the so-called height function transport equation (see Section 5.3.1 for the detailed derivation), i.e.:

$$\frac{\partial \eta}{\partial t}+\frac{\partial}{\partial x}\int_{-h}^{\eta} u\mathrm{d}z+\frac{\partial}{\partial y}\int_{-h}^{\eta} v\mathrm{d}z=0 \qquad (5.61)$$

The above equation can be used to track the movement of the free surface displacement of water waves as long as the free surface does not overturn.

Compared with the original NSEs, the governing equations with the assumption of hydrostatic pressure become simpler. Iteration is no longer needed for pressure solution. The passage of time can be used first to advance the horizontal velocities based on the momentum equations, which will later be used to obtain the vertical velocity from the continuity equation. A model solving the above system of PDEs is usually referred to as a quasi-3D model, which is generally used to simulate large-scale currents or very long waves such as tides and tsunami.

Johns and Jefferson (1980) were one of the early explorers of this approach, though their model was only 2D in a vertical plane. Blumberg and Mellor (1987) developed a 3D ocean circulation model to study various ocean circulation and mixing problems. Casulli and Cheng (1992) also presented a 3D model of this type and used it to simulate coastal flooding by tides. This type of model can also be applied to study various turbulent open channel flows (e.g., Li and Yu, 1996).

Most of these quasi-3D models are constructed on the σ-coordinate, which maps the physical space between a wavy free surface and an uneven bottom to a regular space:

$$\sigma=\frac{z+h}{H} \qquad (5.62)$$

where $H=h+\eta$ and σ is the new coordinate in the vertical direction that varies from 0 (when $z=-h$) to 1 (when $z=\eta$). With the introduction of σ-coordinate transformation, the boundary conditions can be applied exactly on the free surface and bottom.

The introduction of the σ-coordinate transformation modifies the original NSEs to the following forms:

$$\frac{\partial u}{\partial x} + \frac{\partial u}{\partial \sigma}\frac{\partial \sigma}{\partial x^*} + \frac{\partial v}{\partial y} + \frac{\partial v}{\partial \sigma}\frac{\partial \sigma}{\partial y^*} + \frac{\partial w}{\partial \sigma}\frac{\partial \sigma}{\partial z^*} = 0 \tag{5.63}$$

$$\frac{\partial u}{\partial t} + u\frac{\partial u}{\partial x} + v\frac{\partial u}{\partial y} + \omega\frac{\partial u}{\partial \sigma} = -g\frac{\partial \eta}{\partial x} + \frac{\partial \tau_{xx}}{\partial x} + \frac{\partial \tau_{xx}}{\partial \sigma}\frac{\partial \sigma}{\partial x^*} + \frac{\partial \tau_{xy}}{\partial y}$$
$$+ \frac{\partial \tau_{xy}}{\partial \sigma}\frac{\partial \sigma}{\partial y^*} + \frac{\partial \tau_{xz}}{\partial \sigma}\frac{\partial \sigma}{\partial z^*} \tag{5.64}$$

$$\frac{\partial v}{\partial t} + u\frac{\partial v}{\partial x} + v\frac{\partial v}{\partial y} + \omega\frac{\partial v}{\partial \sigma} = -g\frac{\partial \eta}{\partial y} + \frac{\partial \tau_{yx}}{\partial x} + \frac{\partial \tau_{yx}}{\partial \sigma}\frac{\partial \sigma}{\partial x^*} + \frac{\partial \tau_{yy}}{\partial y}$$
$$+ \frac{\partial \tau_{yy}}{\partial \sigma}\frac{\partial \sigma}{\partial y^*} + \frac{\partial \tau_{yz}}{\partial \sigma}\frac{\partial \sigma}{\partial z^*} \tag{5.65}$$

where

$$\omega = \frac{D\sigma}{Dt^*} = \frac{\partial \sigma}{\partial t^*} + u\frac{\partial \sigma}{\partial x^*} + v\frac{\partial \sigma}{\partial y^*} + w\frac{\partial \sigma}{\partial z^*}$$

and

$$\frac{\partial \sigma}{\partial t^*} = -\frac{\sigma}{D}\frac{\partial D}{\partial t} \quad \frac{\partial \sigma}{\partial x^*} = \frac{1}{D}\frac{\partial h}{\partial x} - \frac{\sigma}{D}\frac{\partial D}{\partial x} \quad \frac{\partial \sigma}{\partial y^*} = \frac{1}{D}\frac{\partial h}{\partial y} - \frac{\sigma}{D}\frac{\partial D}{\partial y} \quad \frac{\partial \sigma}{\partial z^*} = \frac{1}{D}$$

The above equations are valid by assuming $x = x^*$ and $y = y^*$.

The free surface tracking equation then becomes:

$$\frac{\partial \eta}{\partial t} + \frac{\partial}{\partial x}\left[(\eta + h)\int_0^1 u\,d\sigma\right] + \frac{\partial}{\partial y}\left[(\eta + h)\int_0^1 v\,d\sigma\right] = 0 \tag{5.66}$$

5.2.2.2   *Wave models: open source codes*

The pioneer work by Blumberg and Mellor (1987) has led to the success of POM, which has been widely used in the simulation of large-scale ocean circulation and ocean mixing processes, storm surge (Minato, 1998), and tidal currents (Liu *et al.*, 2005). Currently, a real-time forecasting system of sea surface height (SSH) and currents is under development with the use of POM (Oey *et al.*, 2005; Figure 5.7), which can be used for the prediction of storm surge in coastal water induced by hurricanes.

Another open source code is the COHERENS (COupled Hydrodynamical Ecological model for REgioNal Shelf seas) model that was developed under the Marine Science and Technology Programme (MAST-III) sponsored by the

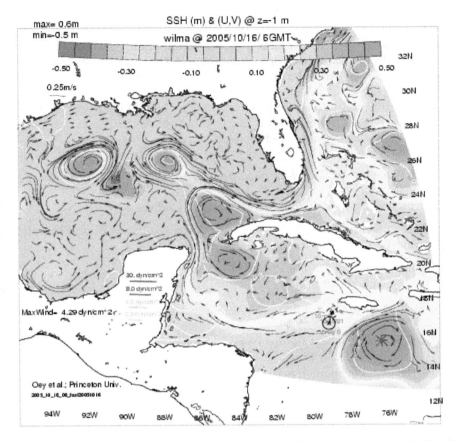

Figure 5.7 Simulation of sea surface height (SSH) and current in Gulf of Mexico during Hurricane Wilma (2005) by using POM. (Courtesy of Dr. Leo Oey, of Princeton University)

European Union. COHERENS is a 3D hydrodynamic multipurpose model for coastal and shelf seas, which is coupled to biological, resuspension, and contaminant models, and resolves mesoscale to seasonal scale processes (Marinov *et al.*, 2006).

Sometimes, the depth-resolved hydrostatic model is also applied to river flows, estuarial flows, or flows in reservoirs or lakes. In these cases, considering that the flow is confined by two lateral sides, the lateral averaging can be performed to further reduce the computational effort. The model CE-QUAL-W2, which was developed by the U.S. Army Corps of Engineers (USACE), is a 2D laterally averaged FD model. This model can be used to model the vertical variation of flow and water quality in rivers, lakes, reservoirs, and estuaries.

5.2.2.3   Wave models: commercial software

Currently, there are two commercial 3D codes based on the σ-coordinate transformation and hydrostatic pressure assumption. One is MIKE 3 developed by the Danish Hydraulic Institute (DHI). MIKE 3 is applicable for simulations of hydrodynamics, water quality, and sediment transport in water bodies (coastal and inland) where 3D effects are important. The other is DELFT3D developed by DELFT Hydraulics. Delft3D provides the integrated modeling environment for hydrodynamics, waves, sediment transport, morphology, water quality, and ecology.

5.2.3   Wave models based on potential flow theory

5.2.3.1   Governing equations and boundary conditions

When the potential flow theory is used to describe a 3D wave problem, the governing equation for the velocity potential is:

$$\nabla^2 \phi = \frac{\partial^2 \phi}{\partial x^2} + \frac{\partial^2 \phi}{\partial y^2} + \frac{\partial^2 \phi}{\partial z^2} = 0 \tag{5.67}$$

in which we have the following definition:

$$\mathbf{u} = u\mathbf{i} + v\mathbf{j} + w\mathbf{k} = -\nabla\phi = -\frac{\partial \phi}{\partial x}\mathbf{i} - \frac{\partial \phi}{\partial y}\mathbf{j} - \frac{\partial \phi}{\partial y}\mathbf{k} \tag{5.68}$$

The general kinematic free surface boundary condition is:

$$\frac{d\mathbf{x}}{dt} = \mathbf{u} \quad \text{on free surface} \tag{5.69}$$

from which the boundary location can be updated. The above general kinematic free surface boundary condition can be reduced to the following form if free surface displacement is a single-value function of the horizontal coordinate:

$$\frac{\partial \eta}{\partial t} = w - u\frac{\partial \eta}{\partial x} - v\frac{\partial \eta}{\partial y} \quad \text{at } z = \eta(x, y, t) \tag{5.70}$$

The corresponding dynamic free surface boundary condition on the free surface is:

$$-\frac{\partial \phi}{\partial t} + \frac{1}{2}\left(u^2 + v^2 + w^2\right) + \frac{p_\eta}{\rho} + gz = 0 \quad \text{at } z = \eta(x, y, t) \tag{5.71}$$

The general boundary condition on a solid surface (e.g., bottom, immersed body surface, and wave paddle surface on the inflow boundary) follows the free-slip condition:

$$-\frac{\partial \phi}{\partial n} = V_n(t) \tag{5.72}$$

where n is the outward unit normal vector from the fluid to the solid surface and V_n is the velocity of the body in the normal direction. For a rigid body and bottom, $V_n = 0$. For a moving body, V_n is a function of time, which can be either prescribed for a forced motion or determined by the dynamic equation of a free body. One application of the above equation is wave generation at the inflow boundary by specifying the wave paddle movement $V_n(t)$ based on wave-maker theory.

Similarly, if the bottom can be represented by a single function of the horizontal coordinate, the generalized bottom boundary condition can be simplified as follows:

$$\frac{\partial h}{\partial t} = -u\frac{\partial h}{\partial x} - v\frac{\partial h}{\partial y} - w \text{ at } z = -h(x, y, t) \tag{5.73}$$

When h is specified as the function of time, waves can be generated from bottom movement, similar to tsunami generation.

To truncate the computational domain in a finite region, the radiation boundary condition must be applied. For normally outgoing waves, the linear radiation boundary condition reads as follows:

$$\frac{\partial \phi}{\partial t} + c\frac{\partial \phi}{\partial n} = 0 \tag{5.74}$$

Theoretically, the potential flow can also be formulated by Euler equations. One can solve Euler equations with the appropriate initial (e.g., irrotational) and boundary conditions for potential flows as well (e.g., Chen and Chiang, 1999). However, this is often more computationally expensive when compared to the numerical solution of the Laplace equation. Furthermore, it is hard to ensure numerically the flow irrotationality near the edge of a structure. Therefore, this approach is not popularly adopted by modelers.

5.2.3.2 *Wave models solving the Laplace equation*

There are essentially two ways of solving the Laplace equation (5.47). One is based on the direct numerical solution to the Laplace equation. Under this category, either FEM or FDM can be used. For example, Wu *et al.* (1998) developed a 3D FEM to solve the Laplace equation for fully nonlinear wave sloshing in a tank. To generate finite elements following the moving free surface, either the σ-coordinate transformation (e.g., Turnbull *et al.*, 2003) or the ALE method (e.g., Nitikitpaiboon and Bathe, 1993) can be used. In contrast, Li and Fleming (1997) solved the Laplace equation in the σ-coordinate with the use of FDM. Frandsen and Borthwick (2003) proposed another σ-transformed FDM model to simulate liquid sloshing in a 2D tank. The major advantage of this approach is that the nonlinear free surface boundary conditions can be imposed directly on the free surface.

The alternative way of solving the Laplace equation is to solve its equivalent form of the boundary integral equation with the use of the BEM (see Section 4.6), using Green's theorem that links volume integration to surface integration. With such treatment, the dimension of the problem is reduced by 1, which may cut down the computational cost in some but not all cases, depending on the balance between the reduced dimension and the resulting full and dense matrix. The main numerical difficulty in the BEM is the application of fully nonlinear kinematic and dynamic free surface boundary conditions (5.70) to (5.71). With the nonlinearity being retained, the matrix being formed will be different at every time step and therefore the computation can be expensive because the large full matrix needs to be inverted at every time step. With all the boundary conditions being linearized, however, the matrix can be formed only once at the beginning of the computation that can drastically reduce the computational time for a dynamic problem. More details of the numerical procedure will be provided in Section 6.2.2.2.

Longuet-Higgins and Cokelet (1976) were the pioneers who successfully developed a 2D BEM model to solve highly nonlinear nearly breaking water waves in deep water. Later, this technique and its variations were adopted in both offshore engineering and coastal engineering. In offshore engineering, the technique is often referred to as the panel method and it was mainly for the analysis of wave loads on structures and structure responses (Isaacson, 1982). In coastal engineering, the development of the BEM is to extend the model to finite depth to study the complex nonlinear wave transformation over changing topography and wave run-up on slopes (e.g., Grilli *et al.*, 1989).

5.2.3.3   Boundary element method wave models: commercial software

Although BEM wave models can be applied to both offshore and coastal engineering, the commercialization of the codes has taken place mainly in offshore engineering. The main reason could be the inherent limitation in extending the BME wave models beyond the breaker line, after which the flow is turbulent and rotational. Besides, the computation of large-scale wave transformation in shallow water, which is often the main objective in coastal engineering, can be too computationally expensive for BEM models. In contrast, in offshore engineering applications, the computation is mainly conducted near an offshore structure, which is usually large and allows potential flow assumption during wave–structure interaction. Furthermore, the local water depth is relatively deep which allows the wave to be linearized in the simulation as long as the wave steepness is not too large.

Currently, most of the commercial BEM codes are developed for the design of offshore oil platforms and ship hulls. The code developers are mainly in two fields, namely the research institutes and universities having offshore industrial partnerships or collaboration, and the ship and offshore classification agencies in different countries (e.g., American Bureau of Shipping (ABS), Bureau Veritas France (BV); Det Norske Veritas (DNV)). Some popular commercial codes in analyzing wave–structure interaction using BEM models include WAMIT (Wave Analysis At Massachusetts Institute of Technology; Lee, 1995), MOSES (MultiOperational Structural Engineering Simulator), UNDA (the Latin word for wave; by the Norwegian University of Science and Technology (NTNU), Norway), Nauticus Hull (by DNV).

Figure 5.8 shows an example of using a BEM model in the simulation of wave interaction with a moving body, which is important in both offshore engineering and ship design. In the simulation, both wave loads on structures and structure responses need to be modeled. In addition, the wave generated by body motion also needs to be simulated. Such a model is applicable as long as the KC number is small so that flow separation is not significant. In Section 6.2, we shall provide more details on using the BEM models in the study of wave–structure interaction.

5.3 Depth-averaged models

All depth-averaged models share the same motivation that the depth-resolved models can be too computationally expensive for large-scale problems. Through depth averaging the computational cost can be greatly reduced. The depth averaging, of course, must be based on some assumptions of the vertical flow structure and therefore the applications of these models are relatively restricted.

In the case where the vertical flow motion is weak and not an important consideration, we can calculate the depth-averaged horizontal velocities only. If we make a further assumption that the horizontal velocities are uniform in the vertical direction, we will end up with the well-known SWEs. The SWEs can deal with the same type of long-wave problems handled by quasi-3D models (Section 5.2.2). Since a SWE model can run much faster than a quasi-3D model, it can be applied to a larger computational domain. To model shorter waves, both vertical acceleration and vertical flow variation must be considered. This leads to the so-called Boussinesq equations. Depending on the way approximation is made in the derivation, there are many types of Boussinesq equations that have different characteristics.

Although Boussinesq equations have been successfully applied to nonlinear dispersive waves in shallow waters, they are not applicable to very deep waters. Alternatively, the MSE, another type of depth-averaged equation with the assumption of a monochromatic linear wave, has been derived and applied in both deep and shallow waters. The derivation of the MSE normally starts from the Laplace equation and its application is restricted to

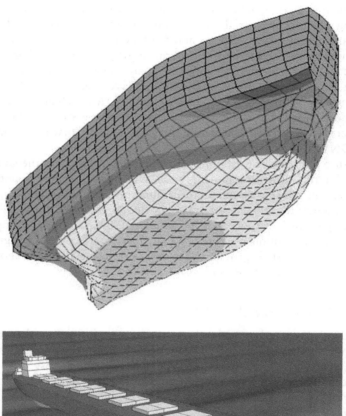

Figure 5.8 Numerical simulation of wave interaction with a moving ship by
Nauticus Hull Wave Load Analysis developed by DNV. (Courtesy
of DNV Software)

periodic water waves only. Because the exact linear dispersion equation is
employed to relate flow property with local water depth, the MSE can be
applied to periodic waves in all water depths.

A wave energy spectral model is not only a depth-averaged, but also
a phase-averaged model. This is different from all other wave models
we have discussed so far, all which are of the phase-resolved type.

The phase-averaging process eliminates the need to resolve the flow change within the wavelength and wave period in the computation. Instead, only the variation of wave height, which is normally slow in both time and space, needs to be resolved. This provides a great advantage in terms of computational efficiency because the mesh size can be so large that it covers many wavelengths. For this reason, this type of model has been used mainly to forecast the wave climate in very large areas (e.g., over the globe). The main limitation of the model is its inability to simulate wave diffraction due to the lack of wave phase information.

5.3.1 Shallow-water equation models

5.3.1.1 Derivation of the shallow-water equations

Let us start with incompressible NSEs:

$$\frac{\partial u_i}{\partial x_i} = 0 \tag{5.75}$$

$$\frac{\partial u_i}{\partial t} + u_j \frac{\partial u_i}{\partial x_j} = -\frac{1}{\rho}\frac{\partial p}{\partial x_i} + g_i + \frac{1}{\rho}\frac{\partial}{\partial x_j}\tau_{ij} \tag{5.76}$$

Taking the vertical integration of the continuity equation from the bottom to the free surface, we have:

$$\int_{-h}^{\eta}\left(\frac{\partial u}{\partial x} + \frac{\partial v}{\partial y} + \frac{\partial w}{\partial z}\right)dz = \int_{-h}^{\eta}\frac{\partial u}{\partial x}dz + \int_{-h}^{\eta}\frac{\partial v}{\partial y}dz + w(x, y, \eta)$$
$$- w(x, y, -h) = 0 \tag{5.77}$$

The above equation can be rewritten as follows:

$$\frac{\partial}{\partial x}\int_{-h}^{\eta}u\,dz - u(x, y, \eta)\frac{\partial \eta}{\partial x} - u(x, y, -h)\frac{\partial h}{\partial x} + \frac{\partial}{\partial y}\int_{-h}^{\eta}v\,dz - v(x, y, \eta)\frac{\partial \eta}{\partial y}$$
$$- v(x, y, -h)\frac{\partial h}{\partial y} + w(x, y, \eta) - w(x, y, -h) = 0 \tag{5.78}$$

In arriving at the above expression, the Leibniz rule of integration was used, i.e.:

$$\frac{\partial}{\partial x}\int_{\alpha(x)}^{\beta(x)}Q(x, y)dy = \int_{\alpha(x)}^{\beta(x)}\frac{\partial}{\partial x}Q(x, y)dy + Q(x, \beta(x))\frac{\partial \beta(x)}{\partial x}$$
$$- Q(x, \alpha(x))\frac{\partial \alpha(x)}{\partial x} \tag{5.79}$$

Recall the kinematic free surface boundary condition:

$$\frac{\partial \eta}{\partial t} + u(x, y, \eta)\frac{\partial \eta}{\partial x} + v(x, y, \eta)\frac{\partial \eta}{\partial y} = w(x, y, \eta) \tag{5.80}$$

and the bottom boundary condition:

$$w(x, y, -h) = -\frac{\partial h}{\partial t} - u(x, y, -h)\frac{\partial h}{\partial x} - v(x, y, -h)\frac{\partial h}{\partial y} \tag{5.81}$$

By substituting (5.80) and (5.81) into (5.78) and defining:

$$U = \frac{1}{H}\int_{-h}^{\eta} u \, dz \quad \text{and} \quad V = \frac{1}{H}\int_{-h}^{\eta} v \, dz \tag{5.82}$$

where $H = \eta + h$ is the total water depth, we have the final form of the integrated continuity equation, i.e.:

$$\frac{\partial H}{\partial t} + \frac{\partial}{\partial x}(UH) + \frac{\partial}{\partial y}(VH) = 0 \tag{5.83}$$

This equation is also called the height function transport equation. It can be used to track the movement of the free surface. The above equation can be rewritten as follows:

$$\frac{\partial \eta}{\partial t} + \frac{\partial}{\partial x}(UH) + \frac{\partial}{\partial y}(VH) = -\frac{\partial h}{\partial t} \tag{5.84}$$

In the case when the still water depth is time-dependent (e.g., in the event of an underwater earthquake), the RHS becomes a source of wave generation.

To derive the vertically integrated momentum equation, let us first rewrite the momentum equation in the x-direction in the conservative form with the application of continuity equation:

$$\frac{\partial u}{\partial t} + \frac{\partial(u^2)}{\partial x} + \frac{\partial(uv)}{\partial y} + \frac{\partial(uw)}{\partial z} = -\frac{1}{\rho}\frac{\partial p}{\partial x} + \frac{1}{\rho}\left(\frac{\partial \tau_{xx}}{\partial x} + \frac{\partial \tau_{xy}}{\partial y} + \frac{\partial \tau_{xz}}{\partial z}\right) \tag{5.85}$$

Assuming hydrostatic pressure and constant density in the vertical direction, i.e., $p = p_a + \rho g(z + \eta)$, we have:

$$-\frac{1}{\rho}\frac{\partial p}{\partial x} = -\frac{1}{\rho}\frac{\partial p_a}{\partial x} - g\frac{\partial \eta}{\partial x} - \frac{\eta g}{\rho}\frac{\partial \rho}{\partial x} \tag{5.86}$$

The first term can be significant when the current and the surface abnormity are driven by an atmospheric pressure difference (e.g., due to storm surge, hurricane, typhoon). The third term can be important for a large-scale ocean circulation driven by density difference. Taking the vertical integration from the bottom to free surface for (5.85) and applying the kinematic free surface and bottom boundary conditions and Leibniz's rule, we have:

$$
\frac{\partial}{\partial t}(UH) + \frac{\partial}{\partial x}(U^2 H) + \frac{\partial}{\partial y}(UVH) + \frac{\partial}{\partial x}\int_{-h}^{\eta}(u-U)^2\,\mathrm{d}z
$$

$$
+ \frac{\partial}{\partial y}\int_{-h}^{\eta}(u-U)(v-V)\,\mathrm{d}z
$$

$$
= -\frac{H}{\rho}\frac{\partial p_a}{\partial x} - gH\frac{\partial \eta}{\partial x} - \frac{\eta H g}{\rho}\frac{\partial \rho}{\partial x} + \frac{1}{\rho}\frac{\partial}{\partial x}\int_{-h}^{\eta}\tau_{xx}\,\mathrm{d}z + \frac{1}{\rho}\frac{\partial}{\partial y}\int_{-h}^{\eta}\tau_{xy}\,\mathrm{d}z
$$

$$
\text{(5.87)}
$$

$$
+ \frac{1}{\rho}\left[-\tau_{xx}(\eta)\frac{\partial \eta}{\partial x} - \tau_{xy}(\eta)\frac{\partial \eta}{\partial y} + \tau_{xz}(\eta)\right]
$$

$$
- \frac{1}{\rho}\left[\tau_{xx}(-h)\frac{\partial h}{\partial x} + \tau_{xy}(-h)\frac{\partial h}{\partial y} + \tau_{xz}(-h)\right]
$$

Depending on how the flow and stress terms are assumed, there are a few variations of SWEs as follows:

Version 1 (Uniform velocity and stresses in the vertical direction): If we assume that the flow is uniformly distributed in the vertical direction (and thus $u = U$, $v = V$) and τ_{xx} and τ_{xy} do not depend on z (thus $\partial\tau_{xx}/\partial x$ and $\partial\tau_{xy}/\partial y$ are constant in the z-direction), equation (5.87) can be simplified as follows:

$$
\frac{\partial}{\partial t}(UH) + \frac{\partial}{\partial x}(U^2 H) + \frac{\partial}{\partial y}(UVH) \tag{5.88}
$$

$$
= -\frac{H}{\rho}\frac{\partial p_a}{\partial x} - gH\frac{\partial \eta}{\partial x} - \frac{\eta H g}{\rho}\frac{\partial \rho}{\partial x} + \frac{H}{\rho}\frac{\partial \tau_{xx}}{\partial x} + \frac{H}{\rho}\frac{\partial \tau_{xy}}{\partial y} + \frac{1}{\rho}\tau_{xz}(\eta) - \frac{1}{\rho}\tau_{xz}(-h)
$$

where $\tau_{xz}(\eta)$ and $\tau_{xz}(-h)$ are surface (e.g., wind) and bottom (e.g., friction) shear stresses, which come into the equation as the source functions. Similarly, the momentum equation in the y-direction is:

$$
\frac{\partial}{\partial t}(VH) + \frac{\partial}{\partial x}(UVH) + \frac{\partial}{\partial y}(V^2 H) = -\frac{H}{\rho}\frac{\partial p_a}{\partial y} - gH\frac{\partial \eta}{\partial y} - \frac{\eta H g}{\rho}\frac{\partial \rho}{\partial y}
$$

$$
+ \frac{H}{\rho}\frac{\partial \tau_{yx}}{\partial x} + \frac{H}{\rho}\frac{\partial \tau_{yy}}{\partial y} + \frac{1}{\rho}\tau_{yz}(\eta) - \frac{1}{\rho}\tau_{yz}(-h)
$$

$$
\text{(5.89)}
$$

The above equations have been used by Dean and Dalrymple (1991).

Version 2 (Varying stresses in vertical direction): The SWEs have another form if we do not make the assumption that τ_{xx}, τ_{xy}, and τ_{yy} are constant, i.e.:

$$\frac{\partial}{\partial t}(UH) + \frac{\partial}{\partial x}(U^2 H) + \frac{\partial}{\partial y}(UVH)$$

$$= -\frac{H}{\rho}\frac{\partial p_a}{\partial x} - gH\frac{\partial \eta}{\partial x} - \frac{\eta Hg}{\rho}\frac{\partial \rho}{\partial x} + \frac{1}{\rho}\tau_{xz}(\eta) - \frac{1}{\rho}\tau_{xz}(-h)$$

$$+ \frac{1}{\rho}\frac{\partial}{\partial x}\int_{-h}^{\eta}\tau_{xx}dz + \frac{1}{\rho}\frac{\partial}{\partial y}\int_{-h}^{\eta}\tau_{xy}dz - \frac{1}{\rho}\tau_{xx}(\eta)\frac{\partial \eta}{\partial x} - \frac{1}{\rho}\tau_{xy}(\eta)\frac{\partial \eta}{\partial y}$$

$$(5.90)$$

$$\frac{\partial}{\partial t}(VH) + \frac{\partial}{\partial x}(UVH) + \frac{\partial}{\partial y}(V^2 H)$$

$$= -\frac{H}{\rho}\frac{\partial p_a}{\partial y} - gH\frac{\partial \eta}{\partial y} - \frac{\eta Hg}{\rho}\frac{\partial \rho}{\partial x} + \frac{1}{\rho}\tau_{yz}(\eta) - \frac{1}{\rho}\tau_{yz}(-h)$$

$$+ \frac{1}{\rho}\frac{\partial}{\partial x}\int_{-h}^{\eta}\tau_{yx}dz + \frac{1}{\rho}\frac{\partial}{\partial y}\int_{-h}^{\eta}\tau_{yy}dz - \frac{1}{\rho}\tau_{yx}(\eta)\frac{\partial \eta}{\partial x} - \frac{1}{\rho}\tau_{yy}(\eta)\frac{\partial \eta}{\partial y} \quad (5.91)$$

In the above equations, the shear stresses τ_{xx}, τ_{xy}, and τ_{yy} on the bottom are all zero because the slip velocities are zero there.

Version 3 (Conversion of stresses following tangents on a sloping free surface and bottom): If we establish the new coordinate (x', y', z') whose (x', y') plane is aligned with the tangential plane of the free surface and z' being normal to the free surface pointing upward, the coordinate transformation can be established between the original coordinate and the new coordinate in the following general form:

$$\begin{pmatrix} x' \\ y' \\ z' \end{pmatrix} = \begin{pmatrix} l_1 & l_2 & l_3 \\ m_1 & m_2 & m_3 \\ n_1 & n_2 & n_3 \end{pmatrix} \begin{pmatrix} x \\ y \\ z \end{pmatrix} \quad (5.92)$$

where l, m, and n are determined by the rotational angles in the new coordinate. This will lead to the following relationship between the stress tensors in the two coordinates:

$$\begin{pmatrix} \tau_{xx}' & \tau_{xy}' & \tau_{xz}' \\ \tau_{yx}' & \tau_{yy}' & \tau_{yz}' \\ \tau_{zx}' & \tau_{zy}' & \tau_{zz}' \end{pmatrix} = \begin{pmatrix} l_1 & l_2 & l_3 \\ m_1 & m_2 & m_3 \\ n_1 & n_2 & n_3 \end{pmatrix} \begin{pmatrix} \tau_{xx} & \tau_{xy} & \tau_{xz} \\ \tau_{yx} & \tau_{yy} & \tau_{yz} \\ \tau_{zx} & \tau_{zy} & \tau_{zz} \end{pmatrix} \quad (5.93)$$

To make the mathematics easier to follow, we shall use a 2D problem in the (x, z) plane as an example where the new coordinate (x', z') rotates in the counterclockwise direction at an angle of θ to follow the surface, from which the following relationship is satisfied:

$$\begin{pmatrix} \tau_{xx}' & \tau_{xz}' \\ \tau_{zx}' & \tau_{zz}' \end{pmatrix} = \begin{pmatrix} \cos\theta & \sin\theta \\ -\sin\theta & \cos\theta \end{pmatrix} \begin{pmatrix} \tau_{xx} & \tau_{xz} \\ \tau_{zx} & \tau_{zz} \end{pmatrix} \quad (5.94)$$

from which we have:

$$\tau_{xz}{}' = \tau_{zx}{}' = -\sin\theta\tau_{xx} + \cos\theta\tau_{zx} = -\sin\theta\tau_{xx} + \cos\theta\tau_{xz} \qquad (5.95)$$

If we further assume that the free surface steepness is small, we would have $\sin\theta \approx \tan\theta = \partial\eta/\partial x$ and $\cos\theta \approx 1$, which leads to the following:

$$\tau_{xz}{}'(\eta) = -\frac{\partial\eta}{\partial x}\tau_{xx}(\eta) + \tau_{xz}(\eta) \qquad (5.96)$$

For a 3D case, with the same assumption of small surface steepness, we would have:

$$\tau_{xz}{}'(\eta) = -\frac{\partial\eta}{\partial x}\tau_{xx}(\eta) - \frac{\partial\eta}{\partial y}\tau_{xy}(\eta) + \tau_{xz}(\eta) \qquad (5.97)$$

Here $\tau_{xz}{}'$ should be regarded as the stress following the free surface orientation such as wind stress. Similarly, we could have the following if the bottom slope is mild:

$$\tau_{xz}{}'(-h) = \frac{\partial h}{\partial x}\tau_{xx}(-h) + \frac{\partial h}{\partial y}\tau_{xy}(-h) + \tau_{xz}(-h) \qquad (5.98)$$

where $\tau_{xz}{}'(-h)$ can be regarded as the bottom shear stress along the sloping surface.

Substituting the above into the original SWE, we have:

$$\begin{aligned}
\frac{\partial}{\partial t}(UH) + \frac{\partial}{\partial x}(U^2 H) + \frac{\partial}{\partial y}(UVH) = &-\frac{H}{\rho}\frac{\partial p_a}{\partial x} - gH\frac{\partial\eta}{\partial x} - \frac{\eta Hg}{\rho}\frac{\partial\rho}{\partial x} \\
&+ \frac{1}{\rho}\frac{\partial}{\partial x}\int_{-h}^{\eta}\tau_{xx}dz + \frac{1}{\rho}\frac{\partial}{\partial y}\int_{-h}^{\eta}\tau_{xy}\,dz \\
&+ \frac{\tau_{xz}{}'(\eta)}{\rho} - \frac{\tau_{xz}{}'(-h)}{\rho}
\end{aligned} \qquad (5.99)$$

$$\begin{aligned}
\frac{\partial}{\partial t}(VH) + \frac{\partial}{\partial x}(UVH) + \frac{\partial}{\partial y}(V^2 H) = &-\frac{H}{\rho}\frac{\partial p_a}{\partial y} - gH\frac{\partial\eta}{\partial y} - \frac{\eta Hg}{\rho}\frac{\partial\rho}{\partial x} \\
&+ \frac{1}{\rho}\frac{\partial}{\partial x}\int_{-h}^{\eta}\tau_{yx}dz + \frac{1}{\rho}\frac{\partial}{\partial y}\int_{-h}^{\eta}\tau_{yy}\,dz \\
&+ \frac{\tau_{yz}{}'(\eta)}{\rho} - \frac{\tau_{yz}{}'(-h)}{\rho}
\end{aligned} \qquad (5.100)$$

The above formulas have been adopted by Kuipers and Vreugdenhil (1973).

Version 4 (On a spherical coordinate with Coriolis forces): For large-scale computations, the SWEs must be constructed on spherical coordinates by including Coriolis forces, i.e.:

$$\frac{\partial H}{\partial t} + \frac{1}{R\cos\phi}\left[\frac{\partial}{\partial\lambda}(UH) + \frac{\partial}{\partial\phi}(\cos\phi VH)\right] = 0 \qquad (5.101)$$

$$\frac{\partial}{\partial t}(UH) + \frac{1}{R\cos\phi}\frac{\partial}{\partial\lambda}(U^2H) + \frac{1}{R}\frac{\partial}{\partial\phi}(UVH) - fHV = -\frac{gH}{R\cos\phi}\frac{\partial\eta}{\partial\lambda}$$
$$(5.102)$$

$$\frac{\partial}{\partial t}(VH) + \frac{1}{R\cos\phi}\frac{\partial}{\partial\lambda}(UVH) + \frac{1}{R}\frac{\partial}{\partial\phi}(V^2H) + fUH = -\frac{gH}{R}\frac{\partial\eta}{\partial\phi} \quad (5.103)$$

where (λ, ϕ) represent the longitude and latitude of Earth, R is the radius of Earth, and f is the Coriolis force parameter due to Earth's rotation. The stress terms are neglected for simplicity. With the introduction of the Coriolis force, there exists another restoring force from the Coriolis effect, besides the conventional gravitational force. The Coriolis restoring force will result in the so-called Rossby wave, which is one type of inertial wave that exists in both atmosphere and ocean. The Rossby wave is also called planetary wave that is different from an ocean surface wind wave in that its wave energy is transmitted through the bulk of the fluid rather than on the surface only. Readers are referred to Dickinson (1978) for more details on Rossby waves.

1D lateral-averaged SWEs (Saint-Venant equations): To apply the SWEs to rivers, the equations can be further simplified by taking the lateral integration across the river section by assuming that the flow is dominated by longitudinal flow. By doing so, the equations are reduced to 1D equations along the river flow direction. The cross-sectional information and river direction will be included in the equation by lateral integration. By neglecting the centrifugal force, we have the well-known Saint-Venant equations in vector form:

$$\frac{\partial \mathbf{U}}{\partial t} + \frac{\partial \mathbf{F}}{\partial x} + \mathbf{S} = 0 \qquad (5.104)$$

where x represents the river flow direction and:

$$\mathbf{U} = \begin{pmatrix} A \\ VA \end{pmatrix} \quad \mathbf{F} = \begin{pmatrix} VA \\ V^2A + gA\bar{y} \end{pmatrix} \quad \mathbf{S} = \begin{pmatrix} -q_l \\ -gA(S_o - S_f) - V_x q_l \end{pmatrix} \qquad (5.105)$$

with V being the mean velocity, A the cross-sectional area, $\bar{y}$ the distance from the water surface to the centroid of the area, q_l the lateral flow, V_x the component of the velocity of the lateral flow in the x-direction, S_o the slope of the river bed, and S_f the slope of the energy line. The main difference of (5.104) from the 1D SWEs is the inclusion of the lateral flow, the cross-sectional area, and the slope of the energy line due to the spatial variation of the mean water level. It is not difficult to prove that for a rectangular channel (thus $A = b + \eta$ for a unit length of the channel) without the lateral flow (and thus $q_l = 0$), equation (5.104) can be reduced to the nonlinear SWEs (5.83) and (5.88) by realizing $S_o = \partial h/\partial x$. Besides, we have also neglected the atmospheric pressure and fluid density variation in the x-direction and viscous effect in (5.88) and assumed that $S_f = -\tau_{xz}(-h)/\rho g H = -\tau_b/\rho g H$. The 1D SWE models are often used in modeling river and reservoir hydraulics (e.g., HEC-2 developed by the USACE).

Conservative and nonconservative SWEs for inviscid fluids: The momentum equations can be simplified as follows if the viscous effects and surface forcing terms are neglected:

$$\frac{\partial}{\partial t}(UH) + \frac{\partial}{\partial x}(U^2 H) + \frac{\partial}{\partial y}(UVH) = -gH\frac{\partial \eta}{\partial x} \qquad (5.106)$$

$$\frac{\partial}{\partial t}(VH) + \frac{\partial}{\partial x}(UVH) + \frac{\partial}{\partial y}(V^2 H) = -gH\frac{\partial \eta}{\partial y} \qquad (5.107)$$

In the above equations, the convection terms are written in conservative form. Making use of the continuity equation (5.83), the equivalent SWEs in nonconservative form can be obtained:

$$\frac{\partial U}{\partial t} + U\frac{\partial U}{\partial x} + V\frac{\partial U}{\partial y} = -g\frac{\partial \eta}{\partial x} \qquad (5.108)$$

$$\frac{\partial V}{\partial t} + U\frac{\partial V}{\partial x} + V\frac{\partial V}{\partial y} = -g\frac{\partial \eta}{\partial y} \qquad (5.109)$$

Linear long-wave equation and linear wave equation: When the wave amplitude is small, i.e., $\eta << h$, and thus the wave nonlinearity is negligible, the SWEs in conservative form can be further simplified to be the linear equations:

$$\frac{\partial}{\partial t}(UH) \approx -gH\frac{\partial \eta}{\partial x} \approx -gh\frac{\partial \eta}{\partial x} \qquad (5.110)$$

$$\frac{\partial}{\partial t}(VH) \approx -gH\frac{\partial \eta}{\partial y} \approx -gh\frac{\partial \eta}{\partial y} \qquad (5.111)$$

By taking the time derivative of the continuity equation and substituting the above momentum equations into it, we are able to merge the three equations into one equation:

$$\frac{\partial^2 H}{\partial t^2} + \frac{\partial}{\partial x}\left(-gh\frac{\partial \eta}{\partial x}\right) + \frac{\partial}{\partial y}\left(-gh\frac{\partial \eta}{\partial y}\right) = 0 \text{ or } \frac{\partial^2 H}{\partial t^2} - \nabla \cdot (c^2 \nabla \eta) = 0$$
$$(5.112)$$

where $c = \sqrt{gh}$. If we first assume that the bottom is fixed, i.e.:

$$\frac{\partial^2 H}{\partial t^2} = \frac{\partial^2 \eta}{\partial t^2}$$

the above equation becomes:

$$\frac{\partial^2 \eta}{\partial t^2} - \nabla \cdot (c^2 \nabla \eta) = 0 \qquad (5.113)$$

For a monochromatic wave train, we can define $\eta(x, y, t) = F(x, y)e^{-i\sigma t}$, where $F(x, y)e^{-i\sigma t}$ is the complex wave amplitude with its modulus representing real wave amplitude and argue the phase angle information. By substituting the definition into (5.113), we would have the steady-state wave equation described by Helmholtz equation:

$$\nabla \cdot \left(c^2 \nabla F\right) + \sigma^2 F = 0 \tag{5.114}$$

In contrast, if the bottom slope is comparably smaller than the free surface steepness, i.e., $|\nabla h| \ll |\nabla \eta|$, the time-dependent equation (5.113) can be further reduced to:

$$\frac{\partial^2 \eta}{\partial t^2} = c^2 \nabla^2 \eta \tag{5.115}$$

This is called linear wave equation, which describes wave propagation on a 2D horizontal plane. For 1D problems, the equation can be split into two convection equations for waves propagating in two opposite directions:

$$\frac{\partial \eta}{\partial t} + c\frac{\partial \eta}{\partial x} = 0$$

$$\frac{\partial \eta}{\partial t} - c\frac{\partial \eta}{\partial x} = 0 \tag{5.116}$$

The two equations are readily solved by the method of characteristics (Erbes, 1993). For 2D SWEs, the numerical methods were discussed by Vreugdenhil (1994) and Durran (1999).

5.3.1.2 Boundary conditions

Since SWEs are depth-integrated equations, only the lateral boundary conditions are needed in computations. The most common lateral boundary conditions include the following:

1. Reflected boundary condition where the wave energy is completely reflected back. This happens on a smooth, rigid, and impermeable vertical wall. Mathematically, it is expressed as a mirror plane, i.e.:

$$U_n = 0 \text{ and } \partial\eta/\partial n = 0 \tag{5.117}$$

2. Outflow boundary condition where waves or currents are allowed to leave the domain freely.
    A. For waves, the radiation boundary condition can be specified:

$$\frac{\partial \phi}{\partial t} + c\frac{\partial \phi}{\partial n} = 0 \tag{5.118}$$

Here ϕ is the variable of the mean flow and free surface displacement and n is the outward unit normal vector.

    B. For flows, different treatments will be used for subcritical and supercritical flows. For subcritical flows, the downstream water level must be provided, whereas for supercritical flows the gradient of total depth must be specified.

3. Inflow boundary condition where the wave or the current is generated:

    A. To generate tides from all open sea boundaries, the time history of water elevation must be provided. The information can be extracted from either the tidal table or gauge record.

    B. To generate long waves, both the flux and the water level need to be provided based on wave theory.

    C. To generate subcritical open channel flows, the flux and the gradient of the wave level need to be provided.

    D. To generate supercritical open channel flows, both the flux and the water level need to be provided.

4. Moving boundary when wave run-up or flow inundation, during which the water wet line changes with time, is modeled. There have been various ways of treating the moving boundary condition in different SWE models, which are summarized below:

    A. In principle, the shoreline-fitted grid system can be generated dynamically at every computational step so that the computation will be carried out on the fluid domain only. Such a technique, however, was rarely used in practical modeling due to the high computational cost.

    B. Tao (1983) proposed the slot technique, in which each computational cell on the possible run-up and run-down area has a narrow and deep slot to allow inside water going up and down, in order to simulate wave run-up and run-down on a beach. This method is like replacing the impermeable beach with a porous one, and thus the water level on each cell can possibly go below the beach surface during the computation. Strictly speaking, Tao's method does not treat a moving boundary because all cells are treated as interior cells and calculated. The advantage of the slot technique is that there is no need to track the wet line of the water front. However, the mass used to fill the slot may cause the reduction of simulated wave run-up. Later, Kennedy *et al.* (2000) tried to improve the method for better mass conservation in their Boussinesq model.

    C. An alternative way is to track the wet line at each time step on a fixed grid system. Certain criteria are used to justify whether the wet line will be moved to the neighborhood cells and, if yes, how much the mass flux is. Examples include Cho (1995) in his SWE model and Lynett *et al.* (2002) in their Boussinesq model (in fact, all the boundary conditions discussed above are equally applicable to Boussinesq models).

5.3.1.3   Modeling of wave and current generation

Waves can be generated in various ways. The most straightforward way of wave and current generation is by specifying the mean flux and water level at the inflow boundary, as discussed in Section 5.3.1.2. By applying a nonuniform wind shear stress on water surface, long waves and currents can also be generated. Similarly, by applying the pressure gradient on water surface, large-scale water surface abnormities can be generated and modeled. Waves can also be generated from the bottom by the change of bathymetry. In an SWE model, this can be easily achieved by making still water depth the function of time, i.e., $h = h(x, y, t)$. Waves will be generated when h changes with time, which can be used to simulate tsunamis generated by an underwater earthquake and landslide (e.g., Heinrich *et al.*, 2001).

An innovative numerical technique was proposed by Larsen and Dancy (1983) to generate waves internally inside the computational domain by specifying a line source function. The idea was practiced in their Boussinesq model to generate waves and it can be equally applied to all SWE models. The source function is added to either the continuity equation or the momentum equation to simulate the wave generation process by the change of mass or momentum inside the domain. When the mass change is adopted, a source function will be added in the continuity equation. This is similar to tsunami generation by the change of $h(x, y, t)$ in SWEs [see (5.84)]. In contrast, when momentum (or energy) change is considered, source functions are added in the momentum equations. This is similar to wave generation by varying the wind speed/pressure during which the stress/pressure gradients are applied in the momentum equations as the forcing terms [see (5.87)].

In using the source function technique, the reflected waves will not interact with the wave generation mechanism. By generating waves internally, all lateral boundaries can become open boundaries that deal with outgoing waves only. The idea of wave generation inside the domain can be implemented in other types of wave models including the NSE solver (Lin and Liu, 1999b), the BEM model (Grilli and Horrillo, 1997), the Boussinesq model (Lee *et al.*, 2001), and the MSE model (Madsen and Larsen, 1987).

5.3.1.4   Turbulence modeling

Most of the SWE models were furnished with some turbulence closure models for the depth-averaged turbulence-induced Reynolds stresses. These turbulence closure models range from the simple mixing-length model to the more advanced $k-\varepsilon$ model. Now let us use the example of the depth-averaged $k-\varepsilon$ model to show how the depth-integrated Reynolds stress:

$$\int_{-h}^{\eta} R_{ij} \, \mathrm{d}z$$

can be assessed. Recall the original $k-\varepsilon$ transport equations:

$$\frac{\partial k}{\partial t} + u_j \frac{\partial k}{\partial x_j} = \frac{\partial}{\partial x_j}\left(\frac{\nu_t}{\sigma_k}\frac{\partial k}{\partial x_j}\right) + P - \varepsilon \qquad (5.119)$$

$$\frac{\partial \varepsilon}{\partial t} + u_j \frac{\partial \varepsilon}{\partial x_j} = \frac{\partial}{\partial x_j}\left(\frac{\nu_t}{\sigma_\varepsilon}\frac{\partial \varepsilon}{\partial x_j}\right) + C_{1\varepsilon}\frac{\varepsilon}{k}P - C_{2\varepsilon}\frac{\varepsilon^2}{k} \qquad (5.120)$$

where

$$P = \nu_t \left(\frac{\partial u_i}{\partial x_j} + \frac{\partial u_j}{\partial x_i}\right)\frac{\partial u_i}{\partial x_j} \text{ and } \nu_t = C_\mu \frac{k^2}{\varepsilon} \text{ with } C_\mu = 0.09, \sigma_k = 1.0,$$

$$\sigma_\varepsilon = 1.3, C_{1\varepsilon} = 1.44, \text{ and } C_{2\varepsilon} = 1.92,$$

Taking the depth average of (5.119) and (5.120), we have:

$$\frac{\partial\left(H\hat{k}\right)}{\partial t} + \frac{\partial\left(HU\hat{k}\right)}{\partial x} + \frac{\partial\left(HV\hat{k}\right)}{\partial y} = \frac{\partial}{\partial x}\left[\frac{\hat{\nu}_t}{\sigma_k}\frac{\partial\left(H\hat{k}\right)}{\partial x}\right] + \frac{\partial}{\partial y}\left[\frac{\hat{\nu}_t}{\sigma_k}\frac{\partial\left(H\hat{k}\right)}{\partial y}\right]$$

$$+ P_h + P_{hV} - \hat{\varepsilon}H \qquad (5.121)$$

$$\frac{\partial\left(H\hat{\varepsilon}\right)}{\partial t} + \frac{\partial\left(HU\hat{\varepsilon}\right)}{\partial x} + \frac{\partial\left(HV\hat{\varepsilon}\right)}{\partial y} = \frac{\partial}{\partial x}\left[\frac{\hat{\nu}_t}{\sigma_\varepsilon}\frac{\partial\left(H\hat{\varepsilon}\right)}{\partial x}\right] + \frac{\partial}{\partial y}\left[\frac{\hat{\nu}_t}{\sigma_\varepsilon}\frac{\partial\left(H\hat{\varepsilon}\right)}{\partial y}\right]$$

$$+ C_{1\varepsilon}\frac{\hat{\varepsilon}}{\hat{k}}P_h + P_{\varepsilon V} - C_{2\varepsilon}\frac{\hat{\varepsilon}^2}{\hat{k}}H \qquad (5.122)$$

where $\hat{k}$ and $\hat{\varepsilon}$ are the depth-averaged turbulence kinetic energy and dissipation rate with:

$$\hat{\nu}_t = C_\mu \frac{\hat{k}^2}{\hat{\varepsilon}} \qquad (5.123)$$

and:

$$P_h = \frac{\hat{\nu}_t}{H}\left\{2\left[\frac{\partial\left(HU\right)}{\partial x}\right]^2 + 2\left[\frac{\partial\left(HV\right)}{\partial y}\right]^2 + \left[\frac{\partial\left(HU\right)}{\partial y} + \frac{\partial\left(HV\right)}{\partial x}\right]^2\right\} \qquad (5.124)$$

which represents the turbulence production by the mean horizontal velocity gradients, and:

$$P_{kV} = c_k u_*^3 \text{ and } P_{\varepsilon V} = c_\varepsilon \frac{u_*^4}{H}, \qquad (5.125)$$

which represents the turbulence production from the turbulent bottom boundary layer. In the above equations, u_* is the friction velocity [e.g., $\tau_b = \rho u_*^2 = \rho c_f \left(U^2 + V^2 \right)$] and $c_k = 1/\sqrt{c_f}$ and $c_\varepsilon = 3.6 \left(C_{2\varepsilon}/c_f^{3/4} \right) \sqrt{C_\mu}$ (Rastogi and Rodi, 1978), with c_f being the friction coefficient that can be determined from either the Manning equation or Chezy formula.

The depth-integrated turbulent stresses can be modeled as follows (Rastogi and Rodi, 1978):

$$\int_{-z_b}^{\eta} R_{ij}\, \mathrm{d}z = \rho H \hat{\nu}_t \left(\frac{\partial U_i}{\partial x_j} + \frac{\partial U_j}{\partial x_i} \right) - \frac{2}{3}\rho H \hat{k} \delta_{ij} \tag{5.126}$$

5.3.1.5   Numerical methods

The numerical solution to the SWEs can be advanced in time from an initial condition. In most of the SWE models, either the FDM or FEM is used. In recent years, the meshless method was extended to solve SWEs. For example, Hon *et al.* (1999) proposed a meshless particle method by using the radial basis function (RBF) to solve SWEs and applied the model to the simulation of a typhoon-induced current. Ata and Soulaimani (2005) extended the SPH technique to solve SWEs.

5.3.1.6   The shallow-water equation models and their applications

Tides: Tides are probably the longest surface water waves on Earth. A tide at a particular location can be estimated from the relative position of Earth, the moon, and the sun as Earth rotates; Earth revolves around the sun, and the moon revolves around Earth. The tidal energy is distributed into many frequencies (or constituents, harmonics). The height of the tide can therefore be represented by a general formula (Schureman, 1958):

$$H_{\text{tide}} = H_0 + \sum H \cos\left[\sigma t - \kappa\right] \tag{5.127}$$

where H_0 is the mean water level above the datum selected, H is the mean amplitude of the particular constituent under the influence of the moon or the sun, σ is the angular frequency of the constituent, and t is the time reckoned from some initial epoch κ.

Tidal analysis is basically the determination of the constants in (5.127) for each tidal constituent. The most important tidal constituents include the semidiurnal constituents generated by the moon (M2) and the sun (S2) with the periods of 12.42 h and 12 h, respectively, and the daily inequality (or diurnal) constituents with the periods of 24.48 h and 24 h, respectively. Besides, there are many other constituents taking account of other factors (Kamphuis, 2000: 117).

The tidal constituent by the moon is the most influential factor to affect tidal height. Two bulges of water on the surface of Earth, one facing the moon and the other on the opposite side, will be drawn by gravitational attraction as the moon revolves around Earth. At a particular location, it will experience the *flood tide* (or incoming tide), during which the water level increases from the low level to the high level, and *ebb tide* (or receding tide, falling tide), during which the water level reduces from its high level to a low level, nearly twice a day. The highest water level is called high tide and the lowest water level is called low tide. Considering the combined effect from the gravitational attraction from the sun, the magnitude of the tidal level will increase from the new moon, when the sun has the maximum counterbalancing effect on the moon-induced tide, to the full moon, during which the sun, the moon, and Earth are nearly in the same line and the tide level will be at its highest level in the month. The tide in the period from the new moon to the full moon is called *spring tide*. From the full moon to the next new moon, the tide is called *neap tide*.

If a tidal gauge is deployed at a particular geographical location, the time history of the tidal height data can be obtained and used to derive the constants in (5.127) using either LS technique or Fourier analysis. The information is used to produce the so-called tidal table after collecting a long history of tidal data in the area. For those harmonics whose period is longer than tidal records, theoretical formulas are used to estimate the corresponding constants. The tidal tables allow us to predict the tidal heights at any time where the tidal tables are derived.

One of the main applications of SWE models is for the simulation of tidal current in coastal waters (e.g., Westerink *et al.*, 1992). The modeling of tidal flows is "probably the oldest form of coastal ocean prediction, and certainly the most accurate" (Parkar *et al.*, 1999). The reason that it is the most accurate compared with other ocean predictions is that the driving force for tides is definite (e.g., astronomical effects) and the tidal flows are of a strict shallow water type.

To employ an SWE model for the simulation of tidal flows in a region, the lateral boundaries are often selected to be in the open sea away from coastlines. Based on the available tidal tables and their interpolated values on the boundaries, the time histories of tidal heights can be specified at all lateral boundaries. The tidal flows inside the computational domain are driven by the difference of tidal heights at different boundaries (e.g., Shankar *et al.*, 1997).

Tsunamis: Tsunami is a particular type of water wave, which often has a very long wavelength and is induced by geophysical effects such as an underwater earthquake or landslide. It may cause severe coastal flooding. The recent Indian Ocean tsunami caused by an earthquake offshore Sumatra, Indonesia, on 26 December 2004, devastated 11 Indian Ocean countries and caused the death of 229,866 people (based on the analysis compiled by the United Nations). Tsunami modeling is another important application of SWE models. To model a tsunami, three main processes need to be simulated, namely tsunami generation due to earthquakes or landslides, tsunami propagation in deep oceans, and tsunami run-up and inundation along coastlines.

Tsunami generation is the process with most of the uncertainties in the simulation because the relation between an earthquake and the displaced ocean surface is difficult to determine precisely. In most of the simulations, the initially displaced volume of water was introduced on the free surface based on the empirical formula that accounts for the crustal movement of the seafloor at the fault of the earthquake. Depending on the relative rupture movement, positive waves can be pushed on one side and negative waves drawn down on the other side. Two waves will then propagate in two opposite directions.

The tsunami propagation simulation often covers a very large area (e.g., across oceans), and thus the SWE model needs to be constructed on spherical coordinates as given in (5.101)–(5.103). During tsunami propagation in deep oceans, the wave amplitude remains small (e.g., a couple of meters). For this reason, the linear SWE is sufficient for the modeling of tsunami propagation. Because the wavelength is long and the wave amplitude is small, little wave energy is consumed during long-distance travel.

When the wave packet comes close to land, the reduction of water depth causes the combined effect of wave shoaling, refraction, and diffraction, the first of which will greatly enhance the incoming wave height and the latter two effects can make the waves warp around an island. As a result, nearly all places along the coastline of an island can possibly face the devastating wave attack. Damage can be more severe in a harbor (or bay) due to the seiching effect that causes the accumulation of wave energy in the partially enclosed basin by wave resonance. As a matter of fact, tsunami is the Japanese term of "great harbor wave" since it can be greatly excited in a harbor. Because the wavelength of a tsunami is rather long, when its front comes to the beach, it will cause continuous coastal flooding (or shoreline receding if the wave front has a negative form) for quite a long time (i.e., a few minutes). This behavior makes a tsunami look like a tide, although the flooding tsunami moves much faster than a tide. In earlier days, without knowing the cause of a tsunami, people called a tsunami a "tidal waves," even though it has no connection with real tide at all.

The actual inundation pattern depends on many factors such as incoming wave characteristics, nearshore bathymetry, and beach configuration. To accurately estimate tsunami run-up and inundation, both wave nonlinearity and moving the water front need to be considered. Although there exist some analytical approaches for the analysis of maximum wave run-up (e.g., Carrier and Greenspan, 1958; Pelinovsky and Mazova, 1992; Carrier and Yeh, 2002), the inundation map for an actual topography can be obtained only from the tsunami run-up modeling. In the numerical modeling, the run-up process of nonlinear waves is simulated. This requires moving the boundary condition, which is handled by various methods to compute the changing wet line caused by wave run-up and drawdown, as discussed in Section 5.3.1.2.

Currently, there are a few existing tsunami models adopted by various tsunami research centers or research institutes. For example, the MOST (Method Of Splitting Tsunami) model, developed by Titov of the Pacific Marine Environmental Laboratory and Synolakis of the University of Southern California (Titov and Synolakis, 1998), was adopted by the National Oceanic and Atmospheric Administration (NOAA), U.S.A. The model uses nested computational grids to telescope down into the high-resolution area of interest. Nested grids are used so there is minimum number of nodes in a wavelength in order to adequately resolve the wave with minimum computational effort. It has been used to investigate the global reach of the 26 December 2004 Indian Ocean tsunami (Titov *et al.*, 2005).

Another well-known model is the TUNAMI-N2 (Tohoku University's Numerical Analysis Model for Investigation of Near-field tsunamis) model, developed by the Disaster Control Research Center of Tohoku University in Japan. It uses a compact leapfrog FD scheme in time and space discretization. The method is efficient and accurate for modeling linear long waves. This model has been widely used in modeling many events of tsunami propagation and run-up (e.g., Shuto *et al.*, 1990; Goto *et al.*, 1997).

Besides the above two models, tsunami models were also developed in many other institutions. To name a few, the tsunami model COMCOT (Cornell Multigrid-COupled Tsunami) developed at Cornell University has been used to simulate solitary wave run-up on a circular island (Liu *et al.*, 1995) and a real tsunami run-up event (Liu *et al.*, 1994). The code NAMI DANCE developed by Efim Pelinovsky, Andrey Kukin, Andrey Zaytsev, and Ahmet Yalciner was used to simulate a few historical tsunami events. To include the wave dispersion effect, the Boussinesq model such as FUNWAVE was also used to model tsunami transformation nearshore (Grilli *et al.*, 2007). To better represent the irregular coastline, Myers and Baptista (1999) developed an FEM tsunami model. Recently, there is also a trend to couple a 2D SWE model with a 3D hydrodynamic model to simulate tsunami transformation and run-up in the nearshore region. Figure 5.9 gives an example of the numerical modeling of the tsunami in the Indian Ocean in 2004 with the use of the SWE module in DELTF3D.

In the tsunami simulation, another interesting issue is the so-called inverse problem. In many cases, the details of initial free surface displacement are unknown even though the magnitude scale and the epicenter of an earthquake have been detected. This triggers an interesting problem, namely, whether we are able to find out the actual seafloor movement by analyzing the observed and collected data of tsunami wave height and run-up on different affected coasts. Mathematically, this is called an inverse problem. Naturally, one would think that the problem can be handled by the reverse modeling that solves the same governing equation but with the negative time step to trace back what happens initially. However, besides the difficulty of collecting sufficient amount of observation data to initiate the simulation, the computation is often numerically unstable. Most of the time, an inverse problem is ill-posed, i.e., it has more unknowns than the number of equations formed and thus no unique solution exists. A reasonable solution must

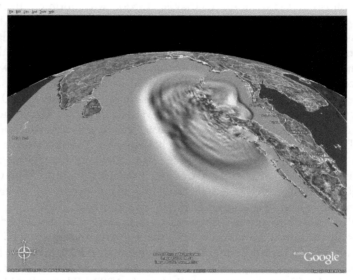

Figure 5.9 Simulation results of the Indian Ocean tsunami on 26 December, 2004 at Indonesian time 8:00, 9:00, and 10:00. (Produced with DELFT3D from WL | Delft Hydraulics; courtesy of Dr. Deepak Vatvani; permission from Google Earth for educational purposes)

Figure 5.9 (Continued).

be obtained under additional constraints. Readers are referred to Tanioka and Satake (2001) and Carl (1996) for the inverse problem in tsunami and ocean circulation problems.

Storm surge and ocean circulation: Wind generates not only waves but also current and ocean surface abnormities. When wind continues blowing toward a shore, water surface will be pushed up along the shoreline and this is called wind-induced setup (in contrast to wave-induced setup discussed in Section 3.1.4.4). Surface set-down results if the wind blows offshore. If the wind direction is parallel to the shoreline, the longshore current can be generated. Besides, a bulge of water surface setup can also be generated under the eye of a storm, where the atmospheric pressure is low. The increase in water surface can be further enhanced by the wave- and wind-induced water setup when the storm is moving toward the shore. The rise of water level caused by the combined wind stress and pressure effect by a storm is termed storm surge. Besides the recent storm surge caused by Hurricane Katrina in August 2005 that left the downtown of New Orleans, U.S.A, under 3 m water, there were many other notorious disasters caused by a storm surge. The greatest disaster of storm surge in the twentieth century occurred in Bangladesh in November 1970. The wind under the cyclone, coupled with the storm surge and heavy rainfall, killed 300,000–500,000 people.

Ocean surface abnormity and ocean current can also be generated from other mechanisms. For example, when a tidal flow or ocean current, which can be generated by wind in the monsoon season, approaches a curved strait, water surface can be pushed up on the outer side of the strait. In contrast, a large-scale ocean circulation can be generated from the mean density difference on a horizontal plane. All the above phenomena can be modeled by using an SWE model.

Other long waves: There are other water waves fitting into the category of long waves. They are tidal bores in the estuarial region, broken waves in surf and swash zones, edge waves on shallow-water beaches, and all long-period waves whose $kh < 0.3$.

Kobayashi and his colleagues at the University of Delaware developed an SWE model for the simulation of various wave phenomena in surf zones, i.e., wave transformation over submerged breakwaters (Kobayashi and Wurjanto, 1989), wave reflection and run-up on slopes (Kobayashi et al., 1990), and wave overtopping of revetments (Kobayashi and Raichle, 1994). Once waves are broken on the beach, they will behave like moving bores, which are similar to hydraulic jumps but moving at the celerity of shallow-water waves. The motion of the bores can be best described by nonlinear SWEs with the inclusion of appropriate energy dissipation terms. This subject has been studied both analytically (e.g., Keller et al., 1960; Hibberd and Peregrine, 1979) and numerically (e.g., Madsen et al., 2005).

Edge waves are a special type of water waves that are either generated locally (e.g., by coastal landslide) or reflected from obliquely incident wind waves or swells, but their further propagation of wave energy is confined nearshore. Under this circumstance, waves will propagate along-shore for a long distance due to negligible energy leakage to the offshore region. Ursell (1952) proposed the analytical solution of the edge wave modes based on the small-amplitude wave theory. At about the same time, Eckart (1951) found that with the use of linear SWEs, the edge wave can also be predicted with good accuracy on a plane beach. This was further confirmed by the laboratory and numerical studies using an SWE model (Liu et al., 1998).

Open channel flows and river flows: Another important application of SWE models is the simulation of open channel flows and river flows. Unsteady open channel flows share a lot of similarity with long waves with varying free surface and near hydrostatic pressure. However, for open channel flows, gravity plays the role of the driving force for flows rather than the restoring force for waves. Depending on whether Fr is greater or less than unity, the flow is classified as a supercritical or subcritical flow. The characteristics and the control section (by various upstream and downstream boundary conditions) of open channel flows are closely related to the flow classification. There are numerous works of numerical modeling of various types of open channel flows with the use of SWE models, e.g., Chapman and Kuo (1985) for shallow-water recirculation flows, Younus and Chaudhry (1994) for supercritical flows in a diverging channel and circular hydraulic jumps, Stelling and Duinmeijer (2003) for dam break flows, and Zhou and Stansby (1999) for hydraulic jumps. Readers are referred to Chow (1973) and Chaudhry (1993) for the theory and computation of open channel flows.

Both commercial software and freeware are available for the modeling of river hydraulics using the SWE approach. For example, MIKE 11 developed by DHI is commercial software that can be used to model unsteady river dynamics, lakes/reservoirs, irrigation canals, and other inland water systems. The software can be used together with data mining for modeling and real-time forecasting (Babovic, 1998). In contrast, HEC-2 developed by the Hydrological Engineering Center of USACE is freeware that can be used for a 1D steady, gradually varied flow in rivers of any cross section. This model can also be used to model the flood in a catchment area (e.g., Greenbaum *et al.*, 1998).

5.3.2   The Boussinesq models

5.3.2.1   The background of the Boussinesq equations

The SWE model is a powerful and efficient numerical tool for the simulation of large-scale long waves and shallow-water flows. However, when it is applied to relatively short waves, i.e., $kh \sim O(1)$, or weakly nonlinear dispersive waves, i.e., solitary waves or N-waves, the errors can be intolerable, especially for the simulation of long-distance propagation. The main reason is that in SWE models, the characteristics of wave dispersion are completely neglected. In reality, however, wave modes with different frequencies will tend to separate in finite water depth. This dispersive property of water waves is the direct result of nonhydrostatic pressure. This means that to model a dispersive wave, a reasonable approximation of the nonhydrostatic pressure is required.

The standard Boussinesq model: The earliest depth-averaged model that included weakly dispersive and nonlinear effects was derived by Boussinesq (1871), in which the nonhydrostatic pressure was approximated and included in the momentum equations. The equations were derived for horizontal bottoms only. Later, Mei and LeMehaute (1966) and Peregrine (1967) derived Boussinesq equations for variable depth. While Mei and LeMehaute (1966) used the velocity at the bottom as the dependent variable, Peregrine (1967) used the depth-averaged velocity and assumed the vertical velocity varying linearly over the depth. Due to wide popularity in the coastal engineering community, the equations derived by Peregrine (1967) are often referred to as the standard Boussinesq equations, which assumed that $h/H = \alpha = O(\varepsilon^2) = O\left((h/L)^2\right) << 1$ for weakly nonlinear and dispersive waves [recall that the Ursell number is defined as $\mathrm{Ur} = \left(L^2 H\right)/h^3 = \alpha/\varepsilon^2$]. The equations, after neglecting the viscous effect, have the following form:

$$\frac{\partial H}{\partial t} + \frac{\partial}{\partial x}(UH) + \frac{\partial}{\partial y}(VH) = 0 \tag{5.128}$$

$$\frac{\partial U}{\partial t} + U\frac{\partial U}{\partial x} + V\frac{\partial U}{\partial y} = -g\frac{\partial \eta}{\partial x} + \frac{1}{2}h\frac{\partial^2}{\partial t \partial x}\left[\frac{\partial (Uh)}{\partial x} + \frac{\partial (Vh)}{\partial y}\right]$$
$$- \frac{1}{6}h^2\frac{\partial^2}{\partial t \partial x}\left[\frac{\partial U}{\partial x} + \frac{\partial V}{\partial y}\right] \tag{5.129}$$

$$\frac{\partial V}{\partial t} + U\frac{\partial V}{\partial x} + V\frac{\partial V}{\partial y} = -g\frac{\partial \eta}{\partial y} + \frac{1}{2}h\frac{\partial^2}{\partial t \partial y}\left[\frac{\partial (Uh)}{\partial x} + \frac{\partial (Vh)}{\partial y}\right]$$
$$- \frac{1}{6}h^2\frac{\partial^2}{\partial t \partial y}\left[\frac{\partial U}{\partial x} + \frac{\partial V}{\partial y}\right] \tag{5.130}$$

The above system of equations has the same continuity equation as that in SWEs. The momentum equations in x- and y-directions, however, contain two more terms that are in the form of third-order mixed derivatives in time and space. The additional terms mathematically represent wave dispersion. These terms result from the vertical acceleration of fluid particles and thus they take into consideration, at least partially, the nonhydrostatic pressure effect. The above equations represent the lowest order of wave dispersion and wave nonlinearity. Peregrine (1967) uses the equation to simulate a solitary wave on a plane beach. Abbott *et al.* (1978) developed a Boussinesq model to simulate regular short waves in shallow water. Freilich and Guza (1984) used the equations to study nonlinear wave interaction between shoaling irregular waves.

The standard Boussinesq equations are valid only for relatively small kh and H/h. To extend the validity range of the equations, researchers have suggested various ways such as (a) improving the linear dispersion characteristics in deeper water and (b) including higher order wave nonlinearity. Witting (1984) attempted to match the dispersion equation using Pade's expansion. The equations make use of free surface velocity and are fully nonlinear, but approximate for dispersion relationship. The equations, however, are valid only for constant water depth. Madsen *et al.* (1991) and Madsen and Sørensen (1992) included higher order terms with adjustable coefficients into the standard Boussinesq equations for constant and variable water depth, respectively. Beji and Nadaoka (1996) presented a formal derivation for Madsen and Sørensen's (1992) improved Boussinesq equations. Their equations were employed later by Li *et al.* (1999) to develop an FEM Boussinesq model for wave simulation.

Nwogu's model: By defining the dependent variable as the velocity at an arbitrary depth, Nwogu (1993) achieved a rational polynomial approximation to the exact linear dispersion relationship without the need to add higher order terms to the equations. The errors in the linear phase speed were minimized over a certain depth range. Nwogu's Boussinesq equations have the following form:

$$
\frac{\partial \eta}{\partial t} + \nabla \cdot [(h + \eta)\, \mathbf{u}_\alpha] + \nabla \left\{ \left(\frac{z_\alpha^2}{2} - \frac{h^2}{6} \right) h \nabla (\nabla \cdot \mathbf{u}_\alpha) \right.
$$

$$
\left. + \left(z_\alpha + \frac{h}{2} \right) h \nabla [\nabla \cdot (h \mathbf{u}_\alpha)] \right\} = 0 \tag{5.131}
$$

$$
\frac{\partial \mathbf{u}_\alpha}{\partial t} + g \nabla \eta + (u_\alpha \cdot \nabla)\, \mathbf{u}_\alpha + \mu^2 \left\{ \frac{z_\alpha^2}{2} \nabla \left(\nabla \cdot \frac{\partial \mathbf{u}_\alpha}{\partial t} \right) + z_\alpha \nabla \left[\nabla \cdot \left(h \frac{\partial \mathbf{u}_\alpha}{\partial t} \right) \right] \right\} = 0
$$

$$
\tag{5.132}
$$

where $\mathbf{u}_\alpha$ is the horizontal velocity vector at an arbitrary depth z_α.

Nwogu's equations have a relatively simple form with the improved ability for representing wave nonlinearity and dispersion. In Nwogu's original paper, an iterative Crank–Nicholson method was employed with a predictor–corrector scheme to solve the Boussinesq equations in 1D space. Wei and Kirby (1995) developed a 2D numerical model using a higher order FD time-stepping scheme. Walkley and Berzins (2002) developed a 2D numerical model with the use of FEM. Lin and Man (2006) developed a 2D FDM model on a staggered grid system, in which the mass is better conserved for long-term computation.

Kirby and his group's model: In the past decade, great advances have been made to further extend the validity range of the Boussinesq equations to deeper water and higher nonlinearity. For example, Kirby and his colleagues at the University of Delaware proposed a few higher order accurate Boussinesq equations extended from Nwogu's equations. The following modified Boussinesq equations improve the nonlinear properties (Wei *et al.*, 1995):

$$
\frac{\partial \eta}{\partial t} = E(\eta, u, v) + E_2(\eta, u, v) \tag{5.133}
$$

$$
\frac{\partial U(u)}{\partial t} = F(\eta, u, v) + \frac{\partial F_1(v)}{\partial t} + F_2 \left(\eta, u, v, \frac{\partial u}{\partial t}, \frac{\partial v}{\partial t} \right) \tag{5.134}
$$

$$
\frac{\partial V(v)}{\partial t} = G(\eta, u, v) + \frac{\partial G_1(u)}{\partial t} + G_2 \left(\eta, u, v, \frac{\partial u}{\partial t}, \frac{\partial v}{\partial t} \right) \tag{5.135}
$$

u and v are the horizontal velocities at $z = z_\alpha = -0.531h$, and U and V are defined as:

$$
U = u + \left[b_1 h \frac{\partial^2 u}{\partial x^2} + b_2 \frac{\partial^2 (hu)}{\partial x^2} \right] \quad \text{and} \quad V = v + \left[b_1 h \frac{\partial^2 v}{\partial y^2} + b_2 \frac{\partial^2 (hv)}{\partial y^2} \right]
$$

The remaining quantities E, E_2, F, F_1, F_2, G, G_1, and G_2 are spatial derivatives of η, u, v, $\dfrac{\partial u}{\partial t}$, or $\dfrac{\partial v}{\partial t}$ that are defined as follows:

$$E = -\frac{\partial}{\partial x}\left[(h+\eta)\,u\right] - \frac{\partial}{\partial y}\left[(h+\eta)\,v\right]$$

$$- \frac{\partial}{\partial x}\left\{ a_1 h^3\left(\frac{\partial^2 u}{\partial x^2} + \frac{\partial^2 v}{\partial x \partial y}\right) + a_2 h^2\left[\frac{\partial^2 (hu)}{\partial x^2} + \frac{\partial^2 (hv)}{\partial x \partial y}\right]\right\}$$

$$- \frac{\partial}{\partial y}\left\{ a_1 h^3\left(\frac{\partial^2 u}{\partial x \partial y} + \frac{\partial^2 v}{\partial y^2}\right) + a_2 h^2\left[\frac{\partial^2 (hu)}{\partial x \partial y} + \frac{\partial^2 (hv)}{\partial y^2}\right]\right\} \tag{5.136}$$

$$F = -g\frac{\partial \eta}{\partial x} - \left(u\frac{\partial u}{\partial x} + v\frac{\partial u}{\partial y}\right) \tag{5.137}$$

$$G = -g\frac{\partial \eta}{\partial y} - \left(u\frac{\partial v}{\partial x} + v\frac{\partial v}{\partial y}\right) \tag{5.138}$$

$$F_1 = -h\left[b_1 h\frac{\partial^2 v}{\partial x \partial y} + b_2\frac{\partial^2 (hv)}{\partial x \partial y}\right] \tag{5.139}$$

$$G_1 = -h\left[b_1 h\frac{\partial^2 u}{\partial x \partial y} + b_2\frac{\partial^2 (hu)}{\partial x \partial y}\right] \tag{5.140}$$

$$E_2 = -\frac{\partial}{\partial x}\left\{\left[a_1 h^2\eta + \frac{1}{6}\eta(h^2-\eta^2)\right]\left(\frac{\partial^2 u}{\partial x^2} + \frac{\partial^2 v}{\partial x \partial y}\right)\right\}$$

$$- \frac{\partial}{\partial x}\left\{\left[a_2 h\eta - \frac{1}{2}\eta(h+\eta)\right]\left(\frac{\partial^2 (hu)}{\partial x^2} + \frac{\partial^2 (hv)}{\partial x \partial y}\right)\right\}$$

$$- \frac{\partial}{\partial y}\left\{\left[a_1 h^2\eta + \frac{1}{6}\eta(h^2-\eta^2)\right]\left(\frac{\partial^2 u}{\partial x \partial y} + \frac{\partial^2 v}{\partial y^2}\right)\right\}$$

$$- \frac{\partial}{\partial y}\left\{\left[a_2 h\eta - \frac{1}{2}\eta(h+\eta)\right]\left(\frac{\partial^2 (hu)}{\partial x \partial y} + \frac{\partial^2 (hv)}{\partial y^2}\right)\right\} \tag{5.141}$$

$$F_2 = -\frac{\partial}{\partial x}\left\{\frac{1}{2}(z_\alpha^2 - \eta^2)\left[u\frac{\partial}{\partial x}\left(\frac{\partial u}{\partial x} + \frac{\partial v}{\partial y}\right) + v\frac{\partial}{\partial y}\left(\frac{\partial u}{\partial x} + \frac{\partial v}{\partial y}\right)\right]\right\}$$

$$- \frac{\partial}{\partial x}\left\{(z_\alpha - \eta)\left[u\frac{\partial}{\partial x}\left(\frac{\partial (hu)}{\partial x} + \frac{\partial (hv)}{\partial y}\right) + v\frac{\partial}{\partial y}\left(\frac{\partial (hu)}{\partial x} + \frac{\partial (hv)}{\partial y}\right)\right]\right\}$$

$$- \frac{1}{2}\frac{\partial}{\partial x}\left\{\left[\frac{\partial (hu)}{\partial x} + \frac{\partial (hv)}{\partial y} + \eta\left(\frac{\partial u}{\partial x} + \frac{\partial v}{\partial y}\right)\right]^2\right\}$$

$$+ \frac{\partial}{\partial x}\left\{\frac{1}{2}\eta^2\left[\frac{\partial}{\partial x}\left(\frac{\partial u}{\partial t}\right) + \frac{\partial}{\partial y}\left(\frac{\partial v}{\partial t}\right)\right] + \eta\frac{\partial}{\partial x}\left[h\frac{\partial u}{\partial t}\right] + \frac{\partial}{\partial y}\left[h\frac{\partial v}{\partial t}\right]\right\} \tag{5.142}$$

$$G_2 = -\frac{\partial}{\partial y}\left\{\frac{1}{2}\left(z_\alpha^2 - \eta^2\right)\left[u\frac{\partial}{\partial x}\left(\frac{\partial u}{\partial x} + \frac{\partial v}{\partial y}\right) + v\frac{\partial}{\partial y}\left(\frac{\partial u}{\partial x} + \frac{\partial v}{\partial y}\right)\right]\right\}$$

$$-\frac{\partial}{\partial y}\left\{\left(z_\alpha - \eta\right)\left[u\frac{\partial}{\partial x}\left(\frac{\partial (hu)}{\partial x} + \frac{\partial (hv)}{\partial y}\right) + v\frac{\partial}{\partial y}\left(\frac{\partial (hu)}{\partial x} + \frac{\partial (hv)}{\partial y}\right)\right]\right\}$$

$$-\frac{1}{2}\frac{\partial}{\partial y}\left\{\left[\frac{\partial (hu)}{\partial x} + \frac{\partial (hv)}{\partial y} + \eta\left(\frac{\partial u}{\partial x} + \frac{\partial v}{\partial y}\right)\right]^2\right\}$$

$$+\frac{\partial}{\partial y}\left\{\frac{1}{2}\eta^2\left[\frac{\partial}{\partial x}\left(\frac{\partial u}{\partial t}\right) + \frac{\partial}{\partial y}\left(\frac{\partial v}{\partial t}\right)\right] + \eta\frac{\partial}{\partial x}\left[h\frac{\partial u}{\partial t}\right] + \frac{\partial}{\partial y}\left[h\frac{\partial v}{\partial t}\right]\right\}$$

$$(5.143)$$

where E_2, F_2, and G_2 are the additional higher order nonlinear terms that would not exist for weakly nonlinear equations. The constants a_1, a_2, b_1, and b_2 are defined as follows:

$$a_1 = \beta^2/2 - 1/6, \quad a_2 = \beta + 1/2, \quad b_1 = \beta^2/2, \quad b_2 = \beta \qquad (5.144)$$

where $\beta = z_\alpha/h$. Later, Gobbi *et al.* (2000) further improved the model's accuracy to $O(kh)$.[4]

Liu and his group's model: Liu and his colleagues at Cornell University (e.g., Liu, 1994; Chen and Liu, 1995) derived fully nonlinear and weakly dispersive Boussinesq equations using velocity potential and free surface elevation. The equations take the following dimensional form (Liu and Wu, 2004):

$$\frac{\partial \eta}{\partial t} + \nabla \cdot [(h + \eta)\mathbf{u}_\alpha] + \nabla \cdot \left\{\left(\frac{z_\alpha^2}{2} - \frac{h^2}{6}\right)h\nabla\left(\nabla \cdot \mathbf{u}_\alpha\right) + \left(z_\alpha + \frac{h}{2}\right)h\nabla\left[\nabla \cdot (h\mathbf{u}_\alpha)\right]\right\}$$

$$+\nabla \cdot \left\{\eta\left[\left(z_\alpha - \frac{\eta}{2}\right)\nabla\left(\nabla \cdot (h\mathbf{u}_\alpha)\right) + \frac{1}{2}\left(z_\alpha^2 - \frac{\eta^2}{3}\right)\nabla\left(\nabla \cdot \mathbf{u}_\alpha\right)\right]\right\} = 0 \quad (5.145)$$

$$\frac{\partial \mathbf{u}_\alpha}{\partial t} + g\nabla\eta + \frac{1}{2}\nabla|\mathbf{u}_\alpha|^2 + \frac{1}{\rho}\nabla p + \left\{\nabla\left[z_\alpha\left(\nabla \cdot \left(h\frac{\partial \mathbf{u}_\alpha}{\partial t}\right)\right)\right] + \nabla\left[\frac{z_\alpha^2}{2}\left(\nabla \cdot \frac{\partial \mathbf{u}_\alpha}{\partial t}\right)\right]\right\}$$

$$+\nabla\left\{\frac{z_\alpha^2}{2}\mathbf{u}_\alpha \cdot \nabla\left(\nabla \cdot \mathbf{u}_\alpha\right) + z_\alpha\mathbf{u}_\alpha \cdot \nabla\left(\nabla \cdot h\mathbf{u}_\alpha\right) + \frac{1}{2}\left[\nabla \cdot (h\mathbf{u}_\alpha)\right]^2 - \eta\nabla \cdot \left(h\frac{\partial \mathbf{u}_\alpha}{\partial t}\right)\right\}$$

$$+\nabla\left\{\eta\left(\nabla \cdot h\mathbf{u}_\alpha\right)\left(\nabla \cdot \mathbf{u}_\alpha\right) - \frac{1}{2}\eta^2\nabla \cdot \left(\frac{\partial \mathbf{u}_\alpha}{\partial t}\right) - \eta\mathbf{u}_\alpha \cdot \nabla\left(\nabla \cdot h\mathbf{u}_\alpha\right)\right\}$$

$$+\nabla\left\{\frac{\eta^2}{2}\left(\nabla \cdot \mathbf{u}_\alpha\right)^2 - \frac{\eta^2}{2}\mathbf{u}_\alpha \cdot \nabla\left(\nabla \cdot \mathbf{u}_\alpha\right)\right\} = 0 \qquad (5.146)$$

where $z_\alpha = -0.531h$, which provides an adequate description of wave propagation in terms of phase speed and group velocity for $0 \leq kh \leq \pi$. Later, Lynett and Liu (2004) presented a two-layer Boussinesq model that is accurate up to $kh = 6$.

Madsen and his group's model: Independently, Madsen and his colleagues at the Technical University of Denmark have proposed a series of new highly nonlinear Boussinesq models, in which either the mean velocity or the velocity at an arbitrary z-level was used to minimize the depth-integrated error of the linear velocity profile. The equations proposed by Schäffer and Madsen (1995) take the following form:

$$\frac{\partial \eta}{\partial t} + \frac{\partial}{\partial x}\left[(h+\eta)\,u_\alpha\right] + \frac{\partial}{\partial y}\left[(h+\eta)\,v_\alpha\right] + \frac{\partial}{\partial x}\left[\left(\frac{z_\alpha^2}{2} - \frac{h^2}{6}\right)h\left(\frac{\partial^2 u_\alpha}{\partial x^2} + \frac{\partial^2 v_\alpha}{\partial x \partial y}\right)\right.$$

$$+ \left(z_\alpha + \frac{h}{2} - \beta_1 h\right) \times h\left(\frac{\partial^2 (hu_\alpha)}{\partial x^2} + \frac{\partial^2 (hv_\alpha)}{\partial x \partial y}\right) + \beta_2 \frac{\partial}{\partial x}\left[h^2 \frac{\partial (hu_\alpha)}{\partial x} + h^2 \frac{\partial (hv_\alpha)}{\partial y}\right]$$

$$- \beta_1 h^2 \frac{\partial^2 \eta}{\partial t \partial x} + \beta_2 \frac{\partial}{\partial x}\left(h^2 \frac{\partial \eta}{\partial t}\right)\right] + \frac{\partial}{\partial y}\left[\left(\frac{z_\alpha^2}{2} - \frac{h^2}{6}\right)h\left(\frac{\partial^2 u_\alpha}{\partial x \partial y} + \frac{\partial^2 v_\alpha}{\partial y^2}\right)\right.$$

$$+ \left(z_\alpha + \frac{h}{2} - \beta_1 h\right) h\left(\frac{\partial^2 (hu_\alpha)}{\partial x \partial y} + \frac{\partial^2 (hv_\alpha)}{\partial y^2}\right) + \beta_2 \frac{\partial}{\partial y}\left[h^2 \frac{\partial (hv_\alpha)}{\partial y} + h^2 \frac{\partial (hu_\alpha)}{\partial x}\right]$$

$$- \beta_1 h^2 \frac{\partial^2 \eta}{\partial t \partial y} + \beta_2 \frac{\partial}{\partial y}\left(h^2 \frac{\partial \eta}{\partial t}\right)\right] = 0 \tag{5.147}$$

$$\frac{\partial u_\alpha}{\partial t} + g\frac{\partial \eta}{\partial x} + u_\alpha \frac{\partial u_\alpha}{\partial x} + v_\alpha \frac{\partial u_\alpha}{\partial y} + \left(\frac{z_\alpha^2}{2} - \gamma_1 h^2\right)\left(\frac{\partial^3 u_\alpha}{\partial t \partial x^2} + \frac{\partial^3 v_\alpha}{\partial t \partial x \partial y}\right)$$

$$+ (z_\alpha + \gamma_2 h)\left[\frac{\partial^2}{\partial x^2}\left(h\frac{\partial u_\alpha}{\partial t}\right) + \frac{\partial^2}{\partial x \partial y}\left(h\frac{\partial v_\alpha}{\partial t}\right)\right]$$

$$- \gamma_1 g h^2 \left\{\frac{\partial^3 \eta}{\partial x^3} + \frac{\partial^3 \eta}{\partial x \partial y^2} + \gamma_2 g h\left[\frac{\partial^2}{\partial x^2}\left(h\frac{\partial \eta}{\partial x}\right) + \frac{\partial^2}{\partial x \partial y}\left(h\frac{\partial \eta}{\partial y}\right)\right]\right\} = 0 \tag{5.148}$$

$$\frac{\partial v_\alpha}{\partial t} + g\frac{\partial \eta}{\partial y} + u_\alpha \frac{\partial v_\alpha}{\partial x} + v_\alpha \frac{\partial v_\alpha}{\partial y} + \left(\frac{z_\alpha^2}{2} - \gamma_1 h^2\right)\left(\frac{\partial^3 u_\alpha}{\partial t \partial x \partial y} + \frac{\partial^3 v_\alpha}{\partial t \partial y^2}\right)$$

$$+ (z_\alpha + \gamma_2 h)\left[\frac{\partial^2}{\partial x \partial y}\left(h\frac{\partial u_\alpha}{\partial t}\right) + \frac{\partial^2}{\partial y^2}\left(h\frac{\partial v_\alpha}{\partial t}\right)\right]$$

$$- \gamma_1 g h^2 \left\{\frac{\partial^3 \eta}{\partial x^2 \partial y} + \frac{\partial^3 \eta}{\partial y^3} + \gamma_2 g h\left[\frac{\partial^2}{\partial x \partial y}\left(h\frac{\partial \eta}{\partial x}\right) + \frac{\partial^2}{\partial y^2}\left(h\frac{\partial \eta}{\partial y}\right)\right]\right\} = 0 \tag{5.149}$$

where $z_\alpha = -0.02907h$, $\beta_1 = -0.13054$, $\beta_2 = -0.53582$, $\gamma_1 = 0.01196$, and $\gamma_2 = 0.00144$. Later, Madsen and Schäffer (1998) and Madsen *et al.* (2002) further improved the Boussinesq models and the latter one is accurate up to $kh = 40$.

5.3.2.2   *Extension of Boussinesq equations in other applications*

Boussinesq models in porous media: Attempts were made to develop Boussinesq equations in porous media so that the model can be used to simulate wave flow over a porous bed or through a porous structure. Cruz *et al.* (1997) presented such a Boussinesq model and used it in the study of wave transformation over a porous bed. Liu and Wen (1997) developed a fully nonlinear and weakly dispersive wave model in porous media. Recently, Hsiao *et al.* (2002) introduced a new set of Boussinesq equations for nonlinear waves over a porous bed. Chen (2006) extended this model to the breaking wave region and used it to study wave interaction with porous structures.

Boussinesq models for stratified flows: Another extension of the development of the Boussinesq equations was to simulate stratified flows. Internal waves may be generated at the interface of two fluid layers with different densities. Lamb (1994) employed a Boussinesq model to investigate the internal wave generated by strong tidal flow. Choi and Camassa (1996, 1999) developed weak and full nonlinear internal wave models in a two-layer fluid system and found that the equations can be reduced to the Boussinesq equations or KdV equations in shallow water and to intermediate long-wave (ILW) equations in deep water. Chen and Liu (1998) proposed a generalized modified the KP equation for interfacial wave propagation. Kataoka *et al.* (2000) derived a fully nonlinear evolution equation for finite amplitude long internal waves in a uniformly stratified fluid. The equation can be reduced to the KP equation in a small-amplitude limit. Lynett and Liu (2002) developed an internal wave model to simulate internal wave evolution in straits and near islands. Strictly speaking, some of the equations used to describe internal wave motion are not Boussinesq equations, rather the extension of the Boussinesq equations to other equation types that share some similar characteristics of wave dispersion and nonlinearity.

Boussinesq models for moving surfaces: The Boussinesq equations were also extended to describe wave generation and propagation by moving surface disturbance. The surface disturbance may come from a free surface, bottom, or a moving object in between. The first scenario is associated with either the large-scale wave generation by wind or the local-scale ship wave generation by a moving hull. In this case, the equations need to be revised to include the surface pressure gradient as in SWEs (5.88) and (5.89). For example, Wu and Wu (1982) proposed a generalized Boussinesq model to simulate nonlinear long waves due to moving surface pressure. To simulate shorter waves, Liu and Wu (2004) employed a nonlinear and dispersive Boussinesq model to simulate ship waves in channels. The second scenario corresponds to tsunami generation by underwater landslides or earthquakes. For example, using the generalized Boussinesq equation, Sander and Hutter (1992) simulated weakly nonlinear dispersive waves generated by submerged moving boundaries. In their approach, the still water depth becomes the prescribed function of time. A similar approach was also adopted in SWE models for simulating tsunami generation (e.g., Heinrich *et al.*, 2001). For an immersed moving body, Lee *et al.* (1989) proposed a forced KdV equation model and a generalized Boussinesq model to study waves generated by a moving body with arched cross section. Their results agreed with experimental data and the numerical results from an NSE-solver model (Zhang and Chwang, 1996).

KdV equation and solitons: Boussinesq models have received a lot of attention in the past few decades not only because of their capability of describing nonlinear dispersive shallow-water waves, but also because of their inherent connection to some important physical phenomena in basic science, e.g., soliton transmission, nonlinear wave–wave interaction. When a wave travels in one direction, the Boussinesq equations can be simplified into the KdV equation as follows:

$$\frac{\partial \phi}{\partial t} + 6\phi \frac{\partial \phi}{\partial x} + \frac{\partial^3 \phi}{\partial x^3} = 0 \tag{5.150}$$

The equation is famous as an example of an "exactly solvable model" for a nonlinear PDE. A few analytical solutions in the form of soliton exist for some well-posed initial and boundary conditions. The analytical expression of the soliton solution can be obtained by the inverse scattering transform, a mathematical procedure for integrating some nonlinear PDEs, e.g., KdV equation, NLS equation, and sine-Gordon equation, by converting them into a system of linear ODEs. All these equations have soliton type of analytical solutions (Ablowitz and Segur, 1981; Drazin and Johnson, 1989).

The soliton phenomenon was first described by John Scott Russell (1808–1882), who observed a solitary wave in the Union Canal and named it as "wave of translation." Later, theoretical investigations were made by Lord Rayleigh and Boussinesq around 1870 and by Korteweg and de Vries in 1895. Compared with the Boussinesq equations that are more restrictedly used in modeling shallow-water waves, the KdV equation has a wider application in general physics such as ion-acoustic waves in plasma and acoustic waves on a crystal lattice. When the KdV equation is generalized in the primary wave propagation direction but with a small spreading angle, the KP equation results. The KP equation is a single PDE and thus is easier to be treated numerically than the Boussinesq equations. This is similar to parabolic approximation made in the elliptic MSE (EMSE), which is to be discussed later.

5.3.2.3   Other nonlinear wave equations

All the previously introduced Boussinesq equations have been derived based on perturbation theory, which allows a different order of approximation for wave nonlinearity and dispersion. Other types of equations for nonlinear water waves can also be obtained by using Luke's variational principle (e.g., Luke, 1967). In contrast, by using the direct approximation to the Euler equations expressed in terms of continuity and energy equation, Green and Naghdi (1976) derived the so-called GN equations for wave propagation in water of variable depth.

Serre equations (1953) are another set of governing equations for highly nonlinear water waves. These equations received less attention than the Boussinesq equations for water wave modeling until recently when some researchers revisited the equations and found some good characteristics in the equations in modeling nonlinear dispersive waves, especially after introducing some higher order correction terms (e.g., Barthelemy, 2004). For a 1D case, the original Serre equations take the following simple form:

$$\frac{\partial \eta}{\partial t} + \frac{\partial [(b+\eta) U]}{\partial x} = 0 \tag{5.151}$$

$$\frac{\partial U}{\partial t} + U\frac{\partial U}{\partial x} + g\frac{\partial \eta}{\partial x} = \frac{1}{3(b+\eta)} \frac{\partial}{\partial x} \left[(b+\eta)^3 \left(\frac{\partial^2 U}{\partial x \partial t} + U\frac{\partial^2 U}{\partial x \partial x} - \left(\frac{\partial U}{\partial x}\right)^2 \right) \right]$$
$$\tag{5.152}$$

The RHS of the momentum equation explicitly contains the nonlinear effect on dispersion. In contrast to the Boussinesq equations, equations (5.151) and (5.152) have the closed-form soliton solution (e.g., Rayleigh solitary wave solution), i.e.:

$$\eta = A\mathrm{sech}^2 \left[\frac{3A}{4b^2(b+A)}(x - Ct) \right] \quad U = C\left(1 - \frac{b}{b+\eta}\right) \tag{5.153}$$

where A is the wave amplitude and $C = \sqrt{gb(1 + A/b)}$. The above solution can be reduced to KdV solitary wave solution for small-amplitude waves, but it may produce more stable solitons for larger amplitude soliton.

Recently, Wu (1999, 2000) attempted to develop a new unified model to allow nonlinear and dispersive wave effects to operate to the same extent as the Euler equations. The model was used to study various nonlinear dispersive gravity–capillary water wave phenomena. These problems include nonlinear wave interaction of solitons during overtaking collisions (e.g., Hirota, 1973; Whitham, 1974) and soliton propagation in variable channels (Shuto, 1974; Miles, 1979; Teng and Wu, 1992, 1997).

Although Boussinesq models and other nonlinear dispersive wave models have been stretched to model waves in much deeper waters in the past decade, such models are still limited to a finite value of kb. This is because the linear dispersion relation is satisfied only in an approximate manner for all Boussinesq models. In deriving a Boussinesq model, the linear dispersion is not explicitly enforced; rather, the momentum equations are manipulated during the integration to ensure that the linear dispersion equation is satisfied to the best possible extent. As a result, in solving the Boussinesq equations, the wave number (or wavelength) is not sought explicitly. This makes the Boussinesq equations not only applicable to shallow-water waves but also to any unsteady shallow-water flow. The price it pays, however, is that the dispersive errors can accumulate and become significant in the simulation of short waves.

5.3.2.4   *Boussinesq models and their applications*

Some widely used and reported Boussinesq models include FUNWAVE (Kirby *et al.*, 1998) developed in the Center for Applied Coastal Research (CACR), University of Delaware; a two-layer Boussinesq model developed at Cornell University and Texas A&M University (Lynett and Liu, 2004; Lynett, 2006); and a higher order Boussinesq model developed in the Technical University of Denmark (Madsen *et al.*, 2006); For commercial software, both DELFT3D and MIKE 21 have modules of Boussinesq models for nonlinear and dispersive water wave modeling.

These Boussinesq models have been used to simulate long-wave breaking and run-up on beaches (e.g., Lynett *et al.*, 2003), wave diffraction around coastal structures (Fuhrman *et al.*, 2005), wave interaction with longshore current (Chen *et al.*, 2003), etc. Figure 5.10 shows an example of the modeling of a surface wave and wave-induced flow by a Boussinesq model (Madsen *et al.*, 1997). Figure 5.11 provides an example of large-scale wave modeling around a port using a Boussinesq model using parallel computation (Sitanggang and Lynett, 2005). Another important application of

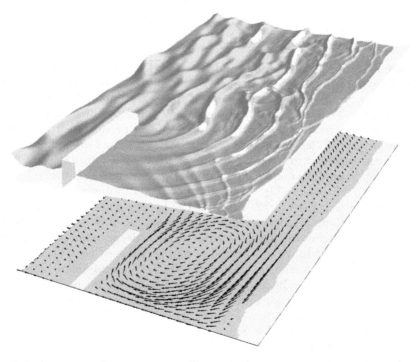

Figure 5.10 Simulation results by a Boussinesq model for surface waves and wave-induced flows around a detached breakwater on a beach subjected to multidirectional irregular waves. (Courtesy of Prof. P.A. Madsen of the Technical University of Denmark)

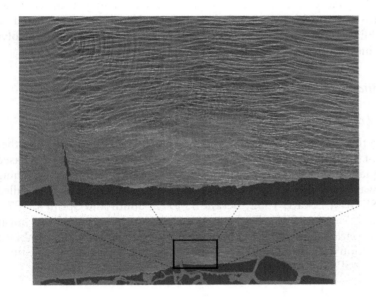

Figure 5.11 Numerical simulation of a quasi-steady-state wave field near Free-
port, Texas, U.S.A. using a Boussinesq model; the simulation area
is 35 km × 8 km with the grid size of 3 m × 3 m; the incoming wave
is irregular with $H_s = 2$ m and $T_p = 12.5$ s; the model was run on
50 parallel Opterons and the wall clock time is about 24 hours
for running 100 wave periods. (Courtesy of Dr. Patrick Lynett of
Texas A&M University)

Boussinesq model is to couple with a sediment transport to calculate sediment transport in the surf zone and the long-term beach profile evolution
(e.g., Rakha *et al.*, 1997; Karambas and Koutitas, 2002).

To apply Boussinesq models to surf zones and swash zones, the correct
modeling of wave breaking and wave–current interaction is inevitable. Two
theoretical barriers need to break through, i.e., the proper modeling of
turbulence in breaking waves that are highly nonuniform in the vertical
direction and the correct inclusion of current-induced vorticity in the potential flow-based Boussinesq equations. Using the eddy viscosity concept,
Zelt (1991) introduced the energy dissipation terms in the Boussinesq
models to simulate breaking waves propagation and run-up. Karambas
and Koutitas (1992) solved the one-equation turbulence kinetic energy (k)
transport model and used the resulting information to calculate the eddy
viscosity under the broken waves. Kabiling and Sato (1994) developed
a breaking wave model and coupled it to the beach evolution model.
Later, Veeramony and Svendsen (2000) and Kennedy *et al.* (2000) also
proposed their Boussinesq models to model wave-breaking processes. The
proper inclusion of current effects into Boussinesq models was pioneered
by Yoon and Liu (1989b) and further extended by Chen *et al.* (2001) for

modeling longshore current and Shen (2000) for modeling stratified flows and surface waves. Although most of Boussinesq models are constructed on FDM or FEM, there is the exception of using the meshless method to solve the Boussinesq equations (Wang and Liu, 2006).

5.3.3 Mild-slope equation models

5.3.3.1 Transient mild-slope equation

To ensure that the wave dispersion relationship is strictly satisfied, another category of depth-integrated wave model was developed. The model is based on the equation that was derived from the potential flow theory assuming that the wave is linear and the bottom slope is mild and thus it is called linear MSE (e.g., Eckart, 1952; Berkhoff, 1972; Smith and Sprinks, 1975; Lozano and Meyer, 1976). The equation is exact in terms of the linear dispersion relationship, and it has been widely used to describe the combined wave refraction and diffraction over a slowly varying topography. The time-dependent MSE is written as (Dingemans, 1997: 254):

$$\frac{\partial^2 \eta}{\partial t^2} - \frac{\partial}{\partial x}(cc_g \frac{\partial \eta}{\partial x}) - \frac{\partial}{\partial y}(cc_g \frac{\partial \eta}{\partial y}) + (\sigma^2 - k^2 cc_g)\eta = 0 \tag{5.154}$$

The relationship among k, h, and σ is determined by the linear dispersion equation:

$$\sigma^2 = gk \tanh kh \tag{5.155}$$

Equation (5.154) is a second-order hyperbolic equation that has second-order derivatives in both time and space. Few numerical attempts have been made to solve (5.154) directly except for Lin (2004a), who proposed a compact FD scheme to solve the equation directly.

Realizing (5.154) has similar characteristics to the linear long-wave equation (5.112), people have tried to split the original second-order MSE into a pair of first-order hyperbolic equations (e.g., Copeland, 1985). The pseudo-fluxes P and Q, both of which are complex variables, were introduced into the equations:

$$\frac{c_g}{c}\frac{\partial \eta}{\partial t} + \frac{\partial P}{\partial x} + \frac{\partial Q}{\partial y} = 0 \tag{5.156}$$

$$\frac{\partial P}{\partial t} + cc_g\frac{\partial \eta}{\partial x} = 0 \tag{5.157}$$

$$\frac{\partial Q}{\partial t} + cc_g\frac{\partial \eta}{\partial y} = 0 \tag{5.158}$$

The above equation system is similar to SWEs but the fluxes do not carry a definite physical meaning. Besides, although the equations are transient, due to the use of pseudofluxes, only their steady-state solution can be sought. Madsen and Larsen (1987) improved the solution procedure by extracting the time harmonic part and employing an efficient ADI algorithm. Introducing a slow coordinate for the time variable, Lee *et al.* (1998) developed a system of hyperbolic MSEs for fast-varying topography. Abohadima and Isobe (1999) proposed a nonlinear time-dependent MSE based on Luke's (1967) variational principle method and employed it to study nonlinear wave diffraction. Zheng *et al.* (2001) adopted the nonlinear dispersion relationship proposed by Kirby and Dalrymple (1986) in their time-dependent MSE model. The inclusion of current and dissipation in the MSE was made by Feng and Hong (2000).

5.3.3.2 *Elliptic mild-slope equation*

For a harmonic wave train, by defining $\eta = F(x, y)e^{-i\sigma t}$, where $F(x, y)$ is the spatially fast-varying complex wave amplitude function, the time derivative term in (5.154) can be eliminated. This leads to the steady-state MSE (Berkhoff, 1972), i.e.:

$$\frac{\partial}{\partial x}\left(cc_g \frac{\partial F}{\partial x}\right) + \frac{\partial}{\partial y}\left(cc_g \frac{\partial F}{\partial y}\right) + k^2 cc_g F = 0. \tag{5.159}$$

The above equation is an elliptic type and is referred to as an EMSE. Radder (1979) suggested that by variable transformation of:

$$\Phi = F\sqrt{cc_g} \text{ and } K^2 = k^2 - \left(\nabla^2 \sqrt{cc_g}\right)/\sqrt{cc_g},$$

the equation can be reduced to Helmholtz equation:

$$\frac{\partial^2 \Phi}{\partial x^2} + \frac{\partial^2 \Phi}{\partial y^2} + K^2 F = 0 \tag{5.160}$$

The numerical solution to an EMSE generally requires a significant number of iterations. Berkhoff (1972) attempted to solve (5.159) numerically but the method is restricted to a small domain. Later, Panchang *et al.* (1991) adopted a preconditional CG method to accelerate the convergence rate and thus the model could be applied to a larger area. Li and Anastasiou (1992) proposed another solution procedure by using the multigrid technique that also aimed at speeding up the convergence rate. Later, various CG versions of EMSE solvers were proposed to improve the efficiency of the numerical solution (e.g., Zhao and Anastasiou, 1996).

There are a few interesting asymptotic behaviors of the original EMSE. In shallow water, i.e., $kh \ll 1$, we have $c = c_g = \sqrt{gh}$ and the above equation can be reduced to (5.114):

$$\frac{\partial}{\partial x}\left(c^2 \frac{\partial F}{\partial x}\right) + \frac{\partial}{\partial y}\left(c^2 \frac{\partial F}{\partial y}\right) + \sigma^2 F = 0 \tag{5.161}$$

In deep water, i.e., $kh \ll 1$, or constant water depth, the spatial gradients of both c_g and c are zero. The EMSE can then be reduced to Helmholtz equation:

$$\frac{\partial^2 F}{\partial x^2} + \frac{\partial^2 F}{\partial y^2} + k^2 F = 0 \tag{5.162}$$

This equation describes the wave diffraction only and was used by Penney and Price (1952a) to develop an analytical solution of wave diffraction behind breakwaters.

5.3.3.3   Eikonal equation and energy equation

If we factor out the fast-varying component in space by further defining $F(x, y) = a(x, y)e^{iS(x,y)}$, where $a(x, y)$ is the slowly varying *real* wave amplitude function and S is the *real* phase function, and substituting into the EMSE, we have:

$$i\left\{ a\frac{\partial}{\partial x}\left(cc_g\frac{\partial S}{\partial x}\right) + 2cc_g\frac{\partial a}{\partial x}\frac{\partial S}{\partial x} + a\frac{\partial}{\partial y}\left(cc_g\frac{\partial S}{\partial y}\right) + 2cc_g\frac{\partial a}{\partial y}\frac{\partial S}{\partial y}\right\}$$
$$+ \left[\frac{\partial}{\partial x}\left(cc_g\frac{\partial a}{\partial x}\right) + \frac{\partial}{\partial y}\left(cc_g\frac{\partial a}{\partial y}\right)\right] - cc_g\left\{\left[\left(\frac{\partial S}{\partial x}\right)^2 + \left(\frac{\partial S}{\partial y}\right)^2\right] - k^2\right\}a = 0 \tag{5.163}$$

Both the real part and the imaginary part in the above equation need to be zero. The real part reduces to the eikonal equation as follows:

$$|\nabla S|^2 = |\mathbf{K}|^2 = \kappa^2 = k^2 + \frac{1}{acc_g}\nabla \cdot (cc_g\nabla a) = k^2 + \frac{1}{a}\nabla^2 a + \frac{\nabla(cc_g)}{cc_g} \cdot \frac{\nabla a}{a} \tag{5.164}$$

which represents the correction of the effective wave number due to the change of topography that causes both wave diffraction (the second term on the RHS) and wave refraction (the third term on the RHS). Note that wave diffraction can occur without wave refraction but wave refraction always occurs with wave diffraction. Generally speaking, for a diffracted/refracted wave field, the wave number k obtained from the linear dispersion equation will be different from the effective wave number κ, from which the local wave propagation direction is determined. Using the identity of $\nabla \times \mathbf{K} = \nabla \times (\nabla S) = 0$ (irrotationality of the effective wave number vector), we are able to uniquely determine the value of S, from which we can trace the wave ray direction.

The imaginary part of (5.163) can be reduced to:

$$\nabla \cdot (cc_g a^2 \nabla S) = 0 \tag{5.165}$$

This is the energy conservation equation. By realizing that wave energy density $E = \rho g a^2/2$ and defining the new wave group velocity vector on varying bathymetry as:

$$\mathbf{v}_g = c_g \nabla S/k = c_g \mathbf{K}/k$$

the above equation can be reduced to the shoaling formula based on the conservation of wave energy flux, i.e.:

$$\nabla \cdot (\mathbf{v}_g E) = 0 \tag{5.166}$$

It shows that the wave energy is propagating in the direction of $\mathbf{K} = \nabla S$, which is orthogonal to wave crests, at the rate of $|\mathbf{v}_g|$.

5.3.3.4 Parabolic mild-slope equation

In the analysis, if we force the phase function S to be a specific form throughout the domain, most of the time being the same as the far-field expression though it can be any other form in principle, the complex wave amplitude function can then be re-expressed as, if the far-field wave propagates in the x-direction with the wave number k_{x0}:

$$F(x, y) = a(x, y)e^{iS(x,y)} = a(x, y)e^{i(K_x x + K_y y)} = a(x, y)e^{i\left[k_{x0} x + (K_x - k_{x0})x + K_y y\right]}$$

$$= a(x, y)e^{i\left[(K_x - k_{x0})x + K_y y\right]}e^{ik_{x0}x} = A(x, y)e^{ik_{x0}x} = A(x, y)e^{iS'} \tag{5.167}$$

In this case, the slowly varying wave amplitude function:

$$A(x, y) = a(x, y)e^{i\left[(K_x - k_{x0})x + K_y y\right]}$$

will have to be a complex function in general when a wave changes its direction from the original x-direction due to either wave refraction and/or wave diffraction. By substituting the above expression into the original EMSE (5.159), we have the following equation:

$$i\left\{ A\frac{\partial}{\partial x}(k_{x0}cc_g) + 2k_{x0}cc_g\frac{\partial A}{\partial x} \right\} + \left[\frac{\partial}{\partial x}\left(cc_g\frac{\partial A}{\partial x} \right) + \frac{\partial}{\partial y}\left(cc_g\frac{\partial A}{\partial y} \right) \right]$$
$$- cc_g\left\{ k_{x0}^2 - k^2 \right\} A = 0 \tag{5.168}$$

This equation is similar to (5.163) but has a complex $A(x, y)$. As a result, the eikonal equation and the energy equation are no longer separable and need to be solved together. In (5.168), the value of k_{x0} can be arbitrary and the deviation of S' from S will be absorbed into $A(x, y)$. Sometimes we can also set $k_{x0} = k(x, y)$ and this makes the last term in (5.168) zero. So far (5.168) is still exact and it is equivalent to (5.163) but expressed differently.

Compared with the coupled eikonal and energy equations, the above equation is decoupled to some extent by absorbing the phase function into the slowly varying complex amplitude function $A(x, y)$. The elliptic equation can be solved to obtain $A(x, y)$, given proper boundary conditions. Once $A(x, y) = A_r + iA_i$ is known, the real wave amplitude can be readily obtained by $a(x, y) = |A(x, y)| = \sqrt{A_r^2 + A_i^2}$. The actual phase function can be obtained by the following equation:

$$S(x, y) = K_x x + K_y y = \arctan\left(\frac{A_i \cos S' + A_r \sin S'}{A_r \cos S' - A_i \sin S'}\right)$$

$$= \arctan\left[\frac{A_i \cos(k_{x0}x) + A_r \sin(k_{x0}x)}{A_r \cos(k_{x0}x) - A_i \sin(k_{x0}x)}\right] \qquad (5.169)$$

The effective wave number can then be obtained by using the definition of $\mathbf{K} = \nabla S$, from which the wave propagation angle can be determined, i.e., $\theta = \arctan(K_y/K_x)$.

If we assume that wave diffraction occurs primarily in the transverse y-direction (perpendicular to the primary wave propagation in the x-direction), equation (5.168) can be simplified to:

$$i2k_{x0}cc_g\frac{\partial A}{\partial x} + \left[(k^2 - k_{x0}^2)cc_g + i\frac{\partial(k_{x0}cc_g)}{\partial x}\right]A + \frac{\partial}{\partial y}\left(cc_g\frac{\partial A}{\partial y}\right) = 0 \quad (5.170)$$

This equation has a parabolic form against the original EMSE in elliptic form. Therefore, it is often called parabolic MSE (PMSE). A similar expression was first obtained by Radder (1979) in an effort to derive the parabolic approximation of the MSE and applied it to the study of wave passage through a submerged shoal during which both wave refraction and diffraction are important. The original motivation of parabolizing the EMSE is to reduce the computational cost of solving the elliptic equation.

When the nonlinear term is kept in the derivation, the above equation can be modified to describe the forward-scattering second-order Stokes waves (Kirby and Dalrymple, 1983b; Liu and Tsay, 1984), i.e.:

$$i2k_{x0}cc_g\frac{\partial A}{\partial x} + \left[(k^2 - k_{x0}^2)cc_g + i\frac{\partial(k_{x0}cc_g)}{\partial x}\right]A + \frac{\partial}{\partial y}\left(cc_g\frac{\partial A}{\partial y}\right)$$

$$- kcc_g G|A|^2 A = 0 \qquad (5.171)$$

where

$$G = k^3 \left(\frac{c}{c_g}\right)\frac{\cosh 4kh + 8 - 2\tanh^2 kh}{8\sinh^4 kh}$$

In constant water depth $k = k_{x0}$, the above equation is reduced to the familiar NLS equation:

$$2ik\frac{\partial A}{\partial x} + \frac{\partial^2 A}{\partial y^2} - kG|A|^2 A = 0 \tag{5.172}$$

By using the Wentzel–Kramers–Brillouin (WKB) expansion, Yue and Mei (1980) arrived at the same expression for weakly nonlinear Stokes waves propagating toward a specific direction in constant water depth. In this case, the wave diffraction is the only mechanism that causes the change of wave propagation direction. In this case although the magnitude $|\nabla S| = \kappa = k_{x0} = |\nabla S'|$, the vector $\nabla S \neq \nabla S' = (k_{0x}, 0)$ in general (consider waves behind a breakwater where waves bend into the shadow area by changing the propagation direction). Therefore, $A(x, y)$ must remain a complex function to account for the difference between the actual phase function $S(x, y)$ and its approximation $S' = k_{x0}x$.

When only linear waves are considered, the above equation can be further reduced to the linear Schrödinger equation (e.g. Lozano and Liu, 1980), i.e.:

$$2ik\frac{\partial A}{\partial x} + \frac{\partial^2 A}{\partial y^2} = 0 \tag{5.173}$$

This is the lowest order parabolic approximation of the original Helmholtz equation. The equation implies that the diffraction process is similar to a diffusion process but with imaginary diffusivity, which causes the change of wave phase rather than the decaying of wave energy that is proportional to $|A|^2$.

Later, Dalrymple and Kirby (1988) extended the validity range of the parabolic approximation and introduced an angular spectrum concept into their model. Liu (1990) summarized the development of parabolic approximation and its application range. The major advantage of the PMSE is that the numerical solution can be marched from deep water to shallow water without iteration, similar to solving an initial value problem but replacing "t" with "x." The research on the PMSE was most active in the 1980s but it became less popular later, partly because of its theoretical limitation to narrowly banded waves only. Nevertheless, the theoretical works of deriving various versions of PMSE provide important theoretical insight on wave refraction and diffraction of a single beam of wave ray, which was later used by Lin *et al.* (2005) to derive the wave energy spectral model capable of handling wave diffraction.

5.3.3.5 *Modified mild-slope equation*

The original MSE is limited by three drawbacks: (1) the equation is applicable only to linear waves; (2) the equation is applicable only to waves on very mild and impermeable bottom geometry; and (3) the equation does not contain energy dissipation. Recently, the advances in theoretical derivation of the MSE are mainly focused on the stretch of its applications to weakly nonlinear waves and steeper bottom slope. The inclusion of various types of energy dissipation was also attempted. This forms the group of so-called modified MSEs (MMSEs) or sometimes extended MSEs.

Nonlinear waves: The inclusion of nonlinear wave effects, even for the lowest order, can have a pronounced effect on wave refraction and diffraction (Kirby and Dalrymple, 1983b; Liu and Tsay, 1984). In principle, the nonlinear wave effects can be considered by including higher order nonlinear terms in the formulation (e.g., Yue and Mei, 1980) and/or the use of the corrected nonlinear dispersion equation (e.g., Booij, 1981). The former approach results in additional nonlinear terms in the MMSE. The recent work on second-order MSEs was developed by Chen and Mei (2006). In the latter approach, the linear dispersion equation is modified as follows (Kirby and Dalrymple, 1986):

$$\sigma^2 = gk \tanh \left(kh + ka \left(\frac{kh}{\sinh(kh)} \right)^4 \right)$$

$$\times \left\{ 1 + \tanh^5(kh)(ka)^2 \frac{\cosh(4kh) + 8 - 2\tanh^2(kh)}{8\sinh^4(kh)} \right\} \tag{5.174}$$

Steep bottom slopes: The original MSE assumes a slowly varying bottom slope, i.e., $|\nabla h| \ll 1$. Booij (1983) investigated numerically the wave reflection from a plane slope and concluded that the MSE model is acceptable up to $\nabla h = 1/3$. To account for a rapidly varying topography effect, the higher order bottom slope must be considered in the derivation. Kirby (1986) extended the EMSE to include a rapidly varying topography. O'Hare and Davies (1992) and Guazzelli *et al.* (1992) included fast-changing bottom undulations by approximating the bed as a series of horizontal shelves and used the MMSE to study Bragg reflection phenomenon. Massel (1993) found that the bottom effect can be more rigorously considered by including higher order bottom effect terms that are proportional to the bottom curvature and the square of the bottom slope. Adopting a similar idea, Chamberlain and Porter (1995) proposed a widely used form of the MMSE that is as follows:

$$\frac{\partial}{\partial x}(cc_g \frac{\partial F}{\partial x}) + \frac{\partial}{\partial y}(cc_g \frac{\partial F}{\partial y}) + \left\{ k^2 cc_g + gu_1 \left(\frac{\partial^2 h}{\partial x^2} + \frac{\partial^2 h}{\partial x^2} \right) \right.$$

$$\left. + gu_2 \left(\left(\frac{\partial h}{\partial x} \right)^2 + \left(\frac{\partial h}{\partial y} \right)^2 \right) \right\} F = 0 \tag{5.175}$$

where

$$u_1 = \frac{\text{sech}^2 kh}{4(2kh + \sinh 2kh)} (\sinh 2kh - 2kh \, \cosh 2kh)$$

$$u_2 = \frac{k\text{sech}^2 kh}{12(2kh + \sinh 2kh)^3} \times \left\{ (2kh)^2 + 4(2kh)^3 \sinh 2kh - 9 \sinh 2kh \sinh 4kh \right.$$

$$\left. + 6kh(2kh + 2 \sinh 2kh)(\cosh^2 2kh - 2 \cosh 2kh + 3) \right\}$$

Later, Chandrasekera and Cheung (1997) and Suh *et al.* (1997) also developed different versions of MMSE to include a rapidly varying bathymetry. Miles and Chamberlain (1998) developed a hierarchy of PDEs and proved that the system can be reduced to various orders of MSE. Agnon and Pelinovsky (2001) performed a systematic derivation and compared their results with various MMSEs. Using the multiple scale method to enforce the seabed boundary condition, Liu and Shi (2008) proposed the so-called complementary MSE for wave propagation over an uneven bottom. It shows that the MMSE in general does provide a more accurate description of wave refraction and diffraction around rapidly varying bottom geometry such as a shoal. The MMSE including a higher order depth effect is especially useful in the simulation of wave propagation problems over depth discontinuity and/or with larger curvature of depth (e.g., ripples and sandbars). Further discussion of the characteristics of MMSEs can be found in Porter (2003) and Hsu *et al.* (2006).

Energy dissipation: Bottom friction and wave breaking are two main sources of energy dissipation when waves approach the shoreline. Dalrymple *et al.* (1984) and Dally *et al.* (1985) proposed the following modified EMSE:

$$\frac{\partial}{\partial x}(cc_g \frac{\partial F}{\partial x}) + \frac{\partial}{\partial y}(cc_g \frac{\partial F}{\partial y}) + \left(k^2 cc_g + i\sigma w + ic_g \sigma \gamma \right) F = 0 \tag{5.176}$$

where w is the damping factor (Dalrymple *et al.*, 1984):

$$w = \frac{2c_f}{3\pi} \frac{ak\sigma}{\sinh 2kh \sinh kh} \tag{5.177}$$

with c_f being the dimensionless friction coefficient that is the function of Re and bottom roughness. γ is the wave-breaking parameter:

$$\gamma = \frac{\chi}{h} \left(1 - \frac{\Gamma^2 h^2}{4a^2} \right) \tag{5.178}$$

where $\chi = 0.15$ (Dally *et al.*, 1985) and $\Gamma = 0.4$ (Demirbilek and Panchang, 1998).

The inclusion of wave-breaking effects in the surf zone was recently attempted and it also resulted in additional nonlinear terms. This requires special treatment in the numerical solution (e.g., Tao and Han, 2001; Zhao *et al.*, 2001; Chen *et al.*, 2005).

Porous beds: Another extension of MSE is to simulate wave propagation over porous beds or through porous structures. Losada *et al.* (1996) presented an MSE model to study wave propagation over porous slopes. Later, Hsu and Wen (2001) proposed a time-dependent parabolic equation to describe wave propagation over permeable beds. Recently, Tsai *et al.* (2006) proposed a new time-dependent MSE to solve wave transformation over porous breakwaters on porous beds.

5.3.3.6   Mild-slope equation models and their applications

The MSE enjoys the theoretical advantage of being capable of modeling waves from deep water to shallow water without the limitation of kh as long as the wave is linear. This is because the linear dispersion relation is explicitly enforced in the equation, and thus the vertical variation of velocity is correctly accounted for in the vertical integration for all kh. The MSE is frequently used to study the combined wave refraction and diffraction above a changing topography (e.g., shoal, pit, or trench) or around an idealized island. Many MSE models were developed in the past few decades based on the various versions of MSE. For example, the REF/DIF model was developed by Dalrymple and Kirby (1985) based on PMSE. A general-purpose coastal wave model CGWAVE was developed at the University of Maine sponsored by USACE based on EMSE (Demirbilek and Panchang, 1998). The model was constructed on FEM formulation.

Waves above a shoal on a beach: The problem the MSE models are often tested against is the wave propagation over an elliptic shoal on a slope (Berkhoff *et al.*, 1982). In this case, the monochromatic wave train has an incident angle of 20° to a plane beach. The combined wave refraction and diffraction around the shoal cause the wave to focus behind the shoal where the wave is amplified and wave nonlinearity becomes significant. Many numerical models based on MSE studied this problem to verify either the model convergence and accuracy (e.g., Panchang *et al.*, 1991) or the contribution from nonlinear terms (e.g., Kirby and Dalrymple, 1984; Zheng *et al.*, 2001). A similar problem of a cusped caustic of the focusing wave behind a shoal was numerically investigated by Liu and Tsay (1984) and compared to Whalin's (1971) experimental data. The computation of wave propagation over an elliptic shoal in constant water depth was originally made by Berkhoff (1972) using FDM and later by Bettess and Zienkiewicz (1977) using FEM and infinite element method and by Zhu (1993) using BEM.

Wave transformation over a changing bathymetry: Another popular problem that is often studied by MSE models is the wave transformation and reflection over a changing bathymetry, e.g., ripple beds, trenches, pits. For example, Dalrymple *et al.* (1989) have investigated wave diffraction over a bottom with slowly varying ridges and troughs. With the use of time-dependent MSE models, Suh *et al.* (1997) investigated the strong Bragg reflection, which occurs when the normally incident wave number is about half of that for the underlying periodic bottom undulation (e.g., sandbar). Their numerical results were compared with the experimental data by Davies and Heathershaw (1984) for a singly sinusoidal bottom and by Guazzelli *et al.* (1992) for a doubly sinusoidal ripple bed. Recently, Porter and Porter (2001) used MMSE to study the interaction of waves with 3D periodic topography.

Combined wave refraction and diffraction around an island: Another commonly studied problem using MSE models is the combined wave refraction and diffraction around an island. For example, Smith and Sprinks (1975) developed an integral technique for modal decomposition to study water waves around a conical island. The trapped edge wave modes around the island were computed numerically. Using the collocation method, Jonsson *et al.* (1976) solved the MSE for wave scattering around a circular island on a parabolic shoal, for which the analytical solution was obtained in a shallow water limit by Homma (1950). The same problem was later studied by Houston (1981) using eigenfunctions. An FD model was developed by Lo (1991) to study the wave condition around an arbitrary island. Zhu (1993) developed an efficient dual-reciprocity boundary element method (DRBEM) model based on MSE to study wave diffraction and refraction around islands. Using the MMSE, Chamberlain and Porter (1999) developed a methodology based on angular decomposition to study the near-trapping of water waves around any axisymmetrical topography. Their results were compared with the earlier modeling results by Xu and Panchang (1993). Lin *et al.* (2002) extended the study to random wave refraction and diffraction around a circular island and conducted experiments to verify their computation based on the MSE model. Lin (2004a) developed a compact FD time-dependent MSE model and used it to study wave scattering around a circular island on a parabolic shoal. Figure 5.12 gives an example of a simulated quasi-steady-state wave field around the island using this time-dependent MSE model.

Harbor resonance: Harbor resonance is an important issue in harbor design. For harbors with simple geometry, an analytical solution based on linear wave theory is possible. For example, Ippen and Goda (1963) presented an analytical solution for harbor resonance in a rectangular harbor. In reality, both the wall reflection coefficient and the bottom friction will contribute to modify the actual resonance pattern. The MSE model is a good candidate for simulating harbor resonance in an actual harbor with a complex boundary configuration, varying bottom bathymetry and bottom friction, partially reflecting walls, etc. Using CGWAVE, Demirbilek and Panchang (1998) simulated rectangular harbor resonance and compared their simulation results with the analytical solution (Ippen and Goda, 1963) and laboratory data (Ippen and Goda, 1963; Lee, 1971).

Random wave simulation: To be practically useful in ocean and coastal engineering to model actual sea state, the inclusion of random waves in MSE models is needed. The linear angular spectrum wave model based on linear MSE was proposed by Dalrymple and Kirby (1988) to study directional

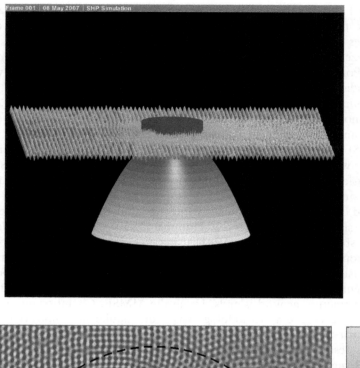

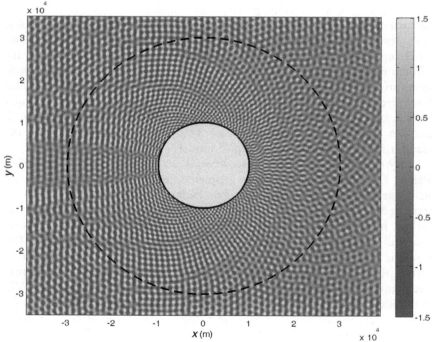

Figure 5.12 Simulated wave pattern around a circular island above a shoal (top for 3D view and bottom for 2D bird's–eye view); the computational domain covers a domain of 100 km × 100 km with 3000 × 3000 uniform cells; the steady wave field is obtained after running Lin (2004a)'s model for 16,000 time steps using 24 CPU hours in a Pentium 4 3.6 GHz PC.

spreading wave propagation. The model was extended to the nonlinear version (Suh *et al.*, 1990) and the inclusion of wave breaking (Chawla *et al.*, 1998). Recently, the use of the parabolic approximation model for spectral waves, OLUCA-SP, in the application of random directional wave breaking over a submerged circular shoal was reported by Liu and Losada (2002). In the model, the energy dissipation due to wave breaking has been modeled by the empirical formulas proposed by Battjes and Janssen (1978), Thornton and Guza (1983), and Rattanapitikon and Shibayama (1998).

Limitations: Although the MSE models have proven to be successful to some extent for simulating nonlinear random breaking waves, the MSE models were rarely used to study the detailed wave phenomenon in swash zones for wave run-up. The main reason is that the wave dispersion equation, even after including nonlinear effects, will fail for large wave nonlinearity beyond the breaking point. At this stage, the explicit enforcement of the dispersion equation, which makes MSEs advantageous over the Boussinesq equations in deep water, becomes its drawback. Furthermore, the nonlinear wave interaction in the surf zone that generates new wave modes and currents cannot be handled by MSEs. For this reason, the MSE models are used less often in modeling waves in coastal waters when compared with the Boussinesq models.

5.3.3.7 Analytical studies based on MSE and its long-wave approximation

Since the EMSE is a linear PDE and has a relatively simple form, analytical approaches have been attempted to obtain the closed-form solution for simple cases. The most common approximation is to assume long waves so that the EMSE reduces to the linear long-wave equation. For example, Homma (1950) proposed an analytical solution of wave transformation around a circular island on a paraboloidal shoal based on the linear long-wave equation in order to find tsunami response near an island. It was found that for a certain incident wave frequency, the wave amplitude around the island can be greatly amplified. Later, Longuet-Higgins (1967) proposed an analytical solution for long-wave propagation over a submerged circular sill, during which the near-trapping phenomenon occurs that causes the amplification of waves above the sill. Summerfield (1972) extended the Longuet-Higgins's study to wave transformation around a small surface-piercing circular cylinder mounted on a submerged larger circular sill. Zhang and Zhu (1994) proposed the series solutions for long-wave scattering around a conical island and over a paraboloidal shoal. The study was later extended to waves around a circular island above a conical shoal (Zhu and Zhang, 1996). Starting with MSE but making the assumption of long-wave approximation, Yu and Zhang (2003) proposed a series-form analytical solution of wave amplification around a circular shoal with the profile of different power functions. Suh *et al.* (2005) proposed the analytical solution for wave transformation and attenuation above a circular pit. Liu and Li (2007) proposed an analytical solution for wave scattering around a circular shoal with the profile of an arbitrary power function and being truncated at any level. Their results can be reduced to that obtained from Longuet-Higgins (1967) for large values of power m. All the above theoretical works are based on linear SWEs, which can be regarded as the limiting factor of the MSE with long-wave approximation.

The above analytical solutions, when compared to both experimental data and analysis based on potential flow theory, overestimated wave amplification above the shoal due to the use of long-wave approximation (Renardy, 1983; Miles, 1986). This calls for a better description of the physics. The original the MSE is apparently a good choice for this purpose. However, since the wave number and the water depth are coupled implicitly through dispersion equation in the MSE, a direct closed-form analytical solution to MSE is impossible, even for rather simple bathymetry. Only recently, by using Hunt's (1979)'s direct approximation to the linear dispersion equation, an approximate analytical solution was obtained by Liu *et al.* (2004) for shorter wave scattering around Homma's Island. The method was then extended to study wave scattering and trapping above a truncated paraboloidal shoal (Lin and Liu, 2007).

5.3.3.8   Unified equations

If we make a comparison between time-dependent MSEs and Boussinesq equations, we find that the two types of equations share some similarities, e.g., both are depth-averaged equations and both can be split into a system of two hyperbolic PDEs. While the MSE has the advantage of representing waves from deep water to shallow water with the explicit use of dispersion equation, the Boussinesq equations are primarily used in relatively shallow water with stronger wave nonlinearity. It is not surprising that people want to find a unified model that contains the advantages from both the MSE model and the Boussinesq model.

The development of a unified wave model started by either extending the Boussinesq equations to include the full dispersive effect or extending the MSE to include the fully nonlinear effect. One of the earliest attempts was made by Witting (1984) to create a unified model for the evolution of nonlinear water waves. Using the series expansion solution, his model can provide accurate results for nearly breaking solitons. However, although he termed his model a "unified model," it is more appropriate to classify it as a higher order Boussinesq model. Later, Karambas (1999) and Wu (2000) developed different "unified models" of the Boussinesq type. In contrast, Tang and Ouellet (1997) proposed a nonlinear MSE and claimed that the model captures the same wave nonlinearity as a Boussinesq model does. A similar extension was made by Li and Fleming (1999) and Huang *et al.* (2001) with the use of variational principle to develop the unified models that can be reduced to MSEs for small-amplitude waves and nonlinear Boussinesq equations in shallow water for finite-amplitude waves.

The difficulty of having a really unified model is the fundamental dilemma of the accurate representation of water waves in both deep and shallow water. In deep water, waves are essentially linear or weakly nonlinear. In this case, the wave form is almost symmetric. The linear and Stokes wave theories, which are governed by dispersion relationship, can precisely describe the wave motion. The MSE inherits the use of the dispersion equation and thus can accurately describe wave motion in deep water. Although the use of dispersion relation ensures the accurate simulation of wave propagation, it limits the application of MSE to purely periodic wave motion only. As the waves get into shallow water with the increase of wave nonlinearity, the wave front becomes steepened, creating an asymmetric wave profile and destroying the dispersion relationship by generating new wave modes and wave-induced current through nonlinear wave interaction. This requires the relaxation of the dispersion equation that is true only for linear periodic waves. Boussinesq models are ideal candidates because the dispersion equation is not explicitly enforced. By abandoning the dispersion equation, there is no need to know the wave number (or wave frequency) in advance. The generation of new wave modes, the steepening of the wave front, and other nonlinear wave phenomena can be automatically handled by the nonlinear terms in the Boussinesq model. The absence of the dispersion equation, however, makes the Boussinesq model restricted to relatively shallow-water region.

Realizing the above facts, a unified model can be successful only if it can generally phase out its dependency on wave number information when waves propagate from deep to shallow water. Such a model, unfortunately, is not available for practical engineering computations but may become needed in the future with the advancement of computer power that makes possible the large-scale phase-resolved computation for wave propagation from very deep water to shallow water.

5.3.4 *Wave spectral models*

5.3.4.1 *Background of wave energy spectrum*

Definition of spectral density function: Waves in the ocean contain random components. A random sea is composed of waves with different frequencies and phases and propagate in different directions. Consider an ocean surface that consists of many wave components slowly varying in time and space. Assuming each wave component is of a linear wave train, the sea state can then be recovered by the integration of waves over frequency and direction spaces as follows:

$$\eta(x, y, t) = \sum_{i=1}^{\infty} \sum_{j=1}^{\infty} \eta_{ij}(x, y, t, \sigma_i, \theta_j) = \int_0^{2\pi} \int_0^{\infty} \eta(x, y, t, \sigma, \theta) d\sigma d\theta$$

$$= \int_0^{2\pi} \int_0^{\infty} a(\sigma, \theta, x, y, t) \cos\left[k(\sigma) \cos\theta x + k(\sigma) \sin\theta y\right.$$
$$\left. -\sigma t + \delta(\sigma, \theta)\right] d\sigma d\theta \qquad (5.179)$$

where $\delta(\sigma, \theta)$ is the random phase information that does not affect the mean wave energy. Each wave component follows a linear dispersion relationship, from which the wave number can be determined locally with the information of local water depth:

$$\sigma^2 = k(\sigma)g \tanh [k(\sigma)h] \tag{5.180}$$

For each wave component, the energy density is related to wave amplitude by the following equation:

$$E(\sigma, \theta, x, y, t) = \frac{1}{2}\rho g a^2 (\sigma, \theta, x, y, t) \tag{5.181}$$

Based on (5.181), we can make the following definition:

$$E(\sigma, \theta, x, y, t) = \rho g \left(\frac{1}{2}a^2 (\sigma, \theta, x, y, t) \right) = \rho g F(\sigma, \theta, x, y, t) \, d\sigma \, d\theta \tag{5.182}$$

where $F(\sigma, \theta, x, y, t)$ is called wave energy spectral density function and can be regarded as the normalized (by ρg) wave energy density per unit frequency and unit directional spreading angle. This gives the following formula for wave amplitude:

$$a(\sigma, \theta, x, y, t) = \sqrt{2F(\sigma, \theta, x, y, t) \, d\sigma \, d\theta} \tag{5.183}$$

The wave frequency spectrum and directional spectrum can be obtained by taking the integration of $F(\sigma, \theta, x, y, t)$ over directional and frequency spaces, respectively:

$$S(\sigma, x, y, t) = \int_0^{2\pi} F(\sigma, \theta, x, y, t) \, d\theta \tag{5.184}$$

$$D(\theta, x, y, t) = \int_0^{\infty} F(\sigma, \theta, x, y, t) \, d\sigma \tag{5.185}$$

and we also have:

$$E_{\text{total}}(x, y, t) = \int_0^{2\pi} \int_0^{\infty} E(\sigma, \theta, x, y, t) \, d\sigma \, d\theta$$

$$= \rho g \int_0^{2\pi} \int_0^{\infty} F(\sigma, \theta, x, y, t) \, d\sigma \, d\theta$$

$$= \rho g \int_0^{2\pi} D(\theta, x, y, t) \, d\theta = \rho g \int_0^{\infty} S(\sigma, x, y, t) \, d\sigma \tag{5.186}$$

where $E_{\text{total}} = \rho g a_{\text{rms}}^2 / 2$, with a_{rms} being the root-mean-square wave amplitude. It is noted that, in principle, the wave spectrum can also be expressed as a function of wave number vector (k_x, k_y), which is mutually convertible to frequency–direction space (σ, θ) by using the linear dispersion equation. Such representation, however, is less popularly used in wave modeling.

Frequency spectrum: For wind waves in oceans, the distribution of wave energy in different frequency and direction components is dependent upon the fetch area of the wind, the strength of the wind, and the relative location in the fetch area. For an infinitely large fetch area, Phillips (1958) found that the equilibrium range of the wave spectrum of a fully developed sea is independent of wind speed, location, and time. Under this circumstance, the simplified spectrum can be expressed as:

$$S_P(\sigma) = \alpha g^2 \sigma^{-5} \tag{5.187}$$

where α is called Phillips constant. Phillips spectrum represents the upper limit of the energy spectrum. Although it is rarely used in any practical computation, it serves as the basis for other spectra that are popularly used in practice.

Pierson–Moskowitz (P–M) spectrum (1964) is an energy spectrum that is widely used in offshore engineering. This spectrum is essentially the modification of Phillips spectrum by considering the additional factor of wind speed for a fully developed sea:

$$S_{\text{P–M}}(\sigma) = \alpha g^2 \sigma^{-5} \exp\left[-0.74(\frac{\sigma U_w}{g})^{-4}\right] \tag{5.188}$$

where $\alpha = 0.0081$ and U_w is the wind speed.

Another popular frequency spectrum is JONSWAP spectrum developed during a JOint North Sea WAve Project (Hasselmann *et al.*, 1973), which essentially is the modified P–M spectrum by further considering the fetch length:

$$S_J(\sigma) = \alpha g^2 \sigma^{-5} \exp\left[-1.25\left(\frac{\sigma}{\sigma_0}\right)^{-4}\right] \gamma^{\exp\left[-\frac{(\sigma-\sigma_0)^2}{2\tau^2\sigma_0^2}\right]} \tag{5.189}$$

where γ is the peak enhancement parameter, τ the shape parameter, and σ_0 the peak angular frequency:

$$\gamma = 3.3 \text{ (varying range } 1-7), \quad \tau = \begin{cases} 0.07(\sigma \le \sigma_0) \\ 0.09(\sigma > \sigma_0) \end{cases} \quad \sigma_0 = \frac{2\pi g}{U_w}\left(\frac{gX}{U_w^2}\right)^{-0.33} \tag{5.190}$$

in which X is the length of the fetch.

In shallow water, the so-called TMA (Texel-Marsen-Arsloe) energy spectrum (Bouws *et al.*, 1985), which is extended from the original JONSWAP spectrum by considering the finite depth, can be used:

$$S_{TMA}(\sigma) = S_J(\sigma)\phi(\sigma, h) \tag{5.191}$$

where

$$\phi(\sigma, h) = \begin{cases} 0.5\omega_h{}^2, & \omega_h \le 1 \\ 1 - 0.5(2 - \omega_h)^2, & 1 < \omega_h \le 2 \\ 1, & \omega_h > 2 \end{cases}$$

and $\omega_h = 2\pi f \sqrt{h/g}$.

There are also other frequency energy spectra, i.e., Neumann spectrum (Neumann, 1953), Bretschneider spectrum (Bretschneider, 1959), ISSC spectrum (International Ship Structures Congress, 1964), ITTC spectrum (ITTC, 1972), Scott Spectrum (Scott, 1965), Liu spectrum (Liu, 1971), Mitsuyasu spectrum (Mitsuyasu, 1972), Ochi–Hubble spectrum (Ochi and Hubble, 1976). Chakrabarti (1987: 120) compared various types of spectra.

Directional spreading spectrum: There are fewer directional spreading spectra when compared with frequency spectra. The two most popular models include cosine-power model and the wrapped-around Gaussian (normal spreading) model, as introduced below. The cosine-power model was first proposed by Pierson *et al.* (1955). It was later refined by Goda (2000: 32) in the following form:

$$D(\theta) = G_0 \cos^{2\beta}\left(\frac{\theta - \theta_0}{2}\right) \tag{5.192}$$

where θ_0 is the principal propagation direction, β is the narrowness coefficient for directional spreading, and G_0 is a constant that normalizes $D(\theta)$:

$$G_0 = \left[\int_{-\pi}^{\pi} \cos^{2\beta}\left[(\theta - \theta_0)/2\right] d\theta\right]^{-1} \tag{5.193}$$

The wrapped-around Gaussian model is essentially an exponential model. It was proposed by Mardia (1972: 239) and refined by Briggs *et al.* (1987) in the following form:

$$D(\theta) = \frac{1}{2\pi} + \frac{1}{\pi}\sum_{n=1}^{N} \exp[-(n\sigma_s)^2/2]\cos[n(\theta - \theta_0)] \tag{5.194}$$

where σ_s is the circular standard deviation in radians and it can be set to 0.6 for most offshore engineering applications. Unlike the cosine-power model, the above wrapped-around Gaussian function does not reduce to zero when $\theta - \theta_0 = \pm\pi$. Briggs *et al.* (1987) suggested the use of $N = 5$.

Besides, there are also the von Mises model based on the modified Bessel function of zeroth order (Abramowitz and Stegun, 1964), the hyperbolic function model (Donelan *et al.*, 1985), and the double-peak spreading model (Zakharov and Shrira, 1990).

5.3.4.2 The transport equation for the wave energy spectrum

Dynamic equation: Based on the concept of energy conservation, a general dynamic equation for the directional wave spectrum can be expressed as follows:

$$\frac{dF}{dt} = \frac{\partial F}{\partial t} + \frac{d\xi}{dt}\frac{\partial F}{\partial \xi} = Q \tag{5.195}$$

where $\xi = (\mathbf{x}, \mathbf{k})$ or $\xi = (\mathbf{x}, \omega, \theta)$ is the multidimensional space variable and Q is the source function that contributes to the local wave energy dissipation, wave energy input, or wave energy exchange between various variable spaces through wave interaction.

In Cartesian coordinates when the divergences of the time derivatives of the space variables are zero, i.e., $\nabla_\xi \cdot (d\xi/dt) = 0$, we have:

$$\frac{\partial F}{\partial t} + \nabla_\xi \cdot \left(\frac{d\xi}{dt}F\right) = Q \tag{5.196}$$

If the wave number space variables are used, the above equation becomes:

$$\frac{\partial F}{\partial t} + \nabla_x \cdot \left(\frac{dx}{dt}F\right) + \nabla_k \cdot \left(\frac{dk}{dt}F\right) = Q \tag{5.197}$$

In contrast, if the wave frequency and direction variables are used, the equation has the following form:

$$\frac{\partial F}{\partial t} + \nabla_x \cdot \left(\frac{dx}{dt}F\right) + \nabla_\omega \cdot \left(\frac{d\sigma}{dt}F\right) + \nabla_\theta \cdot \left(\frac{d\theta}{dt}F\right) = Q \tag{5.198}$$

To apply the above equation in a practical computation, we need to define:

$$\frac{dx}{dt} \& \frac{dk}{dt} \quad \left(\text{or } \frac{dx}{dt} \& \frac{d\omega}{dt} \& \frac{d\theta}{dt}\right)$$

and Q, the former of which can be exactly derived by applying the theory of wave kinematics as shown below and the latter of which must be closed empirically by considering various physical processes.

Wave kinematics: Consider a linear wave train propagating over a changing topography and a nonuniform current $\mathbf{U}$ (Figure 5.13). The free surface displacement of the waves can be described by the following expression:

$$\eta(x, y, t) = a(x, y)e^{iS(x,y,t)} = a(x, y)e^{i(\mathbf{k}\cdot\mathbf{x}-\omega t)} = a(x, y)e^{i(k_x x + k_y y - \omega t)} \tag{5.199}$$

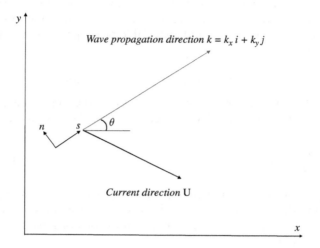

Figure 5.13 Illustration of a steady linear wave train propagating with a current.

where S is the phase function and $k = \sqrt{k_x^2 + k_y^2}$ is the wave number with the local wave propagation angle $\theta = \tan^{-1}(k_y/k_x)$.

In the analysis of wave refraction (Section 3.10), the following conservation equation of wave number has been derived:

$$\frac{\partial k_i}{\partial t} + \frac{\partial \omega}{\partial x_i} = 0 \tag{5.200}$$

and the irrotationality of wave number reads:

$$\frac{\partial k_i}{\partial x_j} - \frac{\partial k_j}{\partial x_i} = 0 \tag{5.201}$$

In contrast, the study of wave–current interaction in Section 3.14.4 provides the relationship between the apparent angular frequency ω and the intrinsic angular frequency σ as follows:

$$\omega = \sigma(k, h) + \mathbf{k} \cdot \mathbf{U} = \sqrt{gk \tanh kh} + \mathbf{k} \cdot \mathbf{U} \tag{5.202}$$

The following interpretation can be obtained by observing (5.200) and (5.202), respectively:

1. The change of apparent angular frequency ω in space will occur only when there is a change of wave field in time; in other words, for steady waves, ω is uniform in space.
2. The intrinsic angular frequency σ can vary in both time and space when the waves propagate over an unsteady nonuniform current.

Expression of dx_i/dt: This term represents the velocity vector at which wave energy propagates. It is the most straightforward term to be evaluated by considering both current and wave effects:

$$\frac{dx_i}{dt} = \frac{\partial \omega}{\partial k_i} = \frac{\partial \sigma}{\partial k_i} + \frac{\partial}{\partial k_i}(k_i U_i) = c_{gi} + U_i \qquad (5.203)$$

To obtain the above equation, we have used the definition of wave group velocity $c_{gi} = \partial \sigma / \partial k_i$.

Expression of dk_i/dt: The total derivative can be written as:

$$\frac{dk_i}{dt} = \frac{\partial k_i}{\partial t} + \frac{dx_j}{dt}\frac{\partial k_i}{\partial x_j} = \frac{\partial k_i}{\partial t} + (c_{gj} + U_j)\frac{\partial k_i}{\partial x_j} \qquad (5.204)$$

The first term on the RHS can be evaluated using (5.200) and (5.202) as:

$$\frac{\partial k_i}{\partial t} = -\frac{\partial \omega}{\partial x_i} = -\frac{\partial \sigma}{\partial h}\frac{\partial h}{\partial x_i} - \frac{\partial \sigma}{\partial k_j}\frac{\partial k_j}{\partial x_i} - U_j\frac{\partial k_j}{\partial x_i} - k_j\frac{\partial U_j}{\partial x_i}$$

$$= -\frac{\partial \sigma}{\partial h}\frac{\partial h}{\partial x_i} - (c_{gj} + U_j)\frac{\partial k_j}{\partial x_i} - k_j\frac{\partial U_j}{\partial x_i} \qquad (5.205)$$

By using the irrotationality of wave number (5.201) and substituting (5.205) into (5.204), we have:

$$\frac{dk_i}{dt} = -\frac{\partial \sigma}{\partial h}\frac{\partial h}{\partial x_i} - k_j\frac{\partial U_j}{\partial x_i} \qquad (5.206)$$

Expression of $d\omega/dt$: As mentioned earlier, instead of formulating the problem using the wave number vector, we can also use the equivalent formulation based on wave frequency and wave direction, which is more popularly used in the wave spectral models. The total derivative of ω can be written as follows by using (5.200) and (5.203):

$$\frac{d\omega}{dt} = \frac{\partial \omega}{\partial t} + \frac{dx_i}{dt}\frac{\partial \omega}{\partial x_i} = \frac{\partial \omega}{\partial t} - (c_{gi} + U_i)\frac{\partial k_i}{\partial t} \qquad (5.207)$$

The first term on the RHS can be evaluated as follows:

$$\frac{\partial \omega}{\partial t} = \frac{\partial \sigma}{\partial h}\frac{\partial h}{\partial t} + \frac{\partial \sigma}{\partial k_i}\frac{\partial k_i}{\partial t} + U_i\frac{\partial k_i}{\partial t} + k_i\frac{\partial U_i}{\partial t} = \frac{\partial \sigma}{\partial h}\frac{\partial h}{\partial t} + (c_{gi} + U_i)\frac{\partial k_i}{\partial t} + k_i\frac{\partial U_i}{\partial t} \qquad (5.208)$$

Substituting the above relationship into (5.207), we have:

$$\frac{d\omega}{dt} = \frac{\partial \sigma}{\partial h}\frac{\partial h}{\partial t} + k_i\frac{\partial U_i}{\partial t} \qquad (5.209)$$

This equation describes the change rate of apparent angular frequency following a wave ray.

Expression of $d\theta/dt$: The kinematics equation for wave propagation angle θ represents the wave refraction process. It can be established by making use of irrotationality of wave number we used in Section 3.10 to derive Snell's law:

$$\frac{\partial (k\sin\theta)}{\partial x} - \frac{\partial (k\cos\theta)}{\partial y} = 0 \tag{5.210}$$

which leads to:

$$k\cos\theta\frac{\partial\theta}{\partial x} + k\sin\theta\frac{\partial\theta}{\partial y} = \cos\theta\frac{\partial k}{\partial y} - \sin\theta\frac{\partial k}{\partial x} \tag{5.211}$$

If we define a new local coordinate (s, n) so that s is in the wave direction and n normal to it, we would have the following relationship from the coordinate transformation:

$$\begin{cases} \dfrac{\partial}{\partial s} = \dfrac{dx}{ds}\dfrac{\partial}{\partial x} + \dfrac{dy}{ds}\dfrac{\partial}{\partial y} = \cos\theta\dfrac{\partial}{\partial x} + \sin\theta\dfrac{\partial}{\partial y} \\ \dfrac{\partial}{\partial n} = \dfrac{dx}{dn}\dfrac{\partial}{\partial x} + \dfrac{dy}{dn}\dfrac{\partial}{\partial y} = -\sin\theta\dfrac{\partial}{\partial x} + \cos\theta\dfrac{\partial}{\partial y} \end{cases} \tag{5.212}$$

Using the above coordinate transformation, equation (5.203) becomes:

$$k\frac{\partial\theta}{\partial s} = \frac{\partial k}{\partial n} \tag{5.213}$$

We then have:

$$\frac{d\theta}{dt} = \frac{\partial\theta}{\partial t} + \frac{dx}{dt}\frac{\partial\theta}{\partial x} = \frac{ds}{dt}\frac{\partial\theta}{\partial s} + \frac{dn}{dt}\frac{\partial\theta}{\partial n} = \frac{ds}{dt}\frac{\partial\theta}{\partial s} = \frac{ds}{dt}\frac{1}{k}\frac{\partial k}{\partial n} = \frac{ds}{dt}\frac{1}{k}\frac{\partial(\omega/c)}{\partial n}$$

$$= \frac{ds}{dt}\frac{\omega}{k}\left(-\frac{1}{c^2}\right)\frac{\partial c}{\partial n} = \frac{(c_g + U_x\cos\theta + U_y\sin\theta)}{c}\left(\sin\theta\frac{\partial c}{\partial x} - \cos\theta\frac{\partial c}{\partial y}\right) \tag{5.214}$$

In the above equation, it is assumed that the apparent frequency is a constant in the n direction, i.e., $\partial\omega/\partial n = 0$.

Transport equation in the Cartesian coordinate: Using the definition of dx/dt and dk/dt found in the earlier section based on wave kinematics, the dynamic equation (5.198) becomes:

$$\frac{\partial F}{\partial t} + \frac{\partial}{\partial x}\left[(c_{gx} + U_x)F\right] + \frac{\partial}{\partial y}\left[(c_{gy} + U_y)F\right]$$

$$+ \frac{\partial}{\partial k_x}\left[\left(-\frac{\partial\sigma}{\partial h}\frac{\partial h}{\partial x} - k_x\frac{\partial U_x}{\partial x} - k_y\frac{\partial U_y}{\partial x}\right)F\right]$$

$$+ \frac{\partial}{\partial k_y}\left[\left(-\frac{\partial\sigma}{\partial h}\frac{\partial h}{\partial y} - k_x\frac{\partial U_x}{\partial y} - k_y\frac{\partial U_y}{\partial y}\right)F\right] = Q \tag{5.215}$$

In contrast, if we use wave frequency and wave direction as the space variables, we have, by substituting $d\omega/dt$ and $d\theta/dt$ into (5.215):

$$
\frac{\partial F}{\partial t} + \frac{\partial}{\partial x}\left[(c_{gx}+U_x)F\right] + \frac{\partial}{\partial y}\left[(c_{gy}+U_y)F\right]
$$

$$
+ \frac{\partial}{\partial \omega}\left[\left(\frac{\partial \sigma}{\partial h}\frac{\partial h}{\partial t} + k_x\frac{\partial U_x}{\partial t} + k_y\frac{\partial U_y}{\partial t}\right)F\right]
$$

$$
+ \frac{\partial}{\partial \theta}\left[\frac{(c_g+U_x\cos\theta+U_y\sin\theta)}{c}\left(\sin\theta\frac{\partial c}{\partial x} - \cos\theta\frac{\partial c}{\partial y}\right)F\right] = Q
$$

$$(5.216)$$

In the case where there is no current and the variation of mean water depth with respect to time is negligible, the above equation can be further simplified to the conventional wave energy spectrum equation in the Cartesian coordinate (e.g., Hasselmann *et al.*, 1988):

$$
\frac{\partial F}{\partial t} + \frac{\partial}{\partial x}\left(c_g\cos\theta F\right) + \frac{\partial}{\partial y}\left(c_g\sin\theta F\right) + \frac{\partial}{\partial \theta}\left[\frac{c_g}{c}\left(\sin\theta\frac{\partial c}{\partial x} - \cos\theta\frac{\partial c}{\partial y}\right)F\right] = Q
$$

$$(5.217)$$

Transport equation in spherical coordinate: The equation can be modified to include the great circle effect on Earth, whose latitude and longitude are defined by ϕ and λ, respectively. In this case, θ is measured clockwise from true north and thus the latitude ϕ is similar to the x-direction and longitude λ to the y-direction in the counterclockwise Cartesian coordinate system (x, y). The modified equation reads:

$$
\frac{\partial F}{\partial t} + (\cos\phi)^{-1}\frac{\partial}{\partial \phi}\left(\frac{d\phi}{dt}\cos\phi F\right) + \frac{\partial}{\partial \lambda}\left(\frac{d\lambda}{dt}F\right) + \frac{\partial}{\partial \theta}\left(\frac{d\theta}{dt}F\right) = Q \quad (5.218)
$$

where

$$
\frac{d\phi}{dt} = c_g R^{-1}\cos\theta
$$

$$
\frac{d\lambda}{dt} = c_g(R\cos\phi)^{-1}\sin\theta
$$

$$
\frac{d\theta}{dt} = c_g\tan\phi R^{-1}\sin\theta + \frac{1}{R}\frac{c_g}{c}\left(\sin\theta\frac{\partial C}{\partial \phi} - \frac{\cos\theta}{\cos\phi}\frac{\partial C}{\partial \lambda}\right)
$$

Source terms: By considering weak nonlinear interactions for wind-induced waves, Hasselmann (1968) proposed the following general form for the source Q, which consists of nine contributions:

$$
Q(k) = \sum_{i=1}^{9} Q_i \qquad\qquad (5.219)
$$

In practical computation, however, there are essentially only three major types of sources affecting wave energy spectrum. The first contribution is the input of wave energy from wind (Q_1 and Q_2), which can be expressed as:

$$Q_{\text{wind}} = \begin{cases} Q_1 = \dfrac{\pi\sigma F_a}{\rho^2 c^3 c_g} & \text{for} \quad t < 1/\mu\sigma \\[2ex] Q_2 = \mu\sigma F & \text{for} \quad t \geq 1/\mu\sigma \end{cases} \tag{5.220}$$

where F_a is the reference wave spectrum (a constant) that is often unknown a priori. Fortunately, it is found that Q_1, which represents a linearly growing sea, contributes only to the initial wave growth and its influence on the future wave evolution is negligible. The continuing wave development is mainly controlled by Q_2, in which μ is called coupling coefficient and the theoretical expression was given by Mitsuyasu and Honda (1982) as:

$$\mu = \frac{0.16}{2\pi}\left(\frac{u_{\text{w}*}}{c}\right)^2 \tag{5.221}$$

where $u_{\text{w}*}$ is the friction velocity induced by wind.

The second contribution is the nonlinear energy transfer among different directional frequency spectral components (Q_5). The complete expression of Q_5 was given by Hasselmann (1968) in the form of triple integration. To facilitate the computation, Hasselmann and Hasselmann (1985) gave the alternative expression based on the parameterization of nonlinear energy transfer as:

$$Q_{nl(i)} = Q_5 = E_{nl} + \sum_{j=1}^{5} C^{(i,j)} H_{nl}^{(j)}, \quad i = 1, \ldots, 18 \tag{5.222}$$

where E_{nl} is the mean value of a set of 18 exact computations for various JONSWAP-type spectra with the enhancement factor γ varying from 1 to 7 and different directional spreading spectra. $H_{nl}^{(j)}$ is a set of five empirical orthogonal functions determined by the Q_{nl} ensemble. $C^{(i,j)}$ is called expansion coefficient.

The third contribution is the energy dissipation due to white capping (deep-water wave breaking Q_7), bottom friction (Q_8), and various shallow-water wave breakings (Q_9), the last two of which are negligible in the deep-water region. Komen *et al.* (1984) based on the P–M spectrum gave the following expression of Q_7:

$$Q_{wc} = Q_7 = -3.33 \times 10^{-5}\bar{\sigma}\left(\frac{\sigma}{\bar{\sigma}}\right)^2 \left(\frac{\hat{\alpha}}{\hat{\alpha}_{\text{PM}}}\right)^2 F(\sigma, \theta) \tag{5.223}$$

where $\hat{\alpha} = \sigma_\eta^2 \bar{\sigma}^4/g^2$, $\hat{\alpha}_{\text{PM}} = 4.57 \times 10^{-3}$, and:

$$\bar{\sigma} = \frac{1}{\sigma_\eta^2}\hat{\alpha}\iint F(\sigma, \theta)\sigma\,\mathrm{d}\sigma\,\mathrm{d}\theta$$

with σ_η^2 being the variance of free surface displacement, which simply equals the total wave energy:

$$\sigma_\eta^2 = a_{\text{rms}}^2/2 = E_{\text{total}}/(\rho g) = \iint F(\sigma, \theta) \, d\sigma \, d\theta$$

if the free surface displacement follows normal distribution.

For bottom friction, Bouws and Komen (1983) proposed the following form:

$$Q_{bt} = Q_{8a} = -\frac{C_{bt}}{g^2} \frac{\sigma^2}{\sinh^2 kh} F(\sigma, \theta) \tag{5.224}$$

where the mean value of C_{bt} is found to be $0.038 \, \text{m}^2\text{s}^{-3}$. Additional wave energy dissipation may result if the seafloor is permeable. This can be modeled as follows if the porous layer has an infinite thickness:

$$Q_{pm} = Q_{8b} = -\frac{2K}{\nu} \frac{\sigma^2}{\sinh(2kh)} F(\sigma, \theta) \tag{5.225}$$

where K is the intrinsic permeability of the porous bed.

For wave breaking in shallow water, Booij *et al.* (1999) proposed the following form:

$$Q_{sb} = Q_9 = -\frac{S_{br}}{E_{\text{total}}/(\rho g)} F(\sigma, \theta) \tag{5.226}$$

where S_{br} is the mean rate of energy dissipation due to wave breaking that can be expressed as:

$$S_{br} = -\frac{1}{4} q_{br} \left(\frac{\bar{\sigma}}{2\pi} \right) H_{\text{max}}^2 \tag{5.227}$$

with q_{br} being determined from the following equation:

$$\frac{1 - q_{br}}{\ln q_{br}} = -8 \frac{E_{\text{total}}/(\rho g)}{H_{\text{max}}^2}$$

and the maximum allowable wave height H_{max} is related to the local still water depth h and beach slope β by $H_m = [0.55 + 0.88 \exp(-0.012 \cot \beta)] h$.

The transport equation in terms of wave action: Strictly speaking, all the above equations have been derived by assuming weak current or no current at all. When waves propagate on a strong unsteady nonuniform current $U(x, t)$, the apparent wave frequency is no longer a constant. This is evident from equation (5.202) where $\omega = \sigma + k \cdot U$ and (5.200) $\partial k/\partial t + \nabla \omega = 0$, from which one can easily deduce that the unsteady current will cause the unsteadiness of **k** that consequently results in the spatial variation of ω. In this case, the wave energy will not be conserved.

Bretherton and Garrett (1969) showed that the so-called wave action N, which is defined as $N = F/\sigma$, is conserved in a moving medium like a current. With the use of wave action, the governing equation expressed in frequency and directional space takes the following form:

$$\frac{\partial N}{\partial t} + \nabla_x \cdot \left(\frac{dx}{dt}N\right) + \nabla_\omega \cdot \left(\frac{d\omega}{dt}N\right) + \nabla_\theta \cdot \left(\frac{d\theta}{dt}N\right) = \frac{Q}{\sigma} \tag{5.228}$$

This equation is used in both deep and shallow waters where the current effect is strong and changes with time and space. For a 2D case, the above equation can be rewritten as:

$$\frac{\partial N}{\partial t} + \frac{\partial(c_x N)}{\partial x} + \frac{\partial(c_y N)}{\partial y} + \frac{\partial(c_\omega N)}{\partial \omega} + \frac{\partial(c_\theta N)}{\partial \theta} = \frac{Q}{\sigma} \tag{5.229}$$

where

$$c_x = \frac{dx}{dt}, c_y = \frac{dy}{dt}, c_\omega = \frac{d\omega}{dt}, \text{ and } c_\theta = \frac{d\theta}{dt}$$

all of which can be evaluated in the same way we introduced earlier in this section.

5.3.4.3   Wave spectral models and their applications

The transport equation for the wave spectrum describes the change of wave spectrum during wave propagation over a changing topography. By including appropriate source terms, the model can be used to simulate wave generation, wave propagation, shoaling and refraction, nonlinear wave energy transfer among different wave components, and wave dissipation. Starting from the so-called first-generation wave model (e.g., Pierson *et al.*, 1966), the model has been improved for better representation of nonlinear wave energy transfer and wave dissipation mechanisms. This leads to the second-generation wave models (e.g., Hasselmann *et al.*, 1976) and the third-generation wave model (Hasselmann *et al.*, 1988) that is a powerful tool for modeling long-term and large-scale wave climates globally. The representative third-generation wave spectral model in deep water includes WAM (Hasselmann *et al.*, 1988) and WaveWatch III (Tolman, 1999), the latter of which was developed by the National Centers for Environmental Prediction (NCEP) of NOAA, U.S.A. The model can cover a very large area (i.e., the entire ocean surface on the globe outside the surf zone) with the mesh size varying from 1 to 10 km. Figure 5.14 shows an example of using this model for forecasting wave height distribution in Pacific Ocean.

For wave transformation in shallow water, the representative wave spectral model is SWAN (Booij *et al.*, 1999; Ris *et al.*, 1999). In this model,

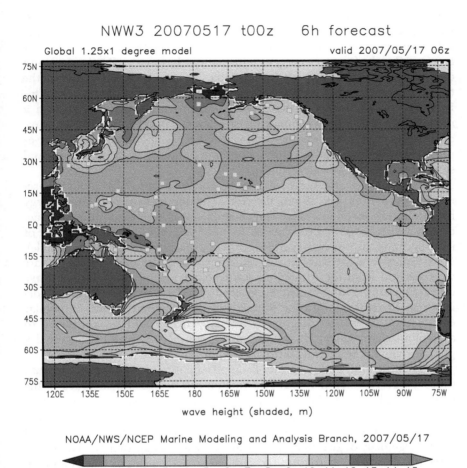

NWW3 20070517 t00z   6h forecast

Global 1.25x1 degree model   valid 2007/05/17 06z

wave height (shaded, m)

NOAA/NWS/NCEP Marine Modeling and Analysis Branch, 2007/05/17

Figure 5.14 Real-time forecast of wave height in Pacific Ocean with the use of the WaveWatch III model; the color scale represents wave height in meters. (Courtesy of Dr. Hendrik Tolman of the Marine Modeling and Analysis Branch (MMAB) of the NCEP of NOAA)

the depth influences on waves, e.g., shoaling, refraction, bottom friction, depth-induced wave breaking, triad wave–wave interactions, are included. For more details on wave spectral models, readers are referred to Sobey (1986) and Massel (1996: 222).

5.3.4.4 *Limitations of wave spectral models*

Wave spectral models possess the great advantage of being able to simulate very large scale problems, i.e., the global wave climate. This is a distinct feature none of the other wave models can possibly have. The reason that wave spectral models can be used to simulate large-scale wave events is that the wave spectrum filters out the wave phase information by retaining only the wave height (or wave energy) information. In most cases, the variation of wave height is slow in both time and space. This means that in solving the transport equation of wave spectrum, one mesh can cross many wavelengths. However, the omission of wave phase information suggests that such a model will not be able to simulate phase-related wave phenomena, i.e., wave diffraction, that may become important in coastal regions when waves propagate over rapidly changing topography, islands, or man-made coastal structures (e.g., breakwaters).

The diffraction effect, which can be easily included by many phase-resolving wave models, is not directly derivable from the spectral energy balance equation. So far, there is no standard way of incorporating diffraction in a directional spectral model. Rivero *et al.* (1997) proposed a way of incorporating the diffraction effect into a conventional spectral wave model. New expressions for wave group velocity and angular energy transfer rate were derived from the eikonal equation that takes into account the wave diffraction effect. In contrast, Booij *et al.* (1997) proposed another model in which an artificial diffusion term is added to correct the term of angular energy transfer rate so that the diffraction effect could be considered to a certain extent. Later, Holthuijsen and Booij (2003) further suggested a phase-decoupled refraction–diffraction approximation.

A revisit to MSE and its parabolic approximation was made by Lin *et al.* (2005), who revealed that wave diffraction can be possibly included in the wave spectral model by introducing the complex wave height spectrum, from which the conventional wave energy spectrum can be recovered. With the use of the complex wave height spectrum, the wave phase information is retained in the argument of the complex variable. By assuming that the change of wave phase is slow away from the singularity point (e.g., the tip of a break water), the wave height spectrum can be solved to simulate wave diffraction.

Regardless of the efforts made so far to extend the wave spectral model for the inclusion of wave diffraction, wave spectral models are generally unable to provide the detailed wave information as accurately as those phase-resolving wave models for local-scale wave simulation. Therefore, a general practise is to use a wave spectral model for large-scale wave propagation simulation that is coupled with a near-field wave model for detailed simulation in coastal areas.

5.3.4.5 *Wave statistics and rogue waves*

A wave spectrum can be extracted from a time history of wave records in a random sea. The data extraction process removes wave phase information, the major component of wave randomness, and retains the important component of wave energy. Therefore, a wave spectrum contains only deterministic information. To recover a random sea state, at least statistically, a random phase $\delta(\sigma, \theta)$ must be introduced to each component of the wave mode in a wave spectrum, i.e.:

$$\eta(x, y, t) = \sum_{i=1}^{\infty} \sum_{j=1}^{\infty} a(\sigma, \theta, x, y, t) \cos\left[k(\sigma)\cos\theta x + k(\sigma)\sin\theta y\right]$$
$$-\sigma t + \delta(\sigma, \theta)] \qquad (5.230)$$

This is a typical way to generate irregular waves in the laboratory to simulate a random sea. A typical wave series for a JONSWAP spectrum using (5.230) is shown in Figure 5.15, from which one may easily identify the randomness of wave height from one wave crest to another. A simple relationship can be established between $F(\sigma)$ and the root-mean-square wave height:

$$H_{\text{rms}} = \sqrt{\frac{1}{N} \sum_{n=1}^{N} H_n^2}$$

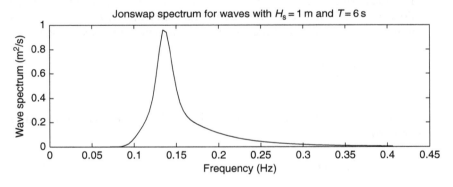

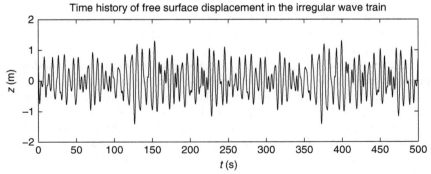

Figure 5.15 An irregular wave train corresponding to a JONSWAP spectrum with $\gamma = 3.3$.

where N is the number of waves in the time-series wave record:

$$\frac{1}{8}H_{rms}^2 = E_{total}/(\rho g) = \int F(\sigma)\, d\sigma \qquad (5.231)$$

An immediate question from the above observation is what could be the possible maximum wave height if the observation is extended to a longer duration, e.g., 5000 s, 1 day, 1 month, or 1 year? The answer to such a question has significance in the design of coastal and offshore structures and ship hulls. The theoretical formulation for this question is available when the following assumptions are true:

1. The waves are narrowly banded (this could be theoretically achieved if we increase the value of γ in the JONSWAP spectrum to have a sharply peaked spectrum, especially swells).
2. The forcing wave condition remains unchanged in the observation period.
3. The sampling volume is large enough to have statistical significance.

Under the above conditions, the distribution of wave height can be proven to follow the Rayleigh distribution (Dean and Dalrymple, 1991). The corresponding probability density function of wave height is then expressed as:

$$f(H) = \frac{2He^{-(H/H_{rms})^2}}{H_{rms}^2} \qquad (5.232)$$

and the wave height that may be exceeded by at least one wave in the entire wave record is:

$$H_{exceed} = \sqrt{\ln N}\, H_{rms} \qquad (5.233)$$

Apparently, with the increase of observation time, the wave number N contained in the wave record also increases, which causes the increase of H_{exceed}. In principle, as $N \to \infty$, $H_{exceed} \to \infty$ too. This implies that, given a sufficiently long time, the linear superposition of many wave modes can possibly create a very large wave height at a particular location.

Such an event, however, will take place less frequently in the real world than what is predicted by the theory. The reason is that in a real sea, the wave-forcing condition keeps changing and the resulting wave field has a widely banded spectrum. This makes the first two basic assumptions of the Rayleigh distribution invalid. In addition, the white capping and depth-induced breaking will further trim down the wave height when the breaking condition is exceeded, which greatly reduce the number of very large-amplitude waves.

Nevertheless, observations were indeed reported for suddenly emerged huge waves on an otherwise quiet and calm background wave field in deep water. Such waves are called rogue waves (or freak waves, monster waves, giant waves, steep waves, etc.). These waves can easily reach a wave height over 10 m without any warning and thus pose great dangers to ships. In shallow waters nearshore, a similar phenomenon has also been observed and the wave is called a sneaker wave.

The real cause of rogue waves is still controversial and even mythical. Most of the reported freak waves were known only through anecdotal evidences from ships that encountered them. The only exception is the so-called Draupner wave that was precisely measured at the Draupner oil platform in the North Sea, offshore of Norway, on 1 January 1995 (see Section 3.4.4). The wave record confirmed a significant wave height of about 12 m and a maximum wave height of 26 m, with the occurrence probability of about 1 in 200,000.

Currently, there are various reasons for the generation of these unexpected giant waves, namely the linear superposition of waves from different sources with their crests coming together at the same time, wave focusing due to a particular incident wave condition and the bathymetry and ambient current, nonlinear wave instability (e.g., Kharif and Pelinovsky, 2003), and strong wind. This mysterious part of nature will remain a challenging problem for both scientists and engineers in the foreseeable future.

5.4 Case studies using model coupling techniques

In modeling realistic waves and currents in oceans, there is a need to couple different wave models into one package to resolve different physical phenomena and/or cover a very large scale domain. In the situation where the entire physical process can be described by the same type of wave model, the coupling mainly takes place by data exchange between meshes of different sizes. A typical example is the simulation of tsunami generation, propagation, and run-up. In contrast, for cases where different wave models are required in different regions or in an overlapped region to correctly represent the coupled physics, the coupling takes place not only for data exchange, but also for the shift of wave models. The examples include the simulation of wind wave generation from deep oceans and the subsequent propagation and transformation in the nearshore region as well as the simulation of wave–current interaction in the same domain.

This section is arranged in the following way: first, some commercial wave and current model packages will be briefly introduced. These commercial software packages normally include different flow and wave modules within one package in order to treat complicated physical problems. Following the introduction of the commercial software, a few examples of numerical simulations that employ the model coupling technique will be presented.

5.4.1   Commercial software for hydrodynamics and wave simulation

Currently, there are a few commercial software packages available for the simulation of waves and coastal and offshore hydrodynamics. They are briefly introduced below.

MIKE 21: The product of DHI. It includes four modules of HD (hydrodynamics for current and water level simulation), waves (wave simulation), AD (advection and dispersion of pollutants), and sediment transport (sediment, mud, and particle transport). In the module of waves, there are models based on the wave action concept for the large-scale wave transformation (MIKE 21 SW: OSW, offshore spectral wind wave module, and NSW, nearshore spectral wind wave module) and models based on the momentum concept for the nearshore waves where the wave diffraction can become significant (BW Boussinesq wave model; EMS, elliptic mild-slope wave model; and PMS, parabolic mild-slope wave model).

DELFT3D: The product of WL | Delft Hydraulics. DELFT3D includes a few modules, e.g., hydrodynamics, waves, sediment transport, morphology, water quality, particle tracking for water quality, and ecology. For Delft Waves, it includes four modules, namely SWAN (spectral wave model for far-field wave simulation), PHAROS (MSE model for middle-field wave simulation), TRITON (Boussinesq model for nearshore wave simulation), and SKYLLA (VOF NSE-solver model for local-scale simulation).

SMS: Product of Environmental Modeling Systems, Inc. Many models have been included in SMS for the simulation of various types of surface flows such as river flows and water waves. Relevant to coastal and ocean engineering problems, for current simulation, there is an SWE model ADCIRC; for wave modeling, there are STWAVE (the wave spectral model), CGWAVE (the MSE model), and BOUSS-2D (the Boussinesq model). Besides, the shoreline change model GENESIS is also included in the package.

5.4.2   Tsunami simulation

The case presented below is the simulation of the 2004 Indian Ocean tsunami using the MIKE 21 software package (simulation results kindly provided by DHI Water • Environment • Health; see also Pedersen *et al.*, 2005). The generation, propagation, and run-up of a tsunami normally cover a very large area that requires different levels of mesh resolution for the optimal simulation results. Since the tsunami can be described by the same SWEs or the Boussinesq equations in both deep oceans and nearshore, the main challenge here is the adequate spatial resolution for the physics in the entire domain. The difficulty, however, arises because the length scales of the tsunami can be very different in deep water and shallow water, say, from hundreds of kilometers in deep oceans to a few kilometers nearshore.

5.4.2.1 Simulation on integrated unstructured mesh

One way of simulating a tsunami from far field to near field is to use an integrated unstructured mesh system that varies its size to adapt to the local shoreline configuration. Usually, coarse meshes are used in the deep ocean whereas fine meshes are deployed in the coastal region. The governing SWEs are solved by either FVM or FEM. By doing so, there is no need for model coupling and data exchange, and a unified model can be applied to model the entire process of tsunami generation, propagation, and run-up. Figure 5.16 gives an example of using the MIKE 21 FM series (flexible mesh) to simulate the 2004 Indian tsunami. Such treatment, however, may exhaust

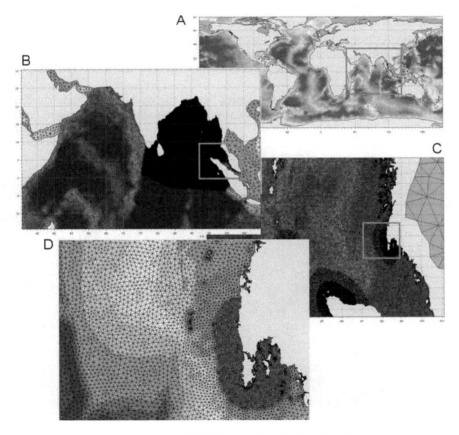

Figure 5.16 The mesh used by MIKE 21 for the 2004 Indian Tsunami simulation; B, C, and D represent various levels of zoom-in regions.

computational resources in the nearshore region before tsunami arrives. At this time, this approach is used less than the nested-mesh approach as elaborated next.

5.4.2.2   Simulation based on the nested-mesh approach

The second way of handling the large-scale simulation of tsunami propagation and run-up is to run the model on different mesh sizes (i.e., large mesh size for far-field tsunami generation and propagation and small mesh size for local wave transformation and run-up). In adopting this approach, it can be advantageous to introduce different approximations to the wave models running on different mesh systems, although these models essentially solve the same basic governing equations. For example, in the far-field simulation, the model is often constructed on spherical coordinates with wave nonlinearity being neglected; for the local-scale run-up simulation, the model can be constructed on the simplified Cartesian coordinate including wave nonlinearity and moving boundary treatment. Figure 5.17 shows an example of the employment of the rectilinear version of MIKE 21 to simulate the 2004 Indian Tsunami, where the regional area is simulated by using the grid size of 9720 m and the local area is simulated by using the grid size of 3240 m.

In this approach, the main difficulty lies in the exchange of data on the boundaries of two mesh systems. Because the large-scale mesh is nested with

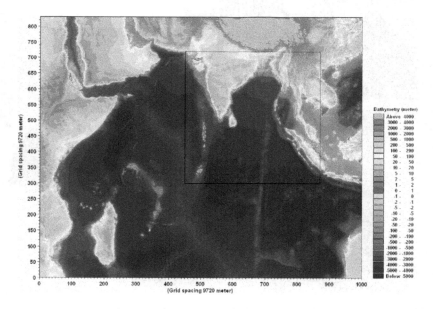

Figure 5.17 Nested regional (left) and local (right) simulation of the 2004 Indian tsunami.

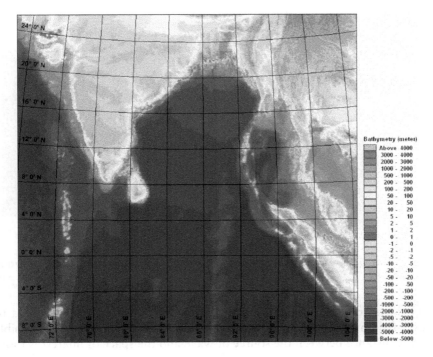

Figure 5.17 (Continued).

the small-scale mesh, this type of modeling is called nested-mesh simulation. For rectilinear grids, the ratio of the large and small grid sizes is often an odd number integer to facilitate the data exchange by interpolation. The data exchange can be either passive (i.e., the data are fed from large-scale simulation results to small-scale simulation as the boundary condition) or dynamic (i.e., the data exchange between the large-scale simulation and the small-scale simulation is allowed so that they are mutually affected). In principle, a multiple-level nested-mesh system can be used to zoom into a very small area for the simulation of detailed wave run-up as long as the local bathymetry data are available (Figure 5.18 for example). Caution, however, must be taken for the data exchange in the nested-mesh approach, because additional numerical errors will be inevitably introduced. This implies that the reliability of the numerical prediction of nearshore wave phenomena will degenerate with the increase of levels of nested meshes.

5.4.3 *Large-scale wind wave simulation*

For wind waves, unlike the tsunami whose wavelength is greatly reduced when approaching the shoreline, the wavelengths from deep water to shallow water are often on the same order of magnitude. This implies that to resolve

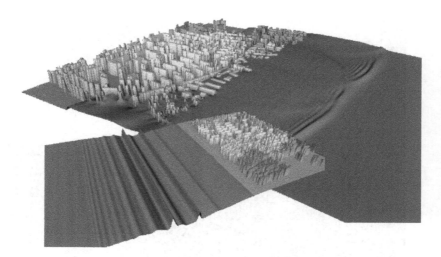

Figure 5.18 An example simulation of wave transformation and inundation nearshore by MIKE 21 BW.

the wave motion, a similar spatial resolution is needed from far-field to near-field modeling if the same wave model is used. This is often infeasible if the distance between the source of wave and the space of interest is thousands of wavelengths away. The effective method of solving this problem is to employ different wave models at different stages and match the boundary conditions at the interface where two wave models are coupled to each other.

For far-field simulation, the most efficient model is the wave spectral model. By removing the wave phase information, a grid space can cover many wavelengths. The wave height, wave period, and wave direction information obtained from the spectral model will be fed into the nearshore wave model, e.g., the Boussinesq model or MSE model, in order to simulate combined wave refraction and diffraction in a relatively shallow-water region. The Boussinesq model can be continued for wave run-up simulation or fed into a more detailed NSE model for wave–structure interaction modeling.

In the following example, a case study of seiche in the Port of Long Beach, California, USA, is induced. The study was originally performed by Kofoed-Hansen *et al.* (2005). Figure 5.19 shows the mesh resolution by the global and regional MIKE 21 SW simulation with the use of the wave spectral model. Figure 5.20 shows the local area near the port, in which the wave field is simulated by the BW model. Figure 5.21 shows the simulated waves in global scale by the wave spectral model, which predicts the significant wave height and main wave propagation direction. On the local scale, the complex wave refraction, diffraction, and reflection pattern can be resolved by the Boussinesq model (Figure 5.22).

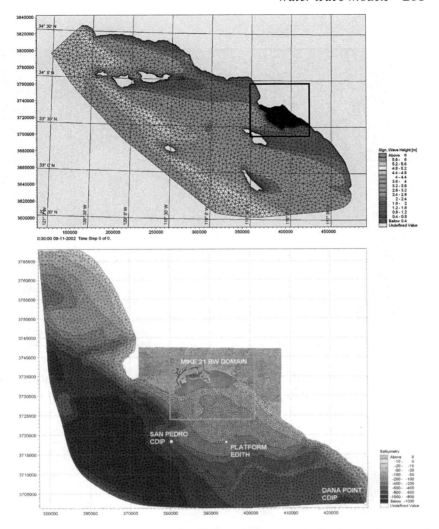

Figure 5.19 The mesh setup in MIKE 21 SW for global (top) and regional
(bottom) simulation. (Images from Figure 5.19 through Figure 5.22
courtesy of DHI Water • Environment • Health and from Kofoed-
Hansen et al., 2005)

5.4.4 Combined wave and current simulation

Unlike the previous two cases where the same wave phenomenon (i.e., long-
wave tsunami or wind wave) is simulated but with either different meshes
or different models being implemented in different regions, the simulation
of combined wave–current interaction is conducted in the same domain. In
this case, the model coupling takes place between two models, normally on

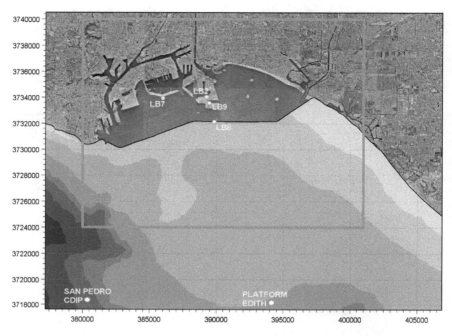

Figure 5.20 Local beach and harbor configuration; waves simulated by MIKE 21 BW.

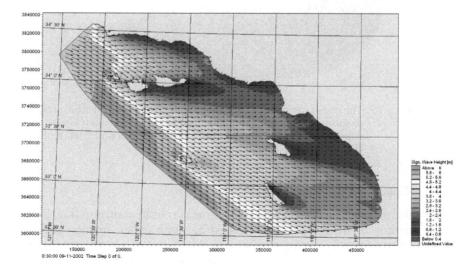

Figure 5.21 Simulated wave height and direction using MIKE 21 SW.

Figure 5.22 Simulated wave fields around the Port of Long Beach.

the same mesh system, by exchanging variable information obtained from the wave and current models, respectively. The data exchange is necessary because wave and current have mutual influences on each other through linear and nonlinear wave–current interaction. The interaction changes both wave and current characteristics. For example, currents can deflect wave direction and change wavelength, whereas waves can affect the current by imposing additional radiation stresses and by changing the effective wind stress and bottom friction experienced by the current.

Different wave and current models can be coupled together, depending on the need of simulation. The commonly adopted wave model is the wave spectral model that is capable of simulating large-scale wave variation in both time and space, e.g., WAM, WaveWatch III, or SWAN. Besides, the phase-resolving MSE model and the Boussinesq model can also be adopted sometimes. The hydrodynamic model for flow simulation can be either the depth-averaged SWE model or the depth-dependent quasi-3D model (e.g., POM or COHERENS).

5.4.4.1 Large-scale modeling

For large-scale modeling of combined wave and current, a typical example is the modeling of storm waves and storm surge under the influence of a large-scale tropical storm (e.g., typhoons, hurricanes, or cyclones, all of which are tropical storms but termed differently based on their origins in the Atlantic Ocean, western Pacific, or near India and Australia, respectively). A storm causes the reduction of atmospheric pressure in its eye where the sea level rises. Accompanying the storm is strong wind that generates storm waves. When the storm approaches the shoreline, the storm waves combines with the storm surge to cause coastal flooding. Sometimes, a high tide will further worsen the situation.

To model this large-scale event, the general practice is to use a wave spectral model to simulate wave generation, propagation, and transformation in deep oceans and use an SWE model to model current and sea surface variation. For example, Zhang and Li (1996) combined a third-generation wave spectral model with a 2D SWE storm surge model to simulate the interaction of wind waves and storm surge. In their work, the current velocity and the sea surface level obtained from the storm surge model are put into the wave model, and the radiation stresses estimated from the wave spectral model results and the modified wind stress are brought back to the storm surge model. Ozer *et al.* (2000) reported a coupled WAM and SWE model to simulate tides, surges, and waves under the MAST III project and employed the model to study North Sea and Spanish coasts. Choi *et al.* (2003) employed a coupled wave–tide–surge model (WAM and SWE) to investigate the tide, storm surge, and wind wave interaction during a winter monsoon in the Yellow Sea and East China Sea.

For a wave spectral model coupled with an SWE model, the wave information is obtained by solving the wave action equation (Section 5.3.4.2), i.e.:

$$\frac{\partial N}{\partial t} + \frac{\partial (c_x N)}{\partial x} + \frac{\partial (c_y N)}{\partial y} + \frac{\partial (c_\sigma N)}{\partial \sigma} + \frac{\partial (c_\theta N)}{\partial \theta} = \frac{Q}{\sigma} \tag{5.234}$$

where c_x, c_y, c_σ, and c_θ are functions of mean water depth (i.e., sum of still water depth h and water setup/set-down η) and mean current velocity U and V. The wind effect is considered in one of the source terms Q. From the wave action information, we are able to compute the root-mean-square wave height, the main wave propagation direction, and the mean wave angular frequency by the following equations:

$$H_{\text{rms}}(x, y, t) = \sqrt{8 E_{\text{total}}(x, y, t)/(\rho g)} = \sqrt{8 \int N(x, y, t, \sigma, \theta)\sigma d\sigma d\theta} \tag{5.235}$$

$$\bar{\theta}(x, y, t) = \tan^{-1}\left\{ \frac{\int_0^{2\pi} D(\theta)\sin\theta d\theta}{\int_0^{2\pi} D(\theta)\cos\theta d\theta} \right\}$$

where $D(x, y, t, \theta) = \int_0^\infty N(x, y, t, \sigma, \theta)\sigma d\sigma \tag{5.236}$

$$\bar{\sigma}(x, y, t) = \frac{\int_0^\infty S(x, y, t, \sigma)\sigma d\sigma}{\int_0^\infty S(x, y, t, \sigma)d\sigma}$$

where $S(x, y, t, \sigma) = \int_0^{2\pi} N(x, y, t, \sigma, \theta)\sigma d\theta$ \hfill (5.237)

To solve (5.234) the mean water level $h + \eta$ and current velocities U and V are needed to estimate c_x, c_y, c_σ, and c_θ. The mean flow information can be obtained by solving the SWEs:

$$\frac{\partial(h + \eta)}{\partial t} + \frac{\partial}{\partial x}(U(h + \eta)) + \frac{\partial}{\partial y}(V(h + \eta)) = 0 \tag{5.238}$$

$$\frac{\partial}{\partial t}(U(h + \eta)) + \frac{\partial}{\partial x}(U^2(h + \eta)) + \frac{\partial}{\partial y}(UV(h + \eta)) - f(h + \eta)V$$

$$= -\frac{(h + \eta)}{\rho}\frac{\partial p_a}{\partial x} - g(h + \eta)\frac{\partial \eta}{\partial x} - \frac{\eta(h + \eta)g}{\rho}\frac{\partial \rho}{\partial x} + \frac{(h + \eta)}{\rho}\frac{\partial \tau_{xx}}{\partial x} + \frac{(h + \eta)}{\rho}\frac{\partial \tau_{xy}}{\partial y}$$

$$+ \frac{1}{\rho}\tau_{xz}(\eta) - \frac{1}{\rho}\tau_{xz}(-h) \tag{5.239}$$

$$\frac{\partial}{\partial t}(V(h + \eta)) + \frac{\partial}{\partial x}(UV(h + \eta)) + \frac{\partial}{\partial y}(V^2(h + \eta)) + f(h + \eta)U$$

$$= -\frac{(h + \eta)}{\rho}\frac{\partial p_a}{\partial y} - g(h + \eta)\frac{\partial \eta}{\partial y} - \frac{\eta(h + \eta)g}{\rho}\frac{\partial \rho}{\partial x} + \frac{(h + \eta)}{\rho}\frac{\partial \tau_{yx}}{\partial x} + \frac{(h + \eta)}{\rho}\frac{\partial \tau_{yy}}{\partial y}$$

$$+ \frac{1}{\rho}\tau_{yz}(\eta) - \frac{1}{\rho}\tau_{yz}(-h) \tag{5.240}$$

where $f = 1.0312 \times 10^{-4}\left(\text{rad s}^{-1}\right)$ is the Coriolis parameter due to the self-rotation of Earth. In this equation, the large-scale change of atmospheric pressure p_a resulting from the moving storm eye is included as the forcing term of the current and the spatial variation of water level. The wind-induced surface stresses $\tau_{xz}(\eta)$ and $\tau_{yz}(\eta)$ are included as other forcing terms for the generation of mean current, which is affected by the surface waves that change the effective water surface roughness (Janssen, 1991). Inside the water column, the total stresses on the horizontal plane include the contribution from viscous effect, turbulence effect, and wave-induced radiation stress:

$$\tau_{xx} = \tau_{Mxx} + \tau_{Txx} + S_{xx} = 2\rho v\frac{\partial U}{\partial x} + \left[2\rho v_t\frac{\partial U}{\partial x} - \frac{2}{3}\rho k\right]$$

$$+ \left[(1 + \cos^2\theta)\frac{kh}{\sinh 2kh} + \frac{1}{2}\cos^2\theta\right]\frac{E_{\text{total}}}{h + \eta} \tag{5.241}$$

$$\tau_{yy} = \tau_{Myy} + \tau_{Tyy} + S_{yy} = 2\rho v\frac{\partial V}{\partial y} + \left[2\rho v_t\frac{\partial V}{\partial y} - \frac{2}{3}\rho k\right]$$

$$+ \left[(1 + \sin^2 \theta) \frac{kh}{\sinh 2kh} + \frac{1}{2} \sin^2 \theta \right] \frac{E_{\text{total}}}{h + \eta} \tag{5.242}$$

$$\tau_{xy} = \tau_{yx} = \tau_{Mxy} + \tau_{Txy} + S_{xy} = \rho \left(\nu + \nu_t \right) \left(\frac{\partial U}{\partial y} + \frac{\partial V}{\partial x} \right)$$

$$+ \sin 2\theta \left(\frac{1}{4} + \frac{kh}{2 \sinh 2kh} \right) \frac{E_{\text{total}}}{h + \eta} \tag{5.243}$$

where $E_{\text{total}} = \rho g H_{\text{rms}}^2 / 8$ is calculated from the wave spectral modeling results. The wave effect in the current is thus reflected in this term by changing the total effective stress. Another influence of wave effect on the current is the change of effective bottom shear stress, i.e.:

$$\tau_{xz}(-h) = \rho c_{\text{wc}} U \sqrt{U^2 + V^2}, \quad \tau_{yz}(-h) = \rho c_{\text{wc}} V \sqrt{U^2 + V^2} \tag{5.244}$$

where the friction coefficient c_{wc} is a function of wave properties, current velocity, and bottom roughness (Grant and Madsen, 1986). Note that when the above approach is used to solve a very large scale problem across the ocean, both wave spectrum transport equation and SWEs should be converted into their equivalent forms in the spherical coordinate.

5.4.4.2 Local-scale modeling

To resolve 3D coastal circulation, it is necessary to employ a quasi-3D hydrodynamic model for current simulation. For example, Xie *et al.* (2001) incorporate the WAM model with the quasi-3D POM to study the 3D wave–current interaction through surface and bottom stresses. In their approach, the 2D wave information is first converted into 2D wave-induced radiation stresses that are subsequently included in the 3D hydrodynamic model by applying the stresses uniformly in the vertical direction. This treatment introduces errors in current computation in the vertical direction because the wave-induced radiation stresses vary in depth and can be significant for large waves. Lin and Zhang (2004) proposed a formula for 3D depth-dependent radiation stresses and incorporated them into the coupled WAM–POM model. The model was used to study wind-driven waves and currents in Singapore coastal water during the monsoon seasons.

Figure 5.23 shows Singapore coastal waters and their surroundings, and Figure 5.24 shows the simulated current profiles on the surface, middle, and bottom layers, from which it is easy to see the vertical flow circulation in the coastal region.

When the wave diffraction effect needs to be considered, the wave spectral model needs to be replaced by the phase-resolving model with the inclusion of current effect, e.g., an MSE model. In this case, wave propagation will be

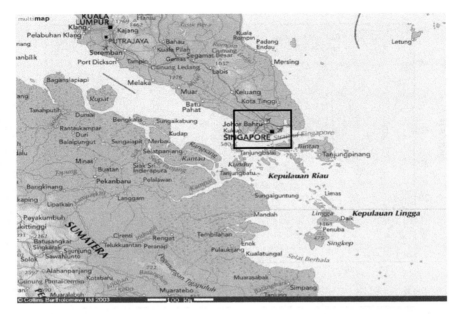

Figure 5.23 Map of Singapore coastal waters and their surroundings; the blocked region is the simulation area for waves and current. (From map © Collins Bartholomew Ltd 2003; reproduced by permission of Harper Collins Publishers)

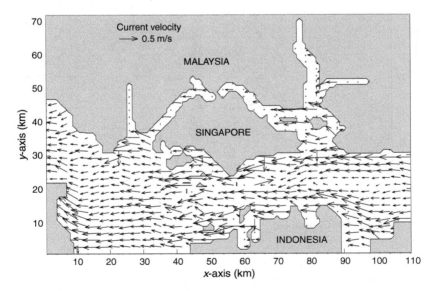

Figure 5.24 Simulation results of current distribution on surface, middle, and bottom layers in Singapore coastal waters by incorporating 3D radiation stresses in the coupled WAM–POM models.

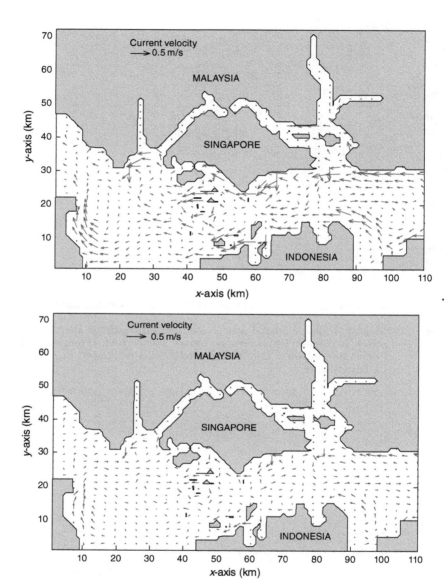

Figure 5.24 (Continued).

simulated by the MSE model, and the water level and current are handled by an SWE or a 3D hydrostatic flow model. The coupling can be done similarly to the coupling of wave spectral–SWE models. Recently, Liu *et al.* (2007) proposed a theoretical formulation for the coupled 3D wave and current flow.

Another alternative is to use Boussinesq models for both wave and current simulations by realizing that a Boussinesq model is similar to an SWE model for current simulation and similar to an MSE model for wave simulation. When properly treated, the Boussinesq model can accurately simulate breaking wave-induced longshore current and circulation in the nearshore region (e.g., Chen *et al.*, 2003). In this case, no model coupling is needed.

To solve the depth-dependent wave and current interaction with the use of a single wave model, we have to turn to the wave model based on NSE solvers. Similar to the Boussinesq equations, the momentum equations of the NSEs represent the contributions from both wave and current. Their interaction has been implicitly included in the nonlinear terms. With the use of a single RANS model, Lin and Liu (2004) simulated the breaking waves and the associated vertical structure of undertow in the surf zone (see Figure 5.25). In this case, the radiation stresses are automatically accounted for when the combined flow of wave and current is simulated, even though they are not explicitly included in the model.

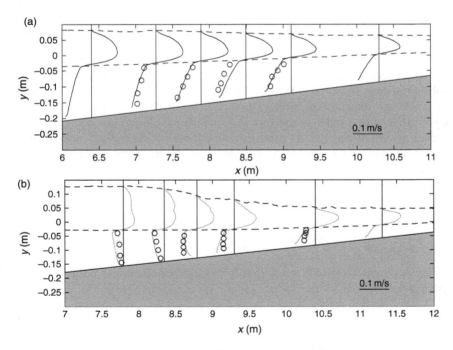

Figure 5.25 Undertow current for spilling (top) and plunging (bottom) breaking waves in the surf zone; solid lines are numerical results and circles are measured data; dashed lines are calculated envelopes that bracket the wave troughs and crests.

5.5   Example wave models and benchmark tests

This section has a twofold purpose: (1) to provide a few examples of wave models with adequate numerical details so that readers can build up their own wave models; and (2) to provide a few benchmark tests simulated by different wave models so that readers can benchmark their own models and understand more thoroughly the behaviors of different wave models under different circumstances.

5.5.1   Examples of wave models

5.5.1.1   The wave models based on the Navier–Stokes equation solvers

As introduced earlier, the wave model based on NSE solvers is the most robust numerical model that can be used to simulate essentially any water wave phenomenon. However, depending on how the NSE and RANS equations are solved numerically, these wave models may display different advantages and limitations. Below we shall introduce three NSE-solver wave models, namely NEWTANK (3D), the σ-coordinate wave model (3D), and the SPH wave model (2D). Focus will be on the introduction of NEWTANK, a 3D numerical wave tank that is currently being developed by the author and his group to simulate general turbulent-free surface flows.

NEWTANK: NEWTANK (Liu and Lin, 2008) is the 3D extension of the earlier 2D NEWFLUME (Lin and Xu, 2006). In addition, NEWTANK also solves the two-phase fluid flows with the use of higher order VOF method to track the interface motion. The NSEs (or RANS equations when the $k-\varepsilon$ turbulence model is used as the turbulence closure model; or SANS equations when the LES is used for turbulence modeling) are solved in a staggered grid system by FDM. Without loss of generality, we shall discuss RANS equations only in this section:

$$\frac{\partial u_i}{\partial x_i} = 0 \tag{5.245}$$

$$\frac{\partial u_i}{\partial t} + u_j \frac{\partial u_i}{\partial x_j} = -\frac{1}{\rho}\frac{\partial p}{\partial x_i} + g_i + \frac{1}{\rho}\frac{\partial \tau_{ij}}{\partial x_j} + \frac{1}{\rho}\frac{\partial R_{ij}}{\partial x_j} \tag{5.246}$$

where R_{ij} is the Reynolds stress that is modeled by the linear eddy viscosity model:

$$R_{ij} = -\rho \langle u_i' u_j' \rangle = 2\rho \nu_t \langle \sigma_{ij} \rangle - \frac{2}{3}\rho k \delta_{ij} \tag{5.247}$$

where the eddy viscosity:

$$\nu_t = C_d \frac{k^2}{\varepsilon}$$

with the turbulence kinetic energy k and its dissipation rate ε being governed by $k-\varepsilon$ transport equations:

$$\frac{\partial k}{\partial t} + u_j \frac{\partial k}{\partial x_j} = \frac{\partial}{\partial x_j}\left[\left(\frac{\nu_t}{\sigma_k} + \nu\right)\frac{\partial k}{\partial x_j}\right] + \nu_t\left(\frac{\partial u_i}{\partial x_j} + \frac{\partial u_j}{\partial x_i}\right)\frac{\partial u_i}{\partial x_j} - \varepsilon \qquad (5.248)$$

$$\frac{\partial \varepsilon}{\partial t} + u_j \frac{\partial \varepsilon}{\partial x_j} = \frac{\partial}{\partial x_j}\left(\frac{\nu_t}{\sigma_\varepsilon}\frac{\partial \varepsilon}{\partial x_j}\right) + C_{1\varepsilon}\frac{\varepsilon}{k}\nu_t\left(\frac{\partial u_i}{\partial x_j} + \frac{\partial u_j}{\partial x_i}\right)\frac{\partial u_i}{\partial x_j} - C_{2\varepsilon}\frac{\varepsilon^2}{k} \qquad (5.249)$$

The coefficients in the above turbulence closure model are as follows:

$$C_d = 0.09,\ C_{1\varepsilon} = 1.44,\ C_{2\varepsilon} = 1.92,\ \sigma_k = 1.0,\ \text{and } \sigma_\varepsilon = 1.3$$

The free surface tracking is accomplished with the use of VOF method, which solves the transport equation of the VOF function, a normalized fluid density function:

$$\frac{\partial F}{\partial t} + u_i \frac{\partial F}{\partial x_i} = 0 \Rightarrow \frac{\partial F}{\partial t} + \frac{\partial (u_i F)}{\partial x_i} = 0 \qquad (5.250)$$

For a two-phase fluid flow, knowing F, the mean density in each computational cell can be recalculated by using the following equation:

$$\rho = \rho_1 F + \rho_2(1 - F) \qquad (5.251)$$

where ρ_1 and ρ_2 represent the density of fluid 1 and fluid 2, respectively. The above equation can also be applied to multiphase fluid problems (e.g., internal wave interaction with free surface wave) when the multiple VOF functions are solved to track the interface between any two adjacent fluids.

To solve the above equation, the two-step projection method (Chorin, 1968) is used. In the first step, the tentative velocity is solved, i.e.:

$$\frac{\tilde{u}_i - u_i^n}{\Delta t} = -u_j^n \frac{\partial u_i^n}{\partial x_j} + \frac{1}{\rho}\frac{\partial \tau_{ij}^n}{\partial x_j} + \frac{1}{\rho}\frac{\partial R_{ij}^n}{\partial x_j} \qquad (5.252)$$

The second step is to project the tentative velocity field onto a divergence-free plane to obtain the final velocity, i.e.:

$$\frac{u_i^{n+1} - \tilde{u}_i}{\Delta t} = -\frac{1}{\rho^n}\frac{\partial p^{n+1}}{\partial x_i} + g_i \qquad (5.253)$$

where:

$$\frac{\partial u_i^{n+1}}{\partial x_i} = 0 \qquad (5.254)$$

Taking the divergence of (5.253) and applying the continuity equation (5.254) to the resulting equation, we obtain get the PPE as follows:

$$\frac{\partial}{\partial x_i}\left(\frac{1}{\rho^n}\frac{\partial p^{n+1}}{\partial x_i}\right) = \frac{1}{\Delta t}\frac{\partial \tilde{u}_i}{\partial x_i} \tag{5.255}$$

After solving (5.255) with appropriate boundary conditions, the pressure information at the $(n+1)$th time step can be obtained and applied to (5.253) for the updated velocity.

The advection terms in the x momentum equation are evaluated at the nth time level:

$$\left(u\frac{\partial u}{\partial x}+v\frac{\partial u}{\partial y}+w\frac{\partial u}{\partial z}\right)^n_{i+1/2,j,k} = u_{i+1/2,j,k}\left(\frac{\partial u}{\partial x}\right)^n_{i+1/2,j,k} + v_{i+1/2,j,k}\left(\frac{\partial u}{\partial y}\right)^n_{i+1/2,j,k}$$

$$+ w_{i+1/2,j,k}\left(\frac{\partial u}{\partial z}\right)^n_{i+1/2,j,k} \tag{5.256}$$

where the combination of a two-point upwind scheme and three-point central difference (for nonuniform grid) is used, e.g.:

$$\left(\frac{\partial u}{\partial x}\right)^n_{i+1/2,j,k} = \left\{\left[1+\alpha\,\mathrm{sgn}\left(u_{i+1/2,j,k}\right)\right]\Delta x_{i+1}\left(\frac{\partial u}{\partial x}\right)^n_{i,j,k}\right.$$

$$\left.+\left[1-\alpha\,\mathrm{sgn}\left(u_{i+1/2,j,k}\right)\right]\Delta x_i\left(\frac{\partial u}{\partial x}\right)^n_{i+1,j,k}\right\}/\Delta x_\alpha \tag{5.257}$$

where:

$$\Delta x_\alpha = \Delta x_{i+1} + \Delta x_i + \alpha\,\mathrm{sgn}\left(u^n_{i+1/2,j,k}\right)\left(\Delta x_{i+1}-\Delta x_i\right) \text{ and } \left(\frac{\partial u}{\partial x}\right)^n_{i,j,k}$$

$$= \frac{u^n_{i+1/2,j,k}-u^n_{i-1/2,j,k}}{\Delta x_i}$$

Similarly, we can have the expression for:

$$\left(\frac{\partial u}{\partial y}\right)^n_{i+1/2,j,k} \text{ and } \left(\frac{\partial u}{\partial z}\right)^n_{i+1/2,j,k}$$

The coefficient α is the weighting factor between the upwind method and the central difference method. When $\alpha = 0$, the FD form becomes the central difference; when $\alpha = 1$, the FD form becomes the upwind difference. In practice, α is selected in the range of 0.3–0.5.

The gradients of the viscous stresses for the x-component equation can be written as:

$$\frac{\partial}{\partial x}\tau_{xx} + \frac{\partial}{\partial y}\tau_{xy} + \frac{\partial}{\partial z}\tau_{xz} \tag{5.258}$$

which can be written in the following FD form:

$$\left(\frac{\partial}{\partial x}\tau_{xx}\right)^n_{i+1/2,j,k} + \left(\frac{\partial}{\partial y}\tau_{xy}\right)^n_{i+1/2,j,k} + \left(\frac{\partial}{\partial z}\tau_{xz}\right)^n_{i+1/2,j,k}$$

$$= \frac{(\tau_{xx})^n_{i+1,j,k} - (\tau_{xx})^n_{i,j,k}}{\Delta x_{i+1/2}} + \frac{(\tau_{xy})^n_{i+1/2,j+1/2,k} - (\tau_{xy})^n_{i+1/2,j-1/2,k}}{\Delta y_j}$$

$$+ \frac{(\tau_{xz})^n_{i+1/2,j,k+1/2} - (\tau_{xz})^n_{i+1/2,j,k-1/2}}{\Delta z_k} \tag{5.259}$$

where $\tau_{ij} = \mu(\partial u_i/\partial x_j + \partial u_j/\partial x_i)$ with $\partial u_i/\partial x_j$ being discretized by the central difference method. The Reynolds stress terms could be solved in a similar way. The momentum equations in the y- and z-directions can be discretized in the same way.

In solving the PPE in the second step, we first write out the following equation:

$$\frac{\partial}{\partial x}\left(\frac{1}{\rho^n}\frac{\partial p^{n+1}}{\partial x}\right) + \frac{\partial}{\partial y}\left(\frac{1}{\rho^n}\frac{\partial p^{n+1}}{\partial y}\right) + \frac{\partial}{\partial z}\left(\frac{1}{\rho^n}\frac{\partial p^{n+1}}{\partial z}\right)$$

$$= \frac{1}{\Delta t}\left(\frac{\partial \tilde{u}}{\partial x} + \frac{\partial \tilde{v}}{\partial y} + \frac{\partial \tilde{w}}{\partial z}\right) \tag{5.260}$$

The central difference is used. For example, the first term on the LHS is discretized as follows:

$$\left\{\frac{\partial}{\partial x}\left(\frac{1}{\rho^n}\frac{\partial p^{n+1}}{\partial x}\right)\right\}_{i,j,k}$$

$$= \frac{1}{\Delta x_i}\left\{\frac{1}{\rho^n_{i+1/2,j,k}}\left(\frac{\partial p}{\partial x}\right)^{n+1}_{i+1/2,j,k} - \frac{1}{\rho^n_{i-1/2,j,k}}\left(\frac{\partial p}{\partial x}\right)^{n+1}_{i-1/2,j,k}\right\}$$

$$= \frac{1}{\Delta x_i}\left\{\frac{1}{\rho^n_{i+1/2,j,k}}\left(\frac{p^{n+1}_{i+1,j,k} - p^{n+1}_{i,j,k}}{\Delta x_{i+1/2}}\right) - \frac{1}{\rho^n_{i-1/2,j,k}}\left(\frac{p^{n+1}_{i,j,k} - p^{n+1}_{i-1,j,k}}{\Delta x_{i-1/2}}\right)\right\} \tag{5.261}$$

with

$$\rho^n_{i+1/2,j,k} = \frac{\rho^n_{i,j,k}\Delta x_{i+1} + \rho^n_{i+1,j,k}\Delta x_i}{\Delta x_i + \Delta x_{i+1}}.$$

Similar FDM can be used for the second and third terms on the LHS of the above PPE. The RHS of the PPE is expressed as follows:

$$\left(\frac{\partial \tilde{u}}{\partial x} + \frac{\partial \tilde{v}}{\partial y} + \frac{\partial \tilde{w}}{\partial z}\right)_{i,j,k} = \frac{\tilde{u}_{i+1/2,j,k} - \tilde{u}_{i-1/2,j,k}}{\Delta x_i} + \frac{\tilde{v}_{i,j+1/2,k} - \tilde{v}_{i,j-1/2,k}}{\Delta y_j}$$

$$+ \frac{\tilde{w}_{i,j,k+1/2} - \tilde{w}_{i,j,k-1/2}}{\Delta z_k} \tag{5.262}$$

Finally, the discretized PPE with the proper boundary conditions yields a set of linear algebraic equations for the pressure field that can be solved by the CG method.

After the updated velocity at the $n+1$ time step is obtained by solving (5.253), the $k-\varepsilon$ equations are solved by FDM as follows:

$$\frac{\varepsilon_{i,j,k}^{n+1} - \varepsilon_{i,j,k}^{n}}{\Delta t} + F\varepsilon X + F\varepsilon Y + F\varepsilon Z = VIS\varepsilon + C_{1\varepsilon}\frac{\varepsilon_{i,j,k}^{n}}{k_{i,j,k}^{n}}P_{i,j,k}^{n+1} - C_{2\varepsilon}\frac{\varepsilon_{i,j,k}^{n}}{k_{i,j,k}^{n}}\varepsilon_{i,j,k}^{n+1}$$

$$(5.263)$$

$$\frac{k_{i,j,k}^{n+1} - k_{i,j,k}^{n}}{\Delta t} + FkX + FkY + FkZ = VISk + P_{i,j,k}^{n+1} - \varepsilon_{i,j,k}^{n+1} \qquad (5.264)$$

where the advection is solved by the upwind scheme to eliminate any oscillation that may bring k or ε to be negative and the diffusion and turbulence production terms are discretized by the central difference method.

The VOF transport equation (5.250) can be discretized in the following FD form for a 3D problem:

$$F_{i,j,k}^{n+1} = F_{i,j,k}^{n} - \frac{\Delta t}{\Delta x_i}\left[(uF)_{i+1/2,j,k}^{n} - (uF)_{i-1/2,j,k}^{n}\right]$$

$$- \frac{\Delta t}{\Delta y_j}\left[(vF)_{i,j+1/2,k}^{n} - (vF)_{i,j-1/2,k}^{n}\right] - \frac{\Delta t}{\Delta z_k}\left[(wF)_{i,j,k+1/2}^{n} - (wF)_{i,j,k-1/2}^{n}\right]$$

$$= F_{i,j,k}^{n} - \Delta F_{\text{east}} + \Delta F_{\text{west}} - \Delta F_{\text{north}} + \Delta F_{\text{south}} - \Delta F_{\text{top}} + \Delta F_{\text{bottom}} \qquad (5.265)$$

Here **u** will be taken from the most updated velocity information. Now the key issue becomes how to determine the VOF fluxes ΔF_{east}, ΔF_{west}, ΔF_{north}, ΔF_{south}, ΔF_{top}, and ΔF_{bottom} across the six cell faces for each computational cell. In NEWTANK, the second-order piecewise linear interface calculation (PLIC) method is used to reconstruct the interface and to determine the VOF fluxes. The detailed procedure is given below for calculating ΔF_{east} and the others can be calculated in the same way.

Step 1 Free surface reconstruction: The linear interface reconstruction is accomplished by knowing both the normal vector of the interface and the intercept of the interface. The modified Young's LS method (e.g., Rider and Kothe, 1998), which is second-order accurate, is employed to estimate the normal vector of the interface first. Define the VOF value in a particular cell $\mathbf{x}_0$ as F_0. The Taylor series expansions of the VOF functions at the neighborhood cells are:

$$F_k^{\text{TSE}}(\mathbf{x}_0 + \Delta\mathbf{x}_k) = F_0 + (\nabla F)\Delta\mathbf{x}_k + o(\Delta\mathbf{x}_k^2) \quad (k = 1, 2, \ldots, n) \qquad (5.266)$$

The sum of $\left(F_k^{\text{TSE}} - F_k\right)^2$ (difference between the estimated and the actual VOF in the neighborhood cells) over all n neighboring cells is then minimized in the LS sense. This L_2 norm minimization will yield the VOF gradient ∇F as the solutions to the linear system:

$$\left(A^{\mathrm{T}}A\right)\mathbf{x} = \left(A^{\mathrm{T}}\mathbf{b}\right) \tag{5.267}$$

where:

$$A = \begin{pmatrix} \omega_1\,(x_1-x_0) & \omega_1\,(y_1-y_0) & \omega_1\,(z_1-z_0) \\ \vdots & \vdots & \vdots \\ \omega_k\,(x_k-x_0) & \omega_k\,(y_k-y_0) & \omega_k\,(z_k-z_0) \\ \vdots & \vdots & \vdots \\ \omega_n\,(x_n-x_0) & \omega_n\,(y_n-y_0) & \omega_n\,(z_n-z_0) \end{pmatrix}, \qquad \mathbf{b} = \begin{pmatrix} \omega_1\,(F_1-F_0) \\ \vdots \\ \omega_k\,(F_k-F_0) \\ \vdots \\ \omega_n\,(F_n-F_0) \end{pmatrix},$$

in which:

$$\omega_k = \frac{1}{(x_k-x_0)^2 + (y_k-y_0)^2 + (z_k-z_0)^2} \quad \text{and} \quad \mathbf{x} = \left(\nabla_x F,\, \nabla_y F,\, \nabla_z F\right)^{\mathrm{T}}$$

The normal vector then can be computed as follows:

$$\mathbf{m} = \left(-\frac{\nabla_x F}{|\nabla F|},\, -\frac{\nabla_y F}{|\nabla F|},\, -\frac{\nabla_z F}{|\nabla F|}\right)^{\mathrm{T}} \tag{5.268}$$

where $|\nabla F| = \sqrt{(\nabla_x F)^2 + (\nabla_y F)^2 + (\nabla_z F)^2}$. In a 3D Cartesian grid system, we choose $n = 26$ surrounding grids to estimate $\mathbf{m}$.

Once $\mathbf{m}$ is known, the normal distance α from the origin of the cell (west–south–bottom corner of the cell in NEWTANK) to the interface plane can be uniquely determined with the given VOF value in the cell. The standard root-finding approach, such as bisection or Newton's method, can be used for this purpose.

Step 2 Determination of donor cell: To determine the VOF flux across a particular cell face, the velocity on the cell face is used to identify the "donor" cell that contributes VOF during the convection. For example, for the cell face on the "east" side (right side), if $u_{i+1/2,j,k} > 0$, the donor cell will be (i, j, k); if $u_{i+1/2,j,k} < 0$, the donor cell becomes $(i+1, j, k)$. In this example, we assume $u_{i+1/2,j,k} > 0$ and thus the reconstructed free surface in the cell (i, j, k) will be used in the following calculation.

Step 3 Lagrangian propagation of the interface: Once $\mathbf{m}$ and α are determined, the interface plane in the cell (i, j, k) can be expressed as:

$$m_1^n x + m_2^n y + m_3^n z = \alpha^n \tag{5.269}$$

where m_1^n, m_2^n, and m_3^n are the x, y, and z components of the unit normal vector $\mathbf{m}$, respectively. During the convection, different parts of the plane

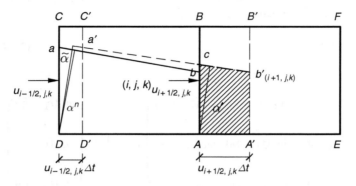

Figure 5.26 Calculation of VOF flux using the PLIC method.

will be moved at different speeds, which can be obtained by the simple linear interpolation within the cell between the two cell faces (e.g., west and east). For the 2D case shown in Figure 5.26, this is done by moving the line ab to a′b′ with $u_{i+1/2,j,k}$ and $u_{i-1/2,j,k}$, respectively. The new interface can be represented by the following equation:

$$\tilde{m}_1 x + \tilde{m}_2 y + \tilde{m}_3 z = \tilde{\alpha} \tag{5.270}$$

where $\tilde{\mathbf{m}} = (\tilde{m}_1, \tilde{m}_2, \tilde{m}_3)$ is the unit normal vector for the new plane interface:

$$\tilde{m}_1 = \frac{m_1^*}{\sqrt{(m_1^*)^2 + (m_2^*)^2 + (m_3^*)^2}} \quad \tilde{m}_2 = \frac{m_2^*}{\sqrt{(m_1^*)^2 + (m_2^*)^2 + (m_3^*)^2}}$$

$$\tilde{m}_3 = \frac{m_3^*}{\sqrt{(m_1^*)^2 + (m_2^*)^2 + (m_3^*)^2}}$$

and the normal distance:

$$\tilde{\alpha} = \frac{\alpha^*}{\sqrt{(m_1^*)^2 + (m_2^*)^2 + (m_3^*)^2}}$$

with:

$$m_1^* = \frac{m_1^n}{1 + \left(\dfrac{u_{i+1/2,j,k} - u_{i-1/2,j,k}}{\Delta x_i}\right) \cdot \Delta t}, \quad m_2^* = m_2^n, \quad m_3^* = m_3^n, \text{ and}$$

$$\alpha^* = \alpha^n + \frac{m_1^n \cdot u_{i-1/2,j,k} \cdot \Delta t}{1 + \left(\dfrac{u_{i+1/2,j,k} - u_{i-1/2,j,k}}{\Delta x_i}\right) \cdot \Delta t}$$

Step 4 Determination of the VOF flux: In the 2D case shown in Figure 5.26, the VOF flux from cell (i, j, k) to cell $(i+1, j, k)$ is the shaded area of AA′b′c. To calculate the area, the information required is the interface function and the normal distance α' in the cell $(i+1, j, k)$. This can be obtained by introducing the simple coordinate transformation from cell (i, j, k), where the updated interface equation is established, to cell $(i+1, j, k)$:

$$x = x' + \Delta x_i \tag{5.271}$$

Substituting (5.271) into (5.270), we have:

$$m_1 x' + m_2 y'' + m_3 z'' = \alpha' \tag{5.272}$$

where $m_1 = \tilde{m}_1$, $m_2 = \tilde{m}_2$, $m_3 = \tilde{m}_3$, and $\alpha' \doteq \tilde{\alpha} - m_1 \Delta x_i$. As expected, the only change is the normal distance to the new origin. The area can then be calculated using the following general formula (Gueyffier *et al.*, 1999):

$$\Delta F_{east} = \text{area} = \frac{1}{2m_1 m_2} \left[\alpha'^2 - \sum_{i=1}^{2} F_2 (\alpha' - m_i \Delta x_i) \right] \tag{5.273}$$

where $F_2(x)$ is the Heaviside step function defined as:

$$F_2(x) = \begin{cases} x^2 & \text{for } x > 0 \\ 0 & \text{for } x \leq 0 \end{cases} \tag{5.274}$$

The convection in the vertical direction can be calculated similarly. For 3D problems, equation (5.273) can be extended to:

$$\Delta F_{east} = \text{volume} = \frac{1}{6m_1 m_2 m_3} \left[\alpha'^3 - \sum_{i=1}^{3} F_3 (\alpha' - m_i \Delta x_i) \right.$$
$$\left. + \sum_{i=1}^{3} F_3 (\alpha' - \alpha'_{max} + m_i \Delta x_i) \right] \tag{5.275}$$

where:

$$\alpha'_{max} = \sum_{i=1}^{3} m_i \Delta x_i \tag{5.276}$$

and $F_3(x)$ is the Heaviside step function and is defined as:

$$F_3(x) = \begin{cases} x^3 & \text{for } x > 0 \\ 0 & \text{for } x \leq 0 \end{cases} \tag{5.277}$$

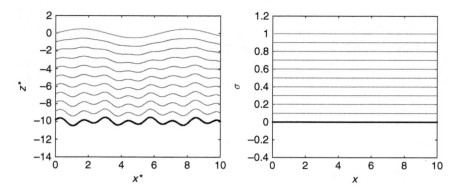

Figure 5.27 Illustration of the transformation from the physical domain (left) to the computational domain (right) using σ-coordinate transformation.

σ-coordinate wave model: This model was constructed on a 3D FD grid system (Lin and Li, 2002). By introducing the new coordinate into the computational domain:

$$t = t^*, \quad x = x^*, \quad y = y^*, \quad \sigma = \frac{z^* + h}{h + \eta} = \frac{z^* + h}{D} \tag{5.278}$$

where the variables with asterisk represent those in the physical domain and not in the computational domain (see Figure 5.27 for illustration of a 2D case). The modified governing equations in the computational domain are the following:

$$\frac{\partial u}{\partial x} + \frac{\partial u}{\partial \sigma}\frac{\partial \sigma}{\partial x^*} + \frac{\partial v}{\partial y} + \frac{\partial v}{\partial \sigma}\frac{\partial \sigma}{\partial y^*} + \frac{\partial w}{\partial \sigma}\frac{\partial \sigma}{\partial z^*} = 0 \tag{5.279}$$

$$\frac{\partial u}{\partial t} + u\frac{\partial u}{\partial x} + v\frac{\partial u}{\partial y} + \omega\frac{\partial u}{\partial \sigma} = -\frac{1}{\rho}\left(\frac{\partial p}{\partial x} + \frac{\partial p}{\partial \sigma}\frac{\partial \sigma}{\partial x^*}\right) + g_x$$

$$+ \frac{\partial \tau_{xx}}{\partial x} + \frac{\partial \tau_{xx}}{\partial \sigma}\frac{\partial \sigma}{\partial x^*} + \frac{\partial \tau_{xy}}{\partial y}$$

$$+ \frac{\partial \tau_{xy}}{\partial \sigma}\frac{\partial \sigma}{\partial y^*} + \frac{\partial \tau_{xz}}{\partial \sigma}\frac{\partial \sigma}{\partial z^*} \tag{5.280}$$

$$\frac{\partial v}{\partial t} + u\frac{\partial v}{\partial x} + v\frac{\partial v}{\partial y} + \omega\frac{\partial v}{\partial \sigma} = -\frac{1}{\rho}\left(\frac{\partial p}{\partial y} + \frac{\partial p}{\partial \sigma}\frac{\partial \sigma}{\partial y^*}\right) + g_y + \frac{\partial \tau_{yx}}{\partial x} + \frac{\partial \tau_{yx}}{\partial \sigma}\frac{\partial \sigma}{\partial x^*}$$

$$+ \frac{\partial \tau_{yy}}{\partial y} + \frac{\partial \tau_{yy}}{\partial \sigma}\frac{\partial \sigma}{\partial y^*} + \frac{\partial \tau_{yz}}{\partial \sigma}\frac{\partial \sigma}{\partial z^*} \tag{5.281}$$

$$\frac{\partial w}{\partial t} + u\frac{\partial w}{\partial x} + v\frac{\partial w}{\partial y} + \omega\frac{\partial w}{\partial \sigma} = -\frac{1}{\rho}\frac{\partial p}{\partial \sigma}\frac{\partial \sigma}{\partial z^*} + g_z + \frac{\partial \tau_{zx}}{\partial x} + \frac{\partial \tau_{zx}}{\partial \sigma}\frac{\partial \sigma}{\partial x^*}$$

$$+ \frac{\partial \tau_{zy}}{\partial y} + \frac{\partial \tau_{zy}}{\partial \sigma}\frac{\partial \sigma}{\partial y^*} + \frac{\partial \tau_{zz}}{\partial \sigma}\frac{\partial \sigma}{\partial z^*} \qquad (5.282)$$

where:

$$\omega = \frac{D\sigma}{Dt^*} = \frac{\partial \sigma}{\partial t^*} + u\frac{\partial \sigma}{\partial x^*} + v\frac{\partial \sigma}{\partial y^*} + w\frac{\partial \sigma}{\partial z^*} \qquad (5.283)$$

and:

$$\frac{\partial \sigma}{\partial t^*} = -\frac{\sigma}{D}\frac{\partial D}{\partial t}, \quad \frac{\partial \sigma}{\partial x^*} = \frac{1}{D}\frac{\partial h}{\partial x} - \frac{\sigma}{D}\frac{\partial D}{\partial x}, \quad \frac{\partial \sigma}{\partial y^*} = \frac{1}{D}\frac{\partial h}{\partial y} - \frac{\sigma}{D}\frac{\partial D}{\partial y}, \quad \frac{\partial \sigma}{\partial z^*} = \frac{1}{D}$$

$$(5.284)$$

In the transformed space, the stresses are calculated as follows:

$$\tau_{xx} = 2(\nu + \nu_t)\left(\frac{\partial u}{\partial x} + \frac{\partial u}{\partial \sigma}\frac{\partial \sigma}{\partial x^*}\right),$$

$$\tau_{xy} = \tau_{yx} = (\nu + \nu_t)\left(\frac{\partial u}{\partial y} + \frac{\partial u}{\partial \sigma}\frac{\partial \sigma}{\partial y^*} + \frac{\partial v}{\partial x} + \frac{\partial v}{\partial \sigma}\frac{\partial \sigma}{\partial x^*}\right)$$

$$\tau_{xz} = \tau_{zx} = (\nu + \nu_t)\left(\frac{\partial u}{\partial \sigma}\frac{\partial \sigma}{\partial z^*} + \frac{\partial w}{\partial x} + \frac{\partial w}{\partial \sigma}\frac{\partial \sigma}{\partial x^*}\right),$$

$$\tau_{yy} = 2(\nu + \nu_t)\left(\frac{\partial v}{\partial y} + \frac{\partial v}{\partial \sigma}\frac{\partial \sigma}{\partial y^*}\right)$$

$$\tau_{yz} = \tau_{zy} = (\nu + \nu_t)\left(\frac{\partial v}{\partial \sigma}\frac{\partial \sigma}{\partial z^*} + \frac{\partial w}{\partial y} + \frac{\partial w}{\partial \sigma}\frac{\partial \sigma}{\partial y^*}\right), \quad \tau_{zz} = 2(\nu + \nu_t)\left(\frac{\partial w}{\partial \sigma}\frac{\partial \sigma}{\partial z^*}\right)$$

$$(5.285)$$

where the eddy viscosity is modeled by LES as:

$$\nu_t = L_s^2\sqrt{2S_{ij}S_{ij}} \qquad (5.286)$$

where S_{ij} is the strain rate of the resolved mean flow and $L_s = 0.15\,(\Delta x \times \Delta y \times \Delta z)^{1/3}$. The governing equation (5.279) for the free surface movement is converted into the following form:

$$\frac{\partial \eta}{\partial t} + \frac{\partial}{\partial x}\left[D\int_0^1 u\,d\sigma\right] + \frac{\partial}{\partial y}\left[D\int_0^1 v\,d\sigma\right] = 0 \qquad (5.287)$$

The time-splitting method is used to solve the momentum equations. For example, in the x-direction, we have:

$$\frac{u_{i,j,k}{}^{n+1/3} - u_{i,j,k}{}^{n}}{\Delta t} + \left(u\frac{\partial u}{\partial x} + v\frac{\partial u}{\partial y} + \omega\frac{\partial u}{\partial \sigma} \right)_{i,j,k} = 0 \tag{5.288}$$

$$\frac{u_{i,j,k}{}^{n+2/3} - u_{i,j,k}{}^{n+1/3}}{\Delta t} = \left(\frac{\partial \tau_{xx}}{\partial x} + \frac{\partial \tau_{xx}}{\partial \sigma}\frac{\partial \sigma}{\partial x^*} + \frac{\partial \tau_{xy}}{\partial y} + \frac{\partial \tau_{xy}}{\partial \sigma}\frac{\partial \sigma}{\partial y^*} + \frac{\partial \tau_{xz}}{\partial \sigma}\frac{\partial \sigma}{\partial z^*} \right)_{i,j,k}^{n+1/3} \tag{5.289}$$

$$\frac{u_{i,j,k}{}^{n+1} - u_{i,j,k}{}^{n+2/3}}{\Delta t} = -\frac{1}{\rho}\left(\frac{\partial p}{\partial x} + \frac{\partial p}{\partial \sigma}\frac{\partial \sigma}{\partial x^*} \right)_{i,j,k}^{n+1} + g_x \tag{5.290}$$

where the updated pressure can be obtained by solving the PPE equation in the transformed plane:

$$\left\{ \frac{\partial^2 p}{\partial x^2} + \frac{\partial^2 p}{\partial y^2} + \left[\left(\frac{\partial \sigma}{\partial x^*}\right)^2 + \left(\frac{\partial \sigma}{\partial y^*}\right)^2 + \left(\frac{\partial \sigma}{\partial z^*}\right)^2 \right]\frac{\partial^2 p}{\partial \sigma^2} \right.$$
$$\left. + 2\left(\frac{\partial \sigma}{\partial x^*}\frac{\partial^2 p}{\partial x \partial \sigma} + \frac{\partial \sigma}{\partial y^*}\frac{\partial^2 p}{\partial y \partial \sigma} \right) + \left(\frac{\partial^2 \sigma}{\partial x^* \partial x} + \frac{\partial^2 \sigma}{\partial y^* \partial y} \right)\frac{\partial p}{\partial \sigma} \right\}_{i,j,k}^{n+1}$$
$$= \frac{\rho}{\Delta t}\left(\frac{\partial u}{\partial x} + \frac{\partial u}{\partial \sigma}\frac{\partial \sigma}{\partial x^*} + \frac{\partial v}{\partial y} + \frac{\partial v}{\partial \sigma}\frac{\partial \sigma}{\partial y^*} + \frac{\partial w}{\partial \sigma}\frac{\partial \sigma}{\partial z^*} \right)_{i,j,k}^{n+2/3} \tag{5.291}$$

For the spatial derivative, the second-order central difference method is used for all terms except for the convection term, which is discretized by the combination of quadratic backward characteristic method and the Lax–Wendroff method for smooth flows without structures inside the computational domain and the combination of central difference and upwind scheme when a structure is present.

To handle wave–structure interaction problems, Lin (2006) extended the above model using the three-layer σ-coordinate system. Considering the problem as shown in Figure 5.28, we can introduce three subdepths, namely h_1, h_2, and $h_3 = h - h_1 - h_2$, where h_1, h_2, and h_3 are function of space (x, y) that can be designed to be conformal to the bottom topography, surface geometry of a solid body, and free surface. Note that the middle layers do not have to be horizontal; instead, they can be designed to follow the body surface configuration. With the problem being thus defined, we then have:

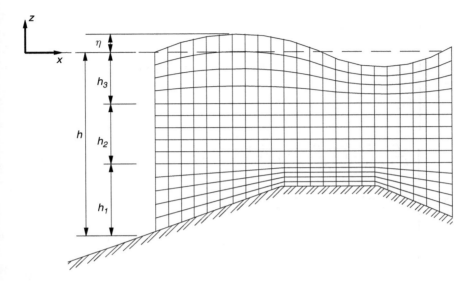

Figure 5.28 Sketch of the three-layer σ-coordinate system.

$$\sigma = \begin{cases} C_1 + C_2 + C_3 \left(\dfrac{h_3(x,y) + z^*}{h_3(x,y) + \eta(x,y)} \right), & -h_3(x,y) \leq z^* \leq \eta(x,y) \\[3mm] C_1 + C_2 \left(\dfrac{z^* + h_2(x,y) + h_3(x,y)}{h_2(x,y)} \right), & -h_2(x,y) - h_3(x,y) \leq z^* < -h_3(x,y) \\[3mm] C_1 \left(\dfrac{z^* + h(x,y)}{h_1(x,y)} \right), & -h(x,y) \leq z^* < -h_2(x,y) - h_3(x,y) \end{cases}$$

$$(5.292)$$

where C_n ($n = 1, 2, 3$) is the weighting coefficients for the nth layer and $C_1 + C_2 + C_3 = 1$. For simplicity, the value of $C_n = N_n / N_t$ can be used where N_n is the number of elements in the nth layer and $N_t = N_1 + N_2 + N_3$ the total number of elements in the vertical discretization.

It is easy to prove that (5.292) can be reduced to the conventional σ-coordinate (5.278) when $h_1 = h_2 = 0$. By arranging the layer system differently, the proposed σ-coordinate can be used to simulate submerged, immersed, and floating structures. Except for the immersed structure, the other two types of structures can be represented by only two layers, the special case of (5.292) with $h_2 = 0$. With the three-layer σ-coordinate, the time derivative and spatial derivatives of σ are revised to:

$$\frac{\partial \sigma}{\partial t^*} = \begin{cases} -\left(\dfrac{\sigma - C_1 - C_2}{D'}\right)\dfrac{\partial(h_3 + \eta)}{\partial t}, & -h_3 \leq z^* \leq \eta, & C_1 + C_2 \leq \sigma \leq 1 \\[2mm] 0, & -h_2 - h_3 \leq z^* < -h_3, & C_1 \leq \sigma < C_1 + C_2 \\[2mm] 0, & -h \leq z^* < -h_2 - h_3, & 0 \leq \sigma < C_1 \end{cases} \tag{5.293}$$

$$\frac{\partial \sigma}{\partial x^*} = \begin{cases} \dfrac{C_3}{h_3 + \eta}\dfrac{\partial h_3}{\partial x} - \left(\dfrac{\sigma - C_1 - C_2}{h_3 + \eta}\right)\dfrac{\partial(h_3 + \eta)}{\partial x}, & -h_3 \leq z^* \leq \eta, & C_1 + C_2 \leq \sigma \leq 1 \\[3mm] \dfrac{C_2}{h_2}\dfrac{\partial h_3}{\partial x} - \left(\dfrac{\sigma - C_1 - C_2}{h_2}\right)\dfrac{\partial h_2}{\partial x}, & -h_2 - h_3 \leq z^* < -h_3, & C_1 < \sigma \leq C_1 + C_2 \\[3mm] -\dfrac{(\sigma - C_1)}{h_1}\dfrac{\partial h_1}{\partial x} + \dfrac{C_1}{h_1}\left(\dfrac{\partial h_2}{\partial x} + \dfrac{\partial h_3}{\partial x}\right), & -h \leq z^* < -h_2 - h_3, & 0 \leq \sigma < C_1 \end{cases}$$

$$\tag{5.294}$$

$$\frac{\partial \sigma}{\partial y^*} = \begin{cases} \dfrac{C_3}{h_3 + \eta}\dfrac{\partial h_3}{\partial y} - \left(\dfrac{\sigma - C_1 - C_2}{h_3 + \eta}\right)\dfrac{\partial(h_3 + \eta)}{\partial y}, & -h_3 \leq z^* \leq \eta, & C_1 + C_2 \leq \sigma \leq 1 \\[3mm] \dfrac{C_2}{h_2}\dfrac{\partial h_3}{\partial y} - \left(\dfrac{\sigma - C_1 - C_2}{h_2}\right)\dfrac{\partial h_2}{\partial y}, & -h_2 - h_3 \leq z^* < -h_3, & C_1 < \sigma \leq C_1 + C_2 \\[3mm] -\dfrac{(\sigma - C_1)}{h_1}\dfrac{\partial h_1}{\partial y} + \dfrac{C_1}{h_1}\left(\dfrac{\partial h_2}{\partial y} + \dfrac{\partial h_3}{\partial y}\right), & -h \leq z^* < -h_2 - h_3, & 0 \leq \sigma < C_1 \end{cases}$$

$$\tag{5.295}$$

$$\frac{\partial \sigma}{\partial z^*} = \begin{cases} \dfrac{C_3}{h_3 + \eta}, & -h_3 \leq z^* \leq \eta, & C_1 + C_2 \leq \sigma \leq 1 \\[3mm] \dfrac{C_2}{h_2}, & -h_2 - h_3 \leq z^* < -h_3, & C_1 < \sigma \leq C_1 + C_2 \\[3mm] \dfrac{C_1}{h_1}, & -h \leq z^* < -h_2 - h_3, & 0 \leq \sigma < C_1 \end{cases} \tag{5.296}$$

SPH wave model: Although the SPH wave model (Shao and Lo, 2003) solved the same NSEs for incompressible fluids, the formulation and evaluation of function derivatives are very different. The governing equations in SPH are written in the Lagrangian form for discrete particles, i.e.:

$$\frac{1}{\rho}\frac{D\rho}{Dt} + \frac{\partial u_i}{\partial x_i} = 0 \tag{5.297}$$

$$\frac{Du_i}{Dt} = \frac{1}{\rho}\left[-\frac{\partial p}{\partial x_i} + \rho g_i + \frac{\partial}{\partial x_j}\tau_{ij}\right] \tag{5.298}$$

The above two equations are equivalent to the mass and momentum conservation equations in NSEs but expressed following particle motion. The mass conservation equation is for general fluid (compressible and incompressible), and the incompressibility condition will be forced later in the computation. The particle location is governed by the following equation:

$$\frac{DX_i}{Dt} = u_i \tag{5.299}$$

Numerically, the above equations are solved by the two-step projection method. In the first step, the tentative velocity is calculated without the inclusion of pressure:

$$\frac{\tilde{u}_i - u_i^n}{\Delta t} = \left\{ \frac{1}{\rho} \left[\rho g_i + \frac{\partial}{\partial x_j} \tau_{ij} \right] \right\}^n \tag{5.300}$$

It must be realized that the velocity on the LHS represents the velocity of the same particle that is located at different positions at different times. After the tentative velocity for each particle is solved, the particle is moved to its tentative location by:

$$\tilde{X}_i = X_i^n + \tilde{u}_i \Delta t \tag{5.301}$$

In the second step, the velocity field is updated based on the updated pressure field that is calculated from the PPE and thus ensures the updated velocity field satisfying the continuity equation:

$$\frac{u_i^{n+1} - \tilde{u}_i}{\Delta t} = -\frac{1}{\rho^{n+1}} \frac{\partial p^{n+1}}{\partial x_i} \tag{5.302}$$

where p^{n+1} can be obtained by solving the PPE:

$$\frac{\partial}{\partial x_i} \left(-\frac{1}{\rho^{n+1}} \frac{\partial p^{n+1}}{\partial x_i} \right) = \frac{\partial}{\partial x_i} \left(\frac{u_i^{n+1} - \tilde{u}_i}{\Delta t} \right) = \frac{1}{\Delta t} \left(\frac{\partial u_i^{n+1}}{\partial x_i} - \frac{\partial \tilde{u}_i}{\partial x_i} \right)$$
$$= \frac{1}{\Delta t} \left(\frac{1}{\rho^{n+1}} \frac{\rho^{n+1} - \rho^n}{\Delta t} - \frac{1}{\rho^{n+1}} \frac{\tilde{\rho} - \rho^n}{\Delta t} \right)$$
$$= \frac{1}{\Delta t^2} \left(\frac{\rho^{n+1} - \tilde{\rho}}{\rho^{n+1}} \right) \tag{5.303}$$

Compared with the mesh-based NSE solvers, the above PPE in the SPH model is different in the sense that instead of forcing the velocity divergence to be zero directly, it forces the density of the particle that is invariant in time, i.e., $\rho^{n+1} = \rho^n = \rho_0$. With the final velocity u_i^{n+1} being obtained from

(5.302), the final particle location is determined by averaging the previous and final velocity:

$$X_i^{n+1} = X_i^n + \frac{\left(u_i^n + u_i^{n+1}\right)}{2}\Delta t \tag{5.304}$$

The above numerical solution involves the assessment of the variable function and the evaluation of function derivatives using a cluster of randomly distributed particle information. In the SPH model, the variable function around any particle a is approximated by the weighted summation of the relevant quantities of its neighborhood particles:

$$\phi(\mathbf{x}_a) = \int_{\mathbf{x}_b \in \Omega} W(\mathbf{x}_a, \mathbf{x}_b)\phi(\mathbf{x}_b)d\Omega \approx \sum_{b=1}^{M} V(\mathbf{x}_b)W(\mathbf{x}_a, \mathbf{x}_b)\phi(\mathbf{x}_b)$$

$$= \sum_{b=1}^{M} \frac{m_b}{\rho(\mathbf{x}_b)} W(\mathbf{x}_a, \mathbf{x}_b)\phi(\mathbf{x}_b) \tag{5.305}$$

where Ω is the integration space around particle a. Usually Ω is taken as a circle for 2D problems and as a sphere for 3D problems with:

$$r = |\mathbf{x}_b - \mathbf{x}_a| \le r_0 = 2h$$

where h is the initial particle spacing (Monaghan, 1994). M is the number of particles within the space Ω (Figure 5.29). $V(\mathbf{x}_b) = m_b/\rho(\mathbf{x}_b)$ is the volume associated with particle b, where m_b is the mass of particle b and $\rho(\mathbf{x}_b)$ is the density around it, which can be readily obtained by employing the same definition above, i.e.:

$$\rho(\mathbf{x}_b) = \sum_{c=1} m_c W(\mathbf{x}_b, \mathbf{x}_c)\phi(\mathbf{x}_c)$$

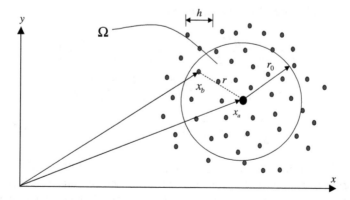

Figure 5.29 Sketch of the particle definition in the SPH model.

Here $W(\mathbf{x}_a, \mathbf{x}_b)$ is the weighting function (sometimes called interpolation kernel), which is used to determine how the neighboring particles are taken into consideration.

There are different ways of defining the kernel, e.g., by Gaussian function, cubic spline, or any other function with similar characteristics of having larger weighting near the particle and diminishing monotonically farther away from it to zero. The employment of different kernels in SPH method is analog to the use of a different FD scheme in FDM, and they have direct impact on the numerical accuracy and stability. Besides, the kernel function should also satisfy the following constraint:

$$\int_{x_b \in \Omega} W(\mathbf{x}_a, \mathbf{x}_b) d\Omega = 1 \tag{5.306}$$

In the discussed SPH wave model, the kernel proposed by Monaghan (1992) is used:

$$W(\mathbf{x}_a, \mathbf{x}_b) = W(|\mathbf{x}_b - \mathbf{x}_a|)$$

$$= W(r) = \begin{cases} \dfrac{10}{7\pi h^2}\left[1 - \dfrac{3}{2}\left(\dfrac{r}{h}\right)^2 + \dfrac{3}{4}\left(\dfrac{r}{h}\right)^3\right] & \text{for} \quad \dfrac{r}{h} \le 1 \\[3mm] \dfrac{10}{28\pi h^2}\left(2 - \dfrac{r}{h}\right)^3 & \text{for} \quad 1 < \dfrac{r}{h} \le 2 \\[3mm] 0 & \text{for} \quad \dfrac{r}{h} > 2 \end{cases}$$

$$\tag{5.307}$$

With the variables being approximated by the above discrete form, various spatial derivatives at particle a can be evaluated as follows. For example, the pressure gradient is evaluated as follows:

$$\left(\frac{1}{\rho}\frac{\partial p}{\partial x_i}\right)_a = \sum_b m_b \left(\frac{p_a}{\rho_a^2} + \frac{p_b}{\rho_b^2}\right) \frac{\partial W(\mathbf{x}_a, \mathbf{x}_b)}{\partial x_i} \tag{5.308}$$

It is seen that the evaluation of the variable gradient is converted to the evaluation of the kernel function, which can be done analytically in general.

Similarly, the divergence, the gradient of stress, Laplacian, etc., can be evaluated as follows (Shao and Ji, 2006):

$$\left(\frac{\partial u_i}{\partial x_i}\right)_a = \rho_a \sum_b m_b \left(\frac{u_{i_a}}{\rho_a^2} + \frac{u_{i_b}}{\rho_b^2}\right) \frac{\partial W(\mathbf{x}_a, \mathbf{x}_b)}{\partial x_i} \tag{5.309}$$

$$\left(\frac{1}{\rho}\frac{\partial \tau_{ij}}{\partial x_i}\right)_a = \sum_b m_b \left(\frac{\tau_{ij_a}}{\rho_a^2} + \frac{\tau_{ij_b}}{\rho_b^2}\right) \frac{\partial W(\mathbf{x}_a, \mathbf{x}_b)}{\partial x_i} \tag{5.310}$$

$$\left[\frac{\partial}{\partial x_i}\left(\frac{1}{\rho}\frac{\partial p}{\partial x_i}\right)\right]_a = \sum_b m_b \frac{8}{(\rho_a + \rho_b)^2} \frac{(p_a - p_b)(x_{i_a} - x_{i_b})}{|\mathbf{x}_b - \mathbf{x}_a|^2} \frac{\partial W(\mathbf{x}_a, \mathbf{x}_b)}{\partial x_i}$$

$$\tag{5.311}$$

The last term is formulated as the hybrid of SPH method and FDM to suppress the instability of the pressure calculation caused by the second-order derivative of kernel function under strong particle disorder.

5.5.1.2   The Boussinesq wave model

Lin and Man (2006) presented a staggered grid FD Boussinesq model that solves Nwogu's Boussinesq equations (Nwogu, 1993):

$$
\frac{\partial \eta}{\partial t} + \frac{\partial}{\partial x} \left[(h + \eta) u \right] + \frac{\partial}{\partial y} \left[(h + \eta) v \right]
$$

$$
+ \frac{\partial}{\partial x} \left\{ a_1 h^3 \left(\frac{\partial^2 u}{\partial x^2} + \frac{\partial^2 v}{\partial x \partial y} \right) + a_2 h^2 \left[\frac{\partial^2 (hu)}{\partial x^2} + \frac{\partial^2 (hv)}{\partial x \partial y} \right] \right\}
$$

$$
+ \frac{\partial}{\partial y} \left\{ a_1 h^3 \left(\frac{\partial^2 u}{\partial x \partial y} + \frac{\partial^2 v}{\partial^2 y} \right) + a_2 h^2 \left[\frac{\partial^2 (hu)}{\partial x \partial y} + \frac{\partial^2 (hv)}{\partial^2 y} \right] \right\} = 0 \qquad (5.312)
$$

$$
\frac{\partial u}{\partial t} + g \frac{\partial \eta}{\partial x} + u \frac{\partial u}{\partial x} + v \frac{\partial u}{\partial y} + b_1 h^2 \frac{\partial}{\partial t} \left(\frac{\partial^2 u}{\partial x^2} + \frac{\partial^2 v}{\partial x \partial y} \right)
$$

$$
+ b_2 h \frac{\partial}{\partial t} \left(\frac{\partial^2 (hu)}{\partial x^2} + \frac{\partial^2 (hv)}{\partial x \partial y} \right) = 0 \qquad (5.313)
$$

$$
\frac{\partial v}{\partial t} + g \frac{\partial \eta}{\partial y} + u \frac{\partial v}{\partial x} + v \frac{\partial v}{\partial y} + b_1 h^2 \frac{\partial}{\partial t} \left(\frac{\partial^2 u}{\partial x \partial y} + \frac{\partial^2 v}{\partial^2 y} \right)
$$

$$
+ b_2 h \frac{\partial}{\partial t} \left(\frac{\partial^2 (hu)}{\partial x \partial y} + \frac{\partial^2 (hv)}{\partial^2 y} \right) = 0 \qquad (5.314)
$$

where the constants a_1, a_2, b_1, and b_2 are given by the following equations:

$$
a_1 = \beta^2/2 - 1/6; \quad a_2 = \beta + 1/2; \quad b_1 = 1/6; b_2 = \beta
$$

with $\beta = z_\alpha / h = -0.531$.

To facilitate the later application of the higher order time-stepping procedure, equations (5.312)–(5.314) are rewritten as follows (e.g., Wei and Kirby, 1995); for convenience, the time and spatial derivatives are represented by subscripts from herein:

$$
\eta_t = E(\eta, u, v) \qquad (5.315)
$$

$$
U_t = F(\eta, u, v) + [F_1(v)]_t \qquad (5.316)
$$

$$
V_t = G(\eta, u, v) + [G_1(u)]_t \qquad (5.317)
$$

where:

$$
U(u) = u + h[b_1 h u_{xx} + b_2 (hu)_{xx}] \qquad (5.318)
$$

$$
V(v) = v + h[b_1 h v_{yy} + b_2 (hv)_{yy}] \qquad (5.319)
$$

The remaining quantities E, F, G, F_1, and G_1 are functions of η, u, and v, which are defined as follows:

$$
\begin{aligned}
E(\eta, u, v) = & -[(h+\eta)u]_x - [(h+\eta)v]_y \\
& - \left\{ \alpha_1 h^3 \left(u_{xx} + v_{xy} \right) + \alpha_2 h^2 \left[(hu)_{xx} + (hv)_{xy} \right] \right\}_x \\
& - \left\{ \alpha_1 h^3 \left(u_{xy} + v_{yy} \right) + \alpha_2 h^2 \left[(hu)_{xy} + (hv)_{yy} \right] \right\}_y
\end{aligned}
\tag{5.320}
$$

$$
F(\eta, u, v) = -g\eta_x - \frac{1}{2}\left(u^2\right)_x - vu_y
\tag{5.321}
$$

$$
G(\eta, u, v) = -g\eta_y - \frac{1}{2}(v^2)_y - uv_x
\tag{5.322}
$$

$$
F_1(v) = -h[b_1 h v_{xy} + b_2 (hv)_{xy}]
\tag{5.323}
$$

$$
G_1(u) = -h[b_1 h u_{xy} + b_2 (hu)_{xy}]
\tag{5.324}
$$

A staggered grid system will be used in the numerical discretization. Figure 5.30 illustrates this staggered grid system in which all scalars such as free surface displacement η and water depth h are defined at the cell center, while vectors such as velocity components u and v are defined at the interfaces of the cell.

The governing equations are matched in time by the fourth-order accurate Adams predictor–corrector method in which the time level n refers to the present time with all the known information. First, the predictor step is implemented for equations (5.315)–(5.317) by the explicit third-order

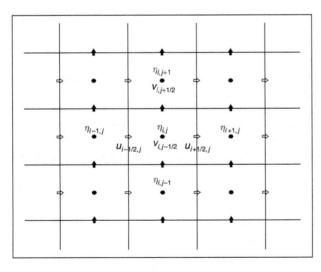

Figure 5.30 Sketch of the staggered grid system in the Boussinesq model.

Adams–Bashforth scheme:

$$\eta_{i,j}^{n+1^*} = \eta_{i,j}^n + \frac{\Delta t}{12}\left[23E_{i,j}^n - 16E_{i,j}^{n-1} + 5E_{i,j}^{n-2}\right] \tag{5.325}$$

$$U_{i+1/2,j}^{n+1^*} = U_{i+1/2,j}^n + \frac{\Delta t}{12}\left[23F_{i+1/2,j}^n - 16F_{i+1/2,j}^{n-1} + 5F_{i+1/2,j}^{n-2}\right]$$

$$+ 2\,(F_1)_{i+1/2,j}^n - 3\,(F_1)_{i+1/2,j}^{n-1} + (F_1)_{i+1/2,j}^{n-2} \tag{5.326}$$

$$V_{i+1/2,j}^{n+1^*} = V_{i+1/2,j}^n + \frac{\Delta t}{12}\left[23G_{i+1/2,j}^n - 16G_{i+1/2,j}^{n-1} + 5G_{i+1/2,j}^{n-2}\right]$$

$$+ 2\,(G_1)_{i+1/2,j}^n - 3\,(G_1)_{i+1/2,j}^{n-1} + (G_1)_{i+1/2,j}^{n-2} \tag{5.327}$$

The evaluation of the predicted velocities:

$$u_{i+1/2,j}^{n+1^*} \text{ and } v_{i,j+1/2}^{n+1^*}$$

requires simultaneous solution of tri-diagonal matrix systems. With the predicted values of $\eta_{i,j}^{n+1^*}$, $u_{i+1/2,j}^{n+1^*}$, and $v_{i,j+1/2}^{n+1^*}$, we can obtain the corresponding values of $E_{i,j}^{n+1^*}$, $F_{i+1/2,j}^{n+1^*}$, $G_{i,j+1/2}^{n+1^*}$, $(F_1)_{i+1/2,j}^{n+1^*}$, and $(G_1)_{i,j+1/2}^{n+1^*}$ based on equations (5.320)–(5.324). These values will be employed in the corrector step with the fourth-order Adams–Moulton method:

$$\eta_{i,j}^{n+1} = \eta_{i,j}^n + \frac{\Delta t}{24}\left[9E_{i,j}^{n+1^*} + 19E_{i,j}^n - 5E_{i,j}^{n-1} + E_{i,j}^{n-2}\right] \tag{5.328}$$

$$U_{i+1/2,j}^{n+1} = U_{i+1/2,j}^n + \frac{\Delta t}{24}\left[9F_{i+1/2,j}^{n+1^*} + 19F_{i+1/2,j}^n - 5F_{i+1/2,j}^{n-1} + F_{i+1/2,j}^{n-2}\right]$$

$$+ (F_1)_{i+1/2,j}^{n+1^*} - (F_1)_{i+1/2,j}^n \tag{5.329}$$

$$V_{i,j+1/2}^{n+1} = V_{i,j+1/2}^n + \frac{\Delta t}{24}\left[9G_{i,j+1/2}^{n+1^*} + 19G_{i,j+1/2}^n - 5G_{i,j+1/2}^{n-1} + G_{i,j+1/2}^{n-2}\right]$$

$$+ (G_1)_{i,j+1/2}^{n+1^*} - (G_1)_{i,j+1/2}^n \tag{5.330}$$

The predictor–corrector procedure will be iterated until the error between two successive results reaches a required limit.

The first-order derivatives of $f = (h + \eta)u$ in the x-direction are discretized by the fourth-order accurate four-point central difference method so that the leading order TE in the form of fifth-order dispersion will not suppress the third-order physical dispersion in the governing equations. For example:

$$\left\{\frac{\partial f}{\partial x}\right\}_i = \frac{1}{24\Delta x}\left[f_{i-3/2} - 27f_{i-1/2} + 27f_{i+1/2} - f_{i+3/2}\right] \tag{5.331}$$

For the first-order derivatives of η in the x-direction, the fourth-order accurate four-point central difference scheme is:

$$\left(\frac{\partial \eta}{\partial x}\right)_{i+1/2,j} = \frac{\eta_{i-1,j} - 27\eta_{i,j} + 27\eta_{i+1,j} - \eta_{i+2,j}}{24\Delta x} \tag{5.332}$$

For the first-order derivatives of u^2 in the x-direction, the four-point finite difference scheme has the following form:

$$\left(\frac{\partial u^2}{\partial x}\right)_{i+1/2,j} = \frac{u^2_{i-3/2,j} - 8u^2_{i-1/2,j} + 8u^2_{i+3/2,j} - u^2_{i+5/2,j}}{12\Delta x} \tag{5.333}$$

For the second-order derivatives in the x-direction, the three-point finite difference scheme is used:

$$(w_{xx})_{i+1/2,j} = \frac{1}{\Delta x^2}\left(w_{i-1/2,j} - 2w_{i+1/2,j} + w_{i+3/2,j}\right) \tag{5.334}$$

where $w_{xx} = hu$ or u. Similar expressions in the y-direction can be obtained for both the first- and second-order derivatives. The cross-differentiation terms are approximated as follows:

$$(u_{xy})_{i,j-1/2} = \frac{u_{i+1/2,j} + u_{i+1/2,j-1} - u_{i-1/2,j} - u_{i+1/2,j-1}}{\Delta x \Delta y} \tag{5.335}$$

$$\left[(hu)_{xy}\right]_{i,j-1/2} = \frac{(hu)_{i+1/2,j} + (hu)_{i+1/2,j-1} - (hu)_{i-1/2,j} - (hu)_{i-1/2,j-1}}{\Delta x \Delta y} \tag{5.336}$$

Again, similar expressions can be obtained for v_{xy} and $(hv)_{xy}$.

The overall accuracy of the above scheme is of second order. It is also proven that the scheme is in conservative form for both mass and momentum and thus allows us to perform long-term simulation with a small level of numerical contamination. Based on the linear von Neumann stability analysis, it can be proven that the predictor scheme is stable when the Courant number:

$$Cr = \sqrt{gh}\left(\frac{\Delta t}{\Delta x}\right) \le 1$$

and the corrector scheme is stable when $Cr \le 0.5$. Considering the nonlinear effect in shallow water, the Courant number is often chosen to be 0.1 to 0.5 in actual computations.

5.5.1.3 The shallow-water equation wave model

The following SWE model is mainly based on Lin and Lin's (2006) SWE model in the simulation of sediment transport and bed morphology, except that here the eddy viscosity is closed by the depth-averaged $k-\varepsilon$ turbulence model. The depth-averaged 2D SWEs have the following forms in the Cartesian coordinate:

$$\frac{\partial H}{\partial t} + \frac{\partial P}{\partial x} + \frac{\partial Q}{\partial y} = 0 \tag{5.337}$$

$$\frac{\partial P}{\partial t} + \frac{\partial}{\partial x}\left(\frac{P^2}{H}\right) + \frac{\partial}{\partial y}\left(\frac{PQ}{H}\right) + gH\frac{\partial \eta}{\partial x} = -\frac{1}{\rho}\tau_{bx} + \frac{1}{\rho}\frac{\partial (HT_{xx})}{\partial x} + \frac{1}{\rho}\frac{\partial (HT_{yx})}{\partial y}$$

$$(5.338)$$

$$\frac{\partial Q}{\partial t} + \frac{\partial}{\partial x}\left(\frac{PQ}{H}\right) + \frac{\partial}{\partial y}\left(\frac{Q^2}{H}\right) + gH\frac{\partial \eta}{\partial y} = -\frac{1}{\rho}\tau_{by} + \frac{1}{\rho}\frac{\partial (HT_{xy})}{\partial x} + \frac{1}{\rho}\frac{\partial (HT_{yy})}{\partial y}$$

$$(5.339)$$

where:

$$\tau_{bx}\left(=\frac{\rho g n^2}{H^{7/3}}P\sqrt{P^2+Q^2}\right) \text{ and } \tau_{by}\left(=\frac{\rho g n^2}{H^{7/3}}Q\sqrt{P^2+Q^2}\right)$$

are bed shear stresses with n being Manning's roughness coefficient, and T_{xx}, T_{yx}, T_{xy}, and T_{yy} are the depth-averaged Reynolds stresses that can be modeled as follows:

$$\frac{1}{\rho}T_{ij} = \hat{\nu}_t\left(\frac{\partial U_i}{\partial x_j} + \frac{\partial U_j}{\partial x_i}\right) - \frac{2}{3}\hat{k}\delta_{ij}$$

$$(5.340)$$

where $\hat{\nu}_t$ is the depth-averaged turbulence eddy viscosity that is the function of depth-averaged turbulent kinetic energy $\hat{k}$ and its dissipation rate $\hat{\varepsilon}$:

$$\hat{\nu}_t = C_\mu \frac{\hat{k}^2}{\hat{\varepsilon}}$$

$$(5.341)$$

in which C_μ is an empirical constant.

The transport of $\hat{k}$ and $\hat{\varepsilon}$ can be determined from the following transport equations for $\hat{k}$ and $\hat{\varepsilon}$:

$$\frac{\partial \left(H\hat{k}\right)}{\partial t} + \frac{\partial \left(HU\hat{k}\right)}{\partial x} + \frac{\partial \left(HV\hat{k}\right)}{\partial y} = \frac{\partial}{\partial x}\left[\frac{\hat{\nu}_t}{\sigma_k}\frac{\partial \left(H\hat{k}\right)}{\partial x}\right] + \frac{\partial}{\partial y}\left[\frac{\hat{\nu}_t}{\sigma_k}\frac{\partial \left(H\hat{k}\right)}{\partial y}\right]$$

$$+ P_h + P_{kV} - \hat{\varepsilon}H \qquad (5.342)$$

$$\frac{\partial (H\hat{\varepsilon})}{\partial t} + \frac{\partial (HU\hat{\varepsilon})}{\partial x} + \frac{\partial (HV\hat{\varepsilon})}{\partial y} = \frac{\partial}{\partial x}\left[\frac{\hat{\nu}_t}{\sigma_\varepsilon}\frac{\partial (H\hat{\varepsilon})}{\partial x}\right] + \frac{\partial}{\partial y}\left[\frac{\hat{\nu}_t}{\sigma_\varepsilon}\frac{\partial (H\hat{\varepsilon})}{\partial y}\right]$$

$$+ C_{1\varepsilon}\frac{\hat{\varepsilon}}{\hat{k}}P_h + P_{\varepsilon V} - C_{2\varepsilon}\frac{\hat{\varepsilon}^2}{\hat{k}}H \qquad (5.343)$$

where σ_k, σ_ε, $C_{1\varepsilon}$, and $C_{2\varepsilon}$ are empirical constants, and the production terms have the following expressions:

$$P_h = \frac{\hat{\nu}_t}{H}\left\{2\left[\frac{\partial (HU)}{\partial x}\right]^2 + 2\left[\frac{\partial (HV)}{\partial y}\right]^2 + \left[\frac{\partial (HU)}{\partial y} + \frac{\partial (HV)}{\partial x}\right]^2\right\} \qquad (5.344)$$

$$P_{kV} = \frac{1}{\sqrt{c_f}} u_*^3 \text{ and } P_{\varepsilon V} = 3.6 \frac{C_{2\varepsilon}}{c_f^{3/4}} \sqrt{C_\mu} \frac{u_*^4}{H} \quad (5.345)$$

where $u_* = \sqrt{c_f (U^2 + V^2)}$ with the friction coefficient $c_f = n^2 g / H^{1/3}$ for rough beds. The empirical constants mentioned above take the following values (Launder and Spalding, 1974):

$$C_\mu = 0.09, \quad \sigma_k = 1.0, \quad \sigma_\varepsilon = 1.3, \quad C_{1\varepsilon} = 1.44, \quad C_{2\varepsilon} = 1.92 \quad (5.346)$$

The FDM constructed in the staggered grid system will be used, where the scalars H, $\hat{k}$, and $\hat{\varepsilon}$ are defined at the cell center and the vectors P and Q are defined on the cell boundaries, respectively. The SWEs are discretized by using the explicit leapfrog FD scheme:

$$\frac{H_{i,j}^{n+1/2} - H_{i,j}^{n-1/2}}{\Delta t} + \frac{P_{i+1/2,j}^n - P_{i-1/2,j}^n}{\Delta x} + \frac{Q_{i,j+1/2}^n - Q_{i,j-1/2}^n}{\Delta y} = 0 \quad (5.347)$$

$$\frac{P_{i+1/2,j}^{n+1} - P_{i+1/2,j}^n}{\Delta t} + \left[\frac{\partial}{\partial x} \left(\frac{P^2}{H} \right) + \frac{\partial}{\partial y} \left(\frac{PQ}{H} \right) \right]^n + g H_{i+1/2,j}^{n+1/2} \frac{\eta_{i+1,j}^{n+1/2} - \eta_{i,j}^{n+1/2}}{\Delta x}$$

$$= \left[-\frac{1}{\rho} \tau_{bx} + \frac{1}{\rho} \frac{\partial (HT_{xx})}{\partial x} + \frac{1}{\rho} \frac{\partial (HT_{yx})}{\partial y} \right]^n \quad (5.348)$$

$$\frac{Q_{i,j+1/2}^{n+1} - Q_{i,j+1/2}^n}{\Delta t} + \left[\frac{\partial}{\partial x} \left(\frac{PQ}{H} \right) + \frac{\partial}{\partial y} \left(\frac{Q^2}{H} \right) \right]^n + g H_{i,j+1/2}^{n+1/2} \frac{\eta_{i,j+1}^{n+1/2} - \eta_{i,j}^{n+1/2}}{\Delta y}$$

$$= \left[-\frac{1}{\rho} \tau_{by} + \frac{1}{\rho} \frac{\partial (HT_{xy})}{\partial x} + \frac{1}{\rho} \frac{\partial (HT_{yy})}{\partial y} \right]^n \quad (5.349)$$

where the nonlinear convection terms are discretized with the upwind scheme and the stress terms with the central difference scheme. In addition, the bottom frictional terms are approximated as follows:

$$\frac{1}{\rho} \tau_{bx} = c_x \left(P_{i+1/2,j}^{n+1} + P_{i+1/2,j}^n \right) \text{ and } \frac{1}{\rho} \tau_{by} = c_y \left(Q_{i,j+1/2}^{n+1} + Q_{i,j+1/2}^n \right) \quad (5.350)$$

in which c_x and c_y are given in terms of Manning's formula:

$$c_x = \frac{1}{2} \frac{gn^2}{\left(H_{i+1/2,j}^n \right)^{7/3}} \left[\left(P_{i+1/2,j}^n \right)^2 + \left(Q_{i+1/2,j}^n \right)^2 \right]^{1/2},$$

$$c_y = \frac{1}{2} \frac{gn^2}{\left(H_{i,j+1/2}^n \right)^{7/3}} \left[\left(P_{i,j+1/2}^n \right)^2 + \left(Q_{i,j+1/2}^n \right)^2 \right]^{1/2} \quad (5.351)$$

The $\hat{k}$ and $\hat{\varepsilon}$ equations can be solved semi-implicitly as follows:

$$\frac{(H\hat{\varepsilon})_{i,j}^{n+1/2} - (H\hat{\varepsilon})_{i,j}^{n-1/2}}{\Delta t} + F\hat{\varepsilon}X + F\hat{\varepsilon}Y = VIS\hat{\varepsilon}X + VIS\hat{\varepsilon}Y$$

$$+ C_{1\varepsilon}\frac{\hat{\varepsilon}_{i,j}^{n-1/2}}{\hat{k}_{i,j}^{n-1/2}}P_{b_{i,j}}^{n+1/2} + P_{\varepsilon V}$$

$$- C_{2\varepsilon}\frac{\hat{\varepsilon}_{i,j}^{n-1/2}\hat{\varepsilon}_{i,j}^{n+1/2}}{\hat{k}_{i,j}^{n-1/2}}H_{i,j}^{n-1/2}$$

$$(5.352)$$

$$\frac{\left(H\hat{k}\right)_{i,j}^{n+1/2} - \left(H\hat{k}\right)_{i,j}^{n-1/2}}{\Delta t} + F\hat{k}X + F\hat{k}Y = VIS\hat{k}X + VIS\hat{k}Y + P_{b_{i,j}}^{n+1/2} + P_{kV}$$

$$- C_{\mu}\frac{\hat{k}_{i,j}^{n-1/2}\hat{k}_{i,j}^{n+1/2}}{\hat{\nu}_{t_{i,j}}^{n-1/2}}H_{i,j}^{n-1/2}$$

$$(5.353)$$

The final FD forms for $\hat{k}-\hat{\varepsilon}$ equations are written as follows:

$$\hat{\varepsilon}_{i,j}^{n+1/2} = \frac{\frac{(H\hat{\varepsilon})_{i,j}^{n-1/2}}{\Delta t} - F\hat{\varepsilon}X - F\hat{\varepsilon}Y + VIS\hat{\varepsilon}X + VIS\hat{\varepsilon}Y + C_{1\varepsilon}\frac{\hat{\varepsilon}_{i,j}^{n-1/2}}{\hat{k}_{i,j}^{n-1/2}}P_{b_{i,j}}^{n+1/2} + P_{\varepsilon V}}{\frac{H_{i,j}^{n+1/2}}{\Delta t} + C_{2\varepsilon}\frac{\hat{\varepsilon}_{i,j}^{n-1/2}}{\hat{k}_{i,j}^{n-1/2}}H_{i,j}^{n-1/2}}$$

$$(5.354)$$

$$\hat{k}_{i,j}^{n+1/2} = \frac{\frac{(H\hat{k})_{i,j}^{n-1/2}}{\Delta t} - F\hat{k}X - F\hat{k}Y + VIS\hat{k}X + VIS\hat{k}Y + P_{b_{i,j}}^{n+1/2} + P_{kV}}{\frac{H_{i,j}^{n+1/2}}{\Delta t} + C_{\mu}\frac{\hat{k}_{i,j}^{n-1/2}}{\hat{\nu}_{t_{i,j}}^{n-1/2}}H_{i,j}^{n-1/2}} \qquad (5.355)$$

The convection and diffusion terms above can be discretized by the upwind scheme and the central difference scheme, respectively. The production term $P_{b_{i,j}}^{n+1/2}$ is evaluated by:

$$P_{b_{i,j}}^{n+1/2} = \frac{\hat{\nu}_{t_{i,j}}^{n-1/2}}{H_{i,j}^{n+1/2}}\left\{2\left[\frac{P_{i+1/2,j}^{n+1/2} - P_{i-1/2,j}^{n+1/2}}{\Delta x}\right] + 2\left[\frac{Q_{i,j+1/2}^{n+1/2} - Q_{i,j-1/2}^{n+1/2}}{\Delta y}\right]\right.$$

$$\left. + \left[\frac{P_{i,j+1/2}^{n+1/2} - P_{i,j-1/2}^{n+1/2}}{\Delta y} + \frac{Q_{i+1/2,j}^{n+1/2} - Q_{i-1/2,j}^{n+1/2}}{\Delta x}\right]\right\} \qquad (5.356)$$

5.5.1.4 The transient mild-slope equation wave model

Lin (2004a) proposed an MSE wave model with the use of compact FD scheme. The model solves the time-dependent MSE:

$$\frac{\partial^2 \eta}{\partial t^2} - \frac{\partial}{\partial x}\left(cc_g \frac{\partial \eta}{\partial x}\right) - \frac{\partial}{\partial y}\left(cc_g \frac{\partial \eta}{\partial y}\right) + (\sigma^2 - k^2 cc_g)\eta = 0 \tag{5.357}$$

The explicit FD scheme for (5.357) is as follows:

$$\frac{\eta_{i,j}^{n+1} - 2\eta_{i,j}^n + \eta_{i,j}^{n-1}}{(\Delta t)^2} - \frac{(cc_g)_{i+1/2,j}\dfrac{\eta_{i+1,j}^n - \eta_{i,j}^n}{\Delta x} - (cc_g)_{i-1/2,j}\dfrac{\eta_{i,j}^n - \eta_{i-1,j}^n}{\Delta x}}{\Delta x}$$

$$- \frac{(cc_g)_{i,j+1/2}\dfrac{\eta_{i,j+1}^n - \eta_{i,j}^n}{\Delta y} - (cc_g)_{i,j-1/2}\dfrac{\eta_{i,j}^n - \eta_{i,j-1}^n}{\Delta y}}{\Delta y}$$

$$+ \left[\sigma^2 - k_{i,j}^2 (cc_g)_{i,j}\right]\eta_{i,j}^n = 0 \tag{5.358}$$

Rearranging it, we have:

$$\eta_{i,j}^{n+1} = \eta_{i,j}^n + \left(\eta_{i,j}^n - \eta_{i,j}^{n-1}\right)$$

$$+ \frac{(\Delta t)^2}{(\Delta x)^2}\left[(cc_g)_{i+1/2,j}\left(\eta_{i+1,j}^n - \eta_{i,j}^n\right) - (cc_g)_{i-1/2,j}\left(\eta_{i,j}^n - \eta_{i-1,j}^n\right)\right]$$

$$+ \frac{(\Delta t)^2}{(\Delta y)^2}\left[(cc_g)_{i,j+1/2}\left(\eta_{i,j+1}^n - \eta_{i,j}^n\right) - (cc_g)_{i,j-1/2}\left(\eta_{i,j}^n - \eta_{i,j-1}^n\right)\right]$$

$$+ (\Delta t)^2\left[\sigma^2 - k_{i,j}^2 (cc_g)_{i,j}\right]\eta_{i,j}^n = 0 \tag{5.359}$$

Note that when we advance the time step from n to $n+1$, the previous time step $(n-1)$ information of η is still needed. The above scheme is compact and second-order accurate in both time and space. The stability condition is found to be the following:

$$\Delta t \leq \min\left(\frac{\Delta x}{\sqrt{cc_g}}, \frac{\Delta y}{\sqrt{cc_g}}\right) \tag{5.360}$$

5.5.1.5 The wave height spectral model for weak diffraction problems

Lin *et al.* (2005) proposed a spectral model based on the complex weight height function that can model weak diffraction problems with the use of a grid size larger than wavelength. For a slowly varying wave group propagating over a changing topography without current, the model solves the following governing equation:

$$\frac{\partial G}{\partial t} + c_x \frac{\partial G}{\partial x} + \frac{G}{2}\frac{\partial c_x}{\partial x} + c_y \frac{\partial G}{\partial y} + \frac{G}{2}\frac{\partial c_y}{\partial y} + c_\theta \frac{\partial G}{\partial \theta} + \frac{G}{2}\frac{\partial c_\theta}{\partial \theta}$$

$$= \frac{\partial}{\partial x}\left(\frac{ic_g}{2k}\frac{\partial G}{\partial x}\right) + \frac{\partial}{\partial y}\left(\frac{ic_g}{2k}\frac{\partial G}{\partial y}\right) + P(G, G^*) \tag{5.361}$$

where $G(f, \theta)$ is the complex wave height spectral function, $c_x = c_g \cos\theta$ and $c_y = c_g \sin\theta$ are group velocities in x- and y-directions:

$$c_\theta = \frac{c_g}{c}\left(\sin\theta \frac{\partial c}{\partial x} - \cos\theta \frac{\partial c}{\partial y}\right)$$

is the refraction velocity, and $P(G, G^*) = Q(GG^*)/(2G^*)$ is the complex source function with $Q(GG^*) = Q(F)$ being the source function of the wave energy spectrum. For example, the bottom friction in the above model can be converted from Booij *et al.*'s (1999) formula as

$$P(G, G^*) = \frac{1}{2G^*}\left[-C_{\text{bottom}}\frac{\omega^2}{g^2 \sinh^2(kh)}GG^*\right] = -C_{\text{bottom}}\frac{\omega^2}{2g^2 \sinh^2(kh)}G \tag{5.362}$$

where $C_{\text{bottom}} = 0.067$ is the bottom friction coefficient.

The model solves the governing equation (5.361) by using the operator-splitting method, which treats the wave propagation, wave refraction, wave diffraction, and wave growth and dissipation (source terms) separately. The wave propagation is represented by the first five terms on the LHS of (5.361). To achieve better accuracy, the equation is regrouped to be the summation of the conservative form minus a source term, i.e.:

$$\frac{G^{n+1/4} - G^n}{\Delta t} + \left[\frac{\partial(c_x G)}{\partial x} - \frac{G}{2}\frac{\partial c_x}{\partial x} + \frac{\partial(c_y G)}{\partial x} - \frac{G}{2}\frac{\partial c_y}{\partial x}\right]^n = 0 \tag{5.363}$$

In this model, equation (5.363) is further split into x- and y-directions and then solved by the conservative Lax–Wendroff method, which is second-order accurate.

The diffraction step is represented by the following equation:

$$\frac{G^{n+2/4} - G^{n+1/4}}{\Delta t} = \left[\frac{\partial}{\partial x}\left(\frac{ic_g}{2k}\frac{\partial G}{\partial x}\right) + \frac{\partial}{\partial y}\left(\frac{ic_g}{2k}\frac{\partial G}{\partial y}\right)\right]^{n+1/4} \tag{5.364}$$

Equation (5.364) is solved by using the Crank–Nicolson method, which is an implicit scheme and unconditionally stable. To improve the efficiency of the solution, the ADI method is used, which further splits the solution procedure into the x- and y-directions. The scheme is again second-order accurate.

The wave refraction is solved next. Similar to the treatment for the wave propagation, the governing equation is first modified to be the summation of the conservative form minus a source term as:

$$\frac{G^{n+3/4} - G^{n+2/4}}{\Delta t} + \left[\frac{\partial(c_\theta G)}{\partial \theta} - \frac{G}{2} \frac{\partial c_\theta}{\partial \theta} \right]^{n+2/4} = 0 \qquad (5.365)$$

The conservative upwind scheme is chosen to discretize (5.365), and it can effectively suppress numerical dispersion. Since the refraction process is to redistribute wave energy among different directional modes with the total energy conserved, the calculated energy is linearly scaled at the end of each step so that the summation of all directional modes at the fixed point and frequency is conserved before and after the computation of refraction. Such treatment will ensure energy conservation during the computation and is especially useful when the directional spreading angle of incident waves is narrow and wave ray bending is strong, which could be encountered in the coastal region.

Finally, the source term is solved by the forward-time scheme as follows:

$$\frac{G^{n+1} - G^{n+3/4}}{\Delta t} = P(G, G^*)^{n+3/4}. \qquad (5.366)$$

Two types of boundary conditions are needed in general. One is the inflow boundary condition that specifies the time-varying complex wave height function F. For the same physical wave height $|H|$, there is an infinite number of valid choices of F by using different combinations of real and imaginary parts, e.g., $G = i|H|$ or $G = (\sqrt{2}/2 - i\sqrt{2}/2)|H|$ is valid. As long as the function F is uniformly specified on the boundary and $GG^* = |H|^2$, the final results in terms of physical wave height $|H|$ will remain the same. Another boundary condition will be specified either on the vertical wall or on the radiation boundary. A reflection coefficient β varying from 0 to 1 is used to represent the completely absorbing boundary to the perfectly reflecting boundary. For the directional wave mode $G(\theta_j)$ incident on the wall aligned with the y-axis, the mathematical representation for such a boundary condition is as follows:

$$\frac{\partial G(\theta_j)}{\partial x} = 0 \quad G(-\theta_j) = \beta G(\theta_j). \qquad (5.367)$$

The wave incident from other angles can be treated similarly. By varying the reflection coefficient, this approach could also be used to represent the partial wave transmission through porous media, wave reflection from uneven bottoms, etc.

The numerical scheme is stable as long as the Courant numbers defined in the propagation and refraction processes are less than unity:

$$\max\left(\frac{c_g\cos\theta\Delta t}{\Delta x},\ \frac{c_g\sin\theta\Delta t}{\Delta y},\ \frac{c_g}{c}(\sin\theta\frac{\partial c}{\partial x}-\cos\theta\frac{\partial c}{\partial y})\frac{\Delta t}{\Delta\theta}\right)\le 1. \qquad (5.368)$$

Once the wave height spectrum is known, the wave height can be recovered:

$$H_{\mathrm{rms}}(x,y,t)=\sqrt{\sum_{i=1}^{n}\sum_{j=1}^{m}\left|G(f_i,\theta_j,x,y,t)\right|^2\Delta f_i\Delta\theta_j} \qquad (5.369)$$

5.5.2 Benchmark tests

A particular physical problem can often be simulated by different wave models with different levels of approximations. In contrast, a particular wave model can simulate different physical phenomena as long as the basic assumptions in the wave model are not violated. In this section, we shall present a few examples that are used to gauge the accuracy of a wave model. For each benchmark test presented, at least two different modeling results will be presented and compared with each other with the additional analytical solution or experimental data. This will help readers know the characteristics of different wave models and be able validate their own wave models.

5.5.2.1 Two-dimensional nonlinear wave transformation above a submerged breakwater

When a nonlinear wave train passes through a submerged breakwater, it may experience dramatic changes of waveform and significant nonlinear wave energy transfer. Beji and Battjes (1993) investigated experimentally the nonlinear periodic wave transformation over a submerged breakwater of trapezoidal shape. In this benchmark test, this problem will be revisited with the use of both NSE-solver model (e.g. the σ-coordinate model by Lin and Li, 2002) and Boussinesq model (Lin and Man, 2007).

The problem setup is shown in Figure 5.31. A regular wave train that has a wave height of 0.04 m and a wave period of 2.86 s was sent from the left boundary $(x=0)$. The open boundary was implemented on the right end of the wave flume. In total, eight wave gauges were deployed at $x=20.44, 24.04, 26.04, 28.04, 30.44, 33.64, 37.04,$ and 41.04 m. For the Boussinesq model, the numerical model was run for 50 s by using the grid size of $\Delta x=0.10$ m and the time step of $\Delta t=0.01$ s. For the σ-coordinate NSE model, $\Delta x=0.05$ m and $\Delta t=0.01$ s are used and 20 grid points are deployed in the vertical direction.

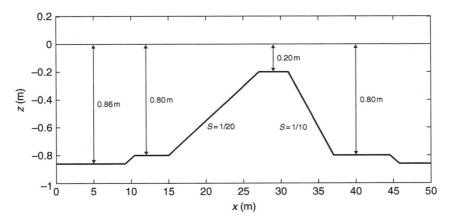

Figure 5.31 Experimental setup of a periodic nonlinear wave train past a submerged breakwater.

Figure 5.32 shows the comparisons between the two numerical results and the gauge data for free surface displacement at eight wave gauge locations. The overall agreement between the measured data and calculated results is reasonably close during wave shoaling ($x = 20.04$–24.04 m) and overtopping above the breakwater crown ($x = 26.04$, 28.04, and 30.44 m). Behind the breakwater, there is the generation of secondary harmonics as evidenced at the last three gauges. The numerical results from the NSE model have an overall better agreement with the experimental data in terms of both phase capturing and wave height prediction, whereas larger discrepancies are observed for Boussinesq model results, especially past the submerged breakwater.

5.5.2.2 Two-dimensional breaking waves on a linear slope

In this benchmark test, the numerical modeling of spilling and plunging breaking waves (Lin and Liu, 1998a,b) is revisited with the use of the improved RANS model NEWFLUME and at various mesh sizes. One of the main objectives of the test is to present a sensitivity study of the numerical results to the change of mesh size for breaking wave simulation. Besides, the NSE model based on the VOF method (NEWFLUME) is also compared with the SPH model based on the Lagrangian particle method.

Spilling breaking waves: The problem setup follows the experiment by Ting and Kirby (1996). The still water depth $d_c = h = 0.4$ m and the linear beach slope $s = 1/35$. The incident wave is of cnoidal shape with the wave period $T = 2.0$ s and the wave height $H = 0.125$ m. When the origin is defined at the toe of the beach, waves break at $x_b = 6.4$ m, where the mean water depth $h_b = 0.199$ m. Four wave gauges are deployed behind the

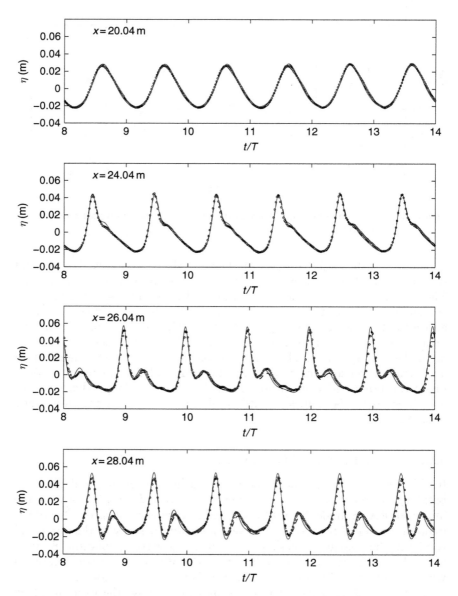

Figure 5.32 Comparisons of measured (circle) and simulated wave profiles by Boussinesq model (solid line) and NSE model (dashed line) for nonlinear wave transformation over a submerged trapezoidal breakwater.

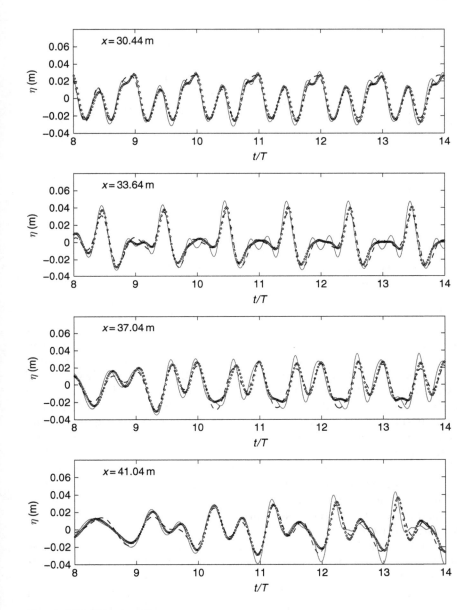

Figure 5.32 (Continued).

breaking point at $(x - x_{\rm b})/h_{\rm b} = 4.397, 7.462, 10.528$, and 13.618, where the mean water depths are $h = 0.175, 0.161, 0.148$, and 0.133 m, respectively (Figure 5.33). The corresponding wave-induced setups at these four sections are $\zeta = 0.006, 0.009, 0.011$, and 0.014 m.

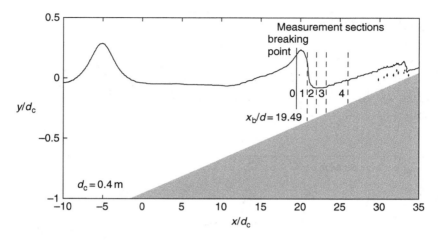

Figure 5.33 Problem setup for spilling and plunging breaking waves on a plane slope.

In the numerical simulation, the original mesh arrangement in Lin and Liu (1998a) uses a uniform $\Delta x = 0.025$ m and a nonuniform mesh for Δy with the fine mesh $\Delta y_{\min} = 0.006$ m near the free surface. In the present test, the uniform mesh of $\Delta x = 0.015$ m and $\Delta y = 0.008$ m is used. Besides, the higher order PLIC method is used for VOF tracking in NEWFLUME (see Section 5.5.1.1) instead of the donor–acceptor method in the original code (Liu and Lin, 1997). The comparisons of free surface displacement at four sections are shown in Figure 5.34 between the numerical simulations (the fifth wave passing the gauges) and the experimental measurements (phase-averaged data). It is seen that with the refinement of mesh in the horizontal direction, the wave amplitude is better predicted right after the breaking point (the first section). However, with the further evolution of the breaking wave into a turbulent bore, the difference between the two numerical results becomes less significant and both of them compare well with the measurements.

Plunging breaking waves: The problem setup is similar to the spilling wave case with the change of wave conditions to $H = 0.128$ m and $T = 5.0$ s (Ting and Kirby, 1995). Waves break at $x_b = 7.795$ m, where the mean water depth $h_b = 0.154$ m. Measurements were made at five sections at $(x - x_b)/h_b = 0$, 3.571, 6.494, 9.740, and 16.883, where the mean water depths are $h = 0.154$, 0.143, 0.132, 0.119, and 0.090 m, respectively. The associated wave-induced set-downs/setups at these sections are $\zeta = -0.002$, 0.001, 0.004, 0.006, and 0.011 m. In the original numerical simulation, the uniform $\Delta x = 0.04$ m and the nonuniform mesh for Δy with the fine mesh $\Delta y_{\min} = 0.0075$ m near the free surface are used. In the present simulation, two uniform mesh systems are used: $\Delta x = 0.015$ m and $\Delta y = 0.008$ m and

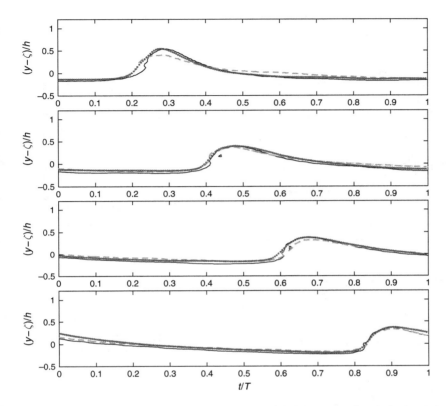

Figure 5.34 Comparison of simulation results (solid: present results; dashed: original results by Lin and Liu, 1998a) and experimental data (circle) for the time history of free surface displacement at four gauge locations under the spilling breaking waves.

$\Delta x = 0.01\,\text{m}$ and $\Delta y = 0.006\,\text{m}$. Comparisons of numerical results (average between the second and third waves passing the gauges) and experimental data (phase-averaged data) are shown in Figure 5.35.

Observe that by reducing the mesh from $\Delta x = 0.04\,\text{m}$ to $\Delta x = 0.015\,\text{m}$, the wave amplitude at the breaking point is captured better. With the further refinement of mesh size to $\Delta x = 0.01\,\text{m}$, not much further improvement is observed. None of the numerical results provided good agreement with the experimental data at the second gauge location, where the aeration on the free surface due to the plunging jet may produce a higher water level. Moving to the downstream gauge locations, the comparisons between the numerical results and experimental data are getting closer and the difference between the coarse mesh results and fine mesh results becomes smaller when the turbulent bore gradually forms. Figure 5.36 shows a

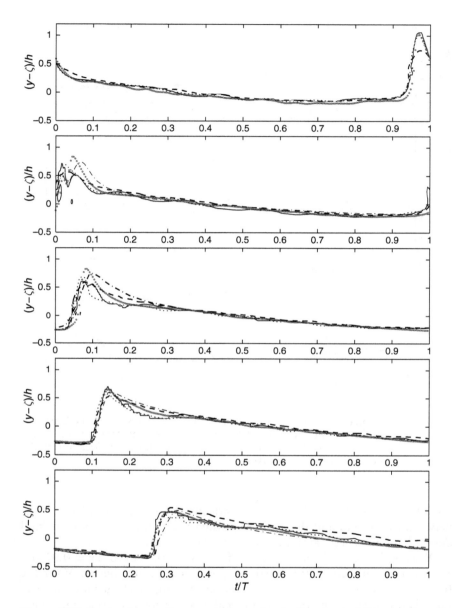

Figure 5.35 Comparisons of numerical results (solid: $\Delta x = 0.01$ m and $\Delta y = 0.006$ m; dotted line: $\Delta x = 0.015$ m and $\Delta y = 0.008$ m; dashed line: original results in Lin and Liu, 1998b; and dashed-dotted line: SPH results using $\Delta x = 0.013$ m by Shao and Ji, 2006) and experimental data (circle) for the time history of free surface displacement at five gauge locations under the plunging breaking waves.

Figure 5.36 Simulated plunging waves from $t/T = 0$ to $t/T = 0.2$ with the mesh resolution of $\Delta x = 0.01$ m and $\Delta y = 0.006$ m. (Courtesy of Dr. Budianto Ontowirjo, JSPS Researcher at Disaster Control Research Control, Tohoku University, Japan)

series of snapshots of the simulated free surface profiles under the plunging breaking waves using the same code NEWFLUME.

5.5.2.3 *Three-dimensional liquid sloshing in a stationary tank*

Consider a closed basin of size $L_x \times L_y$, in which the origin is defined at the left-bottom corner of the basin. The initial free surface displacement is a Gaussian distribution about the center of the basin, i.e.:

$$\eta_0(x, y) = H_0 \exp\left\{-\beta\left[(x - L_x/2)^2 + (y - L_y/2)^2\right]\right\}$$ (5.370)

where H_0 is the initial height of the hump and β is the peak enhancement factor; the linear analytical solution for the free surface evolution is given in Section 3.15.6.2:

$$\eta(x, y, t) = \sum_{n=0}^{\infty} \sum_{m=0}^{\infty} \overline{\eta}_{nm} e^{-i\omega_{nm}t} \cos(n\lambda x) \cos(m\lambda y)$$ (5.371)

in which:

$$\overline{\eta}_{mn} = \frac{4}{(1 + \delta_{n0})(1 + \delta_{m0})L_x L_y} \int_0^{L_x} \int_0^{L_y} \eta_0(x, y) \cos(n\lambda x) \cos(m\lambda y) dx dy$$ (5.372)

where δ_{nm} is the Kronecker delta function and:

$$\lambda = \frac{\pi}{L_x} = \frac{\pi}{L_y}$$ (5.373)

The (n, m) wave modes have the corresponding natural frequency that is determined by the linear dispersion equation:

$$\omega_{nm}^2 = gk_{nm} \tanh(k_{nm}h_0)$$ (5.374)

where h_0 is the still water depth and:

$$k_{nm}^2 = (n\lambda)^2 + (m\lambda)^2 = \left(\frac{\pi}{L_x}\right)^2 (n^2 + m^2)$$ (5.375)

In this benchmark test, we take the following parameters: $L_x = L_y = 10\,\text{m}$, $H_0 = 0.005, 0.05, 0.2\,\text{m}$, $h = 0.50\,\text{m}$, and $\beta = -0.4$. The problem is simulated by Boussinesq model, SWE model, and σ-coordinate NSE model. For Boussinesq model and SWE model, the domain is discretized by uniform grids with $\Delta x = \Delta y = 0.2\,\text{m}$ and the numerical model runs at $\Delta t = 0.01\,\text{s}$. For NSE model, $\Delta x = \Delta y = 0.05\,\text{m}$ and $\Delta t = 0.0025\,\text{s}$ are used. In total, 20 grids are used in the vertical direction. Figure 5.37 shows the simulated free surface displacement at different times from the NSE model.

Figures 5.38–5.40 show the comparison of time histories of free surface displacement at the center and corner of the domain between the linear analytical solution and the numerical results from SWE model, Boussinesq model, and NSE model. Three different initial hump heights are simulated, representing the linear ($H_0/h = 0.01$), weakly nonlinear ($H_0/h = 0.1$), and strongly

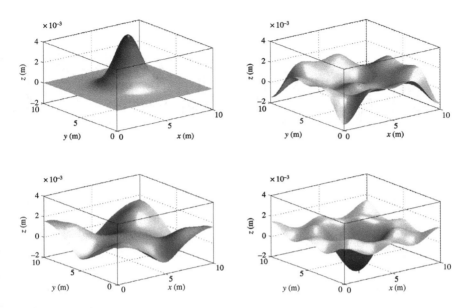

Figure 5.37 Snapshots of free surface profiles of liquid sloshing at $t = 0.0$, 15.0, 20.0, and 25.0 s when $H_0/h = 0.01$.

nonlinear waves ($H_0/h = 0.4$), respectively. For linear waves, both Boussinesq model and NSE model results agree well with the analytical solution, whereas the SWE model overestimates the maximum wave height due to the lack of ability to represent wave dispersion (see Figure 5.38). With the further increase in wave height, differences start to show up among the Boussinesq model, NSE model, and analytical solution, with the NSE model giving slightly better results for the linear analytical solution (see Figure 5.39). When the wave amplitude becomes large, none of the numerical results agree with the analytical solution. Interestingly, the SWE model gives closer results to Boussinesq and NSE models in this strongly nonlinear case (see Figure 5.40).

5.5.2.4 Three-dimensional linear wave diffraction behind a semi-infinite breakwater

Wave diffraction behind a semi-infinite breakwater is a classical problem for testing wave models. The exact theoretical solution was presented by Sommerfeld (1896) by solving the Helmholtz equation. The approximated analytical solution was proposed by Penney and Price (1952a) for the location away from the breakwater (i.e., $> 2L$). In this benchmark test, both the time-dependent MSE model (Lin, 2004a) and the wave height spectral model

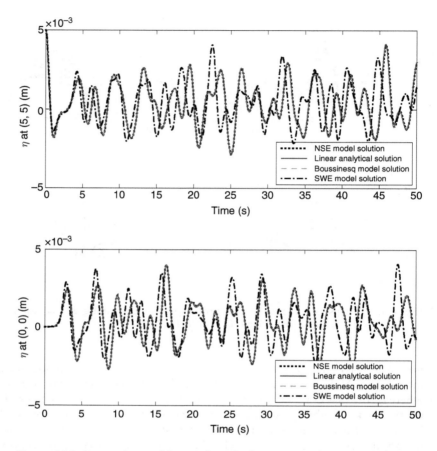

Figure 5.38 Comparisons of free surface displacement at the center and corner of the tank for $H_0/h = 0.01$.

for weakly diffracted waves (Lin *et al.*, 2005) are used. Figure 5.41 shows the problem setup as well as the simulated transient wave field around the breakwater from the MSE model. In the numerical simulation, the computational domain covers an area of 3 km × 3 km with the constant water depth of $h = 10$ m. The incident linear wave has a wave period of $T = 10$ s and a wavelength of $L = 92.32$ m. The MSE model discretizes the domain by 1000 × 1000 uniform cells with $\Delta x = \Delta y = 3$ m and $\Delta t = 0.1$ s; the spectral model discretizes the domain using 100 × 100 uniform cells with $\Delta x = \Delta y = 30$ m and $\Delta t = 1.0$ s.

Figure 5.42 shows the comparisons of the diffraction coefficient $D = H/H_0$ between the two numerical results and the two theories. Although the spectral model provides close agreement with the approximate theory,

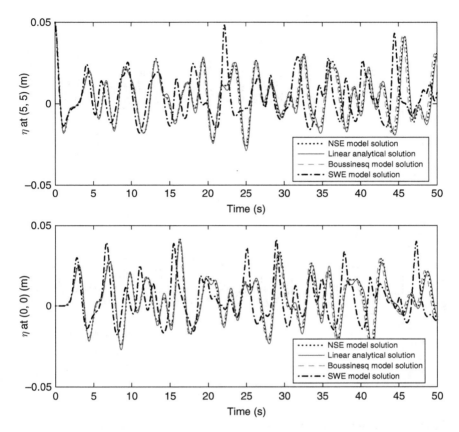

Figure 5.39 Comparisons of free surface displacement at the center and corner of the tank for $H_0/h = 0.1$.

the MSE model has excellent agreement with the exact theory. This is not surprising because both the approximate theory and the spectral model adopt the similar idea of parabolic approximation that underestimates the lateral spreading of wave energy into the shadow area, especially right behind the breakwater. In contrast, the MSE model provides good agreement with the exact analytical solution that gives a slower decay of wave amplitude (and thus faster energy leakage into the shadow area) behind the breakwater. Note also that in the present spectral model, only three grid points are deployed for one wavelength and yet reasonable numerical results are achieved. Lin *et al.* (2005) have shown that for the same incident wave even with the mesh size being increased to 3 km for the spectral model, the numerical results can still show reasonable agreement with the theory in a larger scale (300 km × 300 km) simulation.

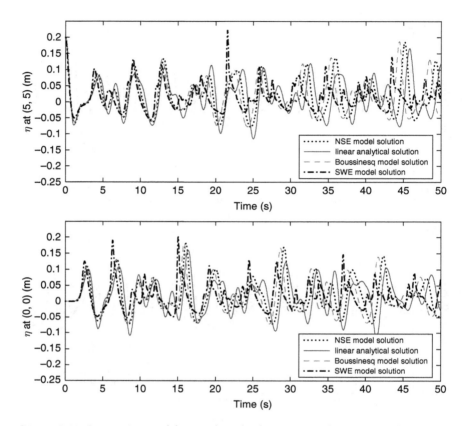

Figure 5.40 Comparisons of free surface displacement at the center and corner of the tank for $H_0/h = 0.4$.

5.5.2.5 Three-dimensional nonlinear wave refraction and diffraction above a shoal on a slope

This benchmark test is for nonlinear wave refraction and diffraction over an elliptical shoal on a linear slope (Berkhoff *et al.*, 1982). Three numerical models will be employed in this test, namely the σ-coordinate NSE model, the MSE model, and the Boussinesq model. Comparisons will be made between the numerical results from these three wave models and the experimental data.

The problem setup is shown in Figure 5.43, in which an elliptic shoal is mounted on a plane beach. The slope-oriented coordinates (x', y') are introduced that are related to the computational coordinates (x, y) as follows:

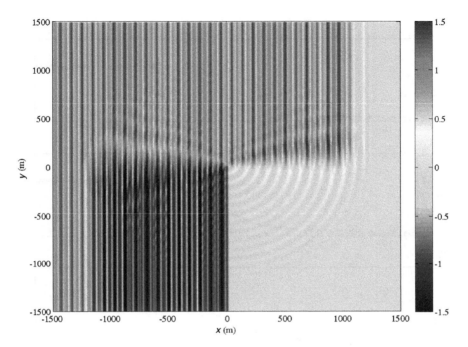

Figure 5.41 Problem setup of wave diffraction behind a semi-infinite breakwater.

$$x' = (x - x_0)\cos 20° - (y - y_0)\sin 20°, \quad y' = (y - y_0)\cos 20° + (x - x_0)\sin 20°$$

$$(5.376)$$

where (x_0, y_0) is the center of the shoal. The water depth on the slope is:

$$h = \begin{cases} 0.45 - 0.02(5.84 - y') & \text{for } y' < 5.84 \\ 0.45 & \text{for } y' \geq 5.84 \end{cases} \qquad (5.377)$$

The water depth above the shoal, i.e., $(x'/4)^2 + (y'/3)^2 < 1$, is modified as:

$$h_{shoal} = h + 0.3 - 0.5\sqrt{1 - \left(\frac{x'}{5}\right)^2 - \left(\frac{y'}{3.75}\right)^2} \qquad (5.378)$$

The incident wave has a wave period of $T = 1\,\text{s}$ and a wave height of $H_0 = 0.0464\,\text{m}$. Experimental measurements were made at five sections behind the shoal (i.e., $y = -1$, -3, -5, -7, and $-9\,\text{m}$) and three sections parallel to the incident wave propagation direction (i.e., $x = -2$, 0, $2\,\text{m}$).

In the numerical simulation, the uniform grid of $\Delta x = \Delta y = 0.1\,\text{m}$ is used for both MSE and Boussinesq models. The constants $\Delta t = 0.002$ and $0.05\,\text{s}$

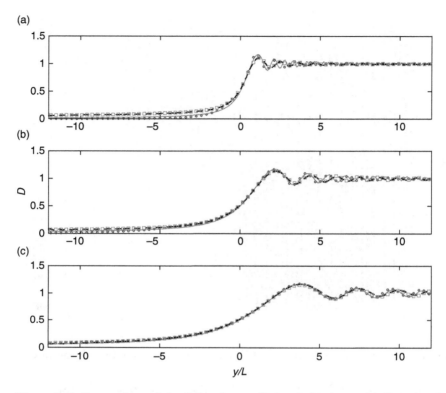

Figure 5.42 Comparison of the diffraction coefficient behind a semi-infinite break-water between numerical results (solid line: spectral model; dashed line: MSE model) and theories (square: exact theory; circle: approximate theory) at (a) $x/L = 1.62$, (b) $x/L = 6.50$, and (c) $x/L = 19.50$.

are used in the Boussinesq model and MSE model, respectively. For the NSE model, $\Delta y = 0.033333$ m, $\Delta x = 0.1$ m, and $\Delta t = 0.005$ s are used. In the σ-coordinate, 20 grid points are used in the vertical direction. All models are run until the wave field becomes stable (e.g., $t > 30$ s) and the wave height is calculated by taking the average of the last fives waves.

The numerical results are compared to the experimental data in Figure 5.44. It can be observed that the comparisons between all numerical results and experimental data are rather good at $y = -1$ m, the section where the nonlinear wave interaction just starts and has not become very significant. However, as we move to the downstream sections, the numerical results from the linear MSE model deviate evidently from the experimental data, due to the absence of wave nonlinearity in the model. Both Boussinesq and NSE models capture the nonlinear

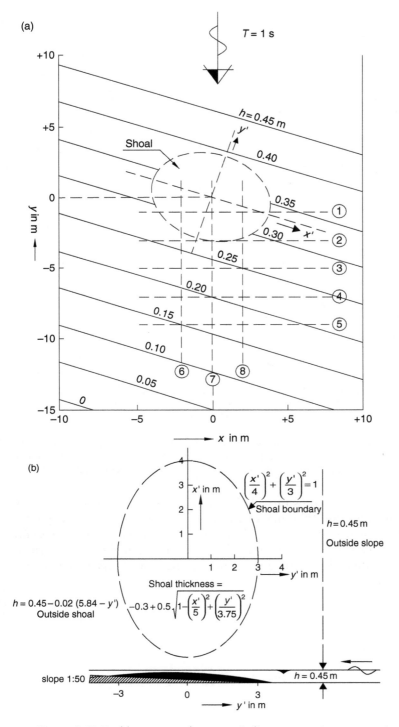

Figure 5.43 Problem setup for a periodic wave train propagating over a submerged elliptical shoal on a slope.

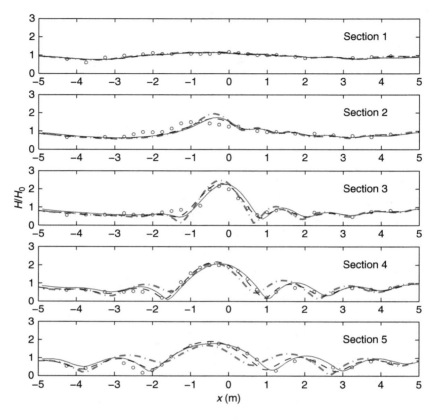

Figure 5.44 Comparisons of the wave amplitude behind the shoal between the numerical results (solid line: NSE model; dashed line: Boussinesq model; and dashed-dotted line: MSE model) and experimental data (circle).

wave interaction and thus provide better agreement with the experimental data. However, the NSE model excels by producing authentic predictions of the wave evolution in all sections behind the shoals, especially at the last section $(y = -11\,\text{m})$, where the accumulative wave dispersion and nonlinearity become so profound that the Boussinesq model is inadequate.

6 Modeling of wave–structure interaction

6.1 Introduction

The study of wave–structure interaction has important applications in coastal and offshore engineering. Although part of the study is to understand the fundamental flow characteristics and wave scattering during wave–structure interaction, most of the study has a goal of finding the wave load on a structure, from which the structure response and stability can be analyzed.

The total flow force on a body can be theoretically obtained by surface integration of pressure and stress on the body, which is, however, difficult to measure directly in practise. In most of the engineering designs, the simple empirical or semiempirical approach such as the Morison equation or the Froude–Krylov (F–K) method is employed. Only in the last two decades, as computer power has rapidly increased, has the direct modeling of wave-induced pressure and/or stress become possible for practical design purposes. This chapter will provide an introduction to various engineering modeling techniques used in a wide range of wave–structure interaction studies.

In general, both pressure and viscous stress under waves are depth-dependent unless the wavelength is very long. Besides, most of the coastal or offshore structures have the surface configuration that is also depth-dependent. This implies that in order to accurately model wave loads on a structure, the depth-resolved wave models must be used. In addition, since the vertical acceleration may become significant during wave–structure interaction along the vertical surface of the body, the models must be capable of simulating nonhydrostatic pressure. Searching the category of wave models discussed in Chapter 5, we find that for general wave–structure interaction modeling only two types of models are valid candidates, namely the potential flow models for inviscid and irrotational flows and the NSE models for viscous and turbulent flows. The depth-averaged models, e.g., MSE models, SWE models, and Boussinesq models, may be used only if the focus of the study is on wave scattering rather than wave loading.

In the following sections, we shall discuss the BEM model, the major type of potential flow models, and the NSE model in connection with the numerical modeling of wave–structure interaction. We will discuss the wave interaction with both fixed and moving bodies. The numerical details of handling irregular body surface and body motion will be highlighted. Some previous numerical studies of wave interaction with structures and structure responses will be reviewed. Finally, the example wave model NEWTANK will be introduced, followed by a series of benchmark tests of wave–structure interaction using the NSE models.

6.2 Models for inviscid and potential flows

Although potential flow theory has been very successful in describing many nonbreaking waves not near a solid boundary, its extension to the modeling of wave–structure interaction must proceed with caution. In Section 3.15.2.1, we have argued that in wave–structure interaction, flow separation may take place around the body. In the flow separation region, vortical flow motion predominates and the flow field is rotational. The low pressure in the vortex results in the form drag, which can be the main contribution of wave force on a small body. Based on these facts, the potential flow theory will be inadequate to give a good approximation for wave loads on a small structure, around which flow separation is considerable.

It is only when the body size is much larger than the fluid trajectory under a wave, i.e., $KC<<1$, that the assumption of potential flow theory is valid because the flow separation is limited to a very small area compared with the body size. Under this circumstance, the wave loads on the structure are mainly caused by the contribution of the irrotational flow part. By solving the Laplace equation, analytically or numerically, one can obtain the velocity potential distribution around the body, from which the total forces on the body can be derived. The closed-form analytical solutions are possible only for simple geometry of the body based on diffraction theory, and they have been discussed in Section 3.15.2.3. In the following sections, we shall focus on various numerical techniques that can be extended to wave interaction with bodies of general shapes.

Although the Laplace equation can be solved directly (e.g., Li and Fleming, 1997 using FDM; Yue *et al.*, 1978 using FEM; Dommermuth and Yue, 1987 using the spectral method; Wu *et al.*, 2006 using the meshless RBF method; and Yan and Ma, 2007 using quasi ALE FEM for waves on 2D moving bodies), it can be computationally costly because a large matrix needs to be inverted. Further difficulty may also arise when the wavy free surface and the irregular body surface geometry are considered. An alternative way of solving the Laplace equation is to solve the velocity potential only on all boundaries, from which the interior velocity potential is obtained based on the special relationship of the variable on the boundaries and in the interior region. The numerical solution is then sought only on the discrete boundary elements that can be represented by lines for 2D problems and polygons for 3D problems. In naval architecture and marine engineering for ship designs, this method is often referred to as (Rankine) panel method because the ship hull surface is broken up into many small panels in the numerical approximation and the problem is solved with the use of Rankine sources. More generally, it is called BEM or sometimes the boundary integral equation method (BIEM). In contrast, depending on how the velocity potential along the solid boundary is solved, the method can be further classified into source–sink method (or source technique) or Green's second identity method, which will be elaborated below.

6.2.1   Source–sink method

6.2.1.1   Introduction of source and sink

It is well known that for irrotational flows governed by the Laplace equation, there are simple solutions of velocity potential functions for elementary flows such as uniform current, point or line source or sink, point or line vortex. The more complex flow can be formed by the superposition of these elementary flow solutions according to the linear property of the Laplace equation (e.g., White, 2003: 526). On a 2D plane (x, z), the uniform current in the x-direction, line source or sink, and line vortex can be represented by the elementary potential function and the corresponding stream function as follows:

$$\phi = -Ux \text{ current; the corresponding stream function} \qquad \psi = Uz$$

$$(6.1)$$

$$\phi = -q \ln r \text{ line source; the corresponding stream function} \quad \psi = q\theta$$

$$(6.2)$$

$$\phi = -k\theta \text{ line vortex; the corresponding stream function} \qquad \psi = -k \ln r$$

$$(6.3)$$

A classical Rankine oval can be formed by superimposing a uniform current with a pair of source and sink in the flow direction:

$$\phi = -Ux - \frac{1}{2} q \ln \frac{(x+a)^2 + z^2}{(x-a)^2 + z^2} \tag{6.4}$$

where a is half the distance between the source and the sink. The corresponding stream function is:

$$\psi = Uz - q \tan^{-1} \frac{2az}{x^2 + z^2 - a^2} \tag{6.5}$$

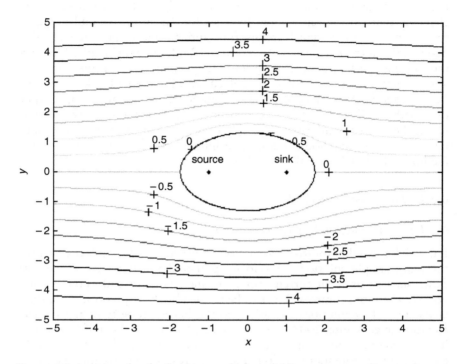

Figure 6.1 An example of a Rankine oval; the solid lines are streamlines.

Figure 6.1 shows the contours of the stream function using the above equation with $a = 1$, $q = 1$, and $U = 1$. The streamlines inside the ellipse are irrelevant and not shown. The Rankine oval represents a uniform potential flow past a closed body with an oval shape. If a vortex is added in the middle of source and sink, a circulation will be introduced that changes the flow pattern around the body and results in a lift force in the transverse direction. In aerodynamics, a series of sources, sinks, and vortices can be superimposed together to represent the flow around an airfoil, from which the drag and lift force can be computed. For wave interaction with a large body, the flow separation is negligible and therefore only sources and sinks are used.

6.2.1.2   Two-dimensional potential flow problems

The Laplace equation governing the potential flow reads as follows:

$$\nabla^2 \phi = \frac{\partial^2 \phi}{\partial x^2} + \frac{\partial^2 \phi}{\partial y^2} = \frac{1}{r} \frac{\partial}{\partial r} \left(r \frac{\partial \phi}{\partial r} \right) + \frac{1}{r^2} \frac{\partial^2 \phi}{\partial \theta^2} = 0 \tag{6.6}$$

A source or a sink on a 2D plane can be expressed as:

$$\phi = -\frac{Q}{2\pi} \ln R = -q \ln R \qquad (6.7)$$

where:

$$R = \sqrt{(x - x_s)^2 + (z - z_s)^2}$$

represents the radial distance from the source or the sink point and Q is the strength of the source that can be obtained by taking the surface integration of ϕ along any circumference of the circle centered at the source location, i.e.:

$$Q = \int_C \left(-\frac{\partial \phi}{\partial R}\right) dC.$$

With the definition of:

$$u_r = -\frac{\partial \phi}{\partial r} \text{ and } u_\theta = -\frac{1}{r}\frac{\partial \phi}{\partial \theta},$$

we have $Q > 0$ for a source and $Q < 0$ for a sink. It is not difficult to prove that for a source or a sink, the Laplace equation is satisfied everywhere except at the source/sink point where the potential function goes to infinity. However, if a continuous and smooth line of sources/sinks is considered, the value of the potential function becomes finite everywhere along the line due to the cancellation effect from the neighborhood sources/sinks. This implies that if we can properly arrange a series of sources/sinks along all boundaries including the solid surface, free surface, and computational boundary, we can correctly simulate the potential function inside the interior flow region.

A moving body in an infinite domain: Now let us consider a simple case of a moving 2D body in an infinite domain of fluid. In this case, the only boundary that needs to be considered is the solid surface, on which the continuous source/sink will be distributed. Assume that the body has a simple translation motion of $u_b(t)$ and the body surface is represented by the surface function $S(x, z, t) = 0$, from which the surface coordinate can be determined as $x_s(S)$ and $z_s(S)$. Therefore, the velocity potential in the entire domain due to one source/sink at a particular location of the body boundary is:

$$\phi_b(x, z, t) = q(x_s, z_s, t) \ln \sqrt{(x - x_s)^2 + (z - z_s)^2} \qquad (6.8)$$

By considering the contribution from all sources/sinks on the boundary, we have the final solution by integrating the sources/sinks along the solid body, i.e.:

$$\phi(x, z, t) = \int_S \phi_b(x, z, t)\, dS = \int_S q(x_s, z_s, t) \ln \sqrt{(x - x_s)^2 + (z - z_s)^2}\, dS$$

$$(6.9)$$

The above resultant velocity potential needs to satisfy the solid boundary condition, i.e., the normal velocity of the fluid particle on the solid surface u_{fn} must be the same as the solid body velocity projected on the normal direction:

$$u_{fn} = -\frac{\partial \phi(x, z, t)}{\partial n_s} = u_{bn} = |u_b| \cos \chi \qquad (6.10)$$

where χ is the angle between the body surface motion and its surface normal direction on the surface.

With the above procedure, we reduce the original 2D problem of the Laplace equation into a 1D problem for solving $q(S)$ along the body surface. For the general shape and motion of the body, however, the closed-form analytical solution does not exist. Thus, the above formulation must be solved numerically in the discrete space. Generally, the approximation is made first by discretizing the curved body surface by N straight line segments and assuming that the source density is constant in each line segment (Figure 6.2). By doing so, at a particular time the total velocity potential becomes:

$$\phi(x, z) \approx \sum_{j=1}^{N} q_j c_j = \sum_{j=1}^{N} q_j \int_{S_j} \ln \sqrt{(x - x_s)^2 + (z - z_s)^2}\, dS \qquad (6.11)$$

For a simple linear segment, the integration c_j in the above expression can be evaluated analytically to generate N constants associated with each line segment. This makes the total velocity potential the sum of N linear terms of unknown q_j. To find q_j, we need to make use of the boundary condition on the body surface. By substituting (6.12) into (6.11) and forcing the boundary condition being satisfied at the center of each segment, we would have total N linear equations in terms of N unknown q_j:

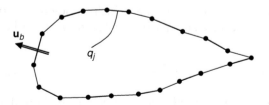

Figure 6.2 Sketch of body surface representation by linear segments in the source/sink method.

$$-\frac{\partial \phi(x,z)}{\partial n_{si}} = \sum_{j=1}^{N} q_j \frac{\partial}{\partial n_{si}} \left[\int_{S_j} \ln \sqrt{(x-x_s)^2 + (z-z_s)^2} \, dS \right]$$

$$= \sum_{j=1}^{N} q_j A_{ij} = |\mathbf{u}_b| \cos \chi_i, \quad i = 1, \ldots, N \quad (6.12)$$

The derivative of the integration, A_{ij}, can be evaluated analytically for a simple line segment. This results in a system of linear equations $A_{ij} q_j = b_i$, from which the numerical values of q_j can be uniquely determined.

Note that the body surface can also be discretized by a series of spline curves rather than straight lines for better representation of body curvature. Within each element, certain variations (linear or higher order polynomial) of source density can be assumed rather than being constant. The closed-form expression of the line integration in (6.11) and its derivative in (6.12) can still be found, but with an increased level of difficulty.

A moving body in a finite domain with free surface: If the body is placed in a finite domain with the presence of bottom, free surface, and side walls, the computation can be carried out in a similar way by placing the source/sink along all boundaries and applying the proper boundary conditions, which specify $\partial \phi / \partial n$, ϕ, or their relationship. For example, the dynamic free surface (where the gauge pressure is zero), bottom, and lateral solid boundary conditions are as follows:

$$-\frac{\partial \phi}{\partial t} + \frac{1}{2}\left(\left(\frac{\partial \phi}{\partial x}\right)^2 + \left(\frac{\partial \phi}{\partial z}\right)^2 \right) + gz = 0 \text{ on free surface} \quad (6.13)$$

$$\frac{\partial \phi}{\partial z} + \frac{\partial \phi}{\partial x}\frac{\partial h}{\partial x} = 0 \text{ at } z = -h \quad (6.14)$$

$$-\frac{\partial \phi}{\partial x} = 0 \text{ at } x = 0, L \text{ (lateral walls)} \quad (6.15)$$

In addition, the kinematic free surface boundary condition is needed to track the motion of the free surface:

$$\frac{D\mathbf{x}_{fs}}{Dt} = \frac{\partial \mathbf{x}_{fs}}{\partial t} - \frac{\partial \phi}{\partial x}\frac{\partial \mathbf{x}_{fs}}{\partial x} - \frac{\partial \phi}{\partial z}\frac{\partial \mathbf{x}_{fs}}{\partial z} = -\nabla \phi \quad (6.16)$$

where $\mathbf{x}_{fs}$ is the free surface location. When the free surface is nonoverturning, the above equation can be simplified as:

$$\frac{\partial \eta}{\partial t} = -\frac{\partial \phi}{\partial z} + \frac{\partial \phi}{\partial x}\frac{\partial \eta}{\partial x} \text{ at } z = \eta(x,t) \quad (6.17)$$

To create a single-connected domain that can be solved by the source technique, an immersed body can be connected to other boundaries (e.g., bottom

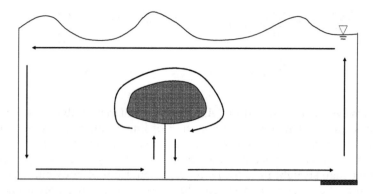

Figure 6.3 Sketch of an immersed body in a finite domain; the loop of arrows suggests the way of making a single-connected domain in the source/sink method.

or solid lateral boundaries) by a straight line, along which both the advance and the return paths are specified so that the source/sink completely cancels out there (this is similar to using scissors to cut out a part from the middle of a paper; see Figure 6.3 for illustration). The idea can be extended to multiple bodies and floating bodies like ships.

Special attention is paid to the two nonlinear free surface boundary conditions. By enforcing the nonlinear boundary condition exactly, a system of nonlinear equations will result that requires iterative procedures to solve, which we shall discuss further in Section 6.2.2.2. For small-amplitude waves, linearization can be made by combining the two free surface conditions into one and applying it on the mean water level:

$$\frac{\partial^2 \phi_s}{\partial t^2} + g \frac{\partial \phi_s}{\partial z} = 0 \text{ at } z = 0 \tag{6.18}$$

As a result, a system of linear equations will be generated that can be solved in the same way as the moving body in an infinite domain discussed earlier.

A body exposed to a linear periodic wave train: If a fixed and rigid body is placed in an open sea with a specified incident wave train, the resulting wave field around the body is composed of both incident waves and diffracted/scattered waves. Rigorously speaking, "wave diffraction" refers to the wave energy *deflected* into the shadow and lower energy regions, while "wave scattering" refers to the wave energy *reflected* from the body (but being modified by the diffracting process it does not necessarily follow exactly the law of reflection). In most cases, however, the use of the terminology of diffraction and scattering is often mixed with any one of them representing the combined wave field. For a moving body, the additional wave can be generated from the moved body and it is often called the "radiation wave." Both diffracted/scattered and radiation waves will have the source location on the surface of the body. In the numerical modeling, they are often treated as single term for the scattered wave.

For a linear periodic wave case, the total velocity potential can then be expressed as the sum of that from the incident wave ϕ_0 and the scattered wave ϕ_s, i.e.:

$$\phi = \phi_0 + \phi_s \tag{6.19}$$

where the incident wave train can be readily expressed in the complex form as:

$$\phi_0 = -\frac{H}{2}\frac{g}{\sigma}\frac{\cosh k(h+z)}{\cosh kh}i e^{i(kx-\sigma t)} = \Phi_0\, e^{-i\sigma t} \tag{6.20}$$

Thus, the solution is needed only for the scattered velocity potential, which is also governed by the Laplace equation:

$$\nabla^2 \phi_s = 0 \tag{6.21}$$

The linkage between the scattered velocity potential and the incident wave potential is made by the solid–surface boundary condition:

$$-\frac{\partial \phi_s}{\partial n} = \frac{\partial \phi_0}{\partial n} + |\mathbf{u}_b| \cos \chi \tag{6.22}$$

For small-amplitude waves, the linearized free surface boundary can be obtained by combining the kinematic and dynamic free surface boundary conditions at the still water level:

$$\frac{\partial^2 \phi_s}{\partial t^2} + g\frac{\partial \phi_s}{\partial z} = 0 \text{ at } z = 0 \tag{6.23}$$

The bottom boundary condition reads:

$$\frac{\partial \phi_s}{\partial z} + \frac{\partial \phi_s}{\partial x}\frac{\partial h}{\partial x} = 0 \text{ at } z = -h \tag{6.24}$$

The radiation boundary condition reads:

$$\lim_{|x|\to\infty}\left(\frac{\partial}{\partial |x|} - ik\right)\phi_s = 0 \tag{6.25}$$

In case the body is fixed and rigid, the scattered wave has the same frequency as that of the incident wave and its velocity potential can be written as:

$$\phi_s(x, y, t) = \Phi_s(x, y)e^{-i\sigma t} \tag{6.26}$$

By substituting the above expression into the governing equation and boundary conditions, we have:

$$\nabla^2 \Phi_s = 0 \tag{6.27}$$

with the solid–surface boundary condition:

$$-\frac{\partial \Phi_s}{\partial n} = \frac{\partial \Phi_0}{\partial n} \tag{6.28}$$

the linearized free surface boundary condition:

$$-\sigma^2 \Phi_s + g \frac{\partial \Phi_s}{\partial z} = 0 \text{ at } z = 0 \tag{6.29}$$

the bottom boundary condition:

$$\frac{\partial \Phi_s}{\partial z} + \frac{\partial \Phi_s}{\partial x} \frac{\partial h}{\partial x} = 0 \text{ at } z = -h \tag{6.30}$$

and the radiation boundary condition:

$$\lim_{|x| \to \infty} \left(\frac{\partial}{\partial |x|} - ik \right) \Phi_s = 0 \tag{6.31}$$

Note that the above is also true if the body has a linear response to the incident wave only (i.e., the body motion has the same frequency as that of the incident wave and thus generates the radiation wave having the same frequency too).

Once the velocity potential is obtained, the pressure can be easily derived from the Bernoulli equation as:

$$p = -\rho g z + \rho \frac{\partial \phi}{\partial t} - \rho \frac{1}{2} \left[\left(\frac{\partial \phi}{\partial x} \right)^2 + \left(\frac{\partial \phi}{\partial y} \right)^2 \right] \tag{6.32}$$

where the first term on the RHS is the hydrostatic pressure, the second term is the linear dynamic pressure, and the third term is the second-order nonlinear dynamic pressure, which can be neglected for small-amplitude linear waves. The total force on the body can be obtained by taking the surface integration of the pressure around the body, i.e.:

$$\mathbf{F} = \int_S p\mathbf{n} \, dS \tag{6.33}$$

6.2.1.3 Three-dimensional potential flow problems

For 3D problems, the major difference is that the functional type of the potential function for source and sink is different, i.e.:

$$\phi = \frac{Q}{4\pi R} \tag{6.34}$$

where:

$$R = \sqrt{(x - x_s)^2 + (y - y_s)^2 + (z - z_s)^2}.$$

Again, it is ready to show that:

$$Q = \int_S \left(-\frac{\partial \phi}{\partial R} \right) dS,$$

where S is the surface of any sphere center at the source location. For a general transient 3D problem, the velocity potential in the fluid domain is:

$$\phi(\mathbf{x}, t) = \int_S \phi_b(\mathbf{x}, \mathbf{x}_s, t)\, dS = \int_S \frac{Q(\mathbf{x}_s, t)}{4\pi R(\mathbf{x}, \mathbf{x}_s)} dS = \frac{1}{4\pi} \int_S Q(\mathbf{x}_s, t) \frac{1}{|\mathbf{x} - \mathbf{x}_s|} dS \tag{6.35}$$

Another major difference is that the radiation condition for the linear scattered wave velocity potential is expressed as:

$$\lim_{r \to \infty} \sqrt{r} \left(\frac{\partial}{\partial r} - ik \right) \phi_s = 0 \tag{6.36}$$

where $r = \sqrt{x^2 + y^2}$ is the radial distance from the source (body) on the horizontal plane. This gives the following approximate condition away from the sources:

$$\phi_s \propto \frac{1}{\sqrt{r}} \tag{6.37}$$

Similar to 2D problems, if the incident wave is monochromatic and linear, the time variation component of the source function can be separated out and represented by a simple harmonic:

$$\phi(\mathbf{x}, t) = \frac{Q(\mathbf{x}_s, t)}{4\pi R(\mathbf{x}, \mathbf{x}_s)} = q(\mathbf{x}_s) G(\mathbf{x}, \mathbf{x}_s) e^{-i\sigma t} \tag{6.38}$$

where $q(\mathbf{x}_s)$ is the source density along the body surface and $G(\mathbf{x}, \mathbf{x}_s)$ is referred to as time-independent Green's function and its analytical expression can be obtained for a body in both infinite water depth (e.g., Havelock,

1942) and finite water depth (Wehausen and Laitone, 1960: 475–9). Green's function, of course, satisfies the Laplace equation and all necessary boundary conditions on the bottom, free surface, and in the far field. Readers can refer to Faltinsen (1990: 109–18) for an example of using the 3D source technique for the analysis of added mass and damping in heave motion and Chakrabarti (1987: 301–22) for a more general numerical solution of wave diffraction around a body.

The source technique introduced in this section is easy to understand and implement in a numerical code for wave–structure interaction. The method takes the advantage of the generalized solution of the Laplace equation that has the special form of Green's function. The BEM based on the source technique solves the source density everywhere on the boundaries, from which the velocity potential and pressure on the boundaries and in the interior domain can be found. Although this method can in principle be used in solving nonlinear water wave problems, its linearized version is more commonly used by treating the incoming waves as linear harmonic waves (or the sum of harmonic waves for irregular waves). This technique is popular in naval architecture and offshore engineering for the design of ship hulls and the analysis of floating platforms, where it is called panel method. In coastal engineering where wave nonlinearity becomes important, the alternative formulation of BEM will often be adopted, which is detailed below.

6.2.2   Green's second identity method

6.2.2.1   Model description

Green's second identity takes the following general form:

$$\int_V \left(\phi \nabla^2 G - G \nabla^2 \phi \right) dV = \int_S \left(\phi \frac{\partial G}{\partial n} - G \frac{\partial \phi}{\partial n} \right) dS \tag{6.39}$$

where G is Green function satisfying the Poisson equation, i.e.:

$$\nabla^2 G = Q \delta(\mathbf{x} - \mathbf{x}_s) \tag{6.40}$$

with $\delta(\mathbf{x} - \mathbf{x}_s)$ being the Dirac delta function and Q the source density coefficient that equals 2π for 2D problems and 4π for 3D problems. Green function satisfies the Laplace equation everywhere except at the source location $\mathbf{x}_s$ where the singularity arises. The free space Green function and the source density take the following forms and values for 2D and 3D problems:

$$G = \begin{cases} \ln R & \text{2D} \\ \frac{1}{R} & \text{3D} \end{cases} \quad \text{where} \quad R = |\mathbf{x} - \mathbf{x}_s| \quad \text{and} \quad Q = \begin{cases} 2\pi & \text{2D} \\ 4\pi & \text{3D} \end{cases} \tag{6.41}$$

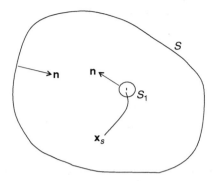

Figure 6.4 Definition sketch of an integration surface used in Green's second identity method where x_s is the singular point.

Consider the domain of interest by excluding a small circle centered at x_s with its radius approaching zero (Figure 6.4). If we substitute (6.40) and (6.41) into (6.39) and force $\nabla^2 \phi = 0$, we would have:

$$0 = \int_V \left(\phi \nabla^2 G - G \nabla^2 \phi \right) dV = \int_S \left(\phi \frac{\partial G}{\partial n} - G \frac{\partial \phi}{\partial n} \right) dS + \int_{S_1} \left(\phi \frac{\partial G}{\partial n} - G \frac{\partial \phi}{\partial n} \right) dS$$

$$\Rightarrow \int_S \left(\phi \frac{\partial G}{\partial n} - G \frac{\partial \phi}{\partial n} \right) dS = -\int_{S_1} \phi \frac{\partial G}{\partial n} \, dS + \int_{S_1} G \frac{\partial \phi}{\partial n} \, S = -\int_{S_1} \frac{\partial G}{\partial r} \, dS = -Q\phi$$

$$\Rightarrow -Q\phi = \int_S \left(\phi \frac{\partial G}{\partial n} - G \frac{\partial \phi}{\partial n} \right) dS \qquad (6.42)$$

In the above equation, Green function G is known as Rankine source and its derivative $\partial G/\partial n$ is called a dipole. The above equation simply states that the velocity potential can be represented by a distribution of Rankine source and its dipole, whose densities are $\partial \phi/\partial n$ and ϕ, over a closed surface.

The above equation connects the velocity potential everywhere inside the computational domain with the velocity potential and its normal derivative on all boundaries of integration. In principle, once ϕ and $\partial \phi/\partial n$ are known on the boundaries, the problem is solved. Caution should be taken when the above equation is applied on the boundary where the value of Q should be reduced by half because the surface integration can be made only on a half space.

To obtain the values of ϕ and $\partial \phi/\partial n$ on all boundaries, discretization will first be made to approximate the actual boundary with a finite number

(N) of linear elements, similar to those in the source technique. When the above equation is written at N nodes or N elements, we have:

$$-Q\phi_i = \int_S \left(\phi \frac{\partial G}{\partial n} - G \frac{\partial \phi}{\partial n} \right) dS =$$

$$\sum_{j=1}^{N} \left[\phi_j \int_{S_j} \left[\frac{\partial G}{\partial n} \right]_i dS + \left[\frac{\partial \phi}{\partial n} \right]_j \times \int_{S_j} (-G)_i \, dS \right], \quad i = 1, \ldots, N$$

(6.43)

This forms a system of N linear equations with 2N unknowns of ϕ and $\partial\phi/\partial n$. The additional information can thus be obtained from N boundary conditions on N nodes or elements. Each boundary condition specifies either ϕ or $\partial\phi/\partial n$ or their relationship and thus can provide another N equation, which may be linear or nonlinear. As a result, we have 2N equations with 2N unknowns, and unique solutions exist if all the 2N equations are independent. Alternatively, the N boundary conditions can be absorbed in the first N linear equations, and thus, the numerical solution is sought only for the N unknowns of ϕ or $\partial\phi/\partial n$ on the boundary. Compared with the source technique that seeks the numerical solution of source density on all boundary elements, Green's second identify method seeks the numerical solution for both velocity potential and its normal derivatives directly. Theoretically speaking, these two methods are equivalent and different only in terms of solution procedures.

6.2.2.2   *Treatment of wave nonlinearity*

Nonlinearity exists in two fully nonlinear free surface boundary conditions (kinematic and dynamic) that need to be incorporated into the solution procedure:

$$\frac{D\mathbf{x}_{fs}}{Dt} = \left(\frac{\partial}{\partial t} - \nabla\phi \cdot \nabla \right) \mathbf{x}_{fs} = -\nabla\phi$$

(6.44)

$$\frac{\partial \phi}{\partial t} = \frac{1}{2} \left(\left(\frac{\partial \phi}{\partial x} \right)^2 + \left(\frac{\partial \phi}{\partial y} \right)^2 + \left(\frac{\partial \phi}{\partial z} \right)^2 \right) + gz + p_a = 0$$

(6.45)

Whereas the first equation is used to update free surface location and the associated velocity potential and its normal derivative, the second equation is needed in the numerical solution procedure of the Laplace equation. There are a few ways of dealing with the nonlinear terms in these free surface boundary conditions and they are summarized as follows.

Iteration for the system of fully nonlinear equations: If the nonlinear boundary condition is involved for either ϕ or $\partial\phi/\partial n$, the resulting system of equations will be nonlinear but the solution is still unique because we have the same number of equations as the number of unknowns. The numerical solution to the system of equations, however, must be obtained by iteration. This approach was adopted by Kim *et al.* (1983). With a similar idea, Nakayama (1983) proposed the error-correcting method to solve the system of nonlinear equations directly.

Successive solution to the Laplace equation: Alternatively, time stepping can also be made based on the Taylor expansion (Dold and Peregrine, 1984) as:

$$\mathbf{x}_{fs}(t+\Delta t) = \mathbf{x}_{fs}(t) + \sum_{k=1}^{n} \frac{(\Delta t)^k}{k!} \frac{D^k \mathbf{x}_{fs}}{Dt^k} + O\left[(\Delta t)^{n+1}\right] \tag{6.46}$$

$$\phi(\mathbf{x}_{fs}(t+\Delta t), t+\Delta t) = \phi(\mathbf{x}_{fs}(t), t) + \sum_{k=1}^{n} \frac{(\Delta t)^k}{k!} \frac{D^k \phi(\mathbf{x}_{fs}(t), t)}{Dt^k} + O\left[(\Delta t)^{n+1}\right] \tag{6.47}$$

Since the Laplace equation is valid for ϕ at any order of its derivative in time, the coefficients in the above Taylor expansion can be obtained by solving the succession of the Laplace equation for various orders of derivatives in time. The solution of the lower order Laplace problem will provide the nonlinear free surface boundary conditions for the next higher order Laplace equation. The details of the treatment can be found in Grilli *et al.* (1989). The methodology can be extended in principle to any order, but most of the time the second-order treatment is sufficient to model strong nonlinear waves (e.g., overturning waves) and their interaction with structures (Grilli and Svendsen, 1990). Later, Celebi *et al.* (1998) developed a numerical wave tank with a similar methodology and used it to study a fully nonlinear wave interaction with vertical cylinders, bottom-mounted or truncated. Since the method solves the velocity potential on the Eulerian frame and tracks the free surface location using the Lagrangian method, it is also called mixed Eulerian–Lagrangian (MEL) approach.

Perturbation technique: For waves in deep water with moderate wave steepness, the perturbation procedure used to derive Stokes wave theory can be similarly applied in the BEM. This method does not need to update the influence matrix corresponding to the system of linear algebraic equations at each time step when the free surface moves to a new position and thus avoids the most expensive part of computation. It is especially useful and efficient for solving 3D wave interaction with structures in relatively deep water, a common case in offshore engineering. Under this circumstance, both velocity potential and free surface displacement can be expanded using Taylor expansion based on the perturbation of a small parameter $\varepsilon = ka$ (wave steepness):

$$\phi(\mathbf{x}, t) = \phi_0(\mathbf{x}, t) + \varepsilon \phi_1(\mathbf{x}, t) + \varepsilon^2 \phi_2(\mathbf{x}, t) + \cdots \tag{6.48}$$

$$\eta(\mathbf{x}, t) = \eta_0(\mathbf{x}, t) + \varepsilon \eta_1(\mathbf{x}, t) + \varepsilon^2 \eta_2(\mathbf{x}, t) + \cdots \tag{6.49}$$

By doing so, the influence matrix for any order of numerical solutions is formed only once in the beginning of the computation, and it remains unchanged in the rest of the computation. The first-order solution is obtained first by imposing the free surface boundary condition at $z = 0$ and the solid boundary condition on the mean location of the body surface. The higher order solution can also be sought once the lower order solutions are known. The detailed procedure can be found in Kim and Kim (1997).

This perturbation method can be used to solve first-, second-, and higher order wave forces on the body, although it is unable to capture the overturning wave. In this approach, the current effect and small structure motion (prescribed or dynamically responding movement) can also be included. There are many reports of using this method for the studies of wave loads or combined wave–current loads on offshore structures (e.g., Isaacson and Cheung, 1992; Teng and Taylor, 1995; Skourup *et al.*, 2000).

6.2.2.3 *Numerical efficiency of boundary element method models*

It generally gives people the illusion that BEM is more computationally efficient than direct NSE solvers because the former only needs to resolve boundary elements and has a reduced dimension by one while the latter solves the problem in the full dimension. However, a more careful analysis reveals the hidden cost for BEM that prevents it from modeling fully nonlinear 3D water waves.

In all approaches of solving water waves using BEM, one needs to solve the following matrix:

$$A_{ij}q_j = b_i, \quad i = j = N \tag{6.50}$$

where N is the number of elements on the boundary. In BEM, since the matrix A is full and dense, the flops of manipulation to solve it is $O(N^3)$. Let us use an example of a 3D problem whose domain can be discretized by $100^3 = 10^6$ small cubes. In the BEM representation, the number of surface elements is reduced to $N = O(100^2) = O(10^4)$. Therefore, the total flops of computation for solving (6.50) using standard Gauss elimination at each time step is around $O(N^3) = O(10^{12})$. When some more efficient matrix decomposition methods (e.g., block LU decomposition, Cholesky decomposition, singular value decomposition) are used, the computational effort can be brought down to $O(N^{2.5}) = O(10^{10})$, still very computationally expensive.

Alternatively, if we directly solve the 3D problem with the use of NSE, we will of course have to use more elements to resolve the domain [e.g., $N = O(10^6)$]. However, if we use the explicit projection method to solve the NSE, the only major part that involves intensive computation is the numerical solution to PPE, the discretized form of which is a matrix with the size of $i = j = N = 10^6$. This matrix, however, is very sparse with only seven nonzeros on each line. The efficient iterative matrix solvers such as the CG method can be used in solving this type of matrix. For most of our tests of water wave problems, the total iteration number can be controlled within 100. This gives the computational efforts of $O(10^8)$ flops for inverting the matrix. Even with the consideration of other manipulations, the total flops at each time step is still cheaper than the BEM. Interestingly, the larger the problem is, the more advantageous an NSE solver will be! This explains why so far there are few reports of the 3D BEM time-domain simulation of fully nonlinear water wave interaction with a structure of complex shape.

Many practical 3D BEM water wave models are based on the perturbation technique introduced earlier. With the use of the perturbation method, the matrix A is no longer time-dependent and only b will be updated at each time step. Therefore, the equation can be rewritten as follows:

$$q_j = A_{ij}^{-1} b_i, \quad i = j = N. \tag{6.51}$$

The inverse matrix A^{-1} will be obtained only once at $t = 0$. By doing so, the boundary condition is always imposed at the mean location of the body and on the still water depth. Therefore, one is unable to simulate fully nonlinear waves that can overturn or surface waves induced by an arbitrary body motion using this approach.

6.2.2.4 *Time-domain analysis and frequency-domain analysis*

All of the fully nonlinear BEM models solve the transient water wave problem in the time domain, as described in the previous section. The numerical results are in the form of time-varying nonlinear free surface displacement, from which the wave transformation and/or wave load on a body can be derived. The solution procedure is called time-domain analysis, which can retain wave nonlinearity as well as its nonlinear interaction with a fixed or moving body. Time-domain analysis can also be performed by using other wave models such as NSE models.

For small-amplitude waves, it is possible to linearize the free surface boundary conditions so that the problem is solved at $z = 0$ and at the mean location of an oscillatory body. This makes possible the frequency-domain analysis, where the irregular wave is first decomposed into a finite number of linear monochromatic waves with discrete frequencies and directions. The numerical solution is then sought for both wave motion and body motion. The resultant wave diffraction and body movements can be obtained by simply summing up the contribution from all wave harmonics. Because the same influence matrix can be used throughout the computation thanks to linearization, the frequency-domain analysis is computationally efficient and thus it is a powerful tool in the study of wave diffraction and structure response in a random sea. However, the fully nonlinear property, for both wave and structure motion, cannot be attained in the analysis.

6.2.2.5   Other numerical issues

Corner: Corner is the place where $\partial\phi/\partial n$ is not continuous. A corner may be present at the intersection of any two boundaries, e.g., wave maker and bottom, structure surface and free surface. Various treatments have been made to develop the compatibility conditions at the corners for better numerical accuracy and robustness. Readers are referred to Grilli *et al.* (1989) for more discussion.

Desingularization: Singularity exists on all source points which coincide with the body surface, free surface, and other boundaries where the boundary conditions are applied. In theory, the singularity is removed after surface integration on a finite length of line segment. However, numerically the singularity may not be completely removed and can lead to instability in certain places (e.g., sharp corners and tips of a breaking wave front). A proposal was made by Kim *et al.* (1998) to move the source distribution outside the fluid domain, so that the source points will not coincide with the fluid collocation points. As a result, the resulting integrals do not contain any singularity, which makes the numerical code more stable.

Leaky panel: When a large and complex 3D body surface (e.g., a ship hull) is represented by plane quadrilateral elements, leaks may appear at the place where the panels do not fit at the junction. This will result in a finite value of normal velocity at the joint of two elements, where it should be zero theoretically. Normally, this will not cause serious problems for the fluid and pressure computation around the body. Readers are referred to Faltinsen (1990: 116) for more discussion.

Sawtooth instability: When breaking waves are simulated with the use of fully nonlinear boundary conditions, sawtooth numerical instability can be developed near the tip of the breaking waves. This phenomenon was first reported by Longuet-Higgins and Cokelet (1976) and solved by applying the smooth procedure. Later, Dold and Peregrine (1984) developed a consistently high-order time-stepping method using Taylor expansion. This minimizes the numerical instability. However, as the tip of the breaking wave becomes very sharp, it is essentially a moving corner. The numerical treatment at this point is in fact a work of art, and different treatments may result in very different details of tip shape. This implies that potential flow theory may start to break down from this point and any endeavor to further the computation is unnecessary.

6.2.3   Strip theory

The strip theory was originally developed for the computation of wave force coefficients on a ship hull, and it was popularly used in naval architecture and ocean engineering for ship design. Sometimes it is also called strip theory method or strip method; readers should not confuse this with "finite strip method" that is used in structural analysis.

The strip theory can be regarded as a simplified version of the BEM methods for slender bodies. Instead of solving the entire problem simultaneously, the strip theory method divides the underwater part of the body into a number of strips along the longitudinal direction. As a result, the 3D body is represented by a series of 2D strips. For each strip, the problem is treated as a 2D problem, from which the BEM can be used to calculate 2D added mass and moment coefficients for the particular incident wave train and ship hull. These coefficients will later be used in combination with equations of motions for the global response analysis or the simpler 1D dynamic analysis of the body motion, in either nonlinear time domain or linear frequency domain.

In strip theory, it is assumed that the flow variation in the longitudinal direction is small, which may not be true at two ends of the ship. Besides, in using strip theory, each strip is solved independently without any interaction or information exchange with other strips. The strip theory can be regarded as a "2.5D" problem solver between the full solution of the 3D problem (e.g., by a 3D BEM model) and the approximated solution of the simplified 2D problem. For a monohull vessel, strip theory can give rather decent results for the heave and roll motions, while it can lead to large errors for surge analysis and multihull vessels. Note that strip theory is also used in aerodynamics (e.g., Delaurier, 1993). Readers are referred to Faltinsen (1990: 50–4) for more discussion.

6.3   Models for viscous and turbulent flows

The model that can solve general viscous and turbulent flows is based on the NSE or its extension (e.g., Reynolds equations), i.e.:

$$\frac{\partial u_i}{\partial x_i} = 0 \tag{6.52}$$

$$\frac{\partial u_i}{\partial t} + u_j \frac{\partial u_i}{\partial x_j} = -\frac{1}{\rho}\frac{\partial p}{\partial x_i} + g_i + \frac{1}{\rho}\frac{\partial \tau_{ij}}{\partial x_j} \tag{6.53}$$

Detailed discussions have been given in Section 5.5.1.1 for the numerical models based on NSE solvers. Most of the NSE models previously introduced have the capability of handling structures with different body shape, although the body surfaces are treated differently. In this section, discussions are mainly made in connection with the treatment of FSI.

When a body is present inside the computational domain, the immediate difficulty for a numerical model is to handle the possibly complex and irregular body surface. This will be treated by either (1) boundary-fitted meshes or (2) Cartesian meshes with special treatment to include irregular boundary effect. If a body moves during the computation, the body surface has the similar characteristics of a free surface, and therefore, it can be handled similarly by (1) using a moving adaptive mesh based on the Lagrangian approach to ensure the boundary conformal mesh at all time or (2) keeping the grid fixed but updating the surface location/information in each cell. The use of boundary-fitted unstructured mesh is often adopted in FEM or FVM, whereas Cartesian mesh is often used in FDM.

For the incompressible NSEs, there is no state equation connecting pressure and local fluid density. As a result, the pressure is coupled with the global velocity field. This means there is no "memory" for pressure, which must adjust itself instantaneously according to the velocity. This is consistent with the fact that the pressure wave transmits at the speed of infinity for incompressible fluid flows. The pressure distribution must satisfy the PPE that can be derived by combining continuity equation and momentum equations. Since the nature of the Poisson equation is elliptic, the numerical solution to the pressure needs to be obtained in an iterative way. There are various fractional-step methods to update velocity and pressure simultaneously for unsteady viscous flows. The simplest yet most accurate one is based on the so-called projection method, which breaks down the numerical procedure into two steps. The first step is to seek the tentative velocity and the second step is to project the tentative velocity on the plane where the continuity equation is satisfied. The projection angle will be determined by the solution of the PPE that incorporates the tentative velocity information as the forcing function. A typical two-step projection method takes the following form:

$$\text{Step 1}: \quad \frac{\tilde{u}_i - u_i^n}{\Delta t} = \left\{ \frac{1}{\rho} \frac{\partial \tau_{ij}}{\partial x_j} - u_j \frac{\partial u_i}{\partial x_j} \right\}^n \tag{6.54}$$

$$\text{Step 2}: \quad \frac{u_i^{n+1} - \tilde{u}_i}{\Delta t} = -\frac{1}{\rho} \frac{\partial p^{n+1}}{\partial x_i} + g_i \tag{6.55}$$

where the pressure is obtained by solving

$$\frac{\partial}{\partial x_i} \left(\frac{1}{\rho} \frac{\partial p^{n+1}}{\partial x_i} \right) = \frac{\partial}{\partial x_i} \left(\frac{\tilde{u}_i}{\Delta t} \right) \tag{6.56}$$

Most of the models discussed below follow similar methodology in terms of time discretization. Differences are mainly in the spatial discretization and numerical approximation of the functions and their derivative terms, which will be discussed below.

6.3.1　Models based on finite element method

Hayashi *et al.* (1991) proposed a comparative study for four different fractional-step FEM NSE models with the use of benchmark tests of solitary wave propagation. Radovitzky and Ortiz (1998) developed a fully Lagrangian FEM for solving NSEs. The mesh is maintained undistorted through continuous and adaptive remeshing of the fluid mass. The model was used to simulate wave breaking to check the accuracy of the adaptive remeshing process. Onate and Garcia (2001) developed a semi-implicit FEM for solving coupled FSI problems with the use of a moving mesh. The model was used for the analysis of totally or partially submerged bodies under surface waves. Since most of the structural analyses were constructed on FEM, the use of FEM for solving fluid motions has the advantage of seamlessly connecting fluid computation to structure response calculation. For this reason, there is an increasing trend of using FEM for solving NSEs in the analysis of FSI. Recently, Aliabadi *et al.* (2003) proposed an FEM model that solved NSEs for fluid motion, coupled with the dynamic equation for nonlinear rigid-body motion and the linear elasticity equations for cable motion. The model was used to simulate mooring forces on floating objects. Zhao *et al.* (2004) used an FEM NSE model to study wave interaction with a submerged pipeline.

Although the remeshing process can provide a boundary-fitted mesh system to both the free surface and the body surface the computation of remeshing itself can be quite involved. This motivates the combined use of Eulerian free surface-tracking techniques, which are constructed on fixed grid systems and have been well established, and unstructured mesh that may provide better resolution to the body surface. For a fixed and rigid body, the hybrid model enjoys the benefit from both the unstructured mesh in resolving the body surface and the Eulerian free surface-tracking techniques that do not require remeshing when the free surface moves. Examples of such models include the FEM VOF model developed by Kim and Lee (2003) and the FEM level set method model by Lin *et al.* (2005b).

Note that besides the conventional formulation of NSEs based on primitive variables, the models constructed on other formulations were also available for solving various wave–structure problems. For example, Lo and Young (2004) developed a FEM ALE model based on the vorticity–velocity formulation of NSEs (see Section 2.1.2) and used the model to simulate solitary wave passage over a submerged structure.

6.3.2 Models based on finite volume method

Similar to FEM, FVM is able to solve the problem in an irregular physical domain with the use of a boundary conformal mesh. In addition, the FVM can explicitly enforce mass and momentum conservation in each computational cell and thus may be more robust and accurate for long-term computation. Many of the commercial software solving 3D NSEs have adopted this methodology. The boundary-conforming mesh is often generated by an automatic mesh generator, some of which have the capability to generate adaptive mesh following the moving surface.

For wave simulation, one of the key issues relevant to an FVM model is the accurate tracking of free surface motion. Some recent developments of FVM-based NSE-solver models are summarized below. Zhang *et al.* (1998)

introduced a multiphase curvilinear FVM model with the free surface being tracked by the level set method and solved by the higher order ENO method. The model was used to study the mold filling and the spreading and solidification of molten droplets. Yue *et al.* (2003) developed a free surface FVM model with the use of level set method for free surface tracking. Alternatively, in their FVM model, Chan and Anastasiou (1999) used the triangular mesh to resolve the curved free surface, and the mesh is updated at each time step based on the updated free surface location solved from the kinematic boundary condition. With the use of combined FDM and FVM, Casulli and Zanolli (2002) proposed a 3D wave model constructed on an unstructured mesh system and the free surface is tracked by solving the height function.

6.3.3 Models based on boundary element method

As pointed out before, BEM has the advantage of converting a domain integration problem to a surface integration problem, and this may improve the computational efficiency. However, the application of BEM is most popular in solving the Laplace equation where the volume–surface transformation, ensured by Green's theorem, is complete. For viscous fluid flows, the governing equations can be converted to Poisson equations under certain special circumstances. Examples include the PPE with the nonlinear source term [e.g., equation (2.16), $\rho Tr(\nabla u \otimes \nabla u) = -\nabla^2 p$ with ρ being a constant], the Poisson stream function equation with the vorticity source for 2D problems [e.g., equation (2.18), $\nabla^2 \psi = -\omega$], and the Poisson velocity equation for Stokes flows with the negligible inertial term (e.g., $\nabla p/\rho - g = \nu \nabla^2 u$). Compared to Laplace equation, Poisson equation requires volume integration of the source term and thus increases the expense of using the BEM. For some cases, however, the complete surface integration can be achieved through the succession of high-order transformation. Wu (1982) proposed a few BEM solutions to viscous flow problems.

The BEM model solving NSEs for water wave problems is rarely reported, mainly because the transformation is too involved, which makes the use of BEM not an attractive option. A limited number of such attempts includes the BEM model presented by Zang *et al.* (2000) for solving the linearized NSEs in the frequency domain to analyze viscous liquid sloshing.

6.3.4 Models based on spectral methods

The spectral methods are efficient and accurate, but they are often used in a relatively simple domain that allows the powerful Fourier transform. The application of spectral methods in solving NSEs for problems with free surfaces and structures of arbitrary shape was not widely reported due to the irregularity of the computational domain. Only a limited number of attempts have been made recently for the development of numerical models using spectral methods for solving free surface flows and fluid interaction with structures. For example, Kevlahan and Ghidaglia (2001) introduced a Brinkman (volume) penalization of the obstacle in their pseudo-spectral method model that solves NSEs on a Cartesian grid. The model was used to simulate flow past an array of cylinders. Chern *et al.* (2005) developed a pseudospectral element model for the simulation of free surface viscous flows.

6.3.5   Models based on finite difference method

Most of the water wave models solving NSEs are based on the FDM formulation. Because FDM is normally constructed on Cartesian grids with small flexibility for dealing with irregular geometry, special schemes are needed on the interface (between two fluids and between a fluid and a solid) that may cross through the grid line at an arbitrary angle. Two immediate difficulties arise, namely the treatment of free surface and the treatment of solid surface, which are elaborated below.

6.3.5.1   Free surface treatment in Cartesian grid

The free surface treatment consists of two parts, namely the tracking of the free surface and the implementation of free surface boundary conditions, which are often interrelated.

MAC method and irregular star technique: The accurate tracking of the free surface can be achieved by the MAC method based on the Lagrangian approach (Harlow and Welch, 1965). In this method, the meshless particles are initially deployed on the free surface, and they are tracked by using the interpolated velocity at the particle location. Since the particles originally on the free surface will remain for a nonbreaking free surface, the locations of the particles will be used to determine the free surface location at any time step. In connection to the MAC method, the normal distance between the free surface cell center and the actual free surface can be determined. To implement a dynamic free surface boundary condition that defines the pressure on the free surface as the atmospheric pressure, the so-called irregular star technique was developed by Chan and Street (1970) to take into account the varying distance of the free surface cell center, where the pressure is defined, to the actual free surface. The wave models in this category include TUMMAC (Miyata *et al.*, 1985), GENSMAC (Tome and Mckee, 1994; Tome *et al.*, 2001), SIMAC (Armenio, 1997), and NS-MAC NWT (Park *et al.*, 1999), all of which are able to simulate wave interaction with structures of complex surface geometry.

VOF method and level set method: The free surface can also be "captured" by using the Eulerian approach. In this approach, two of the most popular methods are the VOF method (Hirt and Nichols, 1981) and the level set method (Osher and Sethian, 1988). In both methods, the free surface cell is treated as a cell with a large gradient of a certain fluid property function, which is the normalized density in the VOF method and the hypothetical interface function in the level set method. By solving this function in the Eulerian frame, the free surface may be approximately reconstructed based on the functional values in the local cell and its neighborhood. The information about the reconstructed free surface in each cell can be used by the "irregular star" technique when the pressure solution is sought. Alternatively, a free surface cell can also be treated as a special fluid cell with the reduced mean density compared to the interior fluid cell. The dynamic free surface boundary condition for pressure is then either imposed on the immediate neighborhood air cell for a single-phase fluid flow computation or not imposed at all for a multiphase fluid flow computation. This alternative treatment is simpler because no free surface orientation information is needed. However, it may generate inaccurate results for pressure and thus for velocity in free surface cells that must be corrected later by applying appropriate velocity boundary conditions (Liu and Lin, 1997).

The earliest practical VOF model was developed by Nichols *et al.* (1980) as SOLA-VOF, which was followed by NASA-VOF2D (Torrey *et al.*, 1985), NASA-VOF3D (Torrey *et al.*, 1987), FLOW3D (Hirt, 1988), RIPPLE (Kothe *et al.*, 1991), and TELLURIDE (Kothe *et al.*, 1997). Some of these codes or their methodology were extended to simulate water wave interaction with structures, and the successful examples include SKYLLA (van der Meer *et al.*, 1992), VOFbreak^2 (Troch, 1997), COBRAS (Liu *et al.*, 1999b), and NEWFLUME (Lin and Xu, 2006). The incorporation of level set method into an NSE-solver wave model is fairly recent. Iafrati *et al.* (2001) proposed a model based on the level set method and demonstrated its capability of capturing a 2D breaking wave front. Gu *et al.* (2005) developed a two-phase 3D model based on the level set method for the simulation of liquid sloshing in a tank. Hong and Doi (2006) presented a 3D level set model to study 3D breaking ship waves.

6.3.5.2   Body surface treatment in Cartesian grid

Compared with the free surface, the body surface in many cases of wave–structure interaction can be regarded as fixed and rigid. Even when it is in motion and deformation, the level of body distortion is generally far less than that of the free surface. Therefore, the tracking of body surface is relatively easy. The main challenge is the accurate representation of body surface geometry, on which the correct boundary conditions must be accurately implemented.

Cut-cell method: One way of representing a body surface in a Cartesian grid is to use the cut-cell method that approximates the body curvature by a series of linear segments, similar to the MAC method approximating the free surface. As a result, the fluid cell closest to the body is modified from a cube to a polygon whose shape is determined by the actual body geometry.

The distances between the body surface to the cell center and cell faces can be calculated precisely. When the velocity and pressure boundary conditions are applied on the solid boundary, interpolation or extrapolation is used so that the distance information is incorporated appropriately, again similar to the "irregular star" method used in the MAC method. Many of NSE solvers for wave problems use the cut-cell method to represent arbitrary body shape, and recent examples include Li *et al.* (2004). The use of cut-cell is similar to the use of unstructured boundary-fitted grid on the body surface, but it avoids the complicated process of mesh generation. However, since the transition between the boundary grid and the nearby Cartesian grid can be abrupt in the cut-cell method, numerical accuracy may be reduced near the body.

Partial-cell treatment: Another alternative of body surface approximation is to use the volume fraction and line fraction of the solid in a cell to represent body effect on the cell crossing the body surface. This approach is similar to the VOF method on one hand due to the use of cell volume fraction information and similar to the MAC method on the other hand due to the use of the line fraction to represent the cell-crossing information. The method is named PCT by Kothe *et al.* (1991). With such a representation, the actual body geometry in a cell was replaced by the cell partially available to the fluid, and the governing equation was reformulated by treating the "partial cell" as a special porous cell with a particular value of porosity (Liu and Lin, 1997). This is different from the traditional cut-cell treatment, in which the FD scheme (not the governing equations) is changed according to the cell shape. The wave models adopting this idea include COBRAS (Liu *et al.*, 1999b), NEWFLUME (Lin and Xu, 2006), and NS-MAC NWT (Park *et al.*, 1999).

Other embedded boundary methods: Sometimes, the cut-cell method, the PCT, and other boundary-fitted methods are also called embedded boundary method, which is the opposite of its counterpart of stair-step representation of the body surface in Cartesian grid system. For a moving solid surface, this method can be extended to adaptive embedded boundary method or hybrid embedded boundary method such as Chimera method introduced before that overlaps the unstructured boundary-fitted mesh around the structure on the background Cartesian grid system.

Porous-cell method: Although the PCT has the advantage of approximating irregular surface geometry by the cell volume fraction, it is not flexible enough due to the introduction of a linear fraction to represent cell boundary-crossing information. This limits its application where the fluid–body interface has a large deformation, e.g., a body with an elastic surface, a bed that evolves in time. Recently, the author and his research group have been developing a so-called porous-cell method to treat fluid interaction with both solid body and porous mediums. The method employs a unified approach of volume fraction of the solid material in the cell to represent the porous medium or the solid body, an analog to the VOF method in representing the free surface. As a result, it is easy to update this fraction to track the deformation of the body surface or the evolution of a porous bed. We shall provide more details in Section 6.5.1.3 for this treatment when NEWTANK is further elaborated.

Virtual boundary force (VBF) method: An alternative way of handling the irregular boundary is to reformulate NSEs by explicitly including a VBF as the source term in the momentum equations. This approach is named as "VBF" method in this book, and a similar version of the method has been called IB method or other names in other studies. The entire idea behind the VBF method is that the presence of a body in fluid flow is equivalent to imposing appropriate reaction forces along the solid boundary, so that the fluid comes to stop on the solid surface. By applying this force, the boundary conditions on the body surface are no longer needed. The force being applied on the solid boundary is a Dirac delta function of the ambient flow and body configuration.

Peskin (1972) was the first person to introduce the concept of "IB method" to solve blood flows. In his treatment, the tension forces are imposed on the surface of the vessel to reflect the influence of vessel stiffness on the blood flow and to calculate the deformation of the vessel. The force is iterated based on the feedback of velocity computation. Later, Mohd-Yusof (1997) introduced the direct forcing immersed boundary (IB) method, in which the force is calculated directly without iteration. The original IB method was revised by Leveque and Li (1994) as the so-called immersed interface (II) method. By introducing Lagrangian multipliers based on the velocity constraint due to the internal solid boundary, Glowinski *et al.* (1994) introduced the so-called fictitious domain method. Recently, Baaijens (2001) coupled the FD method with the mortar element (ME) method (Ben and Maday, 1997) to simulate the deflection of a slender flexible body under fluid flows.

Most of the previously reported VBF models handled only fully immersed bodies. So far, there have been very few reports of VBF models for the simulation of wave–structure interaction where the free surface comes across the body surface. Only recently Lee and Lin (2005) implemented the VBF method in a σ-coordinate model to study wave diffraction around a large circular cylinder. Berthelsen and Ytrehus (2005) also developed a two-phase model based on IB method to solve partially filled pipe flows. Currently, the author and his research group are developing a 3D two-phase flow model incorporating the VBF and VOF methods to study breaking wave interaction with a body of arbitrary shape. More details of the model will be provided later in Section 6.5.1.2.

In principle, the VBF method can also be applied to other types of wave models such as SWE model, Boussinesq model, MSE model, and wave spectral model. The idea of replacing the boundary condition, especially on an irregular geometry of boundary, with an equivalent "virtual force" is applicable to almost all mechanical problems with complex boundary configurations. In fact, the method has been recently incorporated into the solution of wave equation (Bokil and Glowinski, 2005), Helmholtz equation

(Heikkola *et al.*, 2003), and Maxwell equation (Dahmen *et al.*, 2003) for the study of acoustic and electromagnetic wave scattering.

6.3.5.3 Wave models constructed on a transformed plane

Alternatively, the coordinate transformation may be used to map the irregular computational domain to a regular domain so that the FDM can be constructed on regular Cartesian grids in the transformed plane without any special treatment on the interface. The σ-coordinate transformation is the most commonly used method in many water wave models because it is simple yet capable of mapping the space between the uneven bottom and the wavy free surface in a cubic domain (e.g., Casulli, 1999; Lin and Li, 2002). Other transformations (e.g., elliptic grid generation method) are also possible but they are less frequently used in wave modeling (e.g. Hodges and Street, 1999).

The obvious limitation of the σ-coordinate transformation is that it is unable to handle depth discontinuity and structures with sharp corners, both of which result in the singularity of the horizontal gradient of σ that is required in the computation. The use of other coordinate transformations such as the elliptic grid generation method in principle allows the inclusion of sharp corners of the structure but the transformation may be too involved when the changing free surface is considered. So far, there have been few examples of using the wave model on the transformed plane to simulate wave–structure interaction. The exception is the recent work by Lin (2006), who proposed the use of multiple-layer σ-coordinate transformation at the depth discontinuity (e.g., step) or structure corners. In his study, a few demonstrations were made for 2D and 3D wave interaction with submerged, immersed, and floating structures.

6.3.6 Models based on meshless particle methods

6.3.6.1 Smoothed particle hydrodynamics method

SPH is basically a meshless particle Lagrangian method solving NSEs. It is one of the most popular meshless methods employed to solve NSEs. Although this method has been proposed for over three decades, its application to free surface flow problems started quite recently. Monaghan (1994) was the first to apply SPH into solving free surface problems. The method was extended to 3D for modeling casting processes by Cleary *et al.* (2002) and extended to non-Newtonian flows by Shao and Lo (2003). The method was also coupled with turbulence models and used to simulate breaking waves (e.g., Gotoh and Sakai, 2006; Shao and Ji, 2006). So far, SPH has demonstrated good capabilities for modeling 3D fluid flows with highly distorted free surfaces. Considering its flexibility of handling complex body configurations, SPH can serve as a good alternative for modeling wave interaction with structures. However, no advantage in terms of computational efficiency has been reported for this method compared with the conventional mesh-based NSE solvers.

6.3.6.2   Other meshless methods

Besides SPH, NSEs can also be solved with the use of other meshless methods. For example, with the use of MPS, a sister version of SPH, Chikazawa *et al.* (2001) simulated wave propagation and interaction with a breakwater. Idelsohn *et al.* (2004) developed a particle FEM to simulate breaking waves. Ma (2005) proposed a model that solves the Euler equations using the MLPG method for the simulation of nonlinear water waves.

6.3.7   Other types of models

6.3.7.1   Discrete vortex method

There are other equivalent formulations of NSEs in terms of derived variables (Section 2.1.2). One example is the formulation based on stream function and vorticity for 2D problems, i.e.:

$$\nabla^2 \psi = -\omega \tag{6.57}$$

$$\frac{\partial \omega}{\partial t} + \mathbf{u} \cdot \nabla \omega = \nu \nabla^2 \omega \tag{6.58}$$

Such a formulation has the advantage of solving FSI, during which vortices are always generated on the solid boundary that has a constant value of stream function.

When such a formulation is solved numerically, the so-called discrete vortex method (DVM) or other kind of vortex method is often employed (e.g., see Clarke and Tutty, 1994). In the numerical method, the continuous vorticity field is represented by a finite number of discrete vortices that are generated from solid boundaries and decay by diffusion in the process of convection. When the equations are solved in the Eulerian frame, the vortex-in-cell is used so that new vorticity is introduced at each cell on the solid boundary to ensure the local no-slip boundary condition (e.g., Meneghini and Bearman, 1993). When the Lagrangian method is adopted, the vorticity field is represented by many particles with different circulations. The solution procedure is to track these particles' positions and to determine their changing vorticity strengths (e.g., Chang and Chern, 1991).

The DVM is similar to DNS in the sense that it resolves the finest structure of vortices generated from the boundary layer. It has the advantage of simulating fluid flow interaction with moving bodies. The main drawback is that the computational effort will increase rapidly with the increase of flow

Re. Currently, most of the existing applications of DVM are limited to small Re within 1000. There are many other variations of vortex method, e.g., random vortex method, particle strength exchange method, core-spreading method.

6.3.7.2   Lattice Boltzmann method

The LBM solves the discrete Boltzmann equation on the prescribed lattice with the proper particle collision models. By tracking a finite number of particles, the viscous flow behavior will be captured automatically from the intrinsic particle stream and collision process. In fact, through a Chapman–Enskog analysis, one can recover the NSEs from the discrete LBM algorithm, at least on a macroscopic scale. LBM starts from the direct discrete formulation of a mechanical problem based on the fundamental processes of a particle stream and collision. Since LBM does not solve the PDE directly, it is easy to implement, and it is particularly effective in dealing with bodies with complex geometry (e.g., Martys and Chen, 1996). It is also straightforward to incorporate additional microscopic interaction. Besides, since the pressure field is directly available from the density distributions, there is no need to solve the Poisson equation iteratively, a big time-saving step compared with the traditional NSE solvers.

The main challenge in modeling water waves using LBM is the treatment of the free surface. In the last two decades, significant advances have been made to incorporate the LBM in the simulation of multiphase flows, especially those with a large density difference such as gas and liquid. For a small density difference, the collision operator can be modified based on particulate kinetics, and thus, no special treatment is needed to manipulate the interfaces. Such a simplified treatment, however, often results in an interface thickness much larger than that in real fluids. For the multiphase flows with a large density ratio, certain ways of interface tracking are still necessary. One of the popular methods is called free energy method (Swift *et al.*, 1995) that solves a distribution function to track the interface. Such a method is similar to the VOF method (Zheng *et al.*, 2006). Inamuro *et al.* (2004) combined the projection method with the free energy method in their LBM and used the model to simulate capillary waves and air bubbles. In contrast, Ginzburg and Steiner (2003) introduced an antidiffusion algorithm on the free surface to maintain a sharp front and applied the collision only on the nodes for fluids to simulate free surface flows. The model was used to study the filling process in casting. Kurtoglu and Lin (2006) presented a study of the simulation of a 3D air bubble and compared their results with the numerical results from VOF and level set methods.

With a similar idea and formulation, LBM can be extended to describe other fluid motion processes. For example, Gunstensen and Rothman (1993) employed LBM to study flows in porous media. Yan (2000) used LB equation and Chapman–Enskog expansion to obtain 1D and 2D wave equations and used the LBM to simulate wave motion. Marcou *et al.* (2006) validated their two-phase LBM model against the Saint-Venant equations for the open channel flows. There is the potential that the LBM can be proven to be equivalent to other wave models such as a Boussinesq model or an MSE model. Although so far there has been no report of a workable model based on LBM for the simulation of realistic wave–structure interaction, it is expected that it will not take too long for it to appear, given the active research and fast advancement in this area.

6.4   Numerical simulations of wave–structure interaction

In this section, we shall briefly review the numerical studies made previously on the modeling of wave–structure interaction. In coastal engineering, the majority of structures (e.g., breakwaters, seawalls, groins) can be treated as stationary. The main focus of the study is then on the wave transformation caused by the presence of the structure. On the other hand, in offshore engineering and naval architecture, the structures [e.g., tension-leg platforms (TLP), semisubmersibles, FPSOs, ships] are not stationary. The main emphasis of the study then is on the determination of wave loads and the corresponding structure motions.

6.4.1   Wave transmission, dissipation, and reflection over submerged, immersed, or floating structures

In evaluating the performance of a breakwater (porous or impermeable), two main measures are the transmission and reflection coefficients, respectively. These two coefficients define the effectiveness of the breakwater in blocking the wave energy as well as in reflecting back the wave energy to the offshore. The smaller the transmission coefficient is, the less the wave energy can penetrate into the protected area. This will generally in turn increase the reflection coefficient, which sometimes is also unwanted if a waterway is in the vicinity. In this case, a compromise must be made between these two coefficients. In addition, some engineering measures (e.g., change of porous material or shape of breakwater, induction of local wave breaking) can be employed to increase the local energy dissipation.

6.4.1.1   Reflection, transmission, and dissipation coefficients

Although the concept of wave transmission and reflection is very simple, there are no simple formulas to calculate them based on the known information of wave conditions and structure properties. In general, these coefficients are functions of many influential factors, i.e.:

$$\text{RTD} = f\left(\frac{H}{h}, kh, \frac{a}{h}, \frac{b}{h}, \frac{d_{50}}{h}, n, \text{Re}\right) \tag{6.59}$$

where RTD represents reflection, transmission, and dissipation coefficients, a and b are the effective structure length and the height of the breakwater, respectively, and d_{50} and n are the mean porous size and the porosity of the porous breakwater, respectively. The above formula does not include the secondary effects such as the shape of the structure, the packing and gradation characteristics of porous breakwater, etc. Even from the above simplified version of the formula, we may find that the function is composed of seven independent variables, and it is not straightforward to have the closed-form expression. Another reason that prevents the derivation of such a formula is the occurrence of wave breaking in a certain range of parameter combinations.

As a compromise, people only manage to establish empirical formulas for a limited range of parameter values or to derive the analytical expressions based on certain assumptions. For example, Battjes (1974) proposed an empirical formula to estimate the reflection coefficient from a slope; Lin and Liu (2005) derived a closed-form expression for long-wave transmission and reflection from a trapezoidal breakwater. Once the problem is outside the parameter range or the assumed condition, one may have to conduct new experiments (e.g., Chang *et al.*, 2001 for a solitary wave past a submerged block) or numerical modeling (e.g., Isaacson *et al.*, 2000 for wave interaction with slotted barriers).

Recently, with the use of a solitary wave, which reduces the number of wave-related parameters to only one (i.e., H/h for wave nonlinearity), Lin (2004b) and Lin and Karunarathna (2007) conducted numerical experiments for wave transmission and reflection from an impermeable and a porous breakwater, respectively. The full range of a/h, b/h, and d_{50}/h has been investigated to produce the database for engineering usage. Although some factors are still not completely studied and the functional expression of (6.59) is not sought, the simulation covers parameter ranges much larger than previous studies. These data, if properly analyzed and utilized [e.g., by artificial neural networks (ANN) or the genetic algorithms the (GA)], can possibly generate simple and useful formulas for practical designs.

6.4.1.2 *Harmonic generation for periodic waves over a submerged structure*

When a wave train propagates above a submerged obstacle (e.g., breakwater), the wave nonlinearity increases near the crown of the object where the local still wave depth reduces. With the increase of wave nonlinearity, nonlinear wave interaction will be enhanced and accompanied by active harmonic generation. This process redistributes wave energy into the higher frequency range of the transmitted waves. Above the obstacle, the higher harmonics are bound with the incident wave, similar to the bound infragravity wave. Past the obstacle, these bound harmonics are released as free higher frequency waves. The actual amount of energy redistribution depends on the incident wave and the actual shape of the submerged obstacle. The understanding of the nonlinear process has important applications to surf zone dynamics and local structural stability.

For a simplified obstacle shape, an analytical solution may be possible. For example, using the second-order Stokes wave theory, Massel (1983) developed a theoretical work for nonlinear wave decomposition above a deeply submerged finite or infinite rectangular step. In most of the practical problems where the obstacle has a more complex shape or is made of a porous medium, the harmonic generation can be obtained only through either laboratory experiment or numerical computation. For example, Beji and Battjes (1993) conducted laboratory experiments to investigate wave transformation over a submerged bar of trapezoidal shape. Losada *et al.* (1997) extended the experimental study to harmonic generation past submerged porous steps. Ohyama and Nadaoka (1994) developed a BEM model and used it to investigate the nonlinear wave transformation past submerged shelves. Recently, Lin and Li (2002) used a σ-coordinate NSE model to simulate nonlinear wave trains past a submerged impermeable breakwater. Garcia *et al.* (2004) employed COBRAS to simulate wave transformation over porous breakwaters. Shen *et al.* (2004) proposed another VOF model to study wave propogation over a submerged bar.

6.4.1.3 Fission for solitary waves over a submerged structure

When a solitary wave passes over a submerged step or an obstacle, the solitary wave may split into a few smaller solitons. This process is called fission, which is the antonym of "fusion" that is used to describe the merge of a few solitons into a larger one. Solitons are found in many physical phenomena, including nonlinear dispersive water waves. A solitary wave is often used to represent the leading tsunami nearshore. The study of fission problems has a practical implication of interpreting how a tsunami leading front transforms when it passes over a continental shelf. Tanaka (1986) presented a theoretical study of solitary wave instability over a changing topography. Later, Seabra-Santos *et al.* (1987) and Losada *et al.* (1989) conducted experiments to investigate solitary wave transformation over a step or an obstacle. The experiment was furthered by Chang *et al.* (2001) using PIV for the more detailed vortex structure measurements during solitary wave passage over an obstacle. On the other hand, Liu and Cheng (2001) employed COBRAS to study solitary wave fission over a shelf. Lin (2004b) made a more complete study of solitary wave transformation over a rectangular obstacle with its length varying from zero to infinity. Figure 6.5 shows an example of the solitary wave past a submerged obstacle, during which wave reflection occurs when the wave front passes the frontal and rear corners of the obstacle.

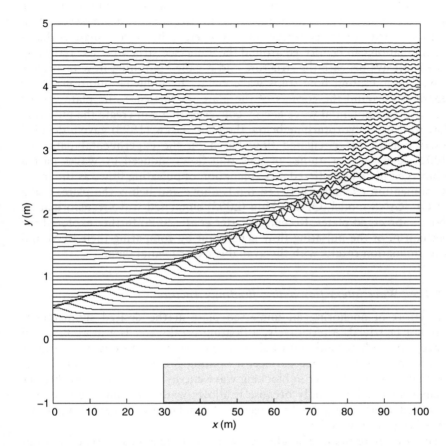

Figure 6.5 Simulation results of a solitary wave ($H = 0.1$ m and $h = 1.0$ m) past a submerged rectangular obstacle (40 m × 0.6 m) using NEWFLUME; the wave profiles from bottom to top have a time interval of 1.0 s.

6.4.1.4 Pulsating flow for periodic waves over an immersed horizontal plate

It has been found that an immersed horizontal plate can behave similarly to a conventional bottom-seated breakwater. The early analytical solution of wave transmission and reflection from a submerged horizontal plate was made by Burke (1964). Patarapanich and Cheong (1989) conducted a series of laboratory experiments to investigate transmission and reflection characteristics of regular and random waves past a submerged horizontal plate. It is found that there is an optimal range of submergence of $0.05h$–$0.15h$, during which wave energy is largely dissipated by wave breaking. Parsons and Martin (1992) formulated the problem with the use of a hypersingular integral equation, from which the transmission and reflection coefficients can be solved. Compared with the traditional bottom-seated breakwater, the plate-type of breakwater can be material-effective because it can be very thin.

Another extraordinary feature of a submerged horizontal plate is that a net circulation can be formed around the immersed structure. The returning flow is always beneath the structure and is called pulsating flow or reverse flow. This pulsating flow traps wave energy by inducing additional flow separation around the corners of the plate. In this case, the plate can be regarded as a potential wave energy converter that can create the pulsating flow to drive a hydroturbine below the plate. Graw (1993) gave a detailed description of the circulated flow based on the experimental data. Yu (2002) reviewed the studies of pulsating flows induced by waves past a submerged plate and offered an explanation of pulsating flows as the result of momentum exchange between fluids above and below the plate. Using a VOF RANS model, Qi and Hou (2003) simulated wave transformation above submerged plates. Recently, Carter *et al.* (2006) further explored the mechanism of the pulsating flow generation by comparing their BEM and NSE simulation results with the available experimental data.

6.4.1.5 *Wave interaction with a floating structure*

The effort of anchoring an immersed plate at the desired submergence may not be trivial. Alternatively, a floating breakwater is a more practical concept in both coastal and offshore engineering. Because wave energy is concentrated near the free surface, a floating breakwater is naturally the most efficient breakwater in blocking wave energy. This type of breakwater is environmentally friendly because it allows water exchange between the protected area and the offshore region. A similar case in nature would be wave interaction with ice sheets.

The concept of floating breakwater was proposed almost two decades ago (e.g., Williams *et al.*, 1991). Most of the floating breakwaters are flexible and moored to the seabed by mooring lines. In this case, the mooring force is one of the major concerns in the design (Sannasiraj *et al.*, 1998). There are many types of floating breakwaters, e.g., Ponton type (Sannasiraj *et al.*, 1998), porous block-type (Wang and Ren, 1993), membrane type (Williams, 1996), cage type (Murali and Mani, 1997), spar buoy type (Liang *et al.*, 2004), hybrid type with both membrane and rigid components [e.g., rapidly installed breakwater system (RIBS), Briggs *et al.*, 2002], pneumatic type (Koo *et al.*, 2006). Recently, with the use of COBRAS, Koftis *et al.* (2006) simulated wave transmission and reflection from a fixed floating breakwater and compared their results with the experimental data. Note that when the structure size becomes much larger than the wavelength, the structure is often called VLFS. During wave interaction with a VLFS, both wave diffraction and structure deformation are significant, and the analysis must include both hydrodynamics and the elastic property of the structure. Such an analysis is called "hydroelastic" analysis.

6.4.2 Wave reflection, run-up, and overtopping a surface-piercing structure

There are generally three scenarios where waves overtopping a structure must be considered. One is for a surface-piercing breakwater that is designed to completely block waves outside the protected area in normal wave and tidal conditions. In the extreme case such as high tide and/or large storm waves, part of the waves can overtop the crown of the breakwater and transmit both mass and energy into the protected area. Another situation is for a seawall that is built along the coastline to prevent a large wave or tsunami attack. The height of a seawall is often designed to allow a certain amount of wave overtopping to take place during the extreme condition. If the waves are not high enough, wave run-up will then take place, followed by run-down and wave reflection, a physical process similar to waves on a beach. The third case would be for a deck above the sea level (e.g., offshore platform or ship deck) subjected to large wave attacks. A great amount of water will overflow the deck, and the process is often called "green water" effect, which is similar to waves overtopping the crown of a breakwater or a seawall.

6.4.2.1 Wave run-up and reflection

Wave run-up and reflection without overtopping occurs when the maximum wave height is less than the crown of a structure. Although a few empirical formulas exist for wave run-up and reflection from a slope (see Sections 3.8.1 and 3.9.5), their reliability is still subject to further verification. Alternatively, a theoretical approach or a numerical simulation can be made. One of the classical analytical solutions was proposed by Carrier and Greenspan (1958) to calculate wave run-up and reflection from a linear slope based on the nonlinear shallow-water approximation. Although progress has been were made since then, the analytical solutions are still limited to relatively simple geometry.

In contrast to the analytical approach, a numerical approach is more flexible in dealing with irregular geometry, permeability, roughness, porosity, etc. of a sloping structure surface. For example, Zelt (1991) used a Boussinesq model to simulate solitary wave run-up on a plane beach. Titov and Synolakis (1998) employed a 2D SWE model to simulate tsunami run-up on beaches. With the use of a RANS model, Lin *et al.* (1999) simulated breaking and nonbreaking solitary wave run-up on a beach. For periodic and irregular wave trains, Kobayashi (1999) used an SWE model to study the run-up and reflection from impermeable slopes. Although a lot of RANS models are capable of simulating regular or irregular wave run-up and reflection from a beach (e.g., Lin and Liu, 2004), so far there have been few reports of dedicated studies of periodic wave run-up and reflection from a slope using such models.

6.4.2.2 Wave overtopping

When the incident wave height exceeds the crown of a structure, wave run-up will continue to cause waves overtopping the structure. This is an important phenomenon in the design of a coastal or an offshore structure. Although there are a few empirical formulas (e.g., Owen, 1980; van der Meer and Janssen, 1995) for estimating the mean overtopping rate, numerical modeling is often a good alternative when a new type of structure profile is considered. Most of the earlier modeling efforts employed SWE models. Examples included Kobayashi and Raichle (1994) for simulating irregular waves overtopping coastal structures, van Gent (1994) for simulating waves action on permeable or impermeable structures, Hu *et al.* (2000) for simulating waves overtopping seawalls, and Hubbard and Dodd (2002) for simulating wave run-up and overtopping using a 2D adaptive mesh model. Although these SWE models have been successful to a certain extent, their prediction accuracy degenerates when the vertical acceleration is significant (e.g., structures with steep frontal slope) or the wave breaking is strong.

The models based on the NSE solvers can better represent the physical processes. Liu *et al.* (1999a) employed a RANS model to simulate waves overtopping a seawall protected by porous armor layers. Figure 6.6

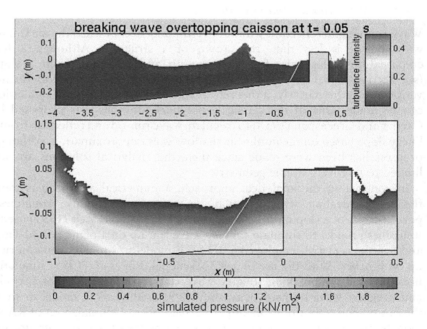

Figure 6.6 Simulated breaking wave overtopping a seawall protected by the porous armor layer (between the seawall and the inclined white line); the top view shows the simulated intensive turbulence and the bottom view shows the pressure distribution.

shows the simulated wave overtopping process using this model. Li *et al.* (2004) developed an LES model to simulate waves overtopping a sea dike. Recently, Reeve *et al.* (2007) employed a RANS model to simulate irregular waves overtopping an impermeable seawall. Besides better representing real physics, the models solving RANS equations have another advantage over the SWE models in that the wave forces on the structure can be directly obtained based on the surface integration of calculated fluid pressure. This is important if the impulsive wave loads are evaluated.

6.4.2.3 Green water

Although green water on an offshore structure is similar to a wave overtopping a coastal structure in nature, differences do exist between these two phenomena. The main difference is that a wave overtopping a coastal structure is often a process continued from wave run-up on a slope or a vertical wall, whereas green water often takes place separately without wave run-up. The main reason is that an offshore structure, below the platform, has an open space that allows waves to pass through. This will effectively reduce the maximum overtopping rate but may induce dangerous uplift force when a large wave attacks the platform. In another event, for ships or other floating vessels, the hull shape is often curved up from the bottom to the deck, which is equivalent to the negative slope angle. This will also reduce the amount of overtopping water.

Compared with the wave overtopping a coastal structure, the detailed characteristics of green water on an offshore platform or a ship deck have been studied less. Only recently, the quantitative study, experimentally or numerically, of green water has become intensive. For example, Cox and Ortega (2002) conducted experiments to measure the free surface and velocity profile in waves overtopping a deck. By employing the SPH technique, Gomez-Gesteira *et al.* (2005) simulated green water overtopping decks of offshore platforms and ships. Figure 6.7 shows a simulation of a deep-water breaking wave impinging on a spar platform using NEWTANK. It is expected that more research will be carried out in this area with the use of NSE-solver models.

6.4.3 Wave diffraction around large bodies

When the structure size is much larger than wavelength, wave diffraction can become important. The wave direction will be modified in the diffraction process and wave energy will be redistributed among different directional components. In the study of wave diffraction, there are two categories of interest. One is of major concern in wave transformation during wave diffraction and the other is of concern in diffracted wave forces on the structure.

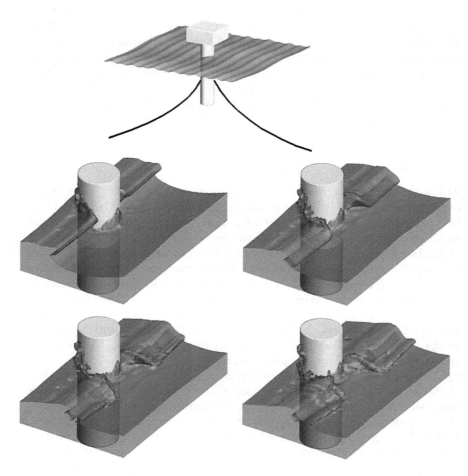

Figure 6.7 Breaking wave impinging on a spar platform; the wave in deep sea
has $H = 15\,\mathrm{m}$, $T = 6.12\,\mathrm{s}$, and $L = 58.5\,\mathrm{m}$ and the spar has the
diameter $D = 20\,\mathrm{m}$ (top). Schematic drawing of the spar platform;
(bottom) simulation by NEWTANK.

6.4.3.1 *Wave diffraction around a rigid body*

When waves pass a fixed structure such as a breakwater, a large cylinder,
or an island, wave diffraction takes place. There have been many theoretical
studies of wave diffraction around large bodies, provided the body geometry
is simple and the incident waves are linear and chromatic. However, for
more general shaped bodies and/or nonlinear waves, numerical modeling may
have to be employed. Many wave models are qualified to provide reliable
results for wave diffraction patterns. For structures with vertical surfaces
(e.g., piles and breakwaters), the depth-averaged models (e.g., MSE model

and Boussinesq model) can be used. For 3D bodies (e.g., ship hulls), only BEM models and NSE models can be employed.

One of the commonly studied problems is to model wave diffraction patterns in a real harbor. For example, for short wave diffraction around an arbitrary island, Lo (1991) proposed an MSE model. Xu *et al.* (1996) developed another MSE model for the study of harbor waves. Li *et al.* (2000) employed a Boussinesq model to simulate random wave diffraction behind breakwaters. Woo and Liu (2004) developed a FEM Boussinesq model to study harbor oscillations. For long-wave (e.g., tsunami) diffraction and refraction around an island, Titov and Synolakis (1998) employed a SWE model.

6.4.3.2 *Diffraction force on a large body*

The calculation of wave force on a large body can be derived from the wave diffraction pattern. In fact, this is one of the most important hydrodynamic measures in designing an offshore structure and a ship hull. Currently, the established way of performing such a calculation is primarily based on the BEM models for potential flows. For a practical 3D problem, almost all the approaches using BEM models introduce various simplifications for the nonlinear free surface boundary conditions. For example, when only the first-order linear wave forces are required, the solution procedure can be linearized (Section 6.2.1.2). When the higher order (mainly second-order) nonlinear wave forces are needed, the perturbation technique (Section 6.2.2.2) can be employed. Although it is possible to recover the time-domain nonlinear wave force on the structure (e.g., Isaacson and Cheung, 1992) based on the perturbation technique, such an approach is unable to solve a fully nonlinear problem in which the free surface may turn over and exert impulsive impact force on the structure. Some of the well-known commercial software of this kind include WAMIT, MOSES, Nauticus Hull, etc. (see Section 5.2.3.3). So far, only a few studies have been made for simulating wave diffraction force on a large body using the NSE models (e.g., Li and Lin, 2001; Lee and Lin, 2005). It is expected, however, that more studies of this kind will take place in the future.

6.4.3.3 *Hydroelastic analysis for wave diffraction around an elastic large body*

Rigid-body assumption is applicable only to relatively small structures with large stiffness. When the body size becomes many times the wavelength's or the structural stiffness is small, the elasticity of the body must be considered in the modeling of wave–structure interaction. Such an analysis is often termed hydroelastic analysis. In the analysis, both the structural deformation and the wave diffraction around the structure must be simultaneously considered. In offshore engineering, one example that requires hydroelastic analysis is VLFS. We have discussed the closed-form analytical solution based on the hydroelastic analysis for some simple VLFS (mainly axisymmetrical thin plate, i.e., circular plate) in Section 3.11.3. For more general shaped structures, numerical analysis is needed. For example, using eigenfunction expansion, Wu *et al.* (1995) analyzed the elastic response of a floating plate to wave action. Ertekin and Kim (1999) studied the hydroelastic response of a mat-type structure in shallow water subject to obliquely incident waves. Combining the BEM model for fluid and the FEM model for structure, Liu and Sakai (2002) simulated the time-domain elastic response of a floating structure to waves. The hydroelastic analysis can also be conducted for other types of structure (e.g., Lu *et al.*, 2000 for the study of hydroelastic response of structures during water impact) or liquid sloshing in an elastic container (e.g., Bermudez and Rodriguez, 1999).

6.4.4   *Wave loads on small fixed structures*

When the body size is small compared with the trajectory motion of fluid particles under waves, the viscous effect becomes important, and the potential flow assumption is no longer valid. Flow separation takes place on the surface of the body and forms the wake region on the lee side, manifested by the presence of vortex generation and shedding. In this case, the correct calculation of wave forces on the structure relies on the reasonable description of the vortical motion in the wake region, which directly affects the net force on the body. For a structure with a simple shape, the empirical Morison equation is a useful tool for calculating such forces. For a structure with a complex shape exposed to a strongly nonlinear wave field, the numerical approach based on NSE models can provide a more accurate computation of wave forces.

To simulate wave action on structures with the presence of strong turbulence, the numerical model based on the RANS equations or LES is the most appropriate choice. In many cases, the wave forces on the structure are always related to structure stability analysis, which in earlier days was mainly done with the use of empirical formulas established on a limited number of experimental tests [e.g., the empirical formulas of van der Meer (1988) for the stability analysis of rubble mount breakwaters]. With the use of NSE models, it is possible to calculate the wave forces acting on the structure directly by integrating the simulated pressure around the body surface. For example, van Gent (1995) developed an NSE model to simulate wave force on both impermeable and permeable structures. Liu *et al.* (1999a) proposed a RANS model to simulate wave action on a seawall. Lin and Li (2003) developed a 3D LES NSE model to simulate the combined wave and current force on a vertical square cylinder. Figure 6.8 shows the example of the simulated vortical structure and the free surface profile around the cylinder at two different instances.

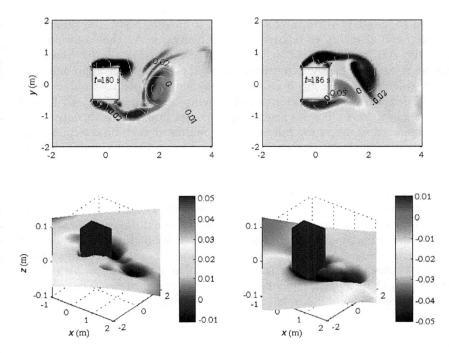

Figure 6.8 Simulation results of the vortical structures on the middle water level (top) and the associated free surface profiles (bottom; color scale for free surface displacement in meters) around a rectangular cylinder (1 m × 1 m) under a combined current ($U = 0.6$ m/s and $h = 1.0$ m) and wave ($H = 0.05$ m and $T = 4.0$ s) action; the time interval between left and right plots is 6.0 s.

6.4.5 *Wave interaction with moving and/or flexible bodies*

The motion of a body during wave–structure interaction adds difficulty to numerical modeling. To handle the body motion in the simulation, there are generally four types of numerical treatments, namely adaptive mesh or its kind, the cut-cell method or its kind (this is mainly applicable to NSE models), moving coordinates, and the perturbation technique (this is mainly applied to potential flow BEM models). In this section, we shall discuss all these methods.

6.4.5.1 *Adaptive mesh approach*

The adaptive mesh method is the most general numerical treatment for a moving body. In this approach, the problem is solved in the physical domain with the use of an adaptive boundary-conformal mesh, at least near the moving body. The method is able to deal with both body motion and body deformation.

Fully nonlinear BEM model: With a fully nonlinear BEM model that is able to simulate wave interaction with a moving body, the generation of adaptive elements is relatively straightforward. As the body surface and the free surface move in time, the new location of the boundary elements can be readily determined using the Lagrangian approach. For 2D problems, this idea has been successfully implemented to simulate wave paddle movement and other types of moving boundaries (e.g., Grilli and Subramanya, 1996). For 3D problems, the major limitation is the large CPU time required to invert the full and dense matrix at each time step. So far, there have been few reports of a BEM model capable of simulating fully nonlinear wave interaction with a 3D moving body with complex surface configuration.

NSE-solver models: With the use of the NSE models, the remeshing process can be computationally expensive for a 3D moving body in fluid. The remeshing could be done in the entire computational domain at each time step with the updated body surface information. Such an approach is called g*lobal remeshing*. Apparently, the full automation for the remeshing process without human interference is needed to simulate a practical problem. This can be very challenging if the body motion is fast and arbitrary. The commercial software FLUENT is able to treat the moving and deforming domain by using the adaptive mesh dynamically generated by the global remeshing process. The technique can be applied to the study of FSI. However, further development and tests for more complicated problems of surface wave interaction with 3D moving bodies are still needed. If the body movement is small, especially periodic, the ALE method can be used to minimize the effort of remeshing process.

Alternatively, one can also use *chimera method* to treat a moving body. In this method, two mesh systems are used, with one being fixed and the other being conformal to body surface and moving with the body. With the use of chimera grid, there is no need to perform remeshing at each time step if the body motion is in simple translation and rotation without deformation because the same boundary conformal mesh can be used at different locations and orientations. The changed location of the body and thus the boundary conformal mesh at each time step will be reflected by different crossing information between the background grid and the moving mesh. Interpolation will be used to exchange the flow information between the two grid systems. This method has been successfully used to simulate a moving ship and the generated ship waves in a harbor (Chen and Chen, 1998). Note, however, that the simplicity of this method is applicable only to rigid bodies without surface deformation.

6.4.5.2 *Fixed Cartesian grid method*

The cut-cell technique can be used to represent an irregular body surface. By using this method, the advantage of the Cartesian grid can be retained and the FD scheme only needs to be modified at the boundary cell. The similar numerical schemes include PCT and the porous-cell method (Section 6.3.5.2). When a moving body is considered, some kind of body surface tracking is needed to update the location of the body and the information of the solid surface crossing through the Cartesian grid. With the use of cut-cell method, Heinrich (1992) employed the modified NASA-VOF2D to simulate water wave generation by landslides. Adopting an idea similar to the VOF method, Xiao (1999) used the color functions to represent a body. The convection equation for the color function was solved in the Eulerian frame to update the body location at each time step. In contrast, with the use of PCT, Lin (2007) proposed the LRS method to handle a moving body in a fixed Cartesian grid, where the rigid-body motion is tracked by the Lagrangian method. Figure 6.9 shows one of the case studies using this method for modeling the impact and entry of a cylinder into water.

Another approach of simulating a moving body in a fixed Cartesian coordinate is based on the VBF approach. In this approach, the numerical solution is sought in the entire computational domain, even inside the solid body. The VBF is applied along the original body surface to enforce the fluid velocity to be zero there (no-slip boundary condition). Some of

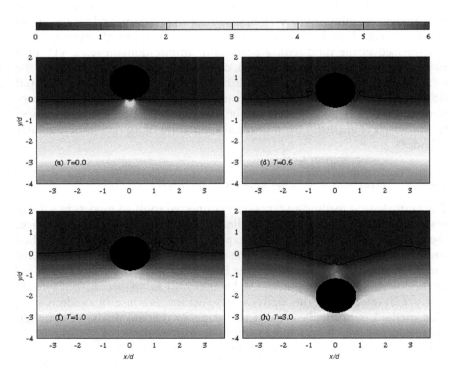

Figure 6.9 Simulation results of a cylinder slamming and entering into water (color scale for pressure).

the representative works of this kind include the IB method (e.g., Lai and Peskin, 2000), the II method (e.g., Leveque and Li, 1994), and the fictitious domain method (e.g., Glowinski *et al.*, 2001), all of which have the capability of treating moving boundaries. Recently, Fadlun *et al.* (2000) developed a second-order-accurate IB method for simulating unsteady 3D incompressible flows and their interaction with moving bodies with complex geometries.

6.4.5.3   *Moving coordinate approach*

Constructing the coordinate frame on a moving body avoids remeshing. This is a natural choice if the body is moving at a constant speed in an infinite domain. In this case, the moving body in quiescent fluid is theoretically equivalent to a stationary body in fluid that moves in the opposite direction. However, constructing the coordinate on an arbitrarily moving body implies that the problem should be recast into a noninertial reference frame that requires the conversion of the original governing equations for the fluid to new equations including the additional noninertial effects. This method loses its advantage when multiple bodies or a single body with deformable surface are present. Usually, this approach is used only under two circumstances, i.e., when the fluid is confined inside a rigid container (e.g., liquid sloshing in a tank) and when a single moving body is immersed in the fluid with a practically infinite domain (e.g., VIV of a riser under current action). In both cases, the moving coordinate can be established following the body motion. For liquid sloshing, both potential flow BEM models and NSE models can be used, whereas for an oscillatory body in fluid only NSE models can be used due to the presence of vortices (e.g., Dütsch *et al.*, 1998 for numerical simulation of an oscillatory circular cylinder).

6.4.5.4   *Perturbation or other simplified methods*

The perturbation method so far is only connected to BEM models in solving wave-induced structure motion. Similar to the treatment of the free surface, the body location and motion are expressed as functions of the perturbed small variables in various orders. The solution is solved with the boundary condition applied on the still water level and mean body location. This methodology is popularly adopted by many BEM (panel method) codes that are developed for the analysis of structure response in random seas. This method is normally applied to frequency-domain analysis, although the results can also be extended to time-domain analysis.

The typical output from this kind of model includes (1) the added mass and moment of inertia coefficients, (2) first-order radiation and wave-drift damping coefficients, (3) first- and second-order (drift) wave forces and moments by each wave component, second-order sum-frequency and difference-frequency wave forces, and moments between any two wave components, and (4) structure responses in terms of first-order RAO, sum-frequency RAO, and difference-frequency RAO (see Sections 3.15.2 and 3.15.3 for the explanation of these forces and motions). One of the representative codes of this kind is WAMIT that can be used to analyze the wave forces on a large floating offshore structure and the corresponding structure responses.

6.4.5.5 Meshless method

In an earlier section, we have introduced the meshless particle method. The main advantage of this type of model is its flexibility in treating highly irregular body surface geometry. It was one of the hot research topics in computational mechanics in the last decade. However, its application in fluid interaction with a moving body just started recently, e.g., Chew *et al.* (2006) developed a meshless GFDM model to simulate incompressible fluid flows around moving bodies. So far, there is no report of a meshless model capable of simulating free surface flow interaction with moving bodies, although the method has a good potential in handling such problems.

6.4.6 Liquid sloshing in partially filled containers

Liquid sloshing is a special type of wave–structure interaction. Compared with the traditional wave–structure interaction where waves are progressive and normally come from a far field source not affected by the local structure, the waves during forced liquid sloshing are confined standing waves that are generated by container motions. We have discussed some analytical approaches for analyzing liquid sloshing in 2D and 3D tanks (Section 3.15.6). Most of these analytical approaches, however, are based on linear wave theory and simple tank geometry.

For the study of nonlinear waves, especially breaking waves, in a tank with complex geometry and excitation, numerical models become an attractive option. For liquid sloshing with nonbreaking free surface, both potential flow models (solved by BEM, FEM, or FDM) and NSE models can be used. For example, Faltinsen (1978) developed a BEM model and compared the numerical results with the linear analytical solution. Based on the potential flow assumption, Wu *et al.* (1998) developed a 3D FEM model and used the model to make a series of 3D simulations of liquid sloshing. Frandsen (2004) developed a FDM that solves the Laplace equation in the σ-coordinate and then employs the model to simulate 2D liquid sloshing in a tank moving horizontally and vertically.

In contrast, Behr and Tezduyar (1994) developed a 3D FEM NSE model to simulate nonbreaking liquid sloshing in a tank subject to vertical vibration. Armenio and La Rocca (1996) developed a 2D FDM RANS model and compared the computational results with those from an SWE model. Celebi

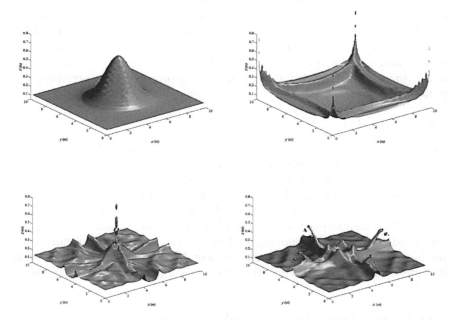

Figure 6.10 Simulated snapshots of broken free surface for liquid sloshing in a square tank.

and Akyildiz (2002) developed a 2D VOF NSE model to simulate liquid sloshing in a tank. Recently, Chen (2005) presented another 2D viscous model to simulate liquid sloshing under the simultaneous excitation of heave, surge, and pitch. Kim *et al.* (2004) developed a 3D viscous model that employs the height function to track the free surface during liquid sloshing. Few reports, however, were made for the NSE models that are capable of accurately simulating 3D liquid sloshing with broken free surfaces except for the recent work by Liu *et al.* (2008). Figure 6.10 shows the simulation results of 3D liquid sloshing in a tank with a broken free surface. The extensive review of liquid sloshing can be found in Ibrahim (2005).

6.5   Benchmark tests

There are many numerical models capable of simulating wave–structure interaction. In this section, we shall present NEWTANK, a generic 3D model for simulating turbulent two-phase fluid flows and their interaction with structures. Details will be provided for the numerical treatment (e.g., porous-cell method and VBF method) of structures in fluids. A series of benchmark tests will be given for the study of various wave–structure interaction problems.

6.5.1 Model description

In Chapter 5, numerical details (e.g., the two-step projection method, k–ε turbulence model, and the VOF method based on PLIC interface reconstruction) have been provided for NEWTANK without the presence of structures (Section 5.5.1.1). In this section, the introduction of NEWTANK is completed by further elaboration of solid boundary treatment and flow computation in a noninertial reference frame.

6.5.1.1 Governing equations

In NEWTANK, the following set of unified equations is solved:

$$\frac{\partial u_i}{\partial x_i} = S \tag{6.60}$$

$$\frac{\partial u_i}{\partial t} + C_c u_j \frac{\partial u_i}{\partial x_j} = -\frac{C_p}{\rho}\frac{\partial p}{\partial x_i} + C_g g_i + \frac{C_v}{\rho}\frac{\partial \tau_{ij}}{\partial x_j} + \frac{C_t}{\rho}\frac{\partial R_{ij}}{\partial x_j} + f_i \tag{6.61}$$

where the coefficients C_c, C_p, C_g, C_v, and C_t are coefficients associated with convection, pressure, gravity, viscous stress, and turbulence stress terms, which are used to account for the porous effect on the flow. In the absence of porous media, $C_c = C_p = C_g = C_v = C_t = 1$. Compared with the conventional RANS equations, the source terms have been included in the RHS of the continuity and momentum equations. The inclusion of the source term in the continuity equation is to generate waves from internal sources (see Section 5.2.1.6). The advantage of generating waves in this way is to avoid the simultaneous wave generation and wave absorption on the same boundary (see Section 5.2.1.6). Besides, it may also be used to simulate other physical processes such as waves generated by underwater explosion. The inclusion of the source term in the momentum equations has multiple purposes (e.g., additional force in the noninertial reference frame, artificial wave damping force in the sponge layer, resistance force in porous media). In this book, we shall only elaborate three types of forces f_i, two of which are related to the solid boundary treatment (VBF and porous-cell methods) and the third related to the force in the noninertial frame.

6.5.1.2 Virtual boundary force method

In the VBF method, the body is removed from the computational domain and replaced by the internal body forces at the original boundary location acting on the fluid flow (see Figure 6.11 for illustration). Now the key issue becomes how the internal VBF can be determined. Apparently, the force is not known a priori before the problem is solved. The obvious physical condition to determine the force is that the velocities at the original location of solid boundaries must vanish (i.e., no-slip boundary condition). In other

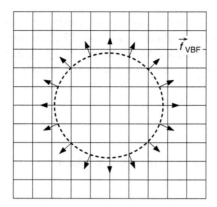

Figure 6.11 Illustration of the VBF method that replaces the boundary condition on the body surface with a virtual boundary force.

words, the VBF should be defined such that the computational results of velocities at all boundaries go to zero at any time. Now let us see how such a force is determined numerically in the model.

In the VBF method, the original NSEs are revised to be:

$$\frac{\partial u_i}{\partial x_i} = 0 \tag{6.62}$$

$$\frac{\partial u_i}{\partial t} + u_j \frac{\partial u_i}{\partial x_j} = -\frac{1}{\rho} \frac{\partial p}{\partial x_i} + g_i + \frac{1}{\rho} \frac{\partial \tau_{ij}}{\partial x_j} + \frac{1}{\rho} \frac{\partial R_{ij}}{\partial x_j} + (f_{\mathrm{VBF}})_i \tag{6.63}$$

where $(f_{\mathrm{VBF}})_i$ is the additional VBF that is used to replace the actual reaction force and is only nonzero on the solid surface. This force is a Dirac delta function in theory but becomes finite in numerical computation that depends on the local discretization, flow characteristics, and boundary configuration. When the above equations are solved on a Cartesian grid, body surface will likely come across the grid lines in various ways. In terms of force computation at each time step based on the no-slip condition for velocity, interpolation is required to connect the information between the grid points where the numerical solution is sought and the nearby body surface. The force can be obtained by iteration with the velocity feedback information, and the method is termed "feedback forcing method" (Peskin, 1972). In contrast, the force can also be obtained by the direct force that will be solved simultaneously with the pressure using the modified PPE. This method is called "direct forcing method" (e.g., Mohd-Yusof, 1997), which is adopted in NEWTANK and elaborated below.

In the first step of the numerical solution, the tentative velocity $\tilde{u}_i$ is sought as usual:

$$\frac{\tilde{u}_i - u_i^n}{\Delta t} = -u_j^n \frac{\partial u_j^n}{\partial x_j} + \frac{1}{\rho^n} \frac{\partial \tau_{ij}^n}{\partial x_j}. \tag{6.64}$$

In the second step, the intermediate velocity field is projected onto a divergence-free plane for the final velocity, i.e.:

$$\frac{u_i^{n+1} - \tilde{u}_i}{\Delta t} = -\frac{1}{\rho^n} \frac{\partial p^{n+1}}{\partial x_i} + g_i + (f_{\mathrm{VBF}})_i \tag{6.65}$$

where:

$$\frac{\partial u_i^{n+1}}{\partial x_i} = 0 \tag{6.66}$$

Taking the divergence of (6.65) and applying (6.66), we have the revised PPE:

$$\frac{\partial}{\partial x_i}\left(\frac{1}{\rho^n}\frac{\partial p^{n+1}}{\partial x_i}\right) = \frac{1}{\Delta t}\frac{\partial \tilde{u}_i}{\partial x_i} + \frac{\partial g_i}{\partial x_i} + \frac{\partial (f_{\mathrm{VBF}})_i}{\partial x_i} \tag{6.67}$$

If $(f_{\mathrm{VBF}})_i$ is known, (6.67) can be solved numerically similar to the NSE model without VBF. By realizing $(f_{\mathrm{VBF}})_i$ is nonzero only at the cells crossing the solid boundary, we can make use of (6.65) and define $(f_{\mathrm{VBF}})_i$:

$$(f_{\mathrm{VBF}})_i = \begin{cases} \dfrac{\hat{u}_i^{n+1} - \tilde{u}_i}{\Delta t} + \dfrac{1}{\rho^n}\dfrac{\partial p^{n+1}}{\partial x_i} - g_i & \text{on or near the virtual boundary} \\ 0 & \text{elsewhere} \end{cases}$$

$$\tag{6.68}$$

Here, we use $\hat{u}_i^{n+1}$ to replace the original u_i^{n+1} in (6.65) to enforce the no-slip velocity boundary constraint at the cells that cross the solid boundary.

Consider the 2D case illustrated in Figure 6.12 where a solid surface crosses through the cell (i,j,k). In this case, the VBF will be applied on the cell face center in the fluid domain nearest to the solid surface (or on the solid boundary if it coincides with the cell face center where the velocity vector is defined). Therefore, the term $\partial (f_{\mathrm{VBF}})_i / \partial x_i$ on the RHS of (6.67) is evaluated as follows:

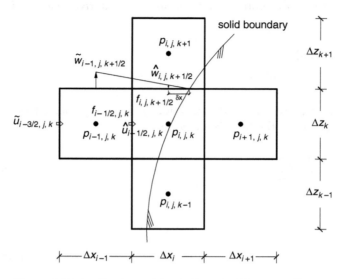

Figure 6.12 Application of the VBF method for a 2D flow computation near a body surface.

$$\left(\frac{\partial f_{\mathrm{VBF}x}}{\partial x}\right)_{i,j,k} + \left(\frac{\partial f_{\mathrm{VBF}z}}{\partial z}\right)_{i,j,k}$$

$$= \frac{(f_{\mathrm{VBF}x})_{i+1/2,j,k} - (f_{\mathrm{VBF}x})_{i-1/2,j,k}}{\Delta x_i} + \frac{(f_{\mathrm{VBF}z})_{i,j,k+1/2} - (f_{\mathrm{VBF}z})_{i,j,k-1/2}}{\Delta z_k}$$

$$= \left[0 - \frac{\dfrac{\hat{u}^{n+1}_{i-1/2,j,k} - \tilde{u}^{n+1}_{i-1/2,j,k}}{\Delta t} + \dfrac{1}{\rho^n_{i-1/2,j,k}}\left(\dfrac{\partial p^{n+1}}{\partial x}\right)_{i-1/2,j,k} - g_{i-1/2,j,k}}{\Delta x_i} \right]$$

$$+ \left[\frac{\dfrac{\hat{w}^{n+1}_{i+1/2,j,k} - \tilde{w}_{i,j,k+1/2}}{\Delta t} + \dfrac{1}{\rho^n_{i,j,k+1/2}}\left(\dfrac{\partial p^{n+1}}{\partial x}\right)_{i,j,k+1/2} - g_{i,j,k+1/2}}{\Delta z_k} - 0 \right]$$

$$(6.69)$$

The other terms in the PPE are discretized as follows:

$$\left\{\frac{\partial}{\partial x}\left(\frac{1}{\rho^n}\frac{\partial p^{n+1}}{\partial x}\right)\right\}_{i,j,k} = \frac{1}{\Delta x_i}\left\{\frac{1}{\rho^n_{i+1/2,j,k}}\left(\frac{\partial p}{\partial x}\right)^{n+1}_{i+1/2,j,k}\right.$$

$$\left. - \frac{1}{\rho^n_{i-1/2,j,k}} \left(\frac{\partial p}{\partial x}\right)^{n+1}_{i-1/2,j,k} \right\}$$

$$= \frac{1}{\Delta x_i} \left\{ \frac{1}{\rho^n_{i+1/2,j,k}} \left(\frac{p^{n+1}_{i+1,j,k} - p^{n+1}_{i,j,k}}{\Delta x_{i+1/2}}\right) \right.$$

$$\left. - \frac{1}{\rho^n_{i-1/2,j,k}} \left(\frac{\partial p}{\partial x}\right)^{n+1}_{i-1/2,j,k} \right\} \tag{6.70}$$

$$\left\{ \frac{\partial}{\partial z} \left(\frac{1}{\rho^n}\frac{\partial p^{n+1}}{\partial z}\right) \right\}_{i,j,k} = \frac{1}{\Delta z_k} \left\{ \frac{1}{\rho^n_{i,j,k+1/2}} \left(\frac{\partial p}{\partial z}\right)^{n+1}_{i,j,k+1/2} \right.$$

$$\left. - \frac{1}{\rho^n_{i,j,k-1/2}} \left(\frac{\partial p}{\partial z}\right)^{n+1}_{i,j,k-1/2} \right\}$$

$$= \frac{1}{\Delta z_k} \left\{ \frac{1}{\rho^n_{i,j,k+1/2}} \left(\frac{p^{n+1}_{i,j,k+1} - p^{n+1}_{i,j,k}}{\Delta z_{k+1/2}}\right) \right.$$

$$\left. - \frac{1}{\rho^n_{i,j,k-1/2}} \left(\frac{p^{n+1}_{i,j,k} - p^{n+1}_{i,j,k-1}}{\Delta z_{k-1/2}}\right) \right\} \tag{6.71}$$

By substituting (6.69), (6.70), and (6.71) into (6.67), we have the discretized form of the revised PPE:

$$\frac{1}{\Delta x_i} \left\{ \frac{1}{\rho^n_{i+1/2,j,k}} \left(\frac{p^{n+1}_{i+1,j,k} - p^{n+1}_{i,j,k}}{\Delta x_{i+1/2}}\right) \right\} + \frac{1}{\Delta z_k} \left\{ -\frac{1}{\rho^n_{i,j,k-1/2}} \left(\frac{p^{n+1}_{i,j,k} - p^{n+1}_{i,j,k-1}}{\Delta z_{k-1/2}}\right) \right\}$$

$$= \frac{1}{\Delta t} \left(\frac{\partial \tilde{u}}{\partial x} + \frac{\partial \tilde{w}}{\partial z}\right)_{i,j,k} + \left(\frac{\partial g_x}{\partial x} + \frac{\partial g_z}{\partial z}\right)_{i,j,k}$$

$$+ \left(0 - \frac{\dfrac{\hat{u}^{n+1}_{i-1/2,j,k} - \tilde{u}_{i-1/2,j,k}}{\Delta t} - g_{i-1/2,j,k}}{\Delta x_i} \right)$$

$$+ \left(\frac{\dfrac{\hat{w}^{n+1}_{i,j,k+1/2} - \tilde{w}_{i,j,k+1/2}}{\Delta t} - g_{i,j,k+1/2}}{\Delta z_k} - 0 \right) \tag{6.72}$$

The only remaining task in the VBF method is to determine $\hat{u}^{n+1}_{i-1/2,j,k}$ and $\hat{w}^{n+1}_{i,j,k+1/2}$. The interpolation is needed in general to obtain these values using the tentative velocity in the interior fluid cell and the no-slip velocity constraint on the solid surface. Taking $\hat{w}^{n+1}_{i,j,k+1/2}$, for example, one can easily obtain:

$$\hat{w}^{n+1}_{i,j,k+1/2} = \frac{\delta x}{(\Delta x_i + \Delta x_{i+1})/2 + \delta x}\tilde{w}_{i,j,k+1/2} \tag{6.73}$$

where δx is the horizontal distance between the solid boundary and the cell top face center where the vertical VBF $f_{\text{VBF}z}$ is applied. Similarly, we can obtain $\hat{u}^{n+1}_{i-1/2,j,k}$. With the interpolated velocities, the revised PPE can be solved to obtain the new pressure field that will enforce the final velocity to be divergence-free and zero on the position of the solid surface. The above methodology has been extended to 3D cases in NEWTANK.

Since $(f_{\text{VBF}})_i$ is actually the reaction force on the fluid by the body, the total fluid force acting on the body is simply the volumetric integration of $(f_{\text{VBF}})_i$ around the body with a minus sign, i.e.:

$$(F_\text{T})_i = -\rho \iiint_\Omega (f_{\text{VBF}})_i \, \mathrm{d}V \tag{6.74}$$

where Ω is the body surface volume, and in the numerical modeling, it represents the volume occupied by the computational cells crossing the body surface.

6.5.1.3   *Flows in porous media*

In Section 5.2.1.7, we have shown that for flows in porous media that can be approximated by uniform spheres with the mean diameter of d_{50}, the general governing equations can be written as:

$$\frac{\partial \bar{u}_i}{\partial x_i} = 0 \tag{6.75}$$

$$\frac{1}{n}\frac{\partial \bar{u}_i}{\partial t} + \frac{\bar{u}_j}{n^2}\frac{\partial \bar{u}_i}{\partial x_j} = -\frac{1}{\rho}\frac{\partial \bar{p}}{\partial x_i} + g_i + \frac{\nu}{n}\frac{\partial^2 \bar{u}_i}{\partial x_j \partial x_j} - \frac{1}{n^2}\frac{\partial \overline{u''_i u''_j}}{\partial x_j} - \frac{C_\text{M}(1-n)}{n^2}\frac{\partial \bar{u}_i}{\partial t}$$

$$-\frac{3}{4}C_\text{D}\frac{(1-n)}{n^3}\frac{1}{d_{50}}\overline{u_c}\,\bar{u}_i \tag{6.76}$$

On rearranging the equations and omitting the overbar for simplicity, we have the following generalized equations:

$$\frac{\partial u_i}{\partial x_i} = 0 \tag{6.77}$$

$$\frac{\partial u_i}{\partial t} + C_\text{c} u_j \frac{\partial u_i}{\partial x_j} = -\frac{C_\text{p}}{\rho}\frac{\partial p}{\partial x_i} + C_\text{g} g_i + \frac{C_\text{v}}{\rho}\frac{\partial \tau_{ij}}{\partial x_j} + \frac{C_\text{t}}{\rho}\frac{\partial R_{ij}}{\partial x_j} + (f_{\text{POR}})_i \tag{6.78}$$

where

$$C_\text{c} = C_\text{t} = \frac{1}{n + C_\text{M}(1-n)}, \quad C_\text{p} = C_\text{g} = \frac{n^2}{n + C_\text{M}(1-n)}, \quad C_\text{v} = \frac{n}{n + C_\text{M}(1-n)}$$

and

$$\tau_{ij} = \rho\nu\left(\frac{\partial u_i}{\partial x_j} + \frac{\partial u_j}{\partial x_i}\right), \quad R_{ij} = -\rho\overline{u_i''u_j''},$$

$$(f_{\text{POR}})_i = -\frac{3}{4d_{50}}\frac{C_{\text{D}}(1-n)}{n^2 + nC_{\text{M}}(1-n)}u_c u_i = -C_f u_c u_i$$

In solving the above equations, the conventional two-step projection method is used, i.e.:

$$\frac{\tilde{u}_i - u_i^n}{\Delta t} = \left\{\frac{C_v}{\rho}\frac{\partial\tau_{ij}}{\partial x_j} + \frac{C_t}{\rho}\frac{\partial R_{ij}}{\partial x_j} - C_c u_j\frac{\partial u_i}{\partial x_j}\right\}^n \tag{6.79}$$

$$\frac{u_i^{n+1} - \tilde{u}_i}{\Delta t} = -\frac{C_p}{\rho}\frac{\partial p^{n+1}}{\partial x_i} + C_g g_i - C_f \tilde{u}_c u_i^{n+1} \tag{6.80}$$

Special attention should be paid to the second step of the method where the nonlinear friction term is linearized by using the product of the tentative total velocity $\tilde{u}_c = \sqrt{\tilde{u}_i\tilde{u}_i}$ and the current velocity u_i^{n+1}. This treatment has two advantages: (1) with linearization, it is possible to form the linear matrix for PPE; (2) with the implicit treatment of the friction term, there is no additional time-step restriction enforced by this source term, even when $C_f \to \infty$ as $n \to 0$ and/or $d_{50} \to 0$. The latter is very important when we want to use special porous media (e.g., $n \to 0$) to represent an impermeable solid body.

The immediate numerical difficulty at this juncture is that we are unable to form the discretised PPE by simply taking the divergence of (6.80) and enforcing the divergence of $u_i^{n+1} = 0$, as we did for nonporous flow cases. The additional terms will result from the friction term, even though it is already linearized. To resolve this problem, an intriguing approach is used to first re-express (6.80) as follows:

$$u_i^{n+1} = \frac{1}{\gamma}\tilde{u}_i - \frac{C_p\Delta t}{\rho\gamma}\frac{\partial p^{n+1}}{\partial x_i} + \frac{C_g\Delta t g_i}{\gamma} \quad \text{where } \gamma = 1 + \Delta t C_f \tilde{u}_c \tag{6.81}$$

Now by taking the divergence of (6.81) and applying $\partial u_i^{n+1}/\partial x_i = 0$, we have:

$$\frac{\partial}{\partial x_i}\left(\frac{C_p}{\rho\gamma}\frac{\partial p^{n+1}}{\partial x_i}\right) = \frac{\partial}{\partial x_i}\left(\frac{\tilde{u}_i}{\Delta t\gamma}\right) + \frac{\partial}{\partial x_i}\left(\frac{C_g g_i}{\gamma}\right) \tag{6.82}$$

Comparing this with the PPE (6.56) we obtained earlier, the differences arise that both the matrix and the RHS are modified. Especially, the additional source term:

$$\frac{\partial}{\partial x_i}\left(\frac{C_g g_i}{\gamma}\right)$$

is required even when the gravitational acceleration is a constant because C_g varies across the porous interface.

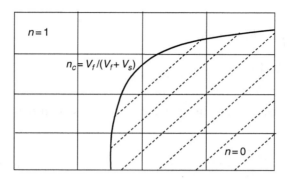

Figure 6.13 Definition sketch the use of porous cell treatment in representing an irregular impermeable body surface.

The method introduced above can be used in principle to solve fluid interaction with a structure with any permeability, which is the function of porous material properties (e.g., mean size d_{50} and porosity n). In the extreme case when n approaches 1.0, it simulates a "transparent" structure, inside which flow can freely move. On the other hand, when n approaches 0.0, it represents an impermeable structure. When the body surface crosses through a computational cell, the mean value of porosity in the cell n_c can be calculated by $n_c = (V_f + nV_s)/(V_f + V_s)$, where V_f and V_s are volumes open to fluid and occupied by the porous medium in the computational cell, respectively (see Figure 6.13 where $n = 0$ for an impermeable body). With such treatment, it is able to simulate wave interaction with a body, porous or impervious, with irregular geometry. The only caution is that as n approaches zero, the resulting PPE can be ill-conditioned. In this case, setting a lower limit of n (e.g. 10^{-4}) often helps resolve the problem.

The VOF method can still be used to track the free surface motion inside and across the porous media. However, correction must be made to account for the porous effect in each cell:

$$\frac{\partial (nF)}{\partial t} + \frac{\partial (u_i F)}{\partial x_i} = 0 \tag{6.83}$$

The porosity does not appear in the convective term because u_i in the porous media represents the mean velocity that already takes into consideration the porosity. As a result, the PLIC established in Section 5.5.1.1 can be equally applied except that the FD equation in (5.265) is revised to:

$$F_{i,j,k}^{n+1} = \left[\left(n_{i,j,k}\, {}^n F_{i,j,k}^n \right) - \Delta F_{\text{east}} + \Delta F_{\text{west}} - \Delta F_{\text{north}} + \Delta F_{\text{south}} - \Delta F_{\text{top}} + \Delta F_{\text{bottom}} \right] / n_{i,j,k}^{n+1}. \tag{6.84}$$

In this case, n can be the function of both time and space to represent the moving body or the evolving porous bed.

6.5.1.4 *Flows in a noninertial reference frame*

For a fluid inside a container that has six DOF motions, the easiest way of solving the fluid flow is to establish the coordinate system following the tank motion. By doing so, the computational domain is fixed. The governing equations, however, need to be modified to take into account the additional virtual forces resulting from the noninertial motions of the reference frame. For general cases, this corresponds to adding a force term $(f_{\text{NIF}})_i$ to the momentum equation:

$$\frac{\partial u_i}{\partial t} + u_j \frac{\partial u_i}{\partial x_j} = -\frac{1}{\rho}\frac{\partial p}{\partial x_i} + g_i + \frac{1}{\rho}\frac{\partial \tau_{ij}}{\partial x_j} + \frac{1}{\rho}\frac{\partial R_{ij}}{\partial x_j} + (f_{\text{NIF}})_i \qquad (6.85)$$

The force $(f_{\text{NIF}})_i$ includes the contribution from the translational and rotational frame motions. The vector form of the force reads as follows:

$$f_{\text{NIF}} = -\frac{d\mathbf{v}}{dt} - \frac{d\dot{\boldsymbol{\theta}}}{dt} \times (\mathbf{r} - \mathbf{R}) - 2\dot{\boldsymbol{\theta}} \times \frac{d(\mathbf{r} - \mathbf{R})}{dt} - \dot{\boldsymbol{\theta}} \times \left[\dot{\boldsymbol{\theta}} \times (\mathbf{r} - \mathbf{R})\right] \qquad (6.86)$$

where $\mathbf{v}$ and $\dot{\boldsymbol{\theta}}$ are the translational and rotational velocity of the noninertial reference frame, respectively; $\mathbf{r}$ and $\mathbf{R}$ are the position vectors for the point of interest in the computational domain and the origin of the rotation that can be either inside or outside the computational domain (Figure 6.14). Following the similar procedure of the two-step projection method, we can write the second step as:

$$\frac{u_i^{n+1} - \tilde{u}_i}{\Delta t} = -\frac{1}{\rho}\frac{\partial p^{n+1}}{\partial x_i} + g_i + \left(f_{\text{NIF}}^{n+1}\right)_i \qquad (6.87)$$

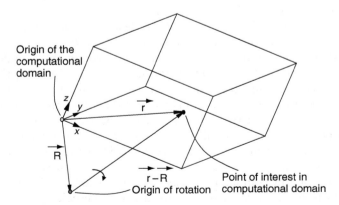

Origin of the computational domain

Point of interest in computational domain

Origin of rotation

Figure 6.14 Definition sketch of simulation of fluid flows in a noninertial reference frame.

This leads to the following equation:

$$\frac{\partial}{\partial x_i}\left(\frac{1}{\rho}\frac{\partial p^{n+1}}{\partial x_i}\right) = \frac{\partial}{\partial x_i}\left(\frac{\tilde{u}_i}{\Delta t}\right) + \frac{\partial \left(f_{\mathrm{NIF}}^{n+1}\right)_i}{\partial x_i}. \tag{6.88}$$

This additional force must be included on the boundaries as well where the velocity comes to zero:

$$\frac{\partial p^{n+1}}{\partial x_i} = \rho\left[g_i + \left(f_{\mathrm{NIF}}^{n+1}\right)_i\right] \tag{6.89}$$

6.5.2   Benchmark tests

In this section, the benchmark tests will be provided for the studies of different wave–structure interaction problems. We will present only the numerical results by the NSE models that are capable of simulating viscous and turbulent flows.

6.5.2.1   Two-dimensional drag and lift forces on a circular cylinder by a steady flow

When a steady flow passes over a circular cylinder, vortices will be formed behind the cylinder. Depending on the flow Re, vortices may be attached with the body or shed from the body alternately. The resulting drag and lift force coefficients are functions of Re, and there are many experimental data for this problem. This is the classical test for a viscous fluid model capable of handling an arbitrary solid surface. In the present test, NEWTANK will be used to simulate the problem. The VBF method is used to handle the cylinder surface in the Cartesian grid.

A circular cylinder with the diameter of $D = 1.0\,\mathrm{m}$ is deployed in the computational domain of $30\,\mathrm{m} \times 10\,\mathrm{m}$. The center of the cylinder is $10\,\mathrm{m}$ away from the left boundary and in the middle of the y-direction. A nonuniform mesh system with a total number of 130×80 is used with the finest grid of $\Delta x = \Delta y = 0.05\,\mathrm{m}$ being arranged near the cylinder. The upstream velocity U is set to be 0.01 m/s and the Re $= UD/\nu$ is changed by adjusting the value of ν. No turbulence closure model is used.

Figure 6.15 shows the simulated vortex structures around the cylinder at Re $= 1$, 10, 100, and 1000. It is observed that for small Re < 10, the vortices being formed are symmetrical and attached to the cylinder. With the increase of Re > 100, the vortices are elongated and shed from the cylinder alternately, causing the asymmetrical flow pattern in the y-direction.

One of the important concerns in this test is whether the numerical model can accurately calculate the drag and lift force coefficients C_D and C_L that are closely related to the flow separation point and are sensitive to the numerical accuracy of flow computation near the cylinder surface. Table 6.1 shows the comparisons of the calculated $\bar{C}_\mathrm{D}$ (mean drag coefficient), $|C_{\mathrm{L\,max}}|$

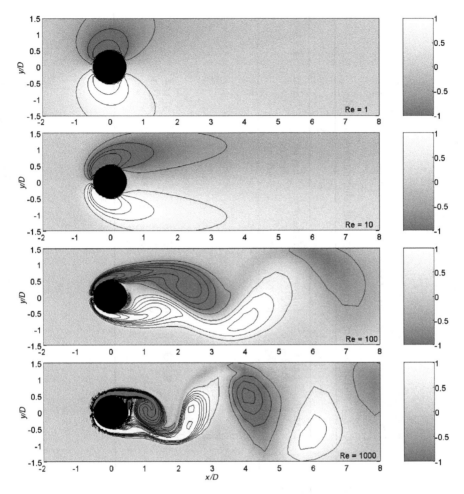

Figure 6.15 Simulated vortex structure behind a circular cylinder for different Re; the contours represent the normalized vorticity $(\omega/(U/D))$ with the interval of 0.5.

(maximum lift coefficient), and Strouhal number St that quantifies the shedding frequency between the present results and other numerical results using the IB method, e.g., Lai and Peskin (2000) and Palma *et al.* (2006). While Lai and Peskin (2000) employed 38 grids near the cylinder without the use of any turbulence model, Palma *et al.* (2006) employed 50 grids with the use of the k–ω turbulence model. It is seen that the present results compare reasonably well with other computations. There is a clear trend of decreased $\bar{C}_D$ and increased $|C_{L\,max}|$ with the increase of Re.

Table 6.1 Comparisons of the calculated $\bar{C}_D$, $|C_{Lmax}|$, and St between the present model and other numerical results

	$\bar{C}_D$			$|C_{Lmax}|$			St		
	Present	Lai and Peskin (2000)	Palma et al. (2006)	Present	Lai and Peskin (2000)	Palma et al. (2006)	Present	Lai and Peskin (2000)	Palma et al. (2006)
1	17.049			0					
2	9.7105			0					
5	5.1831			0					
10	3.4685		2.05	0					
20	2.4504			0					
50	1.655			0					
100	1.4643	1.4473	1.32	0.3231	0.3299	0.331	0.162	0.165	0.163
200	1.4019		1.34	0.5819		0.68	0.184502		0.190
500	1.3985			0.8479			0.208333		
1000	1.4121			0.9617			0.218		

6.5.2.2   *Two-dimensional solitary wave transmission, reflection, and dissipation over an impermeable step*

A solitary wave is often used to test a wave model that is capable of balancing wave dispersion and wave nonlinearity. In this test, the 2D NEWFLUME is validated against the case of solitary waves past a step, a similar case of a tsunami front past a continental shelf. A solitary wave will go through the fission process during which waves split into a number of solitons. The wave nonlinearity and fission process introduces the difficulty of defining the transmission coefficient based on the ratio of wave heights. Alternatively, one can use the definition based on the energy flux concept to define the RTD coefficients (Lin, 2004b):

$$K_R = \sqrt{\frac{-EF_{ref}}{EF_{inc}}}, \quad K_T = \sqrt{\frac{EF_{trans}}{EF_{inc}}}, \quad K_D = \sqrt{\frac{TD}{EF_{inc}}} = \sqrt{1 - K_R^2 - K_T^2}$$

$$(6.90)$$

where EF_{trans}, EF_{ref}, and EF_{inc} are the total energy fluxes integrated across the water depth for the transmitted, reflected, and incident waves. For a nonlinear wave, the energy flux includes both the work done by the dynamic pressure and the nonlinear convective contribution.

By using the linear long-wave approximation, Lamb (1945) gave the analytical solutions of the reflected and transmitted wave heights (H_R and H_T) for wave propagation over a step:

$$\frac{H_R}{H_I} = \frac{1 - \sqrt{1 - b/h}}{1 + \sqrt{1 - b/h}}, \quad \frac{H_T}{H_I} = \frac{2}{1 + \sqrt{1 - b/h}} \tag{6.91}$$

where $H_I = H$ is the incident wave height and b is the height of the step. In the above expressions, we have $1 \geq H_R/H_I \geq 0$ and $2 \geq H_T/H_I \geq 1$. It is easy to prove that the above equations satisfy the total energy conservation among the incident, transmitted, and reflected waves for a linear wave train, i.e., $C_{gd}H_R^2 + C_{gs}H_T^2 = C_{gd}H_I^2$, where $C_{gd} = \sqrt{gh}$ is the group velocity in the deep water while $C_{gs} = \sqrt{g(h-b)}$ is the group velocity on the step. Thus, the above expressions can be modified so that they can fit in the current definition of wave coefficients based on the energy concept:

$$K_R = \frac{H_R}{H_I} = \frac{1 - \sqrt{1 - b/h}}{1 + \sqrt{1 - b/h}}, \quad K_T = \frac{\sqrt{C_{gs}}}{\sqrt{C_{gd}}} \frac{H_T}{H_I} = \frac{2\sqrt[4]{1 - b/h}}{1 + \sqrt{1 - b/h}} \tag{6.92}$$

Later, Sugimoto et al. (1987) introduced an "edge-layer" concept that is based on the potential flow theory and allows vertical motion near the step. The theory incorporates the higher order corrections to Lamb's theory. An analytical solution was found for the wave reflection coefficient:

$$K_R = \frac{H_R}{H_I} = \frac{1}{4} \left\{ \sqrt{1 + 8 \left(\frac{1 - \sqrt{1 - b/h}}{1 + \sqrt{1 - b/h}} \right)} - 1 \right\}^2 \tag{6.93}$$

The number of solitons contained in the transmitted wave packet and the corresponding wave heights were also found analytically. Note that though allowing the vertical velocity in the vicinity of the step, Sugimoto et al.'s result relies on the validity of potential flow theory and cannot represent the vortex shedding mechanism from the step corner. Seabra-Santos et al. (1987) conducted a series of laboratory experiments to study this phenomenon. In the experiments, various values of H/h that range from 0.14 to 0.44 were used, and the results were represented by the ratio of H_R/H_I.

Now let us present the comparisons between the numerical results and the theories and experimental data. The numerical simulations are performed in a domain 100 m long, with the deep-water depth being $h = 1$ m. A uniform grid system of $\Delta x = 0.05$ m and $\Delta y = 0.01$ m is employed. Solitary waves with $H/h = 0.1$ and $H/h = 0.3$, respectively, are sent from the left boundary. The step height varies from 0 to $2H$ (fully block the wave). Figure 6.16 shows the comparisons of the simulated reflection coefficients to the experimental data. The definitions based on both wave height ratio and energy flux integration are presented. It is observed that the differences between two definitions are small for $H/h = 0.1$, but large for $H/h = 0.3$. The simulated results in terms of the wave height ratio compare reasonably well with the experimental data, which are also defined based on the wave height ratio, for both small and large wave height.

Figure 6.17 shows the simulated RTD coefficients together with the theories. For reflection coefficients, the numerical results agree very well with Lamb's theory up to about $b/h \approx 1 - 2H/h$ for both weakly nonlinear waves $H/h = 0.1$ and fully nonlinear waves $H/h = 0.3$. The deviation of the large values of b/h is mainly caused by the fact that K_R actually reaches 1.0 when $b/h = 1 + 2H/h$ rather than $b/h = 1$ as predicted by both theories. On the other hand, Sugimoto et al.'s theory gives consistently smaller results. Since the current numerical model solves the Reynolds equation directly that is free of limitation from either the long-wave approximation or the potential flow assumption, its better agreement with Lamb's theory seems to suggest that the "edge-layer" theory, which enforces the smooth streamlines over the step, may significantly underestimate wave reflection.

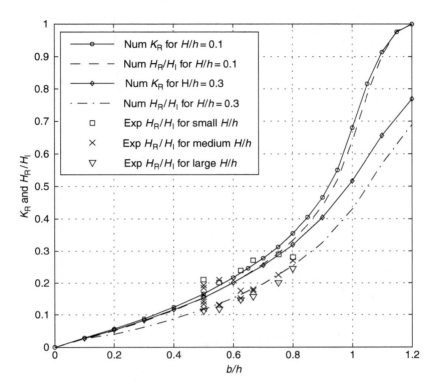

Figure 6.16 Comparison of simulated wave reflection coefficients with the experimental data for small and large incident wave heights. (Sugimoto *et al.*, 1987)

Considering that Lamb's theory is for linear periodic long waves, this agreement also suggests that during the interaction, both the wave dispersion and the periodicity are insignificant and do not affect the wave reflection process. Figure 6.17(b) and (c) shows the results for the transmission and dissipation coefficient. With the increase of step height, wave breaking can be induced that increases the energy dissipation and decreases the wave transmission, whereas the wave reflection is less affected by the breaking process.

With the validated model, Lin (2004b) has made a numerical study of solitary wave interaction with a rectangular block with an arbitrary aspect ratio. The RTD coefficients are produced and tabulated for engineering design. The study was further extended by Lin and Karunarathna (2007) for solitary wave transmission and reflection from a porous breakwater.

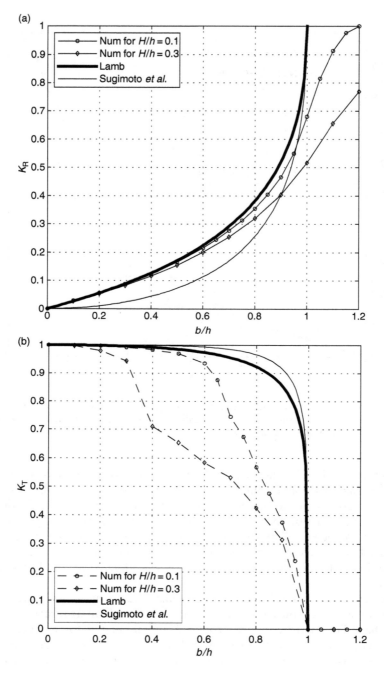

Figure 6.17 Comparisons of (a) K_R, (b) K_T, and (c) K_D between numerical results and theories by Lamb (1945) and Sugimoto *et al.* (1987).

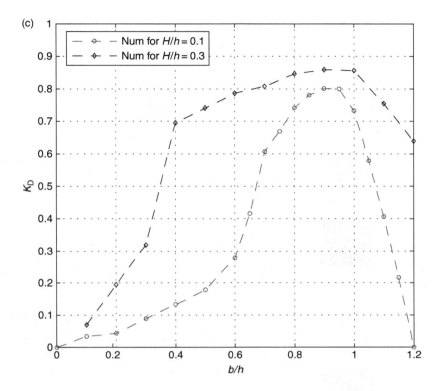

Figure 6.17 (Continued).

6.5.2.3 Two-dimensional solitary wave past a submerged rectangular block

When waves pass above a submerged bluff body, vortices may be formed around the corner of the body and subsequently transported into the region and possibly interact with both the free surface and the bottom. In this test, we shall present the simulation of a solitary wave past a submerged rectangular block. Comparisons will be made among the 2D VOF RANS model (Lin and Liu, 1998a), the 3D multiple-layer σ-coordinate LES model (Lin, 2006), and the experimental measurements (Zhuang and Lee, 1996). The problem setup is shown in Figure 6.18(a), where the still water depth $h = 0.228$ m and the incident wave height $H = 0.069$ m. The rectangular obstacle has height of 0.114 m and length of $L = 0.381$ m. The time histories of horizontal and vertical velocities are measured at two points behind the obstacle, which are Point 1: 0.040 m above the bottom and 0.034 m downstream from the obstacle and Point 2: 0.017 m above Point 1. To simulate this problem, the uniform mesh $\Delta x = 0.0025$ m is used in both models. For vertical resolution, in total 106 meshes are used in the VOF model, whereas

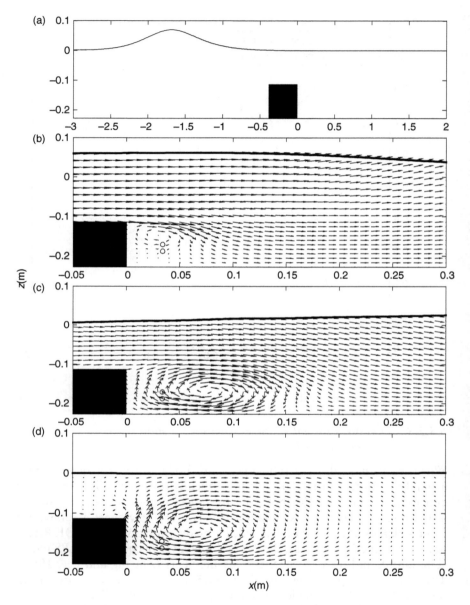

Figure 6.18 (a) Problem setup of a solitary wave past a submerged obstacle and the simulation results of the vertical structure behind the obstacle during the wave passage at (a) $t\sqrt{g/h} = 12.62$, (b) $t\sqrt{g/h} = 15.90$, and (c) $t\sqrt{g/h} = 19.17$; the two circles behind the obstacle represent measurement Points 1 (lower) and 2 (upper).

40 and 100 meshes are used in the multiple-layer σ-coordinate model to represent coarse and fine vertical resolution, respectively.

Figure 6.18(b)–(d) shows the simulation results of the vortex development during the passage of the solitary wave over the obstacle for the coarse vertical resolution case. The vortex is initiated around $t\sqrt{g/h} = 12.62$ when the crest of the solitary wave is above the rear corner of the obstacle. The vortex is further developed and evolved at $t\sqrt{g/h} = 15.90$ after the crest has passed over the obstacle. The size of the vortex grows further and it rises toward the free surface even when the wave has completely left the region (e.g., $t\sqrt{g/h} = 19.17$). The comparisons between numerical results and available experimental data are shown in Figure 6.19. It is found that all numerical results compared reasonably well with the experimental data for both horizontal and vertical velocity at these two points. This is true even for the coarse resolution results, although the fine resolution results are

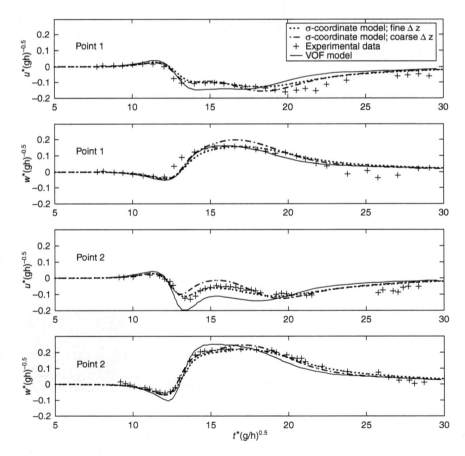

Figure 6.19 Comparisons of the time histories of horizontal and vertical velocities at the two points behind the rectangular obstacle among the σ-coordinate model, VOF model, and experimental data.

even better. Compared to the VOF model, the σ-coordinate model captures the velocity pattern inside the vortical structure better.

6.5.2.4   *Two-dimensional solitary wave run-up on a linear slope*

The main purpose of presenting this benchmark test is to show how the VBF treatment introduced in Section 6.5.1.2 can be used together with the VOF method to simulate the free surface flow interaction with a surface piercing body, e.g., the linear slope in this case. The slope has an angle of 30° and $s = \tan(30°) = 0.577$. The still water depth $h = 0.16$ m and the solitary wave has the wave height of $H = 0.027$ m. The PIV was used to determine the particle velocity and free surface displacement (Lin *et al.*, 1999).

The numerical computations are performed in the domain of 4.49 m $\leq$ $x \leq 6.99$ m and -0.16 m $\leq z \leq 0.11$ m, which is discretized by 250×90 uniform grids with $\Delta x = 0.01$ m and $\Delta z = 0.003$ m. The time step is adjusted dynamically to satisfy the stability constraints. The present numerical results from NEWTANK using the VBF method are compared with the experimental PIV data and the earlier numerical results from NEWFLUME using the PCT method (see Figures 6.20–6.25). The overall comparisons between

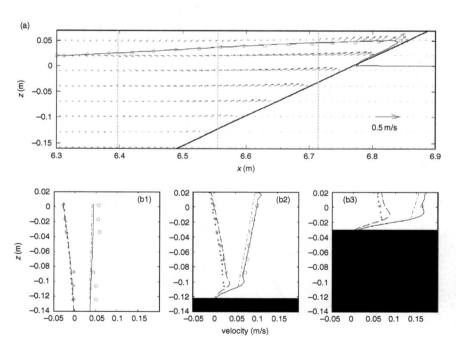

Figure 6.20 Solitary wave run-up and run-down at $t = 6.38$ s: (a) Comparison of free surface profiles (—: VBF model; arrows of velocity field also from VBF model; o: PIV data); (b) Comparisons of velocities at (b1) $x = 6.3972$ m, (b2) $x = 6.5556$ m, and (b3) $x = 6.7146$ m (— and –: u and w by VBF model, -.- and . . . : u and w by NEWFLUME, o and *: u and w by PIV).

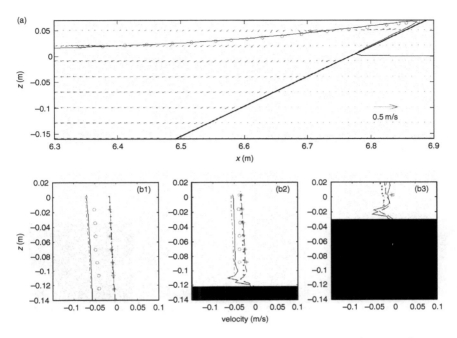

Figure 6.21 Solitary wave run-up and run-down at $t = 6.58$ s; explanatory data are in the caption for Figure 6.20.

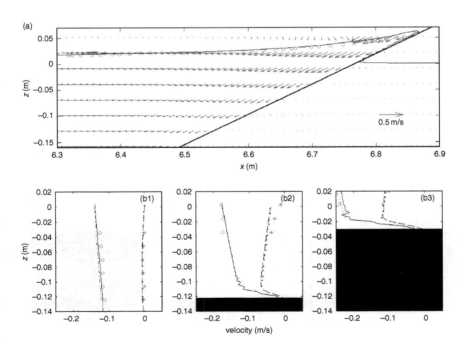

Figure 6.22 Solitary wave run-up and run-down at $t = 6.78$ s; explanatory data are in the caption for Figure 6.20.

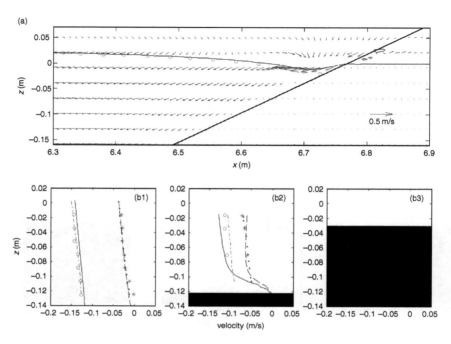

Figure 6.23 Solitary wave run-up and run-down at $t = 7.18$ s; explanatory data are in the caption for Figure 6.20.

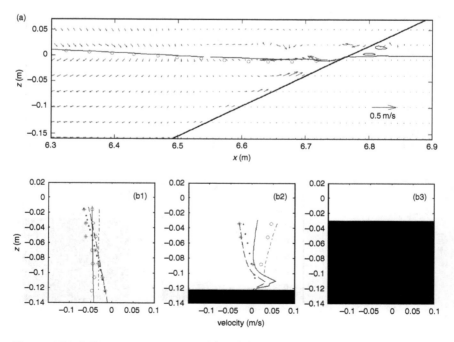

Figure 6.24 Solitary wave run-up and run-down at $t = 7.38$ s; explanatory data are in the caption for Figure 6.20.

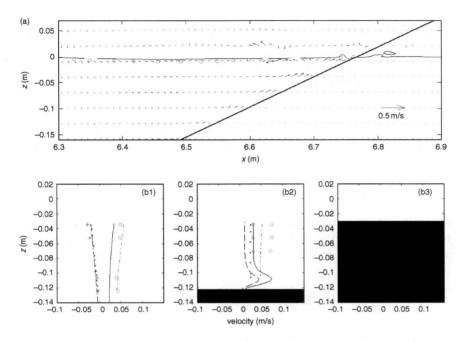

Figure 6.25 Solitary wave run-up and run-down at $t = 7.58$ s; explanatory data are in the caption for Figure 6.20.

the two numerical results and the experimental data are satisfactory. While the present VBF method provides the slightly better prediction during wave run-up process, the PCT is closer to the experimental data during wave run-down.

6.5.2.5 Two-dimensional flows through porous blocks

In this test, we shall present the application of NEWTANK to the simulation of flows in porous media. The objective of this benchmark is to demonstrate the efficacy of the porous-cell method (Section 6.5.1.3) and the validity of the semi-theoretical porous flow model (Section 5.2.1.7). The problems of the transient flow past a porous block filled with rocks or beads (Liu *et al.*, 1999a) are revisited with NEWTANK. The present numerical results are compared to the numerical results using the earlier porous flow model and the laboratory data.

In the laboratory, a glass tank that is 89 cm long, 44 cm wide, and 58 cm high was used with a porous block of $29 \times 44 \times 37$ cm^3 being placed at the center of the tank, i.e., 30.0–59.0 cm from the left glass tank wall. A gate was built 2 cm away from the left edge of the porous block to hold the water in the left-side reservoir. Two porous materials are used in the experiments.

One is the crushed rocks with an equivalent mean diameter of $d_{50} = 1.59$ cm. The porosity of the rocks is 0.49. The other porous material is the uniform glass bead with a diameter of $d_{50} = 0.3$ cm and a porosity of 0.39 cm. The initial water depths are 23.85 cm in the left-side reservoir and 2.4 cm in the downstream for the case of crushed rocks, whereas they are 13.8 cm and 2 cm for the case of beads.

In the numerical computations, a grid system with the uniform $\Delta x = 0.005$ m and $\Delta y = 0.0025$ m is used. In NEWTANK, the drag force coefficient in (6.76) takes the same form as that suggested by Lin and Karunarathna (2007) (Section 5.2.1.7), i.e.:

$$C_D = c_1 \frac{24.0}{Re} + c_2 \left(\frac{3.0}{\sqrt{Re}} + 0.34 \right) \left(1 + \frac{7.5}{KC} \right) \tag{6.94}$$

where we adopt the values of $c_1 = 7.0$ and $c_2 = 4.0$. The porous flow model by Liu et al. (1999a) considered linear friction and nonlinear friction only, and their model can be converted to the equivalent drag force coefficient in (6.76) as:

$$C_D = \frac{800}{3} \frac{(1 - n)}{Re} + \frac{4.4}{3} (1 + \frac{7.5}{KC}) \tag{6.95}$$

For the inertial force coefficient, both of the two models employed $C_M = 0.34$. Because the flow in the experiment is neither steady nor oscillatory, the KC number cannot be precisely defined. By analyzing the time history of the water surface profile in the recorded images, the characteristic time scales T, which are the times for water flowing through the porous dam, are estimated to be 1.0 s and 3.0 s for the rocks and the beads, respectively.

Figure 6.26 shows the comparisons of the flow past the porous block of crushed rocks between the numerical results and the experimental data. It is seen that both the numerical modeling results are in reasonably good agreement with the experimental data. The numerical results from NEWTANK give slightly better results than the earlier porous flow model in the later stage of flow passage (e.g., $t = 2.0$ s). Figure 6.27 gives the comparisons for the case of beads. Again, two numerical results are close to each other, and they provide reasonable comparisons to the measured data, although the comparisons are not as good as in the case of rocks. This example shows that it is possible to use the same set of empirical coefficients to simulate flows in different porous media, although further fine tuning of the model coefficients, especially the proper definition of KC for transient flows and the value of c_2, may still be needed in the future based on larger amount of laboratory data. The influence of medium shape and packing pattern also needs to be investigated further.

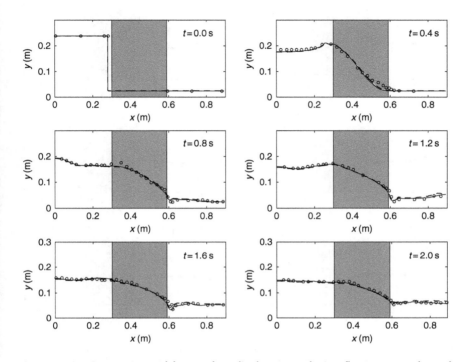

Figure 6.26 Comparison of free surface displacement during flow passage through the porous block of crushed rocks between numerical results (solid: NEWTANK; dashed: Liu *et al.*, 1999a); and experimental data (circle).

6.5.2.6 Three-dimensional liquid sloshing under external excitation

In this benchmark test, NEWTANK will be used to simulate 3D water sloshing in a tank with still water depth h, tank length $2a$, and width $2w$. In the laboratory, the tank is set on a shaking table at an angle φ from the axis of oscillation. The shaking table moves periodically as $u_e = -A \cos \omega t$, where u_e is the excitation velocity, $A = b\omega$ is the velocity amplitude with b being the displacement amplitude, and $\omega = 2\pi f$ is the angular frequency of the excitation (Figure 6.28).

By splitting the velocity of the shaker u_e into x and y components, i.e., $u_x = -A \cos \varphi \cos \omega t$ and $u_y = -A \sin \varphi \cos \omega t$, and treating the problem to have linearly coupled surge and sway motions, we are able to combine the 2D linear analytical solution of Faltinsen (1978) under 1D excitation (surge or sway; see Section 3.14.6.4) to generate the 3D linear analytical solution for liquid sloshing under the combined surge and sway excitations:

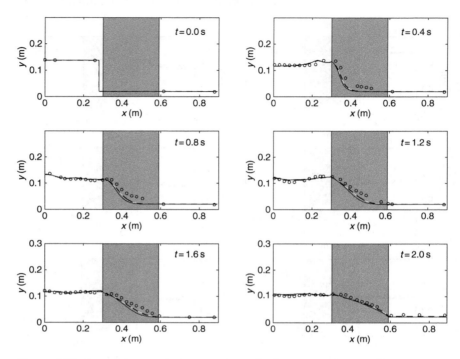

Figure 6.27 Comparison of free surface displacement during flow passage through the porous block of beads between numerical results (solid: NEWTANK; dashed: Liu *et al.*, 1999a); and experimental data (circle).

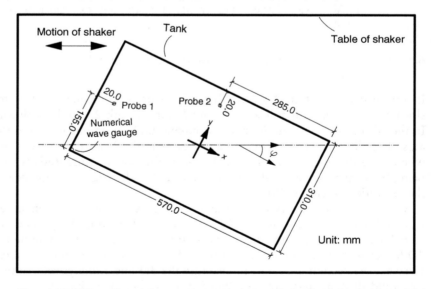

Figure 6.28 Top view of the experimental setup of the tank on the shaking table. (Experimental data provided by Prof. Koh C.G. at the National University of Singapore)

$$\eta = \frac{1}{g} \sum_{n=0}^{\infty} \sin\left\{\frac{(2n+1)\pi}{2a}x\right\} \cosh\left\{\frac{(2n+1)\pi}{2a}h\right\}$$

$$\times \left[-A_{nx}\omega_{nx}\sin\omega_{nx}t - C_{nx}\omega\sin\omega t\right]$$

$$-\frac{1}{g}A\omega x\sin\omega t + \frac{1}{g}\sum_{n=0}^{\infty}\sin\left\{\frac{(2n+1)\pi}{2w}y\right\}\cosh\left\{\frac{(2n+1)\pi}{2w}h\right\}$$

$$\times \left[-A_{ny}\omega_{ny}\sin\omega_{ny}t - C_{ny}\omega\sin\omega t\right] - \frac{1}{g}A\omega y\sin\omega t \tag{6.96}$$

where:

$$\omega_{nx}^2 = g\frac{(2n+1)\pi}{2a}\tanh\left\{\frac{(2n+1)\pi}{2a}h\right\}, \quad C_{nx} = \frac{\omega K_{nx}}{\omega_{nx}^2 - \omega^2},$$

$$A_{nx} = -C_{nx} - \frac{K_{nx}}{\omega}$$

$$K_{nx} = \frac{\omega A}{\cosh\left\{\frac{(2n+1)\pi}{2a}h\right\}}\frac{2}{a}\left[\frac{2a}{(2n+1)\pi}\right]^2 (-1)^n \tag{6.97}$$

and:

$$\omega_{ny}^2 = g\frac{(2n+1)\pi}{2w}\tanh\left\{\frac{(2n+1)\pi}{2w}h\right\}, \quad C_{ny} = \frac{\omega K_{ny}}{\omega_{ny}^2 - \omega^2},$$

$$A_{ny} = -C_{ny} - \frac{K_{ny}}{\omega}$$

$$K_{ny} = \frac{\omega A}{\cosh\left\{\frac{(2n+1)\pi}{2w}h\right\}}\frac{2}{w}\left[\frac{2w}{(2n+1)\pi}\right]^2 (-1)^n \tag{6.98}$$

Note that the origin is still set at the center of the tank and on the still water level.

In this study, we will use the parameters $h = 0.5\,\text{m}$, $a = 0.285\,\text{m}$, and $w = 0.155\,\text{m}$. The tank is set on the shaking table at an angle $\varphi = 30°$. The natural frequencies for different modes are:

$$\omega_{mn}^2 = \sqrt{\left(\frac{mg\pi}{2a}\right)^2 + \left(\frac{ng\pi}{2w}\right)^2}\tanh\sqrt{\left(\frac{m\pi h}{2a}\right)^2 + \left(\frac{n\pi h}{2w}\right)^2}$$

$$(m, n = 0, 1, 2, \ldots) \tag{6.99}$$

The lowest natural frequencies are $\omega_{10} = 6.0578\,\text{s}^{-1}$ and $\omega_{01} = 9.5048\,\text{s}^{-1}$, respectively. In the numerical test, we will choose $\omega = 0.985\omega_{10}$ so that the near resonance phenomenon will occur. In this first simulation, the small amplitude of excitation $b = 0.0005\,\text{m}$ is used, so that the problem remains linear in the entire simulation. In the second simulation, the larger amplitude of excitation $b = 0.005$ m is used, which is also the value used in the laboratory experiment. In this case, strong nonlinearity will be present in the sloshing. Two wave probes are used to measure the free surface displacement and they are located at $(-0.265$ m, 0.0 m$)$ for Probe 1 and at $(0.0$ m, 0.135 m$)$ for Probe 2 (see Figure 6.28).

The numerical simulation is performed in a fixed computational domain with the virtual force being introduced in the noninertial reference frame (Section 6.5.1.4). The mesh system has a uniform horizontal mesh size $\Delta x = \Delta y = 0.005$ m and a nonuniform vertical mesh size Δz with the minimum $\Delta z = 0.001$ m being arranged near the free surface. The time step is automatically adjusted to ensure numerical stability. Figure 6.29 shows the comparisons of the normalized (by displacement amplitude) free surface displacement at two probe locations and one more corner location $(-0.285$ m, -0.155 m$)$. When the excitation amplitude is small, the numerical results agree very well with the linear analytical solution

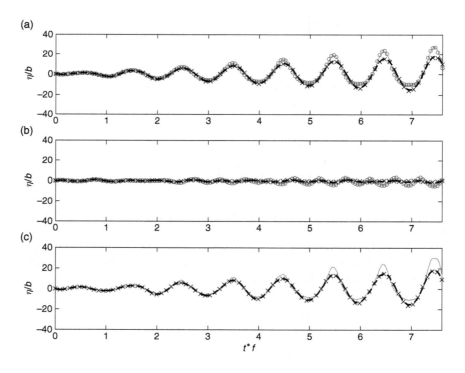

Figure 6.29 Comparisons of the time series of normalized surface elevation at (a) Probe 1 $(0.02, 0.155)$, (b) Probe 2 $(0.285, 0.29)$, and (c) corner of the tank among the linear analytical solution (cross), experimental data (circle), and the numerical results ($b = 0.0005$ m: dashed line; $b = 0.005$ m: solid line).

at all positions. However, as the excitation amplitude increases, the free surface displacement deviates significantly from the linear solution; instead it matches well with the experimental data. This demonstrates the good characteristics of NEWTANK in simulating strongly nonlinear liquid sloshing.

7 Summary

In this chapter, a brief summary will be provided for different wave models introduced in this book. We will also take this opportunity to briefly discuss the branches of wave modeling not adequately covered in this book. The future direction of research in water wave modeling will be highlighted at the end.

7.1 Summaries of wave models and numerical methods

We have introduced different numerical models for water wave modeling. They can be mainly classified into seven types, namely full NSE-solver model, quasi-3D model based on hydrostatic pressure assumption, potential flow model (mainly BEM model), SWE model, Boussinesq model, MSE model, and wave energy (action) spectral model. While one type of wave model can simulate different physical processes, a particular physical phenomenon can also be simulated by more than one type of wave model. Therefore, it is worth summarizing the advantages and limitations of each wave model based on its capability and limitation. In Table 7.1, we provide a summary based on the suitability of various wave models for different physical processes.

7.2 Subjects not covered in this book

It must be pointed out that there have been quite a number of subjects that are relevant to numerical modeling of water waves but not covered, at least adequately, in this book. These subjects often involve multiple physical processes and thus are more challenging in terms of theoretical formulation and numerical computation due to their multidisciplinary nature.

7.2.1 High-speed flows and cavitation

So far all our discussions have been confined to incompressible fluids. Fluid compressibility may become important when the fluid velocity

Table 7.1 Suitability of wave models for simulation of different physical processes

Wave models	Formulation or numerical methods	Wave diffraction	Wave refraction	Wave dispersion	Wave non-linearity	Wave breaking	Wave run-up	Wave over-topping structure	Flow turbulence	Wave-structure interaction	Wave-current interaction	Numerical efficiency	Comments marked by (*)
NSE	FVM or FEM	★★★★	★★★★	★★★★	★★★★	★★★★	★★★★	★★★★	★★★★	★★★★	★★★★	★	*Cut-cell, VBF, or their kinds used
	FDM	★★★★	★★★★	★★★★	★★★★	★★★★	★★★★	★★★★	★★★★	★★★(*)	★★★★	★☆	
	SPH	★★★★	★★★★	★★★★	★★★★	★★★★	★★★★	★★★★	★★★★	★★★★	★★★★	☆	
Quasi-3D	FDM or FVM	★★★★	★★★★		★★				★★★		★★(*)	★★	*Mainly for current simulation only and solved in the σ-coordinate
Potential flow	BEM	★★★★	★★★★	★★★★	★★★	★ (*)	★★★	★		★★★ (**)		★★☆	* For initiation of wave breaking only ** For large structures only
	FEM	★★★★	★★★★	★★★★	★★★	(★★★)	★★★	★		★★★		★★	*** Adaptive mesh or σ-coordinate
	FDM	★★★★	★★★★	★★★★	★★★	(★★★★)	★★★			★★		★★	**** σ-coordinate
SWE	FDM or FEM	★★★★	★★★★		★★★	★★	★★★	★☆	★☆	★☆	★☆(*)	★★★★	*Current module only in a coupled model
Boussinesq	Standard	★★★★	★★★★	★★	★★☆	★★☆	★★★	★☆	★☆	★☆	★★	★★★☆	*For coupled wave–current simulation
	High-order	★★★★	★★★★	★★★	★★★	★★★	★★★	★☆	★☆	★★	★★(*)	★★☆	
MSE	Elliptic (*)	★★★★	★★★★	★★★★	★	★			☆	★☆	★★	★★★	*For steady wave field only
	Hyperbolic (**)	★★★★	★★★★	★★★★	★	★			☆	★★	★★	★★★☆	**For transient wave field
	Parabolic (***)	★★★	★★	★★★★☆	★	★			☆	☆	★	★★★★☆	***Waves with primary propagation direction
Spectral	Wave Energy	☆	★★★★	★★★★	★★	★★			★★			★★★★	
	Wave Action	☆	★★★★	★★★★	★★	★★☆			★★		★★★	★★★★	

Notes:

[1] The number of stars represents the level of suitability of a particular model for the corresponding wave phenomenon; ☆ represents half star.

[2] The "numerical efficiency" in this table has two implications: (a) the size of the computational domain the simulation can possibly cover (e.g., global, regional, or local);
(b) for the same computational domain discretisation the CPU time required.

is high. This can be the case when there is an underwater explosion, an underwater volcano eruption, fluid near a high-speed underwater vehicle or object, fluid near a propeller, etc. In such a case, fluid compressibility can become important. Besides the surface waves, the underwater acoustic (pressure) wave is also important to study. In connection with high-speed fluid flows, the cavitation near the surface of the solid body is another concern. This is one of the major considerations in designing a high-speed underwater device or a ship propeller.

7.2.2 Stratified flows and internal waves

Most of the numerical simulations of water waves are for surface waves, assuming constant fluid density. When the density variation becomes considerable, it increases the complex level of the problem. In oceans or large lakes, the internal wave can take place on the interface of two fluid layers and it is an important physical process for vertical mixing. The studies of internal waves have applications to physical oceanography, coastal and offshore engineering, and submarine design. In principle, all the NSE models have the potential capability of simulating internal waves, provided the interface between two fluid layers can be tracked accurately. The numerical model that tracks both the free surface and internal interface can be used to simulate ocean surface–wave interaction with internal waves.

7.2.3 Sediment transport and bed morphology

Although extensive discussions have been made in this book for wave–structure interaction, few have addressed wave–bottom interaction. On the seafloor, there are various types of bed materials, such as coral reef, mud, fine sediment, and sand. Different bed materials will react differently under wave action. The study of wave–bed interaction involves the study of sediment transport, bed morphology, elastic soil skeleton response to wave action, soil liquefaction under cyclic wave loads, foundation stability, etc. This is an important area of research but requires knowledge from both fluid dynamics and soil mechanics.

7.2.4 Ocean environmental modeling and green energy in oceans

There are many environmental issues related to ocean waves and currents. Water quality modeling in coastal zones and deep oceans requires the hydrodynamic information that can be obtained by the modeling of ocean waves and currents. Examples of environmental modeling in oceans include the oil spill simulation, the modeling of pollutant transport, and fate, the prediction of coastal eutrophication (e.g., "red tides" caused by algal bloom).

On the contrary, green energy can be obtained from the ocean by making use of the wave, current, and tidal power that is clean, sustainable, and renewable. Besides, wind energy and solar energy in oceans are vast and the future exploitation of them requires the deployment of devices in oceans, where the knowledge of waves and currents is needed.

7.3 Future work

The numerical modeling of water waves is still a very active research area. There are many efforts being made for the improvement of wave theory, better numerical technique, and an integrated modeling systems.

7.3.1 *Improvement of wave theory*

7.3.1.1 *NSE-solver models*

Turbulence modeling: A water wave is a complex natural phenomenon, yet it can be described by the very general NSEs. There is an increasing trend to use NSE models to simulate water wave problems. Compared with other models, this type of numerical model has the fewest assumptions and can accurately represent the viscous effect on fluid motion. The major challenge ahead is the better modeling of turbulence for high Re flows, which is the common challenge for most of the fluid flow problems. Besides the Reynolds stress approach, the LES is another promising alternative that is becoming popular in modeling water waves.

Air entrainment: The modeling of flow turbulence near the free surface is closely related to the modeling of air entrainment. Although the MAC, VOF, and level set methods have been developed to track the free surface, the detailed mixing process near the free surface caused by flow turbulence has not been adequately modeled. The aeration process in wave breaking brings air into the water, and the physical property of the mixed two-phase fluid deviates significantly from that of the single-phase fluid. More thorough study is needed in the future to understand the relationship between the free surface mixing mechanism and the flow turbulence.

7.3.1.2 *Unification of depth-averaged wave models*

The depth-averaged models are important for coastal engineers and scientists of physical oceanography because they can cover a much larger simulation area than that by the NSE models. Although efforts have been made in the past two decades to develop a unified depth-averaged wave model, the results are not very encouraging. We basically end up with the Boussinesq-type models that are suitable for highly nonlinear waves in relatively shallow waters or the MSE models that are valid for all water depths but for weak wave nonlinearity only. It can be foreseen that efforts will be continued to extend the Boussinesq models to deeper water and to improve the MSE model for stronger nonlinearity. A really unified depth-averaged model that is applicable from very deep water to very shallow water, however, is still a challenging task.

7.3.1.3   Wave spectral models

The wave spectral models have been very successful in modeling large-scale wave motions. All physical phenomena can be reasonably modeled by the wave spectral models except for wave diffraction. Although efforts have been made to develop the wave spectral model that can partially treat wave diffraction, so far there is no model that is well accepted by the community. In the future, there is the need to further the research in this area, from both the theoretical and practical modeling points of view.

7.3.2   Innovative numerical methods

Most of the wave models have been constructed with conventional numerical methods, e.g., FDM, FEM, or BEM. Recently, there has been an increasing trend to use the meshless particle method to solve water wave problems. Besides, there exist other types of unconventional methods that may possibly be used for wave modeling and wave data analysis, which are briefly introduced below.

7.3.2.1   Semianalytical method

With the advance of computational mathematics, some new methods for solving PDEs semianalytically have been proposed. For example, with the use of homotopy analysis, an explicit series solution to some PDEs, Liao and Cheung (2003) solved nonlinear progressive waves in deep water. Starting from the initial trial solution, the method employed a Maclaurin series expansion that provides successive approximation of the solution by repeated application of a differential operator.

Recently, there is the proposal of the so-called finite analytical method. In this method, the discrete algebraic equation is obtained from the analytical solution for each local element. The method was applied to solve some diffusion problems (e.g., Jin *et al.*, 1996). However, the application of the method is limited to certain types of PDEs only, and the method is not as general as FDM or FEM.

Wavelet method has been primarily used in data analysis. However, recently this method has also been used to solve PDEs. Hong and Kennett (2002) employed the wavelet-based method to solve wave equations for wave propagation.

7.3.2.2 *Numerical methods based on direct problem formulation*

Many of the conventional numerical methods are simply the numerical tools for the PDE, which must be derived a priori to describe a particular physical process mathematically. These numerical methods are therefore equation-based rather than problem-based. The PDE is the bridge to link the numerical method to the physical problem. Alternatively, one can propose a numerical method that solves the problem directly, provided the numerical method faithfully represents the fundamental physics of the problem. By doing so, the PDE is bypassed. The typical example of this approach is the LBM. Although the LBM has not been widely used to model water wave problems, it has good potential, especially for the complex physical processes where the existing governing PDE is not adequate (e.g., air entrainment, sediment transport).

7.3.3 *System integration*

7.3.3.1 *Multidisciplinary problems*

The simulation of water waves is often linked to other physical processes that need to be simulated by different numerical models.

Coupled to a meteorology model: Large-scale wave and current modeling requires knowledge of the instantaneous wind speed and direction as well as the changing atmospheric pressure. Such information can be obtained only from a meteorology simulation model. On the contrary, ocean temperature and evaporation rate information that is obtained from an ocean hydrodynamic model will be fed back to the meteorology model.

Coupled fluid–soil–structure interaction: We have discussed the simulation of coupled wave–structure interaction. However, in both coastal and offshore engineering, the problem often involves soil reaction as well. A complete simulation of this problem must include hydrodynamics, structure dynamics, and soil mechanics. It is an important subject in future research.

7.3.3.2 *Development of a software package*

We have already discussed the merits and limitations of each wave model. Apparently, there is no model that possesses the advantage of both numerical efficiency and complete physical representation. In certain cases, a few models must be employed simultaneously to provide the reliable prediction within the feasible time frame and computational resource. For this reason, the integration of different wave models into one package with the proper linkage among these models for information exchange becomes very important. One of the major efforts in future research is to extend the modeling technique to large-scale engineering computation and management.

7.3.3.3 Data analysis

All the numerical models we discussed in this book are deterministic models. This means that a unique result will be generated with a particular set of initial and boundary conditions. Therefore, one simulation will produce only one result for a particular set of variable combinations. In reality, there are many possible combinations of wave, current, and structure conditions. When the variable number becomes large, simulating all possible cases becomes a mission impossible due to the excessive number of possible variable combinations. In this case, some modern computational techniques have become good tools for reducing the efforts of simulation and providing better interpretation of the generated data. These computational tools for data analysis include the artificial neural network (ANN), genetic algorithm (GA), and fuzzy logic.

Appendices

Appendix I Definition of gradient, divergence, curl, and the Laplacian operator in Cartesian, cylindrical, and spherical polar coordinates

Cartesian coordinates

$$\nabla f = \frac{1}{h_1}\frac{\partial f}{\partial u}\hat{i}_1 + \frac{1}{h_2}\frac{\partial f}{\partial v}\hat{i}_2 + \frac{1}{h_3}\frac{\partial f}{\partial w}\hat{i}_3 \tag{I.1}$$

$$\nabla \cdot Q = \frac{1}{h_1 h_2 h_3}\left[\frac{\partial}{\partial u}(h_2 h_3 Q_1) + \frac{\partial}{\partial v}(h_3 h_1 Q_2) + \frac{\partial}{\partial w}(h_1 h_2 Q_3)\right] \tag{I.2}$$

$$\nabla \times Q = \frac{1}{h_1 h_2 h_3}\begin{vmatrix} h_1\hat{i}_1 & h_2\hat{i}_2 & h_3\hat{i}_3 \\ \partial/\partial u & \partial/\partial v & \partial/\partial w \\ h_1 Q_1 & h_2 Q_2 & h_3 Q_3 \end{vmatrix}$$

$$= \frac{1}{h_2 h_3}\left[\frac{\partial(h_3 Q_3)}{\partial v} - \frac{\partial(h_2 Q_2)}{\partial w}\right]\hat{i}_1 + \frac{1}{h_1 h_3}\left[\frac{\partial(h_1 Q_1)}{\partial w} - \frac{\partial(h_3 Q_3)}{\partial u}\right]\hat{i}_2$$

$$+ \frac{1}{h_1 h_2}\left[\frac{\partial(h_2 Q_2)}{\partial u} - \frac{\partial(h_1 Q_1)}{\partial v}\right]\hat{i}_3 \tag{I.3}$$

where:

$$h_1 = \sqrt{U \cdot U} = \sqrt{x_u^2 + y_u^2 + z_u^2}$$

$$h_2 = \sqrt{V \cdot V} = \sqrt{x_v^2 + y_v^2 + z_v^2} \tag{I.4}$$

$$h_3 = \sqrt{W \cdot W} = \sqrt{x_w^2 + y_w^2 + z_w^2}$$

$\hat{i}_1, \hat{i}_2, \hat{i}_3$ are the normalized U, V, W base vectors:

$$\hat{i}_1 = \frac{U}{\|U\|} = \frac{U}{\sqrt{U \cdot U}} = \frac{U}{h_1}, \quad \hat{i}_2 = \frac{V}{h_2}, \quad \hat{i}_3 = \frac{W}{h_3} \tag{I.5}$$

Combining (I.1) and (I.2), we have the Laplacian operator:

$$\nabla^2 f = \nabla \cdot \nabla f = \frac{1}{h_1 h_2 h_3} \left[\frac{\partial}{\partial u} \left(\frac{h_2 h_3}{h_1} \frac{\partial f}{\partial u} \right) \right.$$
$$\left. + \frac{\partial}{\partial v} \left(\frac{h_1 h_3}{h_2} \frac{\partial f}{\partial v} \right) + \frac{\partial}{\partial w} \left(\frac{h_1 h_2}{h_3} \frac{\partial f}{\partial w} \right) \right] \tag{I.6}$$

Cylindrical coordinates

With $u = r$, $v = \theta$, and $w = z$, we have $\hat{i}_1 = \hat{e}_r$, $\hat{i}_2 = \hat{e}_\theta$, $\hat{i}_3 = \hat{e}_z$ and:

$$h_1 = \sqrt{x_r^2 + y_r^2 + z_r^2} = \sqrt{\cos^2 \theta + \sin^2 \theta + 0} = 1$$

$$h_2 = \sqrt{x_\theta^2 + y_\theta^2 + z_\theta^2} = \sqrt{r^2 \sin^2 \theta + r^2 \cos^2 \theta + 0} = r \tag{I.7}$$

$$h_3 = \sqrt{x_z^2 + y_z^2 + z_z^2} = \sqrt{0 + 0 + 1} = 1$$

It follows that:

$$\nabla f = \frac{\partial f}{\partial r} \hat{e}_r + \frac{1}{r} \frac{\partial f}{\partial \theta} \hat{e}_\theta + \frac{\partial f}{\partial z} \hat{e}_z \tag{I.8}$$

$$\nabla \cdot Q = \frac{1}{r} \frac{\partial}{\partial r} (r Q_r) + \frac{1}{r} \frac{\partial}{\partial \theta} Q_\theta + \frac{\partial}{\partial z} Q_z \tag{I.9}$$

$$\nabla \times Q = \left[\frac{1}{r} \frac{\partial Q_z}{\partial \theta} - \frac{\partial Q_\theta}{\partial z} \right] \hat{e}_r + \left[\frac{\partial Q_r}{\partial z} - \frac{\partial Q_z}{\partial r} \right] \hat{e}_\theta + \frac{1}{r} \left[\frac{\partial (r Q_\theta)}{\partial r} - \frac{\partial Q_r}{\partial \theta} \right] \hat{e}_z \tag{I.10}$$

$$\nabla^2 f = \frac{1}{r} \frac{\partial}{\partial r} \left(r \frac{\partial f}{\partial r} \right) + \frac{1}{r^2} \frac{\partial^2 f}{\partial \theta^2} + \frac{\partial^2 f}{\partial z^2} = \frac{\partial^2 f}{\partial r^2} + \frac{1}{r} \frac{\partial f}{\partial r} + \frac{1}{r^2} \frac{\partial^2 f}{\partial \theta^2} + \frac{\partial^2 f}{\partial z^2} \tag{I.11}$$

where $Q = Q_r \hat{e}_r + Q_\theta \hat{e}_\theta + Q_z \hat{e}_z$.

Spherical polar coordinates

With ρ, θ, and ϕ, we have:

$$\nabla f = \frac{\partial f}{\partial \rho} \hat{e}_\rho + \frac{1}{\rho} \frac{\partial f}{\partial \theta} \hat{e}_\theta + \frac{1}{\rho \sin \theta} \frac{\partial f}{\partial \phi} \hat{e}_\phi \tag{I.12}$$

$$\nabla \cdot Q = \frac{1}{\rho^2} \frac{\partial}{\partial \rho} (\rho^2 Q_\rho) + \frac{1}{\rho \sin \theta} \frac{\partial}{\partial \theta} (\sin \theta Q_\theta) + \frac{1}{\rho \sin \theta} \frac{\partial Q_\phi}{\partial \phi} \tag{I.13}$$

$$\nabla \times Q = \frac{1}{\rho \sin \theta} \left[\frac{\partial Q_\theta}{\partial \phi} - \frac{\partial}{\partial \theta} (\sin \theta Q_\phi) \right] \hat{e}_\rho$$
$$+ \frac{1}{\rho} \left[\frac{\partial (\rho Q_\phi)}{\partial \rho} - \frac{1}{\sin \theta} \frac{\partial Q_\rho}{\partial \phi} \right] \hat{e}_\theta + \frac{1}{\rho} \left[\frac{\partial Q_\rho}{\partial \theta} - \frac{\partial (\rho Q_\theta)}{\partial \rho} \right] \hat{e}_\phi \tag{I.14}$$

and:

$$\nabla^2 f = \frac{1}{\rho^2} \left[\frac{\partial}{\partial \rho} \left(\rho^2 \frac{\partial f}{\partial \rho} \right) + \frac{1}{\sin \theta} \frac{\partial}{\partial \theta} \left(\sin \theta \frac{\partial f}{\partial \theta} \right) + \frac{1}{\sin^2 \theta} \frac{\partial^2 f}{\partial \phi^2} \right] \tag{I.15}$$

where $Q = Q_\rho \hat{e}_\rho + Q_\theta \hat{e}_\theta + Q_\theta \hat{e}_\theta$.

Appendix II Tensor and vector manipulation

Single operators

This section explicitly lists what some symbols mean for clarity.
  Divergence:
  For a vector field V, divergence is generally written as:

$$\nabla \cdot V = \text{div}(V) \tag{II.1}$$

and is a scalar field.
  Curl:
  For a vector field V, curl is generally written as:

$$\nabla \times V = \text{curl}(V) \tag{II.2}$$

and is a vector field.
  Gradient:
  For a vector field V, gradient is generally written as:

$$\nabla V = \text{grad}(V) \tag{II.3}$$

and is a tensor.

Combinations of multiple operators

Curl of the gradient:
  The curl of the gradient of any scalar field ψ is always zero, i.e.:

$$\nabla \times (\nabla \psi) = 0 \tag{II.4}$$

Divergence of the curl:
  The divergence of the curl of any vector field A is always zero, i.e.:

$$\nabla \cdot (\nabla \times A) = 0 \tag{II.5}$$

Divergence of the gradient:
  The Laplacian of a scalar field is defined as the divergence of the gradient, i.e.:

$$\nabla \cdot (\nabla \psi) = \nabla^2 \psi \tag{II.6}$$

Note that the result is a scalar quantity.
Curl of the curl:

$$\nabla \times \nabla \times A = \nabla (\nabla \cdot A) - \nabla^2 A \tag{II.7}$$

Properties

Distributive property:

$$\nabla \cdot (A + B) = \nabla \cdot A + \nabla \cdot B \tag{II.8}$$

$$\nabla \times (A + B) = \nabla \times A + \nabla \times B \tag{II.9}$$

Vector dot product:

$$\nabla (A \cdot B) = (A \cdot \nabla) B + (B \cdot \nabla) A + A \times (\nabla \times B) + B \times (\nabla \times A) \tag{II.10}$$

Vector cross product:

$$\nabla \cdot (A \times B) = B \cdot \nabla \times A - A \cdot \nabla \times B \tag{II.11}$$

$$\nabla \times (A \times B) = A (\nabla \cdot B) - B (\nabla \cdot A) + (B \cdot \nabla) A - (A \cdot \nabla) B \tag{II.12}$$

Product of a scalar and a vector

$$\nabla \cdot (\psi A) = A \cdot \nabla \psi + \psi \nabla \cdot A \tag{II.13}$$

$$\nabla \times (\psi A) = \psi \nabla \times A - A \times \nabla \psi \tag{II.14}$$

More identities

$$\frac{1}{2} \nabla A^2 = A \times (\nabla \times A) + (A \cdot \nabla) A \tag{II.15}$$

Appendix III   Mathematical derivations in hydrodynamics and wave theory

Derivation of Bernoulli equation for a potential flow (Section 2.2.2.2)

Starting from the momentum equation of the NSE (2.10):

$$\frac{\partial u_i}{\partial t} + u_j \frac{\partial u_i}{\partial x_j} = -\frac{1}{\rho} \frac{\partial p}{\partial x_i} + g_i + v \frac{\partial^2 u_i}{\partial x_j^2} \tag{III.1}$$

For a potential flow of an ideal fluid, the above equation can be rewritten as:

$$-\frac{\partial^2 \phi}{\partial t \partial x_i} + \frac{\partial \phi}{\partial x_j} \frac{\partial^2 \phi}{\partial x_j \partial x_i} = -\frac{1}{\rho} \frac{\partial p}{\partial x_i} + \frac{\partial g_i x_i}{\partial x_i} \tag{III.2}$$

The above equation can be rearranged to be:

$$\frac{\partial}{\partial x_i}\left[-\frac{\partial \phi}{\partial t}+\frac{1}{2}\left(\frac{\partial \phi}{\partial x_j}\right)^2+\frac{p}{\rho}-g_j x_j\right]=0 \qquad \text{(III.3)}$$

In a Cartesian coordinate whose vertical axis z is aligned with gravity, we have $g_j x_j = -gz$. Integrating the above equation in space x_i, we have:

$$-\frac{\partial \phi}{\partial t}+\frac{1}{2}\left[\left(\frac{\partial \phi}{\partial x}\right)^2+\left(\frac{\partial \phi}{\partial y}\right)^2+\left(\frac{\partial \phi}{\partial z}\right)^2\right]+\frac{p}{\rho}+gz=C(t) \qquad \text{(III.4)}$$

Apparently, C is a constant in a 3D space but varies in time. Based on this equation, the pressure and the velocity for any two points in a 3D space are related. For a steady flow, the above equation can be reduced to an even more familiar form of Bernoulli equation:

$$\frac{1}{2}\left[u^2+v^2+w^2\right]+\frac{p}{\rho}+gz=C \qquad \text{(III.5)}$$

Proof of the small-amplitude wave requirement (e.g., ka <<1 and a/h <<1) in linear wave theory (Section 3.1.1)

The kinematic free surface boundary condition (3.4) is:

$$w-\frac{\partial \eta}{\partial t}-u\frac{\partial \eta}{\partial x}=0 \quad \text{on } z=\eta \qquad \text{(III.6)}$$

Applying the Taylor series expansion for the kinematic free surface boundary condition at $z=0$, we have:

$$\left(w-\frac{\partial \eta}{\partial t}-u\frac{\partial \eta}{\partial x}\right)_{z=\eta}=\left(w-\frac{\partial \eta}{\partial t}-u\frac{\partial \eta}{\partial x}\right)_{z=0}$$
$$+\eta\frac{\partial}{\partial z}\left(w-\frac{\partial \eta}{\partial t}-u\frac{\partial \eta}{\partial x}\right)_{z=0}+\cdots=0 \qquad \text{(III.7)}$$

To retain only the linear terms, we all the nonlinear product terms must be negligible. For example, we would have to require:

$$\left(w>>u\frac{\partial \eta}{\partial x}\right)_{z=0} \qquad \text{(III.8)}$$

From linear wave theory, we have the expression of u, w, and $\partial \eta/\partial x$ at $z=0$ as:

$$w=-\frac{\partial \phi}{\partial z}=\frac{H}{2}\sigma\frac{\sinh kh}{\sinh kh} \qquad \text{(III.9)}$$

$$u = -\frac{\partial \phi}{\partial x} = \frac{H}{2}\sigma \frac{\cosh kh}{\sinh kh} \tag{III.10}$$

$$\frac{\partial \eta}{\partial x} = \frac{H}{2}k \tag{III.11}$$

Substituting (III.9)–(III.11) into III.8, we have:

$$\frac{H}{2}\sigma\frac{\sinh kh}{\sinh kh} >> \frac{H}{2}\sigma\frac{\cosh kh}{\sinh kh} \cdot \frac{H}{2}k \tag{III.12}$$

Rearranging the above equation, we have:

$$\frac{kH}{2} << \tanh kh \,(\le 1) \tag{III.13}$$

Obviously, we have:

$$ka = \frac{kH}{2} << 1 \tag{III.14}$$

If we divide (III.13) by kh on both sides of the equation, we have:

$$\frac{\frac{kH}{2}}{kh} << \frac{\tanh kh}{kh} \,(\le 1) \tag{III.15}$$

This leads to the proof of another small-amplitude wave constraint:

$$\frac{a}{h} = \frac{H}{2h} << 1 \tag{III.16}$$

Appendix IV   Fortran source code for the fifth-order Stokes wave

```
c
c      Program to calculate fifth-order Stokes waves
c
       PROGRAM MAIN
       implicit real*8 (a-h,o-z)
       common aa,h0,xxt,amp,xxk,xxl,sigma,c0,cnf,gz,pi,c,s,
     & a11,a13,a15,a22,a24,a33,a35,a44,a55,b22,b24,b33,b35,b44,b55

       read(*,*)aa,h0,xxt
       gz=-9.8
       pi=acos(-1.0d0)

       call Stokes5

C.......Calculate free surface displacement within one wave period;
C.......u(z,t) and w(z,t) can be similarly calculated using equations
       do j=1,101
       t=(j-1)/100*xxt
```

```fortran
      eta=amp1*cos(pi/2.0-cnf-sigma*t)+
     &   amp2*cos(2.0*(pi/2.0-cnf-sigma*t))
     &  +amp3*cos(3.0*(pi/2.0-cnf-sigma*t))
     &   +amp4*cos(4.0*(pi/2.0-cnf-sigma*t))
     &  +amp5*cos(5.0*(pi/2.0-cnf-sigma*t))
      end do

      end

C---------------------------------------------------------------------------
      SUBROUTINE Stokes5
      implicit real*8 (a-h,o-z)
      common aa,h0,xxt,amp,xxk,xxl,sigma,c0,cnf,gz,pi,c,s,
     & a11,a13,a15,a22,a24,a33,a35,a44,a55,b22,b24,b33,b35,b44,b55
      dimension bb(2,2)

C.......this program calculates wave steepness (amp) and wave number (xxk) using
C.......given wave height (aa), water depth (h0), and wave period (xxt), based
C.......on fifth-order Stokes wave theory as proposed by Skjelbreia and
C.......Hendrickson (1967) (5-th conf. coastal engng.)

C.......The wave steepness (amp) and wave number (xxk) are obtained by
C.......solving two nonlinear algebraic equations (nonlinear wave
C.......amplitude equation (f1) and nonlinear dispersion equation (f2)).
C.......The Newton's method is used to solve the nonlinear equation system.
C.......f1=pi*aa/h0-2*pi/(xxk*h0)*(amp+amp**3*b33+amp**5*(b35+b55))
C.......f2=(2*pi*h0)/(-gz*xxt**2)
C      &   -xxk*h0/(2*pi)*tanh(xxk*h0)*(1+amp**2*c1+amp**4*c2)

C.......first step of Newton's method: guess the value of amp and xxk.
      xxk=2*pi/(xxt*sqrt(-gz*h0))
      amp=aa/2.0*xxk
      n=0
C.......second step: form matrix B=partial (f1,f2)/partial (xxk,amp)
C.......(Jacobian matrix)
C.......prepare the most frequently used coefficients
800   continue
      c=cosh(xxk*h0)
      s=sinh(xxk*h0)
C.......dc/dk and ds/dk
      ckk=h0*s
      sk=h0*c
      b33=(3.0*(8.0*c**6+1.0))/(64.0*s**6)
      b35=(88128.0*c**14-208224.0*c**12+70848.0*c**10+54000.0*c**8
     &      -21816.0*c**6+6264.0*c**4-54.0*c**2-81.0)
     &      /(12288.0*s**12*(6.0*c**2-1.0))
      b55=(192000.0*c**16-262720.0*c**14+83680.0*c**12+20160.0*c**10
     &  -7280.0*c**8+7160.0*c**6-1800.0*c**4-1050.0*c**2+225.0)
     &  /(12288.0*s**10*(6.0*c**2-1.0)*(8.0*c**4-11.0*c**2+3.0))
C.......d(b33)/dk
      b33k=9.0*c**5*ckk/(4.0*s**6)-
     &      (9.0*(8.0*c**6+1.0))/(32.0*s**7)*sk
      b35k=(14.0*88128.0*c**13*ckk-12.0*208224.0*c**11*ckk
     &      +10.0*70848.0*c**9*ckk+8.0*54000.0*c**7*ckk
     &      -6.0*21816.0*c**5*ckk+4.0*6264.0*c**3*ckk
     &      -2.0*54.0*c**1*ckk)
     &      /(12288.0*s**12*(6.0*c**2-1.0))
     &   -(88128.0*c**14-208224.0*c**12+70848.0*c**10+54000.0*c**8
     &      -21816.0*c**6+6264.0*c**4-54.0*c**2-81.0)*12.0
     &      /(12288.0*s**13*(6.0*c**2-1.0))*sk
     &   -(88128.0*c**14-208224.0*c**12+70848.0*c**10+54000.0*c**8
     &      -21816.0*c**6+6264.0*c**4-54.0*c**2-81.0)*12.0*c*ckk
```

```
     &          /(12288.0*s**12*(6.0*c**2-1.0)**2)
      b55k=(16.0*192000.0*c**15*ckk-14.0*262720.0*c**13*ckk
     &      +12.0*83680.0*c**11*ckk+10.0*20160.0*c**9*ckk
     &      -8.0*7280.0*c**7*ckk+6.0*7160.0*c**5*ckk
     &        -4.0*1800.0*c**3*ckk-2.0*1050.0*c**1*ckk)
     &      /(12288.0*s**10*(6.0*c**2-1.0)*(8.0*c**4-11.0*c**2+3.0))
     &      -(192000.0*c**16-262720.0*c**14+83680.0*c**12+20160.0*c**10
     &       -7280.0*c**8+7160.0*c**6-1800.0*c**4-1050.0*c**2+225.0)*10.0
     &       /(12288.0*s**11*(6.0*c**2-1.0)*(8.0*c**4-11.0*c**2+3.0))*sk
     &      -(192000.0*c**16-262720.0*c**14+83680.0*c**12+20160.0*c**10
     &       -7280.0*c**8+7160.0*c**6-1800.0*c**4-1050.0*c**2+225.0)
     &       *12.0*c*ckk
     &       /(12288.0*s**10*(6.0*c**2-1.0)**2*(8.0*c**4-11.0*c**2+3.0))
     &      -(192000.0*c**16-262720.0*c**14+83680.0*c**12+20160.0*c**10
     &       -7280.0*c**8+7160.0*c**6-1800.0*c**4-1050.0*c**2+225.0)
     &       *(32.0*c**3-22.0*c)*ckk
     &       /(12288.0*s**10*(6.0*c**2-1.0)*(8.0*c**4-11.0*c**2+3.0)**2)
C.......c1 the same as the concentration
      cc1=(8.0*c**4-8.0*c**2+9.0)/(8.0*s**4)
      c2=(3840.0*c**12-4096.0*c**10+2592.0*c**8-1008.0*c**6+
     &     5944.0*c**4-1830.0*c**2+147.0)/(512.0*s**10*(6.0*c**2-1.0))

      c1k=(4.0*8.0*c**3*ckk-2.0*8.0*c**1*ckk)/(8.0*s**4)
     &    -(8.0*c**4-8.0*c**2+9.0)*4.0*sk/(8.0*s**5)
      c2k=(12.0*3840.0*c**11*ckk-10.0*4096.0*c**9*ckk
     &     +8.0*2592.0*c**7*ckk-6.0*1008.0*c**5*ckk+
     &     4.0*5944.0*c**3*ckk-2.0*1830.0*c**1*ckk)
     &     /(512.0*s**10*(6.0*c**2-1.0))
     &    -(3840.0*c**12-4096.0*c**10+2592.0*c**8-1008.0*c**6+
     &     5944.0*c**4-1830.0*c**2+147.0)*10.0*sk/
     &     (512.0*s**11*(6.0*c**2-1.0))
     &    -(3840.0*c**12-4096.0*c**10+2592.0*c**8-1008.0*c**6+
     &     5944.0*c**4-1830.0*c**2+147.0)*12.0*c*ckk
     &     /(512.0*s**10*(6.0*c**2-1.0)**2)

C.......calculate B coefficient
C.......partial f1/partial xxk
      bb(1,1)=2.0*pi/(xxk**2*h0)*(amp+amp**3*b33+amp**5*(b35+b55))
     &        -2.0*pi/(xxk*h0)*(amp**3*b33k+amp**5*(b35k+b55k))
C.......partial f1/partial amp
      bb(1,2)=-2.0*pi/(xxk*h0)*
     &        (1.0+3.0*amp**2*b33+5.0*amp**4*(b35+b55))
C.......partial f2/partial xxk
      bb(2,1)=-h0/(2.0*pi)*tanh(xxk*h0)*(1.0+amp**2*cc1+amp**4*c2)
     &        -xxk*h0/(2.0*pi)*(1.0-(tanh(xxk*h0))**2)*h0*
     &          (1.0+amp**2*cc1+amp**4*c2)
     &        -xxk*h0/(2.0*pi)*tanh(xxk*h0)
     &          *(amp**2*c1k+amp**4*c2k)
C.......partial f2/partial amp
      bb(2,2)=-xxk*h0/(2.0*pi)*tanh(xxk*h0)
     &        *(2.0*amp*cc1+4.0*amp**3*c2)

C.......calculate f1 and f2 (equation 1 and equation 2)
      f1=pi*aa/h0-2.0*pi/(xxk*h0)*(amp+amp**3*b33+amp**5*(b35+b55))
      f2=(2.0*pi*h0)/(-gz*xxt**2)-xxk*h0/(2.0*pi)*tanh(xxk*h0)
     &        *(1.0+amp**2*cc1+amp**4*c2)

C.......check if the criteria are satisfied:
      if ((abs(f1).lt.1.0e-12.and.abs(f2).lt.1.0e-12).or.n.gt.100)
     & goto 1000
```

```
C.......solve the two-equation system Bx=-F
C......bb(1,1)*xk + bb(1,2)*xa = -f1
C......bb(2,1)*xk + bb(2,2)*xa = -f2

      xa=(f1*bb(2,1)-f2*bb(1,1))/(bb(1,1)*bb(2,2)-bb(2,1)*bb(1,2))
      xk=(f2*bb(1,2)-f1*bb(2,2))/(bb(1,1)*bb(2,2)-bb(2,1)*bb(1,2))

C......update xxk and amp
      xxk=xxk+xk
      amp=amp+xa

      n=n+1
      goto 800

1000      continue
      write(*,*)'n=',n,'f1=',f1,'f2=',f2,'xxk=',xxk,'amp=',amp

C......so far, we have obtained the wave number k and the steepness \lambda
C......Now we calculate the coefficients needed for free surface and velocities
      b22=(2.0*c**2+1.0)*c/(4.0*s**3)
      b24=(c*(272.0*c**8-504.0*c**6-192.0*c**4+322.0*c**2+21.0))
     &        /(384.0*s**9)
      b44=(c*(768.0*c**10-448.0*c**8-48.0*c**6+48.0*c**4+106.0*c**2
     &        -21.0))/(384.0*s**9*(6.0*c**2-1.0))
      amp1=amp/xxk
      amp2=(amp**2*b22+amp**4*b24)/xxk
      amp3=(amp**3*b33+amp**5*b35)/xxk
      amp4=amp**4*b44/xxk
      amp5=amp**5*b55/xxk
      a11=1.0/s
      a13=-c**2*(5.0*c**2+1.0)/(8.0*s**5)
      a15=-(1184.0*c**10-1440.0*c**8-1992.0*c**6+2641.0*c**4-249.0*c**2
     &   +18.0)/(1536.0*s**11)
      a22=3.0/(8.0*s**4)
      a24=(192.0*c**8-424.0*c**6-312.0*c**4+480.0*c**2-17.0)/
     &     (768.0*s**10)
      a33=(13.0-4.0*c**2)/(64.0*s**7)
      a35=(512.0*c**12+4224.0*c**10-6800.0*c**8-12808.0*c**6+
     &    16704.0*c**4-3154.0*c**2+107.0)/(4096.0*s**13*(6.0*c**2-1.0))
      a44=(80.0*c**6-816.0*c**4+1338.0*c**2-197.0)
     &    /1536.0/s**10/(6.0*c**2-1.0)
      a55=-(2880.0*c**10-72480.0*c**8+324000*c**6-432000*c**4+
     &    163470.0*c**2-16245.0)/61440.0/s**11/(6.0*c**2-1.0)/
     &    (8.0*c**4-11.0*c**2+3.0)

      write(*,*)'a11=',a11,'a13=',a13,'a15=',a15
      write(*,*)'a22=',a22,'a24=',a24,'a33=',a33
      write(*,*)'a35=',a35,'a44=',a44,'a55=',a55
      write(*,*) 'amp1=',amp1,'amp2=',amp2,'amp3=',amp3
      write(*,*) 'amp4=',amp4,'amp5=',amp5

C......find zero crossing using Newton-Raphson's process so that wave can
C......start from cold (zero)
C     x(n+1)=x(n)-f(x)/f'(x)
C......guess when the free surface displacement is zero
      cnf=pi/4.0
      m=0
4000      continue
      coef=amp1*sin(pi/2.0-cnf)+amp2*2.0*sin(2.0*(pi/2.0-cnf))
```

```
     &    +amp3*3.0*sin(3.0*(pi/2.0-cnf))+amp4*4.0*sin(4.0*(pi/2.0-cnf))
     &    +amp5*5.0*sin(5.0*(pi/2.0-cnf))
          f=amp1*cos(pi/2.0-cnf)+amp2*cos(2.0*(pi/2.0-cnf))
     &    +amp3*cos(3.0*(pi/2.0-cnf))+amp4*cos(4.0*(pi/2.0-cnf))
     &    +amp5*cos(5.0*(pi/2.0-cnf))
          if (abs(f).lt.1.0e-12.or.m.gt.100) goto 3200
          xxx=-f/coef
          cnf=cnf+xxx
          goto 4000

C.......summarize the coefficients wavelength, angular frequency, and phase speed
3200      continue
          xxl=2.0d0*pi/xxk
          sigma=2.*pi/xxt
          c0=xxl/xxt
          write(*,*)'xxl=',xxl,'c0=',c0,'cnf=',cnf

          return
          end
```

Appendix V   Fortran source code for a cnoidal wave

```
c
c         Program to calculate cnoidal waves
c
          PROGRAM MAIN
          implicit real*8 (a-h,o-z)
          common aa,h0,xxt,amp,xxk,xxl,sigma,c0,cnf,gz,pi,ytrough,xmod1

          read(*,*)aa,h0,xxt
          gz=-9.8
          pi=acos(-1.0d0)

          call cnoidal

C.......Calculate free surface displacement within one wave period;
C.......u(z,t) and w(z,t) can be similarly calculated using equations
          do j=1,101
          t=(j-1)/100*xxt
             ccn1=cn(cnf*0.5*ck(xmod1)+2.0*ck(xmod1)*(-t/xxt),xmod1)
             csn1=sqrt(1.0-ccn1**2)
             cdn1=sqrt(1.0-xmod1*(1-ccn1**2))
             ccn2=cn(cnf*0.5*ck(xmod1)+2.0*ck(xmod1)*(-t/xxt),xmod1)
             csn2=sqrt(1.0-ccn2**2)
             cdn2=sqrt(1.0-xmod1*(1-ccn2**2))
             eta=ytrough+aa*cn(cnf*0.5*ck(xmod1)+2.0*ck(xmod1)*
     &        (-t/xxt),xmod1)**2
          end do

          end

C----------------------------------------------------------------
          SUBROUTINE cnoidal
          implicit real*8 (a-h,o-z)
          common aa,h0,xxt,amp,xxk,xxl,sigma,c0,cnf,gz,pi,ytrough,xmod1

C.....This subroutine finds parameter y_t (ytrough) and modulus k (xmod1)
          external ck,ce,cn
```

```
C.....Newton-Raphson iteration to determine the parameter m
      xmod=0.99999999d0
      n=0
40    n=n+1
      xa=xmod*h0+2.0d0*aa-xmod*aa-3.0d0*aa*
     & ce(xmod)/ck(xmod)-16.0d0*h0**3*xmod**2*ck(xmod)**2/
     & 3.0d0/(-gz)/aa/xxt**2
      if (abs(xa).le.1.0e-8.or.n.gt.1000) goto 50

      xb=h0-aa+3.0d0*aa/2.0d0/xmod/(1.0d0-xmod)/ck(xmod)**2*
     & ((1.0d0-xmod)*ck(xmod)**2+ce(xmod)**2-2.0d0*(1.0d0-xmod)*
     & ck(xmod)*ce(xmod))-16.0d0*h0**3*xmod*ck(xmod)/3.0d0/(-gz)/
     & (1.0d0-xmod)/aa/xxt**2*((1.0d0-xmod)*ck(xmod)+ce(xmod))
      xmod=xmod-xa/xb
      goto 40
50    continue
C.....after sobey el at (1987, J. Waterway)
      xxl=4.0d0*ck(xmod)*h0*sqrt(xmod*h0/aa/3.0)

C.....after Mei (1983) or simply c=L/T
      xtemp=-xmod+2.0d0-3.0d0*ce(xmod)/ck(xmod)
      c0=sqrt(-gz*h0*(1.0d0+aa/h0/xmod*xtemp))
      ytrough=aa/xmod*(1.0d0-xmod-ce(xmod)/ck(xmod))
      write(*,*)'ytrough=',ytrough,'xmod=',xmod,'c0=',c0

      xmod1=xmod
C.....find the wave parameters and check the validity of theory
       xxk=2.*pi/xxl
       sigma=2.*pi/xxt
      if (xxl.gt.1.0e6) then
            write(*,*)'no cnoidal wave exists for given parameters.'
      stop
      endif

C.....iterative method to find zero-cross point of the cnoidal wave
              zup=1.0
              zlow=0.0
              zmid=(zup+zlow)/2.0d0
              nzero=0
3000           nzero=nzero+1
              zero1=ytrough+aa*cn(zmid*0.5*ck(xmod1),xmod1)**2
              if (abs(zero1).le.1.0e-6) goto 3100
              if (nzero.gt.1000) then
                    write(*,*)'too many iterations; stop'
                    stop
              endif
              if (zero1.lt.0.0) then
                    zup=zmid
                    zmid=(zup+zlow)/2.0d0
                    goto 3000
              else
                    zlow=zmid
                    zmid=(zup+zlow)/2.0d0
                    goto 3000
              endif
3100          continue
      cnf=zmid
      write(*,*) 'xxl=',xxl,'ytrough=',ytrough,'xmod1=',xmod1
      write(*,*) 'xxk=',xxk,'c0=',c0
```

```fortran
      write(*,*) 'cnf=',cnf,'zero1=',zero1

      return
      end

c ................................................................
      function ck(xmod)
      implicit real*8 (a-h,o-z)
       common aa,h0,xxt,amp,xxk,xxl,sigma,c0,cnf,gz,pi,ytrough,xmod1
      xmod2=1.0d0-xmod
      ck=0.0d0
      a0=1.0d0
      b0=sqrt(xmod2)
      c00=sqrt(xmod)
      n=1
15    if (abs(c00).lt.1.0e-15.or.n.gt.1000) then
          goto 20
      else
          n=n+1
          a1=(a0+b0)/2.0d0
          b1=sqrt(a0*b0)
          c11=(a0-b0)/2.0d0
          a0=a1
          b0=b1
          c00=c11
          goto 15
      endif
20    ck=pi/2.0d0/a0

      return
      end

c ................................................................
      function ce(xmod)
      implicit real*8 (a-h,o-z)
       common aa,h0,xxt,amp,xxk,xxl,sigma,c0,cnf,gz,pi,ytrough,xmod1
       dimension ccc(1000)
      ce=0.0d0
      xmod2=1.0d0-xmod
      a0=1.0d0
      b0=sqrt(xmod2)
      c00=sqrt(xmod)
      n=1
      ccc(n)=c00

15    if (abs(c00).lt.1.0e-15.or.n.gt.1000) then
          goto 20
      else
            n=n+1
            a1=(a0+b0)/2.0d0
            b1=sqrt(a0*b0)
            c11=(a0-b0)/2.0d0
            a0=a1
            b0=b1
            c00=c11
            ccc(n)=c00
            goto 15
      endif
20    ck=pi/2.0d0/a0
      sum=0.0d0
```

```fortran
      do 30 k1=1,n
        sum=sum+2.0d0**(k1-2)*ccc(k1)**2
30    continue
      ce=ck*(1.0d0-sum)

      return
      end

c .........................................................
      function cn(u,xmod)
      implicit real*8 (a-h,o-z)
       common aa,h0,xxt,amp,xxk,xxl,sigma,c0,cnf,gz,pi,ytrough,xmod1
       dimension yy(1000),c(1000),a(1000)
      xmod2=1.0d0-xmod
      a0=1.0d0
      b0=sqrt(xmod2)
      c00=sqrt(xmod)
      n=1
      a(n)=a0
      c(n)=c00

15    if (abs(c00).lt.1.0e-15.or.n.gt.1000) then
          goto 20
        else
          n=n+1
          a1=(a0+b0)/2.0
          b1=sqrt(a0*b0)
          c11=(a0-b0)/2.0
          a0=a1
          b0=b1
          c00=c11
          a(n)=a0
          c(n)=c00
          goto 15
        endif
20    yy(n)=2.0**(n-1)*a(n)*u

      do 10 i=n-1,1,-1
        yy(i)=(yy(i+1)+asin(c(i+1)/a(i+1)*sin(yy(i+1))))/2.0
10    continue
      cn=cos(yy(1))

      return
      end
```

Appendix VI Matlab script for wave diffraction behind a semi-infinite breakwater

```matlab
clear;
[XX]=load('Yspec.dat');
N=100;
mm=N;
X=(XX(1:mm)+XX(2:mm+1))/2-1500; %change y according to equation
Y=[1.62, 6.50, 19.50]; %set the section position x/L
k=0.068059;
L=2*pi/k;
Y=Y*L;
```

```
for ii=1:N
  for jj=1:3
    x=X(ii);        %position of x
    y=Y(jj);
    r=sqrt(x*x+y*y);
    Beta =sqrt(4/L*(r-y));
    Beta_=sqrt(4/L*(r+y));
    i=sqrt(-1);
    if(x>=0 && y>0)
    Fxy=(1+i)/2*(exp(-i*k*y)*(0.5-0.5*i-(mfun('FresnelC',Beta)...
    -i*mfun('FresnelS',Beta)))+exp(i*k*y)*(0.5-0.5*i-...
    (mfun('FresnelC',Beta_)-i*mfun('FresnelS',Beta_))));
    elseif(x<0 && y>0)
    Fxy=(1+i)/2*(exp(-i*k*y)*(0.5-0.5*i+(mfun('FresnelC',Beta)...
    -i*mfun('FresnelS',Beta)))+exp(i*k*y)*(0.5-0.5*i-...
    (mfun('FresnelC',Beta_)-i*mfun('FresnelS',Beta_))));
    end
  HH(jj,ii)=abs(Fxy);
  end
end

for ii=1:N
  for jj=1:3
    H(jj,N-ii+1)=HH(jj,ii);
  end
end
plot(X/L,H)
save('ExactSolution.dat','H','-ASCII')
%============================================================%
%The above calculation makes use of the following conversion%
```

$$\int\limits_{-\infty}^{-\beta} e^{-i(\pi/2)u^2}\,du = \int\limits_{-\infty}^{0} e^{-i(\pi/2)u^2}\,du - \int\limits_{-\beta}^{0} e^{-i(\pi/2)u^2}\,du$$

$$\% = \int\limits_{-\infty}^{0} \cos\left(\frac{\pi}{2}u^2\right)du - i*\int\limits_{-\infty}^{0} \sin\left(\frac{\pi}{2}u^2\right)du - \left(\int\limits_{-\beta}^{0} \cos\left(\frac{\pi}{2}u^2\right)du - i*\int\limits_{-\beta}^{0} \sin\left(\frac{\pi}{2}u^2\right)du\right)$$

$$= \frac{1}{2} - \frac{1}{2}i - \left(FresnelC(\beta) - i*FresnelS(\beta)\right)$$

Appendix VII   Matlab script for wave diffraction around a large vertical circular cylinder

```
%This script calculates time history of wave profile around a large vertical
%circular cylinder, from which the diffraction coefficients are derived.
format long

%USER INPUT AREA%
  %DISC PARAMETERS IN SI UNITS%
  r0=5.0;         %Disc radius in meters%
  teta0=0.0;

  %WAVE PROPERTIES (LINEAR WAVE ONLY)%
  g=9.81;          %Gravitational acceleration%
  rho2=1000;       %Density of water%
  h=80;            %Still water depth%
```

```
    period=2.0;      %Wave period%
    WH=0.5;          %Wave height; insignificant for diffraction coef.%
%END OF USER INPUT%


%TO DETERMINE WAVE NUMBER%
    omega=2*pi/period;
    k=0.0;
    beta2=0.0;
    beta=omega^2/g;
    while (k<1000)
          if (beta2<beta+1E-4)&(beta2>beta-1E-4)
            break
          else
            beta2=k*tanh(k*h);
            k=k+0.00001;
          end
    end
kh=k*h;   %Check for intermediate to deep water range%


%START OF MAIN LOOP OF COMPUTATION%
    num1=1;
    for teta=pi:-pi/200:0
        Fmax(num1)=0.0;
        Fmin(num1)=0.0;
        num1=num1+1;
    end
    num=1;
    for t=period/50:period/50:period
        r=r0;
        num1=1;
        for teta=pi:-pi/200:0
           bwc=0.0;
             for m=1:1:100;
               if (m==1)
                  delta=1;
                else
                  delta=2;
                end
               JP=(k/2)*(besselj(m-2,k*r0)-besselj(m,k*r0));
               HP=(k/2)*(besselh(m-2,k*r0)-besselh(m,k*r0));
               bwc=bwc+(delta*(i^(m))*(besselj(m-1,k*r)-((JP/HP)*...
                  besselh(m-1,k*r)))*cos((m-1)*(teta)));
              end
              eta=bwc;
              F(num1)=(eta)*WH/2*exp((-i)*omega*t);
          num1=num1+1;
        end


       num1=1;
       for teta=pi:-pi/200:0
         if(F(num1)>Fmax(num1))
             Fmax(num1)=F(num1);
         end
         if(F(num1)<Fmin(num1))
             Fmin(num1)=F(num1);
          end
         num1=num1+1;
        end
     progress=t/period*100
     num=num+1;
```

```
      end
%END OF COMPUTATION LOOP%

%Determine x-axis for plotting graph%
num1=1;
for teta=pi:-pi/200:0
   xp(num1)=teta;
   num1=num1+1;
end
num1=1;
for teta=pi:-pi/200:0
   Falt(num1)=Fmax(num1)-Fmin(num1);
   num1=num1+1;
end

%PLOTTING FOR RESULTS%
Fl=real(Falt);
Fplot=Fl/WH;
plot(xp,Fplot')
axis([0 pi 0 2.1])
xlabel '\theta (rad)'
ylabel 'Diffraction coefficient H/H_0'

%END OF FILE%
```

Appendix VIII   Matlab script for long-wave reflection and transmission through a submerged breakwater or a trench

```
% This Matlab script calculates reflected and transmitted wave heights over
% a submerged trapezoidal breakwater or trench (Lin and Liu, 2005)
% using long wave approximation. The input data include incident
% wave height and period, wave depths in front of, above and behind the
% crown, the length of the crown, and the frontal and back slopes.
clear ;
h0=2;          %Water depth in front of the breakwater
h1=4;          %Water depth on the crown (or h1>h0 for trench)
h2=2;          %Water depth behind the breakwater
T=50 ;         %Incident wave period
FrontSlope = .1;    %Frontal slope tan(theta1)
BackSlope  = .1;    %Back slope tan(theta2)
CrownLength= 10;    %Length of the Crown
HI=1 ;         %Incident wave height.
g=9.81;        %Gravity
Lambda=T*sqrt(g*h0); %incident wave Length in front of the breakwater

M1= abs(h1-h0)/FrontSlope/Lambda;
M2= CrownLength/Lambda;
M3= abs(h2-h1)/BackSlope/Lambda;
s01=sqrt(h0/h1);
s21=sqrt(h2/h1);

a0=4*pi*M1*s01^2/(s01^2-1);
a1=4*pi*M1*s01/(s01^2-1);
b1=4*pi*M3*s01/(s21^2-1);
b2=4*pi*M3*s01*s21/(s21^2-1);

FI=2*pi*M2*s01;
c00a=det([besselj(0,a0),besselj(0,a1);bessely(0,a0),bessely(0,a1)]);
c01a=det([besselj(0,a0),besselj(1,a1);bessely(0,a0),bessely(1,a1)]);
```

```
c10a=det([besselj(1,a0),besselj(0,a1);bessely(1,a0),bessely(0,a1)]);
c11a=det([besselj(1,a0),besselj(1,a1);bessely(1,a0),bessely(1,a1)]);
c00b=det([besselj(0,b1),besselj(0,b2);bessely(0,b1),bessely(0,b2)]);
c01b=det([besselj(0,b1),besselj(1,b2);bessely(0,b1),bessely(1,b2)]);
c10b=det([besselj(1,b1),besselj(0,b2);bessely(1,b1),bessely(0,b2)]);
c11b=det([besselj(1,b1),besselj(1,b2);bessely(1,b1),bessely(1,b2)]);

P=c01b+i*c00b;
Q=c11b+i*c10b;

z1=c11a*(P*cos(FI)+Q*sin(FI))+c10a*(Q*cos(FI)-P*sin(FI));
z2=c01a*(P*cos(FI)+Q*sin(FI))+c00a*(Q*cos(FI)-P*sin(FI));

HR_HI=-(z1-i*z2)/(z1+i*z2)*exp(-i*a0);
KR=abs(HR_HI)           %Reflection coefficient
HR=KR*HI                %Reflection wave height
Cgd=sqrt(g*h0);            %Group veloctiy at h0
Cgs=sqrt(g*h2);            %Group velocity at h1
HT=sqrt((Cgd*HI^2-Cgd*HR^2)/Cgs) %Transmitted wave height
HT_HI=HT/HI;              %Transmitted coefficient
```

Appendix IX   Abbreviation list

ABS	American Bureau of Shipping
ACM	artificial compressibility method
ADI	alternating direction implicit
ALE	arbitrary Lagrangian–Eulerian
ANN	artificial neural network
ASM	algebraic stress model
BEM	boundary element method
BGK	Bhatnagar–Gross–Krook
BIEM	Boundary integral equation method
BV	Bureau Veritas France
CFD	computational fluid dynamics
CFL	Courant–Friedrichs–Lewy
CG	conjugate gradient
COHERENS	COupled Hydrodynamical Ecological model for REgioNal Shelf seas
COMCOT	COrnell Multigrid COupled Tsunami
CSM	Chebyshev spectral method
DEM	discrete element method
DES	detached eddy simulation
DHI	Danish Hydraulic Institute
DNS	direct numerical simulation
DNV	Det Norske Veritas
DOF	degree of freedom
DRBEM	dual reciprocity boundary element method
DVM	discrete vortex method
EBM	energy balance model

EFGM	element-free Galerkin method
EMSE	elliptic mild-slope equation
ENO	essentially nonoscillatory
FD	finite difference
FDM	finite difference method
FEM	finite element method
F–K	Froude–Krylov
FPM	finite point method
FPSO	floating production storage and offloading
FSI	fluid–structure interaction
FSM	Fourier spectral method
FTCS	forward-time and central-space
FVM	finite volume method
GA	Genetic algorithm
GFDM	generalized finite difference method
GLM	generalized Lagrangian mean
GRA	global response analysis
IB	immersed boundary
ICCG	incomplete Cholesky conjugate gradient
ISSC	International Ship Structures Congress
JONSWAP	JOint North Sea WAve Project
KC	Keulegan–Carpenter
KdV	Korteweg–de Vries
KP	Kadomtsev–Petviashvili
LBE	Lattice Boltzmann equation
LBM	Lattice Boltzmann method
LDA	Laser Doppler anemometer
LES	Large eddy simulation
LG	lattice gas
LHS	left-hand side
LRS	locally relative stationary
LS	least-squares
LSM	Legendre spectral method
MAC	marker-and-cell
ME	mortar element
MLPG	meshless local Petrov–Galerkin
MLS	moving least-squares
MMSE	modified mild-slope equation
MOSES	Multioperational Structural Engineering Simulator
MOST	Method Of Splitting Tsunami
MPS	Moving Particle Semi-implicit
MSE	mild-slope equation
NCEP	National Centers for Environmental Prediction
NEM	natural element method
NEWTANK	NumErical Wave TANK

NLS	nonlinear Schrödinger equation
NOAA	National Oceanic & Atmospheric Administration
NSE	Navier–Stokes-equation
NSEs	Navier–Stokes equations
ODE	ordinary differential equation
PCT	partial-cell treatment
PDE	partial differential equation
PIC	particle-in-cell
PISO	pressure implicit with splitting of operators
PIV	particle image velocimetry
P–M	Pierson–Moskowitz
PMSE	parabolic mild-slope equation
POM	Princeton Ocean model
PPE	Poisson pressure equation
QUICK	Quadratic Upwind Interpolation for Convective Kinematics
QUICKEST	Quadratic Upstream Interpolation for Convective Kinematics with Estimated Streaming Terms
RANS	Reynolds-averaged Navier–Stokes
RAO	response amplitude operator
RBF	radial basis function
RHS	right-hand side
RKPM	reproducing kernel particle method
RNG or RG	renormalization group
RTD	reflection, transmission, and dissipation
SANS	spatially-averaged Navier–Stokes
SGS	subgrid scale
SHG	second harmonic generation
SIMPLE	Semi-Implicit Method for Pressure-linked Equations
SIMPLEC	SIMPLE Consistent
SIMPLEM	SIMPLE Modified
SIMPLER	SIMPLE Revised
SIMPLEST	SIMPLE ShorTened
SM	spectral method
SOLA	A numerical SOLution Algorithm for transient fluid flows
SOR	Successive over-relaxation
SPH	smoothed particle hydrodynamics
SSH	sea surface height
SUMMAC	the Stanford-University-Modified-Marker-And-Cell method
SVD	singular-value decomposition
SWAN	Simulating WAves Nearshore
SWE	shallow water equation
TDV	total variation diminishing
TE	truncation error
TLP	tension-leg platform
TUMMAC	Tokyo-University-Modified-Marker-And-Cell method

ULTIMATE	Universal Limiter for Transport Interpolation Modeling of the Advective Transport Equation
USACE	the United States Army Corps of Engineers
VBF	virtual boundary force
VIV	vortex-induced vibration
VLFS	very large floating structure
VOF	volume of fluid
WAM	WAve Simulation Model
WAMIT	Wave Analysis at Massachusetts Institute of Technology
WENO	Weighted ENO
WKB	Wentzel–Kramers–Brillouin

References

Abbott, M.B., Petersen, H.M., and Skovgaard, P., 1978. On the numerical modeling of short waves in shallow water, *J. Hydraul. Res.*, 16, 173–203.

Ablowitz, M. and Segur, H., 1981. *Solitons and the Inverse Scattering Transform*, SIAM, Philadelphia, PA.

Abohadima, S. and Isobe, M., 1999. Linear and nonlinear wave diffraction using the nonlinear time-dependent mild slope equations, *Coast. Eng.*, 37, 175–92.

Abramowitz, M. and Stegun, I.A., 1964. *Handbook of Mathematical Functions, Applied Mathematics Series*, 55, New York: Dover.

Acharya, S. and Moukalled, F., 1989. Improvements to incompressible flow calculation on a non-staggered curvilinear grid, *Numer. Heat Transf. B – Fundam.*, 15, 131–52.

Agnon, Y. and Pelinovsky, E., 2001. Accurate refraction–diffraction equations for water waves on a variable-depth rough bottom, *J. Fluid Mech.*, 449, 301–11.

Ahrens, J.P, 1981. Irregular wave runup on smooth slopes, *CETA No. 81-17*, US Army Corps of Eng., Coastal Eng. Res. Center, Ft. Belvoir, VA.

Akylas, T.R., 1984. On the excitation of long nonlinear water waves by a moving pressure distribution, *J. Fluid Mech.*, 141, 455–66.

Aliabadi, S., Abedi, J., and Zellars, B., 2003. Parallel finite element simulation of mooring forces on floating objects, *Int. J. Numer. Meth. Fluids*, 41, 809–22.

Allsop, N.W.H. and Channell, A.R., 1988. Wave reflections in harbours, The reflection performance of wave screens, *Report OD 102*, Wallingford.

Allsop, W., Bruce, T., Pearson, J., and Besley, P., 2005. Wave overtopping at vertical and steep seawalls, *Proc. Ice-Marit. Eng.*, 158, 103–14.

Andersen, O.H., 1994. Flow in porous media with special reference to breakwater structures. Ph.D. Thesis, Hydraulics and Coastal Engineering Laboratory, Aalborg University.

Andrews, D.G. and McIntyre, M.E., 1978. An exact theory of nonlinear waves on a Lagrangian mean flow, *J. Fluid Mech.*, 89, 609–46.

Andrianov, A.I. and Hermans, A.J., 2005. Hydroelasticity of a circular plate on water of finite or infinite depth, *J. Fluids Struct.*, 20, 719–33.

Apsley, D., Chen, W.L., Leschziner, M., and Lien, F.S., 1998. Non-linear eddy-viscosity modeling of separated flows, *J. Hydraul. Res.*, 35, 723–48.

Armenio, V., 1997. An improved MAC method (SIMAC) for unsteady high-Reynolds free surface flows, *Int. J. Numer. Meth. Fluids*, 24, 185–214.

Armenio, V. and La Rocca, M., 1996. On the analysis of sloshing of water in rectangular containers: numerical and experimental investigation, *Ocean Eng.*, 23, 705–39.

Arnskov, M.M., Fradsoe, J., and Sumer, B.M., 1993. Bed shear stress measurements over a smooth bed in three-dimensional wave–current motion, *Coast. Eng.*, 20, 277–316.

Ata, R. and Soulaimani, A., 2005. A stabilized SPH method for inviscid shallow water flows, *Int. J. Numer. Meth. Fluids*, 47, 139–59.

Atluri, S.N. and Zhu, T., 1998. A new meshless local Petrov–Galerkin (MLPG) approach in computational mechanics, *Comput. Mech.*, 22, 117–27.

Austin, D.I. and Schlueter, R.S., 1982. A numerical model of wave breaking/breakwater interactions, *Proc. 18th Int. Conf. Coast Eng.*, Cape Town, Republic of South Africa, pp. 2079–96.

Baaijens, F.P.T., 2001. A fictitious domain/mortar element method for fluid–structure interaction, *Int. J. Numer. Meth. Fluids*, 35, 743–61.

Babovic, V., 1998. A data mining approach to time series modelling and forecasting, *Hydroinformatics, '98. Proc. Third Int. Conf. Hydroinformatics*, Copenhagen, Denmark, August 1998, pp. 24–26.

Babuska, I. and Melenk, J.M., 1997. The partition of unity method, *Int. J. Numer. Meth. Eng.*, 40, 727–58.

Barthelemy, E., 2004. Nonlinear shallow water theories for coastal waves, *Surv. Geophys.*, 25, 315–37.

Bartholomeusz, E.F., 1958. The reflection of long waves at a step, *Proc. Camb. Philos. Soc.*, 54, 106–18.

Battjes, J.A., 1974. Surf similarity parameter, *Proc. 14th Int. Conf. Coast Eng.*, pp. 69–85.

Battjes, J.A. and Janssen, J.P.F.M., 1978. Energy loss and set-up due to breaking of random waves, *Proc. 16th Conf Coast. Eng.*, Hamburg, Germany, pp. 569–87.

Behr, M. and Tezduyar, T.E., 1994. Finite-element solution strategies for large-scale flow simulations, *Comput. Meth. Appl. Mech. Eng.*, 112, 3–24.

Beji, S. and Battjes, J.A., 1993. Experimental investigation of wave propagation over a bar, *Coast. Eng.*, 19, 151–62.

Beji, S. and Nadaoka, K., 1996. A formal derivation and numerical modelling of the improved Boussinesq equations for varying depth, *Ocean Eng.*, 23(8), 691–704.

Belytschko, T. and Chen, J.S., 2007. *Meshfree and Particle Methods*, New York: John Wiley and Sons Ltd.

Belytschko, T., Lu, Y.Y., and Gu, L., 1994. Element-free Galerkin methods, *Int. J. Numer. Meth. Fluids*, 37, 229–56.

Belytschko, T., Krongauz, Y., Organ, D., and Fleming, M., 1996. Meshless methods: an overview and recent developments, *Comput. Meth. Appl. Mech. Eng.*, 119, 3–47.

Ben, B.F. and Maday, Y., 1997. The mortar finite element method for three dimensional finite elements, *RAIRO Numer. Anal.*, 31, 289–302.

Benek, J.A., Buning, P.G., and Steger, J.L., 1985. A 3D chimera grid embedding technique, *AIAA J.*, 85–1523, 322–31.

Benjamin, T.B. and Ursell, F., 1954. The stability of the plane free surface of a liquid in a vertical periodic motion, *Proc. R. Soc. Load. Ser. A*, 225, 505–15.

Benjamin, T.B. and Feir, J.E., 1967. The disintegration of wave trains on deep water Part 1. Theory, *J. Fluid Mech.*, 27, 417–30.

Benzi, R., Succi, S., and Vergassola, M., 1992. The lattice Boltzmann equation theory and applications, *Phys. Rep.*, 222, 145–97.

Berger, V. and Kohlhase, S., 1976. Mach-reflection as a diffraction problem, *Proc. 15th Int. Coast. Eng. Conf.*, pp. 796–814.

Berkhoff, J.C.W., 1972. Computation of combined refraction diffraction, *Proc. 13th Int. Coast. Eng. Conf.*, New York, USA, pp. 471–90.

Berkhoff, J.C.W., Booy, N., and Radder, A.C., 1982. Verification of numerical wave propagation models for simple harmonic linear water waves, *Coast. Eng.*, 6, 255–79.

Bermudez, A. and Rodriguez, R., 1999. Finite element analysis of sloshing and hydroelastic vibrations under gravity, *RAIRO Numer. Anal.*, 33, 305–27.

Berthelsen, P.A. and Ytrehus, T., 2005. Calculations of stratified wavy two-phase flow in pipes, *Int. J. Multiph. Flow*, 31, 571–92.

Bettess, P. and Zienkiewicz, O.C, 1977. Diffraction and refraction surface waves using finite and infinite elements, *Int. J. Numer. Methods Eng.*, 11, 1271–90.

Bhatnagar, P.L., Gross, E.P., and Krook, M., 1954. A model for collision processes in gases. I. Small amplitude processes in charged and neutral one-component system, *Phys. Rev.*, 94, 511–25.

Bicanic, N., 2004. Discrete element methods, in: Stein, de Borst and Hughes, *Encyclopedia of Computational Mathematics*, Vol. 1, New York: Wiley.

Blumberg, A.F. and Mellor, G.L., 1987. A description of a three-dimensional coastal ocean circulation model, in: N. Heaps (Ed.), *Three-dimensional Coastal Ocean Models*, Washington, DC: American Geophysical Union, p. 208.

Bokil, V.A. and Glowinski, R., 2005. An operator splitting scheme with a distributed Lagrange multiplier based fictitious domain method for wave propagation problems, *J. Comput. Phys.*, 205, 242–68.

Booij, N., 1981. Gravity waves on water with non-uniform depth and current, *Report No. 81-1*, Delft Univ. of Tech., Holland.

Booij, N., 1983. A note on the accuracy of the mild-slope approximation, *Coast. Eng.*, 7, 191–203.

Booij, N., Holthuijsen, L.H., Doorn, N., and Kieftenburg, A.T.M.M., 1997. Diffraction in a spectral wave model, *Proc. 3rd Int. Symp. Ocean Wave Measurement and Analysis*, WAVES'97, Virginia Beach, VA, USA, pp. 243–55.

Booij, N., Ris, R.C., and Holthuijsen, L.H., 1999. A third-generation wave model for coastal regions. 1. Model description and validation, *J. Geophys. Res. Ocean*, 104, 7649–66.

Boussinesq, M.J., 1871. Theorie de l'intumescence appelee onde solitaire ou. de translation se propageant dans un canal rectangulaire., *C. R. Acad. Sci. Paris*, 72, 755–9.

Bouws, E. and Komen, G.J., 1983. On the balance between growth and dissipation in an extreme depth-limited wind-sea in the southern North Sea, *J. Phys. Oceanogr.*, 13, 1653–8.

Bouws, E., Gunther, H., Rosenthal, W., and Vincent, C., 1985. Similarity of the wind wave spectrum in finite depth water, *J. Geophys. Res. Ocean*, 90, 975–86.

Bretherton, F.P. and Garret, C.J.R., 1969. Wave trains inhomogeneous moving media, *Proc. R. Soc. Lond.*, A302, 529–54.

Bretschneider, 1959. Wave variability and wave spectra for wind-generated gravity waves, Beach Erosion Board, *US Army Corps Eng.*, *Tech. Memo.*, 118.

Bridges, T.J. and Mielke, A., 1995. A proof of the Benjamin–Feir instability, *Arch. Ration. Mech. Anal.*, 133, 145–98.

Briggs, M.J., Borgman, L.E., and Outlaw, D.G., 1987. Generation and analysis of directional spectral waves in a laboratory basin, OTC 5416, *Proc. Offshore Tech. Conf.*, Houston, TX, USA.

Briggs, M., Ye, W., Demirbilek, Z., and Zhang, J., 2002. Field and numerical comparisons of the RIBS floating breakwater, *J. Hydraul. Res.*, 40, 289–301.

Burcharth, H.F. and Andersen, O.H., 1995. On the one-dimensional steady and unsteady porous flow equations, *Coast. Eng.*, 24, 233–57.

Burcharth, H.F. and Hughes, S.A., 2002. Fundamentals of design, in: S. Hughes (Ed.), *Coastal Engineering Manual, Part VI, Design of Coastal Project Elements, Chapter VI-5, Engineering Manual 1110-2-1100*, Washington, DC: US Army Corps of Engineering.

Burke, J.E., 1964. Scattering of surface waves on an infinitely deep fluid, *J. Math. Phys.*, 5, 805–19.

Canuto, C., Hussaini, M.Y., Quarteroni, A., and Zhang, T.A., 1987. *Spectral Methods in Fluid Dynamics*, Berlin: Springer-Verlag.

Cao, Y., Beck, R.F., and Schultz, W.W., 1993. Numerical computations of two-dimensional solitary waves generated by moving disturbances, *Int. J. Numer. Meth. Fluids*, 17, 905–20.

Carl, W., 1996. *The Ocean Circulation Inverse Problem*, Cambridge: Cambridge University Press.

Carman, P.C., 1937. Fluid flow through granular beds, *Trans. Inst. Chem. Eng.*, 15, 150–66.

Carrier, G.F. and Greenspan, H.P., 1958. Water waves of finite amplitude on a sloping beach, *J. Fluid Mech.*, 4, 97–109.

Carrier, G.F. and Yeh, H., 2002. Exact long wave runup solution for arbitrary offshore disturbance, *27th General Assembly of European Geophysical Society (EGS)*, Nice, France, EGS02-01939.

Carrier, G.F., Wu, T.T., and Yeh, H., 2003. Tsunami run-up and draw-down on a plane beach, *J. Fluid Mech.*, 475, 79–99.

Carter, R.W., Ertekin, R.C., and Lin, P, 2006. On the reverse flow beneath a submerged plate due to wave action, *OMAE 2006 Proceedings*, OMAE2006-92623.

Casulli, V., 1999. A semi-implicit finite difference method for non-hydrostatic, free-surface flows, *Int. J. Numer. Meth. Fluids*, 30, 425–40.

Casulli, V. and Cheng, R.T., 1992. Semi-implicit finite difference methods for three-dimensional shallow water flow, *Int. J. Numer. Meth. Fluids*, 15, 629–48.

Casulli, V. and Zanolli, P., 2002. Semi-implicit numerical modeling of nonhydrostatic free-surface flows for environmental problems, *Math. Comput. Model.*, 36, 1131–49.

Celebi, M.S. and Akyildiz, H., 2002. Nonlinear modeling of liquid sloshing in a moving rectangular tank, *Ocean Eng.*, 29, 1527–53.

Celebi, M.S., Kim, M.H., and Beck, R.F., 1998. Fully nonlinear 3-D numerical wave tank simulation, *J. Ship Res.*, 42, 33–45.

CERC, 1984. Shore protection manual, *Coastal Eng. Res. Ctr.*, US Army Corps of Eng., Vicksburg.

Chakrabarti, S.K., 1971. Nonlinear wave forces on vertical cylinder, *J. Hydraul. Div. ASCE*, 98 (HY11), Paper 9333, 1895–909.

Chakrabarti, S.K., 1978. Wave forces on multiple vertical cylinders, *J. Waterw. Port Coast. Ocean Eng. ASCE*, 104, 147–61.

Chakrabarti, S.K., 1987. *Hydrodynamics of Offshore Structures*, WIT Press, UK.

Chakrabarti, S.K., 1994. *Offshore Structures Modeling*, World Scientific, Singapore.

Chakrabarti, S.K., 2002. *The Theory and Practice of Hydrodynamics and Vibration*, World Scientific, Singapore.

Chakrabarti, S.K. and Naftzger, R.A., 1974. Nonlinear wave forces on half cylinder and hemisphere, *J. Waterw. Port Coast. Ocean Eng. Div. ASCE*, 100, 189–204.

Chamberlain, P.G. and Porter, D., 1995. The modified mild-slope equation, *J. Fluid Mech.*, 291, 393–407.

Chamberlain, P.G. and Porter, D., 1999. Scattering and near-trapping of water waves by axisymmetric topography, *J. Fluid Mech.*, 388, 335–54.

Chan, C.T. and Anastasiou, K., 1999. Solution of incompressible flows with or without a free surface using the finite volume method on unstructured triangular meshes, *Int. J. Numer. Meth. Fluids*, 29, 35–57.

Chan, K.C. and Street, R.L., 1970. A computer study of finite amplitude water wave, *J. Comput. Phys.*, 6, 68–94.

Chandrasekera, C.N. and Cheung, K.F., 1997. Extended linear refraction–diffraction model, *J. Waterw. Port Coast. Ocean Eng. ASCE*, 123, 280–6.

Chang, C.C. and Chern, R.L., 1991. A numerical study of flow around an impulsively started circular cylinder by a deterministic vortex method, *J. Fluid Mech.*, 233, 243–63.

Chang, K.A., Hsu, T.J., and Liu, P.L.F., 2001. Vortex generation and evolution in water waves propagating over a submerged rectangular obstacle. Part I. Solitary waves, *Coast. Eng.*, 44, 13–36.

Chang, K.A., Hsu, T.J., and Liu, P.L.F., 2005. Vortex generation and evolution in water waves propagating over a submerged rectangular obstacle. Part II. Cnoidal waves, *Coast. Eng.*, 52, 257–283.

Chang, Y.C., Hou, T.Y., Merriman, B., and Osher, S., 1996. A level set formulation of Eulerian interface capturing methods for incompressible fluid flows, *J. Comput. Phys.*, 124, 449–64.

Chapman, R.S. and Kuo, C.Y., 1985. Application of the two-equation *k*–epsilon turbulence model to a two-dimensional steady, free surface flow problem with separation, *Int. J. Numer. Meth. Fluids*, 5, 257–68.

Chau, F.P. and Taylor, R.E., 1992. Second-order wave diffraction by a vertical cylinder, *J. Fluid Mech.*, 240, 571–99.

Chaudhry, M.H., 1993. *Open-channel Flow*, Englewood Cliffs, NJ: Prentice-Hall.

Chawla, A., Özkan-Haller, H.T., and Kirby, J.T., 1998. Spectral model for wave transformation over irregular bathymetry, *J. Waterw. Port Coast. Ocean Eng. ASCE*, 124, 189–98.

Chen, B.F., 2005. Viscous fluid in tank under coupled surge, heave, and pitch motions, *J. Waterw. Port Coast. Ocean Eng. ASCE*, 131, 239–56.

Chen, B.F. and Chiang, H.W., 1999. Complete 2D and fully nonlinear analysis of ideal fluid in tanks, *J. Eng. Mech. ASCE*, 125, 70–8.

Chen, H.C. and Chen, M., 1998. Chimera RANS simulation of a berthing DDG-51 Ship in translational and rotational motions, *Int. J. Offshore Polar Eng.*, 8, 182–191.

Chen, M.Y. and Mei, C.C., 2006. Second-order refraction and diffraction of surface water waves, *J. Fluid Mech.*, 552, 137–66.

Chen, Q., 2006. Fully nonlinear Boussinesq-type equations for waves and currents over porous beds, *J. Eng. Mech. ASCE*, 132, 220–30.

Chen, Q., Kirby, J.T., Dalrymple, R.A., Kennedy, A.B., Thornton, E.B., and Shi, F., 2001. Boussinesq modeling of waves and longshore currents under field conditions, *Proc. 27th Int. Conf. Coast Eng.*, Sydney, Australia, pp. 651–63.

Chen, Q., Kirby, J.T., Dalrymple, R.A., Shi, F., and Thornton, E.B., 2003. Boussinesq modeling of longshore currents, *J. Geophys. Res.*, 108, 3362–80.

Chen, S. and Doolen, G.D., 1998. Lattice Boltzmann method for fluid flows, *Annu. Rev. Fluid Mech.*, 30, 329–64.

Chen, W., Panchang, V., and Demirbilek, Z., 2005. On the modeling of wave-current interaction using the elliptic mild-slope wave equation, *Ocean Eng.*, 32, 2135–64.

Chen, X.B., Duan, W.Y., and Lu, D.Q., 2006. Gravity waves with effect of surface tension and fluid viscosity, *Proc. Conf. Global Chinese Scholars on Hydrodynamics*, Shanghai, China, pp. 171–6.

Chen, Y. and Liu, P.L.F., 1995. Modified Boussinesq equations and associated parabolic models for water wave propagation, *J. Fluid Mech.*, 288, 351–81.

Chen, Y. and Guza, R.T., 1998. Resonant scattering of edge waves by longshore periodic topography, *J. Fluid Mech.*, 369, 91–123.

Chen, Y. and Liu, P.L.F., 1998. A generalized modified Kadomtsev–Petviashvili equation for interfacial wave propagation near the critical depth level, *Wave Motion*, 27, 321–39.

Chern, M.J., Borthwick, A.G.L., and Taylor, R.E., 2005. Pseudospectral element model for free surface viscous flows, *Int. J. Numer. Meth. Heat Fluid Flow*, 15, 517–54.

Chew, C.S., Yeo, K.S., and Shu, C., 2006. A generalized finite-difference (GFD) ALE scheme for incompressible flows around moving solid bodies on hybrid meshfree-Cartesian grids, *J. Comput. Phys.*, 218, 510–48.

Chikazawa, Y., Koshizuka, S., and Oka, Y., 2001. A particle method for elastic and visco-plastic structures and fluid–structure interactions, *Comput. Mech.*, 27, 97–106.

Cho, Y.S., 1995. Numerical simulations of tsunami propagation and run-up, Ph.D. Dissertation, Cornell University.

Cho, Y.S. and Lee, C., 2000. Resonant reflection of waves over sinusoidally varying topographies, *J. Coast. Res.*, 16, 870–6.

Choi, B.H., Eum, H.M., and Woo, S.B., 2003. A synchronously coupled tide–wave–surge model of the Yellow Sea, *Coast. Eng.*, 47, 381–98.

Choi, W. and Camassa, R., 1996. Weakly nonlinear internal waves in a two-fluid system, *J. Fluid Mech.*, 313, 83–103.

Choi, W. and Camassa, R., 1999. Fully nonlinear internal waves in a two-fluid system, *J. Fluid Mech.*, 396, 1–36.

Chorin, A.J., 1967. A numerical method for solving incompressible viscous flow problems, *J. Comput. Phys.*, 2, 12–26.

Chorin, A.J., 1968. Numerical solution of the Navier–Stokes equations, *Math. Comput.*, 22, 745–62.

Chorin, A.J., 1969. On the convergence of discrete approximations of the Navier–Stokes equations, *Math. Comput.*, 23, 341–353.

Chou, T., 1998. Band structure of surface flexural-gravity waves along periodic interfaces, *J. Fluid Mech.*, 369, 333–50.

Chow, V.T., 1973. *Open-channel Hydraulics*, Int. ed., New York: McGraw-Hill.

Clarke, N.R. and Tutty, O.R., 1994. Construction and validation of a discrete vortex method for the two-dimensional incompressible Navier–Stokes equations, *Comput. Fluids*, 23, 751–83.

Cleary, P., Ha, J., Alguine, V., and Nguyen, T., 2002. Flow modelling in casting processes, *Appl. Math. Model.*, 26, 171–90.

Cokelet, E.D., 1977. Steep gravity waves in water of arbitrary uniform depth, *Philos. Trans. R. Soc. Lond., Ser. A*, 286, 183–230.

Cole, S.L., 1985. Transient waves produced by flow past a bump, *Wave Motion*, 7, 579–87.

Copeland, G.J.M., 1985. A practical alternative to the "mild-slope" wave equation, *Coast. Eng.*, 9, 125–49.

Cox, D.T. and Ortega, J.A., 2002. Laboratory observations of green water overtopping a fixed deck, *Ocean Eng.*, 29, 1827–40.

Craik, A.D.D., 2004. The origins of water wave theory. *Annu. Rev. Fluid Mech.*, 36, 1–28.

Cruz, E.C., Isobe, M., and Watanabe, A., 1997. Boussinesq equations for wave transformation on porous beds, *Coast. Eng.*, 30, 125–56.

Cundall, P.A. and Strack, O.D.L., 1979. A discrete numerical model for granular assemblies, *Geotechnique*, 29, 47–65.

Dahmen, W., Klint, T., and Urban, K., 2003. On fictitious domain formulations for Maxwell's equations, *Found. Comput. Math.*, 3, 135–60.

Dally, W.R., Dean, R.G., and Dalrymple, R.A., 1985. Wave height variation across beaches of arbitrary profile, *J. Geophys. Res.*, 90(C6), 11917–27.

Dalrymple, R.A. and Kirby, J.T., 1985. Wave modification in the vicinity of islands. *REF/DIF 1 Documentation Manual*, Newark, DE: Coastal and Offshore Engineering and Research, Inc.

Dalrymple, R.A. and Kirby, J.T., 1988. Models for very wide-angle water waves and wave diffraction, *J. Fluid Mech.*, 192, 33–50.

Dalrymple, R.A. and Rogers, B.D., 2006. Numerical modeling of water waves with the SPH Method, *Coast. Eng.*, 53, 141–7.

Dalrymple, R.A., Kirby, J.T., and Hwang, P.A., 1984. Wave diffraction due to areas of energy dissipation, *J. Waterw. Port Coast. Ocean Eng. ASCE*, 110, 67–79.

Dalrymple, R.A., Suh, K.D., Kirby, J.T., and Chae, J.W., 1989. Models for very wide angle water waves and wave diffraction, Part 2. Irregular bathymetry, *J. Fluid Mech.*, 201, 299–322.

Dalrymple, R.A., Losada, M.A., and Martin, P.A., 1991. Reflection and transmission from porous structures under oblique wave attack, *J. Fluid Mech.*, 224, 625–44.

Daly, B.J. and Harlow, F.H., 1970. Transport equations of turbulence, *Phys. Fluids*, 8, 2634–49.

Davidson, M.A., Bird, P.A., Bullock, G.N., and Huntley, D.A., 1996. A new non-dimensional number for the analysis of wave reflection from rubble mound breakwaters, *Coast. Eng.*, 28, 93–129.

Davies, A.G. and Heathershaw, A.D., 1984. Surface-wave propagation over sinusoidal varying topography, *J. Fluid Mech.*, 291, 419–33.

De Waal, J.P. and Van der Meer, J.W., 1992. Wave runup and overtopping on coastal structures, *Proc. 23rd Int. Conf. Coast Eng.*, 2, 1758–71.

Dean, R.G., 1964. Long wave modification by linear transitions, *J. Waterw. Harbors Div.-ASCE*, 90, 1–29.

Dean, R.G., 1965. Stream function representation of nonlinear ocean waves, *J. Geophys. Res.*, 70, 4561–72.

Dean, R.G., 1970. Relative validity of water wave theories, *J. Waterw. Harbors Div.- ASCE*, 96, 105–19.

Dean, R.G. and Dalrymple, R.A., 1991. *Water Wave Mechanics for Engineers and Scientists*, Singapore: World Scientific.

Delaunay, B., 1934. Sur la sphère vide and Izvestia Akademii Nauk SSSR, *Otdelenie Matematicheskikh i Estestvennykh Nauk*, 7, 793–800.

Delaurier, J.D., 1993. An aerodynamic model for flapping-wing flight, *Aeronaut. J.*, 97, 125–30

Demirbilek, Z. and Panchang, V., 1998. CGWAVE: A coastal surface water wave model of the mild slope equation, *Technical Report*, US Army Engineer Research and Development Center (ERDC), Vicksburg, MS.

Demuren, A.O. and Sarkar, S., 1993. Perspective: systematic study of Reynolds stress closure models in the computations of plane channel flows, *J. Fluids Eng., Trans. ASME*, 115, 5–12.

Devillard, P., Dunlop, F., and Souillard, B., 1988. Localization of gravity waves on a channel with random bottom, *J. Fluid Mech.*, 186, 521–38.

Dickinson, R.E., 1978. Rossby waves-long-period oscillations of oceans and atmospheres, *Annu. Rev. Fluid Mech.*, 10, 195.

Dingemans, M.W., 1997. *Water Wave Propagation Over Uneven Bottoms*, Singapore: World Scientific.

Dingemans, M.W., Van Kester, J.A.Th.M., Radder, A.C., and Uittenbogaard, R.E., 1996. The effect of the CL-vortex force in 3D wave–current interaction, *Proc. 25th Int. Conf. Coast Eng.*, Orlando, FL, USA, pp. 4821–32.

Dold, W. and Peregrine, D.H., 1984. Steep unsteady water waves: an efficient computational scheme, *Proc. 19th Int. Conf. Coast Eng.*, pp. 955–67.

Dommermuth, D.G. and Yue, D.K.P., 1987. A high-order spectral method for the study of nonlinear gravity-waves, *J. Fluid Mech.*, 184, 267–88.

Donelan, M.A., Hamilton, J., and Hui, W.H., 1985, Directional spectra of wind-generated waves, *Philos. Trans. R. Soc. Lond. Ser. A*, 315, 509–62.

van Doormaal, J.P. and Raithby, G.D., 1984. Enhancement of the SIMPLE method for predicting Incompressible fluid flows, *Numer Heat Transf.*, 7, 147–63.

Doorn, N. and van Gent, M.R.A., 2004. Pressure by breaking waves on a slope computed with a VOF model, *Proc. 3rd Int. Conf. Coastal Structures, 2003*, Portland, OR, New York, NY, USA, pp. 728–39.

Drazin, P.G. and Johnson, R.S., 1989. *Solitons: An Introduction*, Cambridge: Cambridge University Press.

Duarte, C.A. and Oden, J.T., 1996. An h–p adaptive method using clouds, *Comput. Meth. Appl. Mech. Eng.*, 139, 237–62.

Durran, D.R., 1999. *Numerical Methods for Wave Equations in Geophysical Fluid Dynamics*, New York: Springer.

Dütsch, H., Durst, F., Becker, S., and Lienhart, H., 1998. Low-Reynolds-number flow around an oscillating circular cylinder at low Keulegan–Carpenter numbers, *J. Fluid Mech.*, 360, 249–71.

Dyke, P., 2007. *Modeling Coastal and Offshore Processes*, Singapore: World Scientific.

Eckart, C., 1951. Surface waves on water of variable depth, Wave Rep., Scripps Institution of Oceanography, 100, *Ref. No. 51-12*.

Eckart, C., 1952. The propagation of gravity waves from deep to shallow water, Circular 20, *National Bureau of Standards*, pp. 165–73.

Engelund, F.A., 1953. *On the Laminar and Turbulent Flows of Ground Water Through Homogeneous Sand*, Danish Academy of Technical Sciences, Denmark.

Engquist, B. and Majda, A., 1977. Absorbing boundary conditions for the numerical simulation of waves, *Math. Comput.*, 31, 629–51.

Erbes, G., 1993. A semi-Lagrangian method of characteristics for the shallow-water equations, *Mon. Weather Rev.*, 121, 3443–52.

Ergun, S., 1952. Fluid flow through packed columns, *Chem. Eng. Prog.*, 48, 89–94.

Eriksson, L.E., 1982. Generation of boundary-conforming grids around wing-body configurations using transfinite interpolation, *AIAA J.*, 20, 1313–20.

Ertekin, R.C. and Kim, J.W., 1999. Hydroelastic response of a floating mat-type structure in oblique, shallow-water waves, *J. Ship Res.*, 43, 241–54.

Ertekin, R.C., Webster, W.C., and Wehausen, J.V., 1986. Waves caused by a moving disturbance in a shallow channel of finite width, *J. Fluid Mech.*, 169, 275–92.

Ertekin, R.C., Riggs, H.R., Che, X.L., and Du, S.X., 1993. Efficient methods for hydroelastic analysis of very large floating structures, *J. Ship Res.*, 37, 58–76.

Evans, D.V. and Linton, C.M., 1994. On step approximations for water wave problems, *J. Fluid Mech.*, 278, 229–49.

Fadlun, E.A., Verzicco, R., Orlandi, P., and Mohd-Yusof, J., 2000. Combined immersed-boundary finite-difference methods for three dimensional complex flow simulations, *J. Comput. Phys.*, 161, 35–60.

Fair, G.M., Geyer, J.C., and Okum, D.A., 1968. *Waste Water Engineering, Vol. 2, Water Purification and Wastewater Treatment and Disposal*, New York: Wiley.

Faltinsen, O.M., 1978. A numerical nonlinear method of sloshing in tanks with two-dimensional flow, *J. Ship Res.*, 22, 193–202.

Faltinsen, O.M., 1990. *Sea Loads on Ships and Offshore Platforms*, Cambridge: Cambridge University Press.

Faltinsen, O.M. and Timokha, A.N., 2002. Asymptotic modal approximation of nonlinear resonant sloshing in a rectangular tank with small fluid depth, *J. Fluid Mech.*, 470, 319–57.

Faltinsen, O.M., Newman, J.N., and Vinje, T., 1995. Nonlinear wave loads on a slender vertical cylinder, *J. Fluid Mech.*, 289, 179–98.

Faltinsen, O.M., Rognebakke, O.F., Lukovsky, I.A., and Timokha, A.N., 2000. Multidimensional modal analysis of nonlinear sloshing in a rectangular tank with finite water depth, *J. Fluid Mech.*, 407, 201–34.

Faraday, M., 1831. On a peculiar class of acoustical figures, and on certain forms assumed by groups of particles upon vibrating elastic surfaces, *Philos. Trans. R. Soc. Lond.*, 121, 299–340.

Farmer, J., Martinelli, L., and Jameson, A., 1994. Fast multigrid method for solving incompressible hydrodynamic problems with free surfaces, *AIAA J.*, 32, 1175–85.

Feng, W. and Hong, G., 2000. Numerical modeling of wave diffraction–refraction in water of varying current and topography, *Chin. Ocean Eng.*, 14, 45–58.

Floryan, J.M. and Rasmussen, H., 1989. Numerical methods for viscous flows with moving boundaries, *Appl. Mech. Rev.*, 42, 323–41.

Forchheimer, P., 1901. Wasserbewegung durch Boden, *Zeits. V. Deutsch. Ing.*, 45, 1782–88.

Franco, C. and Franco, L., 1999. Overtopping formulas for caisson breakwaters with nonbreaking 3D waves, *J. Waterw. Port Coast. Ocean Eng. ASCE*, 125, 98–108.

Frandsen, J.B., 2004. Sloshing motions in excited tanks, *J. Comput. Phys.*, 196, 53–87.

Frandsen, J.B. and Borthwick, A.G.L., 2003. Simulation of sloshing motions in fixed and vertically excited containers using a 2-D inviscid sigma-transformed finite difference solver, *J. Fluids Struct.*, 18, 197–214.

Freilich, M.H. and Guza, R.T., 1984. Nonlinear effects on shoaling surface gravity waves, *Philos. Trans. R. Soc. Lond., Ser. A*, 311, 1–41.

Freitas, C.J., 1995. Perspective: selected benchmarks from commercial CFD codes, *J. Fluids Eng., Trans. ASME*, 117, 210–18.

Frey, P.J. and George, P.L., 2000. *Mesh Generation Application to Finite Elements*, Paris: Hermes Science Europe Ltd.

Fringer, O.B., Gerritsen, M., and Street, R.L., 2006. An unstructured-grid, finite-volume, nonhydrostatic, parallel coastal ocean simulator, *Ocean Model*, 14, 139–73.

Frisch, U., Hasslacher, B., and Pomeau, Y., 1986. Lattice gas automata for the Navier–Stokes equations, *Phys. Rev. Lett.*, 56, 1505–08.

Fuhrman, D.R., Bingham, H.B., and Madsen, P.A., 2005. Nonlinear wave–structure interactions with a high-order Boussinesq model, *Coast. Eng.*, 52, 655–72.

Gao, X. and Zhao, Z., 1995. Interaction between waves, structure and sand beds, *J. Hydrodyn.*, 7, 103–10.

Garcia, N., Lara, J.L., and Losada, I.J., 2004. 2-D numerical analysis of near-field flow at low-crested permeable breakwaters, *Coast. Eng.*, 51, 991–1020.

Garret, C.R.J., 1971. Wave forces on a circular dock, *J. Fluid Mech.*, 46, 129–39.

Garrison, T., 2002. *Oceanography: An Invitation to Marine Science*, 4th ed., Singapore: Thomson Learning, Inc. S.A.

van Gent, M.R.A., 1994. The modeling of wave action on and in coastal structures, *Coast. Eng.*, 22 (3–4), 311–39.

van Gent, M.R.A., 1995. *Wave Interaction With Permeable Coastal Structures*, Ph.D. thesis, Delft University of Technology, ISBN 90-407-1182-8, Delft University Press, Delft, The Netherlands.

van Gent, M.R.A., Tonjes, P., Petit, H.A.H., and Van den Bosch, P., 1994. Wave action on and in permeable coastal structures, *Proc. 24th Int. Conf. Coast. Eng.*, Kobe, Japan, pp. 1739–53.

Gerstner, F.J.v., 1809. Theorie der wellen. *Ann. der Physik*, 32, 412–40.

Ghalayini, S.A. and Williams, A.N., 1991. Nonlinear wave forces on vertical cylinder arrays, *J. Fluids Struct.*, 5, 1–32.

Gingold, R. and Monaghan, J., 1977. Smoothed particle hydrodynamics – theory and application to non-spherical stars, *Mon. Not. R. Astron. Soc.*, 181, 375–89.

Ginzburg, I. and Steiner, K., 2003. Lattice Boltzmann model for free-surface flow and its application to filling process in casting, *J. Comput. Phys.*, 185, 61–99.

Givoli, D., 1991. Nonreflecting boundary-conditions, *J. Comput. Phys.*, 94, 1–29.

Glowinski, R., Pan, T.W., and Periaux, J., 1994. A fictitious domain method for Dirichlet problem and applications, *Comput. Meth. Appl. Mech. Eng.*, 111, 283–303.

Glowinski, R., Pan, T.W., Hesla, T.I., Joseph, D.D., and Periaux, J., 2001. A fictitious domain approach to the direct numerical simulation of incompressible viscous flow past moving rigid bodies: application to particulate flow, *J. Comput. Phys.*, 169, 363–426.

Gobbi, M.F., Kirby, J.T., and Wei, G., 2000. A fully nonlinear Boussinesq model for surface waves Part 2. Extension to O(kh), *J. Fluid Mech.*, 405, 181–210.

Goda, Y., 1970. Numerical experiments on wave statistics with spectral simulation, *Port Harbour Res. Inst., Jpn*, 9, 1–57.

Goda, Y., 2000. *Random Seas and Design of Maritime Structures*, Singapore: World Scientific.

Gomez-Gesteira, M., Cerqueiroa, D., Crespoa, C., and Dalrymple, R.A., 2005. Green water overtopping analyzed with a SPH model, *Ocean Eng.*, 32, 223–38.

Goto, C., Ogawa, Y., Shuto, N., and Imanura, F., 1997. Numerical method of tsunami simulation with the leap-frog scheme (IUGG/IOC Time Project), *IOC Manual*, UNESCO, No. 35.

Gotoh, H. and Sakai, T., 2006. Key issues in the particle method for computation of wave breaking, *Coast. Eng.*, 53, 171–9.

Gottlieb, D. and Orszag, S.A., 1977. *Numerical Analysis of Spectral Methods: Theory and Applications*, Philadelphia, PA: SIAM-CBMS.

Grant, W.D. and Madsen, O.S., 1986. The continental shelf bottom boundary layer, *Annu. Rev. Fluid Mech.*, 18, 265–305.

Graw, K.U., 1993. Shore protection and electricity by submerged plate wave energy converter, *European Wave Energy Symposium*, Edinburgh, UK, pp. 1–6.

Greaves, D.M. and Borthwick, A.G.L., 1999. Hierarchical tree-based finite element mesh generation, *Int. J. Numer. Meth. Eng.*, 45, 447–71.

Green, A.E. and Naghdi, P.M., 1976. A derivation of equations for wave propagation in water of variable depth, *J. Fluid Mech.*, 78, 237–46.

Greenbaum, N., Margalit, A., Schick, A.P., Sharon, D., and Baker, V.R., 1998. A high magnitude storm and flood in a hyperarid catchment, Nahal Zin, Negev Desert, Israel, *Hydrol. Process.*, 12, 1–23.

Greenhow, M. and Li, Y., 1987. Added masses for circular cylinders near or penetrating fluid boundaries–review, extension and application to water-entry, exit and slamming, *Ocean Eng.*, 14, 325–48.

Grilli, S.T. and Svendsen, I.A., 1990. Wave interaction with steeply sloping structures, *Proc. 22nd Int. Conf. Coast. Eng.*, Delft, The Netherlands, pp. 1200–13.

Grilli, S.T. and Subramanya, R., 1996. Numerical modeling of wave breaking induced by fixed or moving boundaries, *Comput. Mech.*, 17, 374–91.

Grilli, S.T. and Horrillo, J., 1997. Numerical generation and absorption of fully nonlinear periodic waves, *J. Eng. Mech. ASCE*, 123, 1060–9.

Grilli, S.T., Skourup, J., and Svendsen, I.A., 1989. An efficient boundary element method for nonlinear water waves, *Eng. Anal. Bound. Elem.*, 6, 97–107.

Grilli, S.T., Ioualalen, M., Asavanant, J., Shi, F., Kirby, J.T., and Watts, P., 2007. Source constraints and model simulation of the December 26, 2004 Indian Ocean tsunami, *J. Waterw. Port Coast. Ocean Eng. ASCE*, 133 (6), 414–28.

Gu, D. and Phillips, O.M., 1994. On narrow V-like ship wakes, *J. Fluid Mech.*, 275, 301–21.

Gu, H., Li, Y., and Lin, P., 2005. Modeling 3D fluid sloshing using level set method, *Mod. Phys. Lett. B*, 19, 1743–6.

Gu, Z. and Wang, H., 1991. Gravity waves over porous bottoms, *Coast. Eng.* 15, 497–524.

Guazzelli, E., Rey, V., and Belzons, M., 1992. Higher order Bragg reflection of gravity surface waves by periodic beds, *J. Fluid Mech.*, 245, 301–17.

Gueyffier, D., Li, J., Nadim, A., Scardovelli, R., and Zaleski, S., 1999. Volume-of-fluid interface tracking with smoothed surface stress methods for three-dimensional flows, *J. Comput. Phys.*, 152, 423–56.

Gunstensen, A.K. and Rothman, D.H., 1993. Lattice-Boltzmann studies of immiscible 2-phase flow through porous-media, *J. Geophys. Res.*, 98(B4), 6431–41.

Hamamoto, T. and Tanaka, Y., 1992. Coupled free vibrational characteristics of artificial floating islands. Fluid–structure interaction analysis of floating elastic circular plate Part 1, *J. Struct. Constr. Eng.*, AIJ, 438, 165–77 (in Japanese).

Hanjalic, K. and Launder, B.E., 1972. A Reynolds stress model of turbulence and its application to thin shear flows, *J. Fluid Mech.*, 52, 609–38.

Hansen, J.B. and Svendsen, I.A., 1984. A theoretical and experimental study of undertow, *Proc. 19th Int. Conf. Coast Eng.*, pp. 2246–62.

Hara, T. and Mei, C.C., 1991. Frequency downshift in narrow-banded surface waves under the influence of wind, *J. Fluid Mech.*, 230, 42–77.

Harlow, F.H. and Welch, J.E., 1965. Numerical calculation of time-dependent viscous incompressible flow, *Phys. Fluids*, 8, 2182–9.

Harten, A., 1984. On a class of high resolution total-variation stable finite difference schemes, *SIAM J. Numer. Anal.*, 21, 1–21.

Harten, A., Engquist, B., Osher, S., and Chakaravathy, S., 1987. Uniformly high order essentially non-oscillatory schemes, *J. Comput. Phys.*, 71, 231–303.

Hassan, O., Probert, E.J., Morgan, K., and Weatherill, N.P., 2000. Unsteady flow simulation using unstructured meshes, *Comput. Meth. Appl. Mech. Eng.*, 189, 1247–75.

Hasselmann, K., 1968. Weak-interaction theory of ocean waves, in: M. Holt (Ed.), *Basic Developments in Fluid Dynamics*, Vol. 2, London: Academic Press, pp. 117–82.

Hasselmann, K., Barnett, T.P., Boyn, E., Carlson, H., Cartwright, D.E., Erick, K., Ewing, J.A., Gienapp, H., Hasselmann, D.E., Kruseman, P., Meerburg, A., Muller, P., Olbers, D.J., Richter, K., Sell, W., and Walden, H., 1973. Measurements of wind-wave growth and swell decay during the Joint North Sea Project (JONSWAP), *Dtsch. Hydrogr. Z Suppl.*, 12, 1–95.

Hasselmann, K., Ross, D.B., Muller, O., and Sell, W., 1976. A parametric wave prediction model, *J. Phys. Oceanogr.*, 6, 200–28.

Hasselmann, S. and Hasselmann, K., 1985. Computations and parameterization of the nonlinear energy transfer in a gravity-wave spectrum. 1. A new method for efficient computations of the exact nonlinear transfer integral, *J. Phys. Oceanogr.*, 15, 1369–77.

Hasselmann, S., Hasselmann, K., Bauer, E., Janssen, P.A.E.M., and Komen, G.J., 1988. The WAM model – a third generation ocean wave prediction model, *J. Phys. Oceanogr.*, 18, 1775–810.

Havelock, T.H., 1908. The propagation of groups of waves in dispersive media, with application to waves on water produced by a travelling distance, *Proc. R. Soc. A*, 81, 398–430.

Havelock, T.H., 1942. The damping of the heaving and pitching motion of a ship, *Philos. Mag.*, 33, 666–73.

Hayashi, M., Hatanaka, K., and Kawahara, M., 1991. Lagrangian finite element method for free surface Navier–Stokes flows using fractional step methods, *Int. J. Numer. Meth. Fluids*, 13, 805–40.

Hayse, T., Humphery, J.A.C., and Grief, R., 1992. A consistently formulated QUICK scheme for fast and stable convergence using finite-volume iterative calculation procedures, *J. Comput. Phys.*, 93, 108–18.

He, X.Y. and Luo, L.S., 1997. Theory of the lattice Boltzmann method: from the Boltzmann equation to the lattice Boltzmann equation, *Phys. Rev. E*, 56, 6811–17.

Hedges, T.S. and Reis, M.T., 1998. Random wave overtopping of simple sea walls: a new regression model, *Proc. ICE: Water Marit. Energy*, 130, 1–10.

Heikkola, E., Rossi, T., and Toivanen, J., 2003. A parallel fictitious domain method for the three-dimensional Helmholtz equation, *SIAM J. Sci. Comput.*, 24, 1567–88.

Heinrich, P., 1992. Nonlinear water-waves generated by submarine and aerial landslides, *J. Waterw. Port Coast. Ocean Eng. ASCE*, 118, 249–66.

Heinrich, P., Piatanesi, A., and Hebert, H., 2001. Numerical modelling of tsunami generation and propagation from submarine slumps: the 1998 Papua New Guinea event, *Geophys. J. Int.*, 145, 97–111.

Hibberd, S. and Peregrine, D.H., 1979. Surf and run-up on a beach: a uniform bore, *J. Fluid Mech.*, 95, 323–45.

Hill, D.F. and Frandsen, J., 2005, Transient evolution of weakly-nonlinear sloshing waves: an analytical and numerical comparison, *J. Eng. Math.*, 53, 187–98.

Hirota, R., 1973. Exact *n*-soliton solutions of the wave equation of long waves in shallow water and in nonlinear lattices, *J. Math. Phys.*, 14, 810–4.

Hirt, C.W., 1988. *Flow-3D User's Manual*, Santa Fe, NM: Flow Science, Inc., FSI-91-00-1.

Hirt, C.W. and Nichols, B.D., 1981. Volume of fluid (VOF) method for the dynamics of free boundaries, *J. Comput. Phys.*, 39, 201–25.

Hirt, C.W., Amsden, A.A., and Cook, J.L., 1974. An arbitrary Lagrangian–Eulerian computing method for all speeds, *J. Comput. Phys.* 14, 227–53.

Hirt, C.W., Nichols, B.D., and Romero, N.C., 1975. SOLA: A numerical solution algorithm for transient fluid flows, *Los Alamos National Laboratory Report LA-5852*.

Hodges, B.R. and Street, R.L., 1999. On simulation of turbulent nonlinear free-surface flows, *J. Comput. Phys.*, 151, 425–57.

Holthuijsen, L.H. and Booij, N., 2003. Phase-decoupled refraction–diffraction for spectral wave models, *Coast. Eng.*, 49, 291–305.

Homma, S., 1950. On the behavior of seismic sea waves around circular island, *Geophys. Mag.*, 21, 199–208.

Hon, Y.C., Cheung, K.F., Mao, X.Z., and Kansa, E.J., 1999. Multiquadric solution for shallow water equations, *J. Hydraul. Eng. ASCE*, 125, 524–33.

Hong, C.B. and Doi, Y., 2006. Numerical and experimental study on ship wash including wave-breaking on shore, *J. Waterw. Port Coast. Ocean Eng. ASCE*, 132, 369–78.

Hong, T.K. and Kennett, B.L.N., 2002. On a wavelet-based method for the numerical simulation of wave propagation, *J. Comput. Phys.*, 183, 577–622.

Houston, J.R., 1981. Combined refraction and diffraction of short waves using the finite element method, *Appl. Ocean Res.*, 3, 163–70.

Hsiao, S.C., Liu, P.L.F., and Chen, Y., 2002. Nonlinear water waves propagating over a permeable bed, *Proc. R. Soc. Lond. A*, 458, 1291–322.

Hsu, T.W. and Wen, C.C., 2001. A parabolic equation extended to account for rapidly varying topography, *Ocean Eng.*, 28, 1479–98.

Hsu, T.-W., Lin, T.-Y., Wen, C.-C., and Ou, S.-H., 2006. A complementary mild-slope equation derived using higher-order depth function for waves obliquely propagating on sloping bottom, *Phys. Fluids*, 18, 087106.

Hu, K., Mingham, C.G., and Causon, D.M., 2000. Numerical simulation of wave overtopping of coastal structure using the non-linear shallow water equation, *Coast. Eng.*, 41, 433–65.

Huang, H., Ding, P.X., and Lu, X.H., 2001. Nonlinear unified equations for water waves propagating over uneven bottoms in the nearshore region, *Prog. Natl. Sci.*, 11, 746–53.

Huang, J.B. and Taylor, R.E., 1996. Semi-analytical solution for second-order wave diffraction by a truncated circular cylinder in monochromatic waves, *J. Fluid Mech.*, 319, 171–96.

Huang, Z.H. and Mei, C.C., 2003. Effects of surface waves on a turbulent current over a smooth or rough seabed, *J. Fluid Mech.*, 497, 253–87.

Hubbard, M.E. and Dodd, N., 2002. A 2D numerical model of wave run-up and overtopping, *Coast. Eng.*, 47, 1–26.

Hudspeth, R.T., 2006. *Waves and Wave Forces On Coastal and Ocean Structures*, Singapore: World Scientific.

Hughes, S.A., 1993. *Physical Models and Laboratory Techniques in Coastal Engineering*, Singapore: World Scientific.

Hughes, S.A., 2004. Estimation of wave run-up on smooth, impermeable slopes using the wave momentum flux parameter, *Coast. Eng.*, 51, 1085–104.

Hunt, I.A., 1959a. Design of seawalls and breakwaters, *J. Waterw. Harbours Division, ASCE*, 85, 123–52.

Hunt, J.N., 1959b. On the damping of gravity waves propagated over a permeable surface, *J. Geophys. Res.*, 64(4), 437–42.

Hunt, J.N., 1979. Direct solution of wave dispersion equation, *J. Waterw. Port Coast. Ocean Eng. ASCE*, 105, 457–9.

Iafrati, A., Di Mascio, A., and Campana, E.F., 2001. A level set technique applied to unsteady free surface flows, *Int. J. Numer. Meth. Fluids*, 35(3), 281–97.

Ibrahim, R.A., 2005. *Liquid Sloshing Dynamics: Theory and Applications*, Cambridge: Cambridge University Press.

Idelsohn, S.R., Onate, E., and Del, P.F., 2004. The particle finite element method: a powerful tool to solve incompressible flows with free-surfaces and breaking waves, *Int. J. Numer. Meth. Eng.*, 61(7), 964–89.

Ijima, Y., Tabuchi, M., and Yumura, Y., 1972. On the motion of a floating circular cylinder in water of finite depth, *Trans. JCE*, 206, 71–84.

Inamuro, T., Ogata, T., Tajima, S., and Konishi, N., 2004. A lattice Boltzmann method for incompressible two-phase flows with large density differences, *J. Comput. Phys.*, 198(2), 628–44.

Ippen, A.T. and Goda, Y., 1963. Wave induced oscillations in harbors: the solution for a rectangular harbor connected to the open-sea, *Hydrodynamic Laboratory Report. No. 59*, Cambridge: MIT Press.

Iribarren, R. and Nogales, C., 1949. Protection des ports, *Proc. 17th Int. Navigation Congress (Lisbon)*, II-4, 31–82.

Isaacson, M., 1977. Nonlinear wave forces on large offshore structures, *J. Waterw., Ports, Coastal and Ocean Division ASCE*, 103 (WW2), 299–300.

Isaacson, M., 1982. Nonlinear wave effects on fixed and floating bodies, *J. Fluid Mech.*, 120, 267–81.

Isaacson, M. and Cheung, K.F., 1992. Time-domain second-order wave diffraction in three dimensions, *J. Waterw. Port Coast. Ocean Eng., ASCE*, 118, 496–516.

Isaacson, M., Baldwin, J., Allyn, N., and Cowdell, S., 2000. Wave interactions with perforated breakwater, *J. Waterw. Port Coast. Ocean Eng. ASCE*, 126(5), 229–35.

Issa, R.I., 1982. Solution of the implicit discretized fluid flow equations by operator splitting, *Mechanical Eng. Rep. FS/82/15*, Imperial College, London.

ISSC, 1964. *Proc. Sec. Int. Ship Structures Congress*, Delft, Netherlands.

ITTC, 1972. Technical decision and recommendation of the seakeeping committee, *Proc. 13th Int. Towing Tank Conf.*, Berlin, Germany.

Iwata, K., Kawasaki, R.C., and Kim, D., 1996. Breaking limit, breaking and post-breaking wave deformation due to submerged structures, *Proc. 25th Conf. Coast. Eng.*, Orlando, FL, USA, pp. 2338–51.

Jaluria, Y. and Torrance, K.E., 2003. *Computational Heat Transfer*, 2nd ed., London: Taylor & Francis.

Janssen, P.A.E.M., 1991. Quasi-linear theory of wind wave generation applied to wave forecasting, *J. Phys. Oceanogr.*, 21, 1631–42.

Jin, B.K., Qian, W.J., Zhang, Z.X., and Shi, H.S., 1996. Finite analytic numerical method – a new numerical simulation method for electrochemical problems, *J. Electroanal. Chem.*, 411, 19–27.

Johns, B. and Jefferson, R.J., 1980. The numerical modeling of surface wave propagation in the surf zone, *J. Phys. Oceanogr.*, 10, 1061–69.

Johnson, R.S., 1972. Some numerical solutions of a variable coefficient Kortweg–de Vries equation (with application to solitary wave development on a shelf), *J. Fluid Mech.*, 54, 81–91.

Jonsson, I.G, Skovgaard, O., and Brink-Kjaer, O., 1976. Diffraction and refraction calculations for waves incident on an island, *J. Mar. Res.*, 34(3), 469–96.

Jung, K.H., Chang, K.A., and Huang, E.T., 2004. Two-dimensional flow characteristics of wave interactions with a fixed rectangular structure, *Ocean Eng.*, 31, 975–98.

Kabiling, M. and Sato, S., 1994. A numerical model for nonlinear waves and beach evolution including swash zone, *Coast. Eng. Jpn*, 37, 67–86.

Kadomtsev, B.B. and Petviashvili, V.I., 1970. On the stability of solitary waves in weakly dispersive media, *Sov. Phys. Dokl.*, 15, 539–41.

Kagemoto, H. and Yue, D.KP., 1986. Interactions among multiple three-dimensional bodies in water waves: an exact algebraic method, *J. Fluid Mech.*, 166, 189–209.

Kajiura, K., 1961. On the partial reflection of water waves passing over a bottom of variable depth, *I.U.G.G. Monogr.*, 24, 206–30.

Kamphuis, J.W., 1975. Friction factor under oscillatory waves, *J. Waterw. Harbors Coast. Eng. Div. ASCE*, 101, 135–44.

Kamphuis, J.W., 1991. Incipient wave breaking, *Coast. Eng.*, 15, 185–203.

Kamphuis, J.W., 2000. *Introduction to Coastal Engineering and Management*, Singapore: World Scientific.

Kanoglu, U. and Synolakis, C.E., 1998. Long wave runup on piecewise linear topographies, *J. Fluid Mech.*, 374, 1–28.

Karambas, T.V., 1999. A unified model for periodic non-linear dispersive waves in intermediate and shallow water, *J. Coast. Res.*, 15(1), 128–39.

Karambas, T.V. and Koutitas, C.A., 1992. A breaking wave propagation model based on the Boussinesq equations, *Coast. Eng.*, 18(1–2), 1–19.

Karambas, T.V. and Koutitas, C., 2002. Surf and swash zone morphology evolution induced by nonlinear waves, *J. Waterw. Port Coast. Ocean Eng. ASCE*, 128(3), 102–13.

Karunarathna, S.A.S.A. and Lin, P., 2006. Numerical simulation of wave damping over porous seabeds, *Coast. Eng.*, 53(10), 845–55.

Kashiwagi, M., 2000. Research on hydroelastic responses of VLFS: recent progress and future work, *Int. J. Offshore Polar Eng.*, 10(2), 81–90.

Kataoka, T., Tsutahara, M., and Akuzawa, A., 2000. Two-dimensional evolution equation of finite-amplitude internal gravity waves in a uniformly stratified fluid, *Phys. Rev. Lett.*, 84(7), 1447–50.

Keller, H.B., Levine, D.A., and Whitham, G.B., 1960. Motion of a bore over a sloping beach, *J. Fluid Mech.*, 7, 302–16.

Kemp, P.H. and Simons, R.R., 1982. The interaction between waves and a turbulent current: waves propagating with the current, *J. Fluid Mech.*, 116, 227–50.

Kemp, P.H. and Simons, R.R., 1983. The interaction of waves and turbulent current: waves propagating against the current, *J. Fluid Mech.*, 130, 73–89.

Kennedy, A.B., Chen, Q., Kirby, J.T., and Dalrymple, R.A., 2000. Boussinesq modeling of wave transformation, breaking and runup, I: 1D, *J. Waterw. Port Coast. Ocean Eng. ASCE.*, 126, 39–47.

Keulegan, G.H. and Carpenter, L.H., 1958. Forces on cylinders and plates in an oscillating fluid, *J. Res. Natl. Bur. Stand.*, 60(5), 423–40.

Kevlahan, N.K.R. and Ghidaglia, J.M., 2001. Computation of turbulent flow past an array of cylinders using a spectral method with Brinkman penalization, *Eur. J. Mech. B Fluids*, 20(3), 333–50.

Kharif, C. and Pelinovsky, E., 2003. Physical mechanisms of the rogue wave phenomenon, *Eur. J. Mech. B Fluid*, 22, 603–34.

Kim, D.J. and Kim, M.H., 1997. Wave–current–body interaction by a time-domain high-order boundary element method, *Proc. 7th Int. Offshore and Polar Eng. Conf.*, Honolulu, HI, USA, pp. 107–115.

Kim, M.H. and Yue, D.K.P., 1989. The complete second-order diffraction solution for an axisymmetric body. 1. Monochromatic incident waves, *J. Fluid Mech.*, 200, 235–64.

Kim, M.H., Celebi, M.S., and Kim, D.J., 1998, Fully nonlinear interactions of waves with a three-dimensional body in uniform currents, *Appl. Ocean Res.*, 20, 309–21.

Kim, M.S. and Lee, W.I., 2003. A new VOF-based numerical scheme for the simulation of fluid flow with free surface. Part I: New free surface-tracking algorithm and its verification, *Int. J. Numer. Meth. Fluids*, 42(7), 765–90.

Kim, S.K., Liu, P.L.F., and Liggett, J.A., 1983. Boundary integral equation solutions for solitary wave generation, propagation and runup, *Coast. Eng.*, 7, 299–317.

Kim, Y., 2001. Numerical simulation of sloshing flows with impact load, *Appl. Ocean Res.*, 23, 53-62.

Kim, Y., Sin, Y.S., and Lee, K.H., 2004. Numerical study on slosh-induced impact pressures on three-dimensional prismatic tanks, *Appl. Ocean Res.*, 26, 213–26.

Kirby, J.T., 1986. A general wave equation for waves over rippled beds, *J. Fluid Mech.*, 162, 171–86.

Kirby, J.T. and Dalrymple, R.A., 1983a. Propagation of oblique incident water waves over a trench, *J. Fluid Mech.*, 133, 47–63.

Kirby, J.T. and Dalrymple, R.A., 1983b. A parabolic equation for the combined refraction–diffraction of Stokes waves by mildly-varying topography, *J. Fluid Mech.*, 136, 453–66.

Kirby, J.T. and Dalrymple, R.A., 1984. Verification of a parabolic equation for propagation of weakly-nonlinear waves, *Coast. Eng.*, 8, 219–32.

Kirby, J.T. and Dalrymple, R.A., 1986. An approximate model for nonlinear dispersion in monochromatic wave-propagation models, *Coast. Eng.*, 9, 545–61.

Kirby, J.T., Wei, G., Chen, Q., Kennedy, A.B., and Dalrymple, R.A., 1998. FUNWAVE 1.0. Fully nonlinear Boussinesq wave model. Documentation and user's manual, *Report CACR-98-06*, Department of Civil and Environmental Engineering, Center for Applied Coastal Research, University of Delaware.

Klopman, G., 1994. Vertical structure of the flow due to waves and currents, *Report H840*, Delft Hydraulics Laboratory, Delft, The Netherlands.

Kobayashi, N., 1999. Numerical modeling of wave runup on coastal structures and beaches, *Mar. Technol. Soc. J.*, 33(3), 33–7.

Kobayashi, N. and Wurjanto, A., 1989. Wave transmission over submerged breakwater, *J. Waterw. Port Coast. Ocean Eng. ASCE.*, 115(5), 662–80.

Kobayashi, N. and Wurjanto, A., 1990. Numerical model for waves on rough permeable slopes, *J. Coast. Res.*, 7, 149–66.

Kobayashi, N. and Raichle, A., 1994. Irregular wave overtopping of revetments in surf zones, *J. Waterw. Port Coast. Ocean Eng. ASCE*, 120(1), 56–73.

Kobayashi, N., Otta, A.K., and Roy, I., 1987. Wave reflection and run-up on rough impermeable slopes, *J. Waterw. Port Coast. Ocean Eng. ASCE*, 113(3), 282–98.

Kobayashi, N., Cox, D.T., and Wurjanto, A., 1990. Irregular wave reflection and run-up on rough impermeable slopes, *J. Waterw. Port Coast. Ocean Eng. ASCE*, 116(6), 708–26.

Kofoed-Hansen H, Kerper, D.R., Sørensen, O.R., and Kirkegaard, J., 2005. Simulation of long wave agitation in ports and harbors using a time-domain Boussinesq model, *Proc. 5th Int. Symp. Ocean Wave Measurement and Analysis – WAVES 2005*, Madrid, Spain, pp. 32–32.

Koftis, T.H., Prinos, P., and Koutandos, E., 2006. 2D-V hydrodynamics of wave–floating breakwater interaction, *J. Hydraul. Res.*, 44(4), 451–69.

Kolmogorov, A.N., 1942. Equations of turbulent motion of an incompressible fluid, *Izv. Akad. Nauk SSR Ser. Phys.*, 6, 56.

Kolmogorov, A.N., 1962. A refinement of previous hypotheses concerning the local structure of turbulence in a viscous incompressible fluid at high Reynolds number, *J. Fluid Mech.*, 13, 82–5.

Komen, G.J., Hasselmann, S., and Hasselmann, K., 1984. On the existence of a fully developed wind-sea spectrum, *J. Phys. Oceanogr.*, 14, 1271–85.

Kondo, H. and Toma, S., 1972. Reflection and transmission for a porous structure, *Proc. 13th Int. Conf. Coast. Eng. ASCE*, New York, NY, USA, pp. 1847–65.

Koo, W., Kim, M.H., Lee, D.H., and Hong, S.Y., 2006. Nonlinear time-domain simulation of pneumatic floating breakwater, *Int. J. Offshore Polar Eng.*, 16(1), 25–32.

Korteweg, D.J. and de Vries, F., 1895. On the change of form of long waves advancing in a rectangular canal, and on a new type of long stationary waves, *Philos. Mag.*, 39, 422–43.

Koshizuka, S., Nobe, A., and Oka, Y., 1998. Numerical analysis of breaking waves using the moving particle semi-implicit method, *Int. J. Numer. Meth. Fluids*, 26(7), 751–69.

Kothe, D.B. and Mjolsness, R.C., 1991, RIPPLE: a new model for incompressible flows with free surfaces, *AIAA J.*, 30(11), 2694–700.

Kothe, D.B., Mjolsness, R.C., and Torrey, M.D., 1991. RIPPLE: a computer program for incompressible flows with free surfaces, *Report LA-12007-MS*, Los Alamos National Laboratory.

Kothe, D.B., Ferrell, R.C., Turner, J.A., and Mosso, S.J., 1997. A high resolution finite volume method for efficient parallel simulation of casting processes on unstructured meshes, *Report LA-UR-97-30*, Los Alamos National Laboratory.

Kuipers, J. and Vreugdenhil, C.B., 1973. Calculations of two-dimensional horizontal flows, *Report S 163, Part I*, Delft Hydraulics Laboratory, Delft, The Netherlands.

Kurtoglu, I.O. and Lin, C.L., 2006. Lattice Boltzmann study of bubble dynamics, *Numer. Heat Transf. Part B Fundam.*, 50(4), 333–51.

Lai, M.C. and Peskin, C.S., 2000. An immersed boundary method with formal second-order accuracy and reduced numerical viscosity, *J. Comput. Phys.*, 160, 705–19.

Lamb, H., 1945. *Hydrodynamics*, 6th ed., New York: Dover.

Lamb, K.G., 1994. Numerical experiments of internal wave generation by strong tidal flow across a finite-amplitude bank edge, *J. Geophys. Res., Oceans*, 99(C1), 843–64.

Lanzano, P., 1991. Waves generated by a moving point-source within a finite-depth ocean, *Earth, Moon, and Planets*, 52, 233–252.

Larsen, J. and Dancy, H., 1983. Open boundaries in short wave simulations–a new simulation, *Coast. Eng.*, 7, 285–97.

Launder, B.E. and Spalding, D.B., 1974. The numerical computation of turbulent flow, *Computer Meth. Appl. Mech. Eng.*, 3, 269–89.

Launder, B.E., Reece, G.T., and Rodi, W., 1975. Progress in development of a Reynolds stress turbulence closure, *J. Fluid Mech.*, 68, 537–66.

Le Méhauté, B. and Wang, S., 1996. *Water Waves Generated by Underwater Explosion*, Singapore: World Scientific.

Lee, C., Park, W.S., Cho, Y.S., and Suh, K.D., 1998. Hyperbolic mild-slope equations extended to account for rapidly varying topography, *Coast. Eng.*, 34, 243–57.

Lee, C., Cho, Y.S., and Yum, K., 2001. Internal generation of waves for extended Boussinesq equations, *Coast. Eng.*, 42(2), 155–62.

Lee, C.H., 1995. WAMIT theory manual, *Report No. 95–2*, Dept. of Ocean Eng., Massachusetts Institute of Technology.

Lee, J.J., 1971. Wave-induced oscillations in harbors of arbitrary geometry, *J. Fluid Mech.*, Part 2, 45, 375–93.

Lee, J.J., Skjelbreia, J.E., and Raichlen, F., 1982. Measurement of velocities in solitary waves, *J. Waterw. Port Coast. Ocean Eng. ASCE*, 108, 200–18.

Lee, S.J., Yates, G.T., and Wu, T.Y, 1989. Experiments and analyses of upstream-advancing solitary waves generated by moving disturbances, *J. Fluid Mech.*, 199, 569–93.

Lee, Y.J. and Lin, P., 2005. Modeling flow–structure interaction with the immersed boundary method, *Proc. 3rd Int. Conf. Asian and Pacific Coast*, Korea, pp. 371–4.

Leibovich, S., 1983. The form and dynamics of Langmuir circulations, *Annu. Rev. Fluid Mech.*, 15, 391–427.

Lemos, C.M., 1992. Wave breaking, a numerical study, *Lecture Notes in Engineering No. 71*. Berlin: Springer-Verlag.

Leonard, B.P., 1979. A stable and accurate convective modeling procedure based on quadratic upstream interpolation, *Comput. Meth. Appl. Mech. Eng.*, 19, 59–98.

Leonard, B.P., 1988, Elliptic systems: finite-difference method IV, in: W.J. Minkowycz, E.M. Sparrow, G.E. Schneider, and R.H. Pletcher (Eds.), *Handbook of Numerical Heat Transfer*, New York: Wiley, pp. 347–78.

Leonard, B.P., 1991. The ULTIMATE conservative difference scheme applied to unsteady one–dimensional advection, *Comput. Meth. Appl. Mech. Eng.*, 88, 17–74.

Leveque, R.J. and Li, Z., 1994. The immersed interface method for elliptic equations with discontinuous coefficients and singular sources, *Numer. Anal.*, 31, 1019–44.

Li, B. and Anastasiou, K., 1992. Efficient elliptic solvers for the mild-slope equation using the multigrid technique, *Coast. Eng.*, 16, 245–66.

Li, B. and Fleming, C.A., 1997. A three-dimensional multigrid model for fully nonlinear water waves, *Coast. Eng.*, 30, 235–58.

Li, B. and Fleming, C.A., 1999. Modified mild-slope equations for wave propagation, *Proc. Inst. Civil Engrs Water Marit. Energy*, 136(1), 43–60.

Li, C.W. and Yu, T.S., 1996. Numerical investigation of turbulent shallow recirculating flows by a quasi-three-dimensional k–ε model, *Int. J. Numer. Meth. Fluids*, 23, 485–501.

Li, C.W. and Lin, P., 2001. A numerical study of three-dimensional wave interaction with a square cylinder, *Ocean Eng.*, 28, 1545–55.

Li, H.J., Hu, S.L.J., and Takayama, T., 1999. The optimal design of TMD for offshore structures, *China Ocean Eng.*, 13(2), 133–44.

Li, S. and Liu, W.K., 2004. *Meshfree Particle Methods*, Berlin: Springer-Verlag.

Li, T.Q., 2003. Computation of turbulent free-surface flows around modern ships, *Int. J. Numer. Meth. Fluids*, 43, 407–30.

Li, T.Q., Troch, P., and De Rouck, J., 2004. Wave overtopping over a sea dike, *J. Comput. Phys.*, 198, 686–726.

Li, Y. and Raichlen, F., 2001. Solitary wave runup on plane slopes, *J. Waterw. Port Coast. Ocean Eng. ASCE*, 127, 33–44.

Li, Y. and Raichlen, F., 2003. Energy balance model for breaking solitary wave runup, *J. Waterw. Port Coast. Ocean Eng. ASCE*, 129, 47–59.

Li, Y.C., Liu, H.J., Teng, B., and Sun, D.P., 2002. Reflection of oblique incident waves by breakwaters with partially-perforated wall, *Chin. Ocean Eng.*, 16(3), 329–342.

Li, Y.S., Liu, S.X., Yu, Y.X., and Lai, G.Z., 1999. Numerical modeling of Boussinesq equations by finite element method, *Coastal Eng.*, 37, 97–122.

Li, Y.S., Liu, S.X., Yu, Y.X., and Lai, G.Z., 2000. Numerical modelling of multidirectional irregular waves through breakwaters, *Appl. Math. Model.*, 24, 551–74.

Li, Z. and Johns, B., 2001. A numerical method for the determination of weakly non-hydrostatic non-linear free surface wave propagation, *Int. J. Numer. Meth. Fluids*, 35, 299–317.

Lian, J.J., Zhao, Z.D., and Zhang, Q.H., 1999. A nonlinear viscoelastic model for bed mud transport due to waves and currents, *Chin. Sci. Bull.*, 44(17), 1597–600.

Liang, N.K., Huang, J.S., and Li, C.F., 2004. A study of spar buoy floating breakwater, *Ocean Eng.*, 31, 43–60.

Liao, S.J. and Cheung, K.F., 2003. Homotopy analysis of nonlinear progressive waves in deep water, *J. Eng. Math.*, 45, 105–16.

Lighthill, M.J., 1978. *Waves in Fluids*, Cambridge: Cambridge University Press.

Lilly, D.K., 1967. The representation of small-scale turbulence in numerical simulation experiments, *Proc. IBM Sci. Comput. Symp. Environmental Sciences*, 195–210.

Lilly, D.K., 1992. A proposed modification of the Germano subgrid-scale closure method, *Phys. Fluids*, A4, 633–5.

Lin, C., Ho, T.C., Chang, S.C., Hsieh, S.C., and Chang, K.A., 2005. Vortex shedding induced by a solitary wave propagating over a submerged vertical plate, *Int. J. Heat Fluid Flow*, 26, 894–904.

Lin, C.L., Lee, H., Lee, T., and Weber, L.J., 2005. A level set characteristic Galerkin finite element method for free surface flows, *Int. J. Numer. Meth. Fluids*, 49, 521–47.

Lin, M.C., Hsu, C.M., Wang, S.C., and Ting, C.L., 2002. Numerical and experimental investigations of wave field around a circular island with the presence of weak currents, *Chin. J. Mech. Ser. A*, 18, 35–42.

Lin, P., 2004a. A compact numerical algorithm for solving the time-dependent mild slope equation, *Int. J. Numer. Meth. Fluids*, 45, 625–42.

Lin, P., 2004b. A numerical study of solitary wave interaction with rectangular obstacles, *Coast. Eng.*, 51, 35–51.

Lin, P., 2006. A multiple-layer σ-coordinate model for simulation of wave–structure interaction, *Comput. Fluids*, 35(2), 147–67.

Lin, P, 2007. A fixed-grid model for simulating a moving body in fluids, *Comput. Fluids*, 36(3), 549–61.

Lin, P. and Liu, P.L.F., 1998a. Turbulence transport, vorticity dynamics, and solute mixing under plunging breaking waves in surf zone, *J. Geophys. Res*, 103(C8), 15677–94.

Lin, P. and Liu, P.L.F., 1998b. A numerical study of breaking waves in the surf zone, *J. Fluid Mech.*, 359, 239–64.

Lin, P. and Liu, P.L.F., 1999a. Free surface tracking methods and their applications to wave hydrodynamics, in: P.L.F. Liu (Ed.), *Advances in Coastal and Ocean Engineering*, Vol. 5, Singapore: World Scientific, pp. 213–40.

Lin, P. and Liu, P.L.F., 1999b. An internal wave-maker for Navier–Stokes equations models, *J. Waterw. Port Coast. Ocean Eng. ASCE*, 125, 207–15.

Lin, P. and Li, C.W., 2002. A σ-coordinate three-dimensional numerical model for surface wave propagation, *Int. J. Numer. Meth. Fluids*, 38, 1045–68.

Lin, P. and Li, C.W., 2003. Wave–current interaction with a vertical square cylinder, *Ocean Eng.*, 30(7), 855–76.

Lin, P. and Liu, P.L.F., 2004. Discussion of "Vertical variation of the flow across the surf zone" [Coastal Engineering 45(2002) 169–198], *Coast. Eng.*, 50, 161–4.

Lin, P. and Zhang, D., 2004. The depth-dependent radiation stresses and their effect on coastal currents, *Proc. 6th Int. Conf. Hydrodynamics: Hydrodynamics VI Theory and Applications*, Perth, West Australia, pp. 247–53.

Lin, P. and Liu, H.W., 2005. Analytical study of linear long-wave reflection by a two-dimensional obstacle of general trapezoidal shape, *J. Eng. Mech. ASCE*, 131(8), 822–30.

Lin, P. and Man, C.J., 2006. A staggered-grid numerical algorithm for the extended Boussinesq equations, *Appl. Math. Model.*, 31(2), 349–68.

Lin, P. and Xu, W., 2006. NEWFLUME: a numerical water flume for two-dimensional turbulent free surface flows, *J. Hydraul. Res.*, 44(1), 79–93.

Lin, P. and Karunarathna, S.A.S.A., 2007. A numerical study of solitary wave interaction with porous breakwaters, *J. Waterw. Port Coast. Ocean Eng. ASCE*, 133(5), 352–63.

Lin, P. and Liu, H.W., 2007. Scattering and trapping of wave energy by a submerged truncated paraboloidal shoal, *J. Waterw. Port Coast. Ocean Eng. ASCE*, 133(2), 94–103.

Lin, P. and Zhang, W., 2008. Numerical simulation of wave-induced laminar boundary layers, *Coast. Eng.*, 55, 200–208.

Lin, P., Chang, K.A., and Liu, P.L.F., 1999. Runup and rundown of solitary waves on sloping beaches, *J. Waterw. Port Coast. Ocean Eng. ASCE*, 125, 247–55.

Lin, P., Li, C.W., and Liu, H.W., 2005. A wave height spectral model for simulation of wave diffraction and refraction, *J. Coast. Res.*, SI 42, 448–59.

Lin, Q. and Lin, P., 2006. Modeling of sediment transport and morphological change in open channels, *15th Congress Asia and Pacific Division of the International Association for Hydraulic Research*, Chennai, India, pp. 195–200.

Linton, C.M. and Evans, D.V., 1990. The interaction of waves with arrays of vertical circular cylinders, *J. Fluid Mech.*, 215, 549–69.

Liszka, T. and Orkisz, J., 1980. The finite-difference method at arbitrary irregular grids and its application in applied mechanics, *Comput. Struct.*, 11, 83–95.

Liu, D. and Lin, P., 2008. A numerical study of three-dimensional liquid sloshing in tanks, *J. Comput. Phys.*, 227, 3921–3939.

Liu, G.R., 2002, *Mesh Free Methods – Moving Beyond the Finite Element Method*, Boca Raton, FL: CRC Press.

Liu, H., Yin, B.S., Xu, Y.Q., and Yang, D.Z., 2005b. Numerical simulation of tides and tidal currents in Liaodong Bay with POM, *Prog. Nat. Sci.*, 15(1), 47–55.

Liu, H.W. and Lin, P., 2005. Discussion for wave transformation by two dimensional bathymetric anomalies with sloped transitions, *Coast. Eng.*, 52, 197–200.

Liu, H.W. and Li, Y.B., 2007. An analytical solution for long-wave scattering by a submerged circular truncated shoal, *J. Eng. Math.*, 57 (2), 133–4.

Liu, H.W., Lin, P., and Shankar, N.J., 2004. An analytical solution of the mild-slope equation for waves around a circular island on a paraboloidal shoal, *Coast. Eng.*, 51(5–6), 421–37.

Liu, P.C., 1971. Normalized and equilibrium spectra of wind waves in Lake Michigan, *J. Phys. Oceanogr.*, 1, 249–57.

Liu, P.L.F., 1973. Damping of water waves over porous bed, *J. Hydraul. Div.*, 99(12), 2263–71.

Liu, P.L.F., 1983. Wave-current interactions on a slowly varying topography, *J. Geophys. Res.*, 88(c7), 4421–6.

Liu, P.L.F., 1990. Wave transformation, *The Sea, Ocean Engineering Science*, Vol. 9A, New York: Wiley, pp. 27–63.

Liu, P.L.F., 1994. Model equations for wave propagation from deep to shallow water, in: P.L.F. Liu (Ed.), *Advances in Coastal Engineering*, Vol. 1, Singapore: World Scientific, pp. 125–57.

Liu, P.L.F. and Dalrymple, R.A., 1984. The damping of gravity waves due to percolation, *Coast. Eng.*, 8(1), 33–49.

Liu, P.L.F. and Tsay, T.K., 1984. Refraction–diffraction model for weakly nonlinear water waves, *J. Fluid Mech.* 141, 265–74.

Liu, P.L.F. and Cho, Y.S., 1993. Bragg reflection of infragravity waves by sandbars, *J. Geophys. Res.*, 98(C2), 22733–41.

Liu, P.L.F., and Lin, P., 1997. A numerical model for breaking wave: the volume of fluid method, Research *Rep. No. CACR-97-02*, Center for Applied Coastal Research, Ocean Eng. Lab., Univ. of Delaware, Newark, Delaware, 19716.

Liu, P.L.F. and Wen, J.G., 1997. Nonlinear diffusive surface waves in porous media, *J. Fluid Mech.*, 347, 119–39.

Liu, P.L.F and Cheng, Y.G., 2001. A numerical study of the evolution of a solitary wave over a shelf, *Phys. Fluids*, 13(6), 1660–7.

Liu, P.L.F and Losada, I.J., 2002. Wave propagation modeling in coastal engineering, *J. Hydraul. Res. IAHR*, 40(3), 229–40.

Liu, P.L.F and Wu, T.R., 2004. Waves generated by moving pressure disturbances in rectangular and trapezoidal channels, *J. Hydraul. Res.*, 42(2), 163–71.

Liu, P.L.F., Cho, Y.S., Kostense, J.K., and Dingemans, M.W. 1992. Propagation and trapping of obliquely incident wave groups over a trench with currents, *Appl. Ocean Res.*, 14, 201–12.

Liu, P.L.F., Cho, Y.S., Yoon, S.B., and Seo, S.N., 1994. Numerical simulations of the 1960 Chilean tsunami propagation and inundation at Hilo, Hawaii, in: M.I. El-Sabh (Ed.), *Recent Developments in Tsunami Research*, Dordrecht, Netherlands: Kluwer Academic Publishers, pp. 99–115.

Liu, P.L.F., Cho, Y.S., Briggs, M.J., Kanoglu, U., and Synolakis, C.E., 1995. Runup of solitary waves on a circular island, *J. Fluid Mech.*, 302, 259–85.

Liu, P.L.F., Yeh, H.H., Lin, P., Chang, K.T., and Cho, Y.S., 1998. Generation and evolution of edge-wave packet, *Phys. Fluids*, 10(7), 1635–57.

Liu, P.L.F., Lin, P., Chang, K.A., and Sakakiyama, T., 1999a. Numerical modeling of wave interaction with porous structures, *J. Waterw. Port Coast. Ocean Eng. ASCE*, 125(6), 322–30.

Liu, P.L.F, Hsu, T.J., Lin, P., Losada, I., Vidal, C., and Sakakiyama, T., 1999b. The Cornell Breaking Wave and Structure (COBRAS) model, *Proc. Conf. Wave Structures 1999*, Santander, Spain, pp. 169–74.

Liu, W.K., Jun, S.F., Adee, J., and Belytschko, T., 1995. Reproducing kernel particle methods for structural dynamics, *Int. J. Numer. Meth. Eng.*, 38(10), 1655–79.

Liu, X.D. and Sakai, S., 2002. Time domain analysis on the dynamic response of a flexible floating structure to waves, *J. Eng. Mech. ASCE*, 128(1), 48–56.

Liu, X.D., Osher, S., and Chan, T. 1994. Weighted essentially nonoscillatory schemes, *J. Comput. Phys.*, 115, 200–12.

Liu, Y.M. and Yue, D.K.P, 1998. On generalized Bragg scattering of surface-waves by bottom ripples, *J. Fluid Mech.*, 356, 297–326.

Liu, Y.Z. and Shi, J.Z., 2008. A theoretical formulation for wave propagations over uneven bottom, *Ocean Eng.*, 35(3–4), 426–32.

Liu, Y.Z., Shi, J.Z., and Perrie, W., 2007. A theoretical formulation for modeling 3D wave and current interations in estuaries, *Adv. Water Resour.*, 30(8), 1737–45.

Lo, D.C. and Young, D.L., 2004. Arbitrary Lagrangian–Eulerian finite element analysis of free surface flow using a velocity–vorticity formulation, *J. Comput. Phys.*, 195(1), 175–201.

Lo, J.M., 1991. A numerical model for combined refraction–diffraction of short waves on an island, *Ocean Eng.*, 18(5), 419–34.

Lohner, R., Yang, C., Baum, J., Luo, H., Pelessone, D., and Charman, C. M., 1999. The numerical simulation of strongly unsteady flow with hundreds of moving bodies, *Int. J. Numer. Meth. Fluids*, 31, 113–20.

Longuet-Higgins, M.S., 1953. Mass transport in water waves, *Philos. Trans. R. Soc. Lond.*, 245(A), 535–81.

Longuet-Higgins, M.S., 1967. On the trapping of wave energy around islands, *J. Fluid Mech.*, 29, 781–821.

Longuet-Higgins, M.S., 1977. The mean forces exerted by waves on floating or submerged bodies with applications to sand bars and wave power machines, *Proc. R. Soc. Lond. A*, 352, 463–80.

Longuet-Higgins, M.S. and Stewart, R.W., 1961. The changes in amplitudes of short gravity waves on steady non-uniform currents, *J. Fluid Mech.*, 10, 529–49.

Longuet-Higgins, M.S. and Stewart, R.W., 1964. Radiation stresses in water waves: a physical discussion with applications, *Deep-Sea Res.*, 11, 529–62.

Longuet-Higgins, M.S. and Cokelet, E.D., 1976. The deformation of steep surface waves on water. I. A numerical method of computation, *Proc. R. Soc.*, A 350, 1–25.

Losada, I.J., Silva, R., and Losada, M.A., 1996. 3-D non-breaking regular wave interaction with submerged breakwaters, *Coast. Eng.*, 28(1–4), 229–48.

Losada, I.J., Patterson, M.D., and Losada, M.A., 1997. Harmonic generation past a submerged porous step, *Coast. Eng.*, 31(1–4), 281–304.

Losada, M.A., Vidal, C., and Medina, R., 1989. Experimental study of the evolution of a solitary wave at an abrupt junction, *J. Geophys. Res.*, 94(10), 14557–66.

Lozano, C. and Meyer, R.E., 1976. Leakage and response of waves trapped by round islands, *Phys. Fluids*, 19, 1075–88.

Lozano, C. and Liu, P.L.F., 1980. Refraction–diffraction model for linear surface water, *J. Fluid Mech.*, 101(4), 705–20.

Lu, C.H., He, Y.S., and Wu, G.X., 2000. Coupled analysis of nonlinear interaction between fluid and structure during impact, *J. Fluids Struct.*, 14(1), 127–46.

Lucy, L., 1977. A numerical approach to the testing of the fission hypothesis, *Astron. J.*, 82, 1013–24.

Luke, J.C., 1967. A variational principle for a fluid with a free surface, *J. Fluid Mech.*, 27, 395–7.

Lynett, P.J., 2006. Nearshore wave modeling with high-order Boussinesq-type equations, *J. Waterw. Port Coast. Ocean Eng. ASCE*, 132(5), 348–57.

Lynett, P.J. and Liu, P.L.F., 2002. A two-dimensional, depth-integrated model for internal wave propagation over variable bathymetry, *Wave Motion*, 36(3), 221–40.

Lynett, P.J. and Liu, P.L.F., 2004. Linear analysis of the multi-layer model, *Coast. Eng.*, 51(5–6), 439–54.

Lynett, P.J, Wu, T.R., and Liu, P.L.F., 2002. Modeling wave runup with depth-integrated equations, *Coast. Eng.*, 46(2), 89–107.

Lynett, P.J., Borrero, J.C., Liu, P.L.F., and Synolakis, C.E., 2003. Field survey and numerical simulations: a review of the 1998 Papua New Guinea tsunami, *Pure Appl. Geophys.*, 160(10–11), 2119–46.

Ma, Q.W., 2005. Meshless local Petrov–Galerkin method for two-dimensional nonlinear water wave problems, *J. Comput. Phys.*, 205(2), 611–25.

MacCamy, R.C. and Fuchs, R.A., 1954. Wave forces on piles: a diffraction theory, *Tech. Memo 69*, US Army Corps of Engineers, Beach Erosion Board.

McCowan, J., 1894. On the highest wave of permanent type, *Philos. Mag. J. Sci.*, 5(38), 351–8.

McIver, M. and McIver, P., 1990. Second-order wave diffraction by a submerged circular cylinder, *J. Fluid Mech.*, 219, 519–29.

Madsen, O.S., 1974. Wave transmission through porous structures, *J. Waterw. Harbour and Coast. Eng. Div. ASCE*, 100(3), 169–88.

Madsen, O.S. and White, S.M., 1975. Reflection and transmission characteristics of porous rubble mound breakwaters, *Rep. No. 207*, The US Army, Coast. Engrg. Res. Ctr.

Madsen, P.A. and Larsen, J., 1987. An efficient finite-difference approach to the mild-slope equation, *Coast. Eng.*, 11, 329–51.

Madsen, P.A. and Sørensen, O.R., 1992. A new form of the Boussinesq equations with improved linear dispersion characteristics. Part 2. A slowly varying bathymetry, *Coast. Eng.*, 18, 183–204.

Madsen, P.A. and Schäffer, H.A., 1998. Higher order Boussinesq-type equations for surface gravity waves – derivation and analysis, *Philos. Trans. R. Soc. Lond. A*, 356, 3123–86.

Madsen, P.A., Murray, R., and Sørensen, O.R., 1991. A new form of the Boussinesq equations with improved linear dispersion characteristics, *Coast. Eng.*, 15, 371–88.

Madsen, P.A., Sørensen, O.R., and Schäffer, H.A., 1997. Surf zone dynamics simulated by a Boussinesq type model. Part I. Model description and cross-shore motion of regular waves, *Coast. Eng.*, 32, 225–87.

Madsen, P.A., Bingham, H.B., and Liu, H., 2002. A new Boussinesq method for fully nonlinear waves from shallow to deep water, *J. Fluid Mech.*, 462, 1–30.

Madsen, P.A., Simonsen, H.J., and Pan, C.H., 2005. Numerical simulation of tidal bores and hydraulic jumps, *Coast. Eng.*, 52(5), 409–33.

Madsen, P.A., Fuhrman, D.R., and Wang, B.L., 2006. A Boussinesq-type method for fully nonlinear waves interacting with a rapidly varying bathymetry, *Coast. Eng.*, 53(5–6), 487–504.

Marcou, O., Yacoubi, S.E.L., and Chopard, B., 2006. A BI-fluid lattice Boltzmann model for water flow in an irrigation channel, *Lect. Notes Comput. Sci.*, 4173, 373–82.

Mardia, K.V., 1972. *Statistics of Directional Data*, London: Academic Press.

Marinov, D., Norro, A., and Zaldivar, J.M., 2006. Application of COHERENS model for hydrodynamic investigation of Sacca di Goro coastal lagoon (Italian Adriatic Sea shore), *Ecol. Modell.*, 193, 52–68.

Martys, N.S. and Chen, H.D., 1996. Simulation of multicomponent fluids in complex three-dimensional geometries by the lattice Boltzmann method, *Phys. Rev. E*, 53(1), 743–50.

Mase, H., Memita, T., Yuhi, M., and Kitano, T., 2002. Stem waves along vertical wall due to random wave incidence, *Coast. Eng.*, 44, 339–50.

Massel, S.R., 1983. Harmonic generation by waves propagating over a submerged step, *Coast. Eng.*, 7, 357–80.

Massel, S.R., 1993. Extended refraction–diffraction equation for surface waves, *Coast. Eng.*, 19, 97–126.

Massel, S.R., 1996. *Ocean Surface Waves: Their Physics and Prediction, Advanced Series On Ocean Engineering*, 11. Singapore: World Scientific.

van der Meer, J.W., 1988. Deterministic and probabilistic design of breakwater armour layers, *J. Waterw. Port Coast. Ocean Eng. ASCE*, 114 (1), 66–80.

van der Meer, J.W., Petit, H.A.H., Van den Bosch, P., Klopman, G., and Broekens, R.D., 1992. Numerical simulation of wave motion on and in coastal structures, *Proc. 23 rd Int. Conf. Coast. Eng.*, Venice, Italy, pp. 1172–84.

van der Meer, J.W. and Janssen, J.P.F., 1995. Wave run-up and wave overtopping at dikes, in: N. Kobayashi and Z. Demirbilek (Eds.), *Wave Forces On Inclined and Vertical Structures*, ASCE, Reston, VA, pp. 1–27.

Mei, C.C., 1985. Resonant reflection of surface water waves by periodic sandbars, *J. Fluid Mech.*, 152, 315–35.

Mei, C.C., 1986. Radiation of solitons by slender bodies advancing in a shallow channel, *J. Fluid Mech.*, 162, 53–67.

Mei, C.C., 1989. *The Applied Dynamics of Ocean Surface Waves*, Singapore: World Scientific.

Mei, C.C. and LeMehaute, B., 1966. Note on the equations of long waves over an uneven bottom, *J. Geophys. Res.*, 71, 393–400.

Mei, C.C. and Black, J.L., 1969. Scattering of surface wave by rectangular obstacles in waters of finite depth, *J. Fluid Mech.*, 38, 499–511.

Mei, C.C., Stiassnie, M., Yue, D., 2005. *Theory and Applications of Ocean Surface Waves*, Singapore: World Scientific.

Mellor, G.L. and Herring, H.J., 1973. A survey of mean turbulent field closure, *AIAA J.*, 11, 590–9.

Melville, W.K., 1980. On the Mach reflection of solitary waves, *J. Fluid Mech.*, 98, 285–97.

Meneghini, J.R. and Bearman, P.W., 1993. Numerical simulation of high amplitude oscillatory-flow about a circular cylinder using a discrete vortex method. *AIAA*, Paper No. 93-3288.

Miche, R., 1944. Mouvement endulatoires de la mer en pro-fondeur constante ou decroissante, *Ann. Ponts et Chaussees*, 121, 285–318.

Miles, J.W., 1967. Surface-wave scattering matrix for a shelf, *J. Fluid Mech.*, 28, 755–67.

Miles, J.W., 1979. On the Korteweg–de Vries equation for a gradually varying channel, *J. Fluid Mech.*, 91, 181–90.

Miles, J.W., 1981. Oblique surface-wave diffraction by a cylindrical obstacle, *Dyn. Atmos. Oceans*, 6, 207–32.

Miles, J.W., 1986. Resonant amplification of gravity waves over a circular still, *J. Fluid Mech.*, 167, 169–79.

Miles, J.W. and Chamberlain, P.G., 1998. Topographical scattering of gravity waves, *J. Fluid Mech.*, 361, 175–88.

Minato, S., 1998. Storm surge simulation using POM and a revisitation of dynamics of sea surface elevation short-term variation, *Meteorol. Geophys.*, 48(3), 79–88.

Mindlin, R.D., 1951. Influence of rotatory inertia and shear on flexural motions of isotropic, elastic plates, *J. Appl. Mech. ASME*, 18, 31–8.

Mitsuyasu, H., 1972. The one-dimensional wave spectra at limited fetch, *Proc. 13th Coast. Eng. Conf.*, 1289–306.

Mitsuyasu, H. and Honda, T., 1982. Wind-induced growth of water waves, *J. Fluid Mech.*, 123, 425–42.

Miyata, H., 1986. Finite-difference simulation of breaking waves, *J. Comput. Phys.*, 65, 179–214.

Miyata, H., Nishimura, S., and Masuko, A., 1985. Finite-difference simulation of nonlinear-waves generated by ships of arbitrary 3-dimensional configuration, *J. Comput. Phys.*, 60, 391–436.

Mohd-Yusof, J., 1997. Combined immersed boundaries/B-splines methods for simulations of flows in complex geometries, *CTR Annual Research Briefs*, NASA Ames/Stanford University.

Molin, B., 1979. Second-order diffraction loads upon three-dimensional bodies, *Appl. Ocean Res.*, 1, 197–202.

Monaghan, J.J., 1992. Smoothed particle hydrodynamics, *Annu. Rev. Astron. Astrophys.*, 30, 543–74.

Monaghan, J.J., 1994. Simulating free surface flows with SPH, *J. Comput. Phys.*, 110, 399–406.

Morison, J.R., O'Brien, M.P., Johnson, J.W., and Schaaf, S.A., 1950. The force exerted by surface wave on piles, *Petrol. Trans. AIME*, 189, 149–54.

Moukalled, F. and Darwish, M., 2000, A unified formulation of the segregated class of algorithms for fluid flow at all speeds, *Numer. Heat Transf.*, Part B, 37, 103–39.

Munk, W.H., 1949. Surf beats, *EOS Trans.*, 30, 849–54.

Murali, K. and Mani, J.S., 1997. Performance of cage floating breakwater, *J. Waterw. Port Coast. Ocean Eng. ASCE*, 123, 172–9.

Murray, J.D., 1965. Viscous damping of gravity waves over a permeable bed, *J. Geophys. Res.* 70, 2325–31.

Myers, E.P. and Baptista, A.M., 1999. Finite element modeling of potential Cascadia subduction zone tsunamis, *Sci. Tsunami Hazards*, 17, 3–18.

Nakayama, T., 1983. Boundary element analysis of nonlinear water wave problems, *Int. J. Numer. Meth. Eng.*, 19, 953–70.

Nayroles, B., Touzot, G., and Villon, P., 1992. Generalizing the finite element method: diffuse approximation and diffuse elements, *Comput. Mech.* 10, 301–18.

Nepf, H.M., 1999. Drag, turbulence, and diffusion in flow through emergent vegetation, *Water Resour. Res.* 35, 479–89.

Neumann, G., 1953. On ocean wave spectra and a new method of forecasting wind-generated sea, *Tech. Mem., No. 43*, Beach Erosion Board.

Newman, J.N., 1965a. Propagation of water waves past long dimensional obstacles, *J. Fluid Mech.*, 23, 23–9.

Newman, J.N., 1965b. Propagation of water waves over an infinite step, *J. Fluid Mech.*, 23, 399–415.

Newman, J.N., 1977. *Marine Hydrodynamics*, Cambridge: MIT Press.

Nichols, B.D., Hirt, C.W., and Hotchkiss, R.S., 1980. SOLA-VOF: A solution algorithm for transient fluid flow with multiple free-boundaries, *Rep. LA-8355*, Los Alamos Scientific Laboratory.

Nielsen, P., 1992. *Coastal Bottom Boundary Layers and Sediment Transport*, Singapore: World Scientific.

Nitikitpaiboon, C. and Bathe, K.J., 1993. An arbitrary Lagrangian–Eulerian velocity potential formulation for fluid–structure interaction, *Comput. Struct.*, 47, 871–91.

Nwogu, O., 1993. Alternative form of Boussinesq equations for nearshore wave propagation, *J. Waterw. Port Coast. Ocean Eng. ASCE*, 119, 618–38.

Ochi, M. and Hubble, E., 1976. On six-parameter wave spectra, *Proc. 15th Int. Conf. Coastal Eng.*, pp. 301–28.

Oey, L.Y., Ezer, T., Forristall, G., Cooper, C., DiMarco, S., and Fan, S., 2005. An exercise in forecasting loop current and eddy frontal positions in the Gulf of Mexico, *Geophys. Res. Lett.*, 32, L12611.

O'Hare, T.J. and Davies, A.G., 1992. A new model for surface wave propagation over rapidly-varying topography, *Coast. Eng.*, 18, 251–66.

Ohkusu, M. and Namba, Y., 2004. Hydroelastic analysis of a large floating structure, *J. Fluids Struct.*, 19, 543–55.

Ohyama, T. and Nadaoka, K., 1994. Transformation of a nonlinear-wave train passing over a submerged shelf without breaking, *Coast. Eng.*, 24, 1–22.

Onate, E. and Garcia, J.A., 2001. Finite element method for fluid–structure interaction with surface waves using a finite calculus formulation, *Comput. Meth. Appl. Mech. Eng.*, 191, 635–60.

Orszag, S.A. and Patterson, G.S., 1972. Numerical simulation of three-dimensional homogeneous isotropic turbulence, *Phys. Rev. Lett.*, 28, 76–69.

Orszag, S.A., Staroselsky, I., Flannery, W.S., and Zhang, Y., 1996. Introduction to renormalization group modeling of turbulence, in: T.B. Gatski, M.Y. Hussaini, and J.L. Lumley (Eds.), *Simulation and Modeling of Turbulent Flows*, Oxford: Oxford University Press, pp. 155–83.

Osborne, A.R., Onorato, M., and Serio, M., 2000. The nonlinear dynamics of rogue waves and holes in deep-water gravity wave trains, *Phys. Lett. A*, 275, 386–93.

Osher, S. and Sethian, J.A., 1988. Fronts propagating with curvature-dependent speed-algorithms based on Hamilton–Jacobi formulations, *J. Comput. Phys.*, 79, 12–49

Oumeraci, H. and Kortenhaus, A., 1994. Analysis of the dynamic response of caisson breakwaters, *Coastal Eng.*, 22, 159–83.

Owen, M.W., 1980. Design of seawalls allowing for wave overtopping, *Technical Report EX-924*, HR-Wallingford, UK.

Ozer, J., Padilla-Hernandez, R., Monbaliu, J., Alvarez, F.E., Carretero, A.J.C., Osuna, P., Yu, J.C.S., and Wolf, J., 2000. A coupling module for tides, surges and waves, *Coast. Eng.*, 41, 95–124.

Palma, P.D., Tullio, M.D., Pascazio, G., and Napolitano, M., 2006. An immersed-boundary method for compressible viscous flows, *Comput. Fluids*, 35, 693–702.

Panchang, V.G., Pearce, B.R., Wei, G., and Cushman-Roisin, B., 1991. Solution of the mild slope wave problem by iteration, *Appl. Ocean Res.*, 13, 187–99.

Papaspyrou, S., Valougeorgis, D., and Karamanos, S.A., 2004. Sloshing effects in half-full horizontal cylindrical vessels under longitudinal excitation, *J. Appl. Mech. Trans. ASME*, 71, 255–65.

Park, J.C., Kim, M.H., and Miyata, H., 1999. Fully non-linear free-surface simulations by a 3D viscous numerical wave tank, *Int. J. Numer. Meth. Fluids*, 29, 685–703.

Park, K.Y., Borthwick, A.G.L., and Cho, Y.S., 2006. Two-dimensional wave–current interaction with a locally enriched quadtree grid system, *Ocean Eng.*, 33, 247–63.

Parkar, B.B., Davies, A.M., and Xing, J., 1999. Tidal height and current prediction, *Coastal Ocean Prediction*, Washington, DC: AGU Publications, pp. 277–327.

Parsons, N.F. and Martin, P.A., 1992. Scattering of water waves by submerged plates using hypersingular integral equations, *Appl. Ocean Res.*, 14, 313–21.

Patankar, S.V., 1981. *Numerical Heat Transfer and Fluid Flow*, New York: Hemisphere.

Patankar, S.V. and Spalding, D.B., 1972. A calculation procedure for heat, mass and momentum transfer in three-dimensional parabolic flows, *Int. J. Heat Mass Transf.*, 15, 1787–806.

Patarapanich, M. and Cheong, H.F., 1989. Reflection and transmission characteristics of regular and random waves from a submerged horizontal plate, *Coast. Eng.*, 13, 161–82.

Péchon, P. and Teisson, C., 1994. Numerical modeling of three-dimensional wave-driven currents in the surf-zone, *24th ICCE, ASCE*, Kobe, Japan, pp. 2503–12.

Pedersen, N.H., Rasch, P.S., and Sato, T., 2005. Modeling of Asian tsunami off the coast of northern Sumatra, *Proc. 3rd DHI Asia-Pacific Software Conference*, Kuala Lumpur, Malaysia.

Pelinovsky, E.N. and Mazova, R.K.H., 1992. Exact analytical solutions of nonlinear problems of tsunami wave runup on slopes with different profiles, *Nat. Hazards*, 5, 227–49.

Penney, W.G. and Price, A.T., 1952a. The diffraction theory of sea waves and the shelter afforded by breakwaters, *Philos. Trans. R. Soc. Lond.*, A244, 236–53.

Penney, W.G. and Price, A.T., 1952b. Some gravity wave problems in the motion of perfect liquids. Part II. Finite periodic stationary gravity waves in a perfect liquid, *Philos. Trans. R. Soc. Lond. Ser. A*, 244, 254–84.

Peregrine, D.H., 1967. Long waves on a beach, *J. Fluid Mech.*, 27, 815–27.

Peregrine, D.H. and Svendsen, I.A., 1978. Spilling breakers, bores and hydraulic jumps, *Proc. 16th Int. Conf. Coastal Eng. ASCE*, Hamburg, Germany, pp. 540–50.

Perlin, M. and Hammack, J., 1991. Experiments on ripple instabilities, Part 3: Resonant quartets of the Benjamin–Feir type, *J. Fluid Mech.*, 229, 229–68.

Perroud, P.H., 1957. The solitary wave reflection along a straight vertical wall at oblique incidence, Ph.D. Thesis, University of California, Berkeley, CA.

Peskin, C.S., 1972. Flow patterns around heart valves: a numerical method, *J. Comput. Phys.*, 10, 252–71.

Phillips, O.M., 1958. The equilibrium range in the spectrum of wind-generated waves, *J. Fluid Mech.*, 4, 426–434.

Phillips, O.M., 1977. *The Dynamics of the Upper Ocean*, 2nd ed., Cambridge: Cambridge University Press.

Pierson, W.J. and Moskowitz, L., 1964. A proposed spectral form for fully developed wind seas based on the similarity theory of S.A. Kitaigorodskii, *J. Geophys. Res.*, 69, 5181–90.

Pierson, W.J., Jr., Neumann, G., and James, R.W., 1955. *Practical Methods For Observing and Forecasting Ocean Waves by Means of Wave Spectra and Statistics*, US Navy Hydrographic Office, H.O. Pub.No.603.

Pierson, W.J., Tick, L.J., and Bear, L., 1966. Computer-based procedure for preparing global wave forecasts and wind field analyses capable of using wave data obtained from a spacecraft, *Proc. of 6th Symp. Naval Hydrodynamics*, Washington, DC, pp. 499–529.

Pope, S.B., 1975. A more general effective-viscosity hypothesis, *J. Fluid Mech.*, 72, 331–40.

Pope, S.B., 2000. *Turbulent Flows*, Cambridge: Cambridge University Press.

Porter, D., 2003. The mild-slope equations. *J. Fluid Mech.*, 494, 51–63.

Porter, R. and Porter, D., 2001. Interaction of water waves with three-dimensional periodic topography, *J. Fluid Mech.*, 434, 301–35.

Press, W.H., Teukolsky, S.A., Vetterling, W.T., and Flannery, B.P., 2007. *Numerical Recipes: The Art of Scientific Computing*, 3rd ed., Cambridge University Press.

Putnam, J.A., 1949. Loss of wave energy due to percolation in a permeable sea bottom transactions, *Am. Geol. Union*, 30(3), 349–56.

Putnam, J.A. and Johnson, J.W., 1949. The dissipation of wave energy by bottom friction transactions, *Am. Geol. Union*, 30(1), 67–74.

Qi, P. and Hou, Y.J., 2003. Numerical wave flume study on wave motion around submerged plates, *Chin. Ocean Eng.*, 17(3), 397–406.

Qian, Y.H., Succi, S., and Orszag, S.A., 1995. Recent advances in lattice Boltzmann computing, *Annu. Rev. Comp. Phys.*, 3, 195–242.

Quartapelle, L., 1993. *Numerical Solution of the Incompressible Navier–Stokes Equations*, Basel: Birkhäuser Verlag.

Radder, A.C., 1979. On the parabolic equation method for water-wave propagation, *J. Fluid Mech.*, 95(1), 159–76.

Radovitzky, R. and Ortiz, M., 1998. Lagrangian finite element analysis of Newtonian fluid flows, *Int. J. Numer. Meth. Eng.* 43, 607–19.

Rakha, K.A., Deigaard, R., and Broker, I., 1997. A phase-resolving cross shore sediment transport model for beach profile evolution, *Coast. Eng.*, 31(1–4), 231–61.

Raman, H., Prabhakararao, G.V., and Venkatanarasaiah, P., 1975. Diffraction of nonlinear surface waves by circular cylinder, *Acta Mech.*, 23, 145–58.

Raphael, E. and de Gennes, P.G., 1996. Capillary gravity waves caused by a moving disturbance: wave resistance, *Phys. Rev. E*, 53, 3448–55.

Rastogi, A.K. and Rodi, W., 1978. Predictions of heat and mass transfer in open channels, *J. Hydraul. Div. ASCE*, 104(HY3), 397–420.

Rattanapitikon, W. and Shibayama, T., 1998. Energy dissipation model for regular and irregular breaking waves, *Coast. Eng. J.*, 40, 327–46.

Raupach, M.R., Finnigan, J.J., and Brunet, Y., 1996. Coherent eddies and turbulence in vegetation canopies: the mixing-layer analogy, *Bound.-Layer Meteor.*, 78, 351–82.

Reeve, D., Chadwick, A., and Fleming, C., 2004. *Coastal Engineering: Processes, Theory and Design Practice*, London: Son Press, Taylor & Francis Group.

Reeve, D.E., Soliman, A. and Lin, P., 2007. Numerical study of combined overflow and wave overtopping over a smooth impermeable seawall, *Coast. Eng.*, doi:10.1016/j.coastaleng.2007.09.008.

Reeve, D.E., Soliman, A., and Lin, P., 2008. Numerical study of combined overflow and wave overtopping over a smooth impermeable seawall, *Coast. Eng.*, 55, 155–66.

Renardy, Y., 1983. Trapping of water waves above a round sill, *J. Fluid Mech.*, 132, 105–18.

Rey, V., Belzons, M., and Guazzelli, E., 1992, Propagation of surface gravity waves over a rectangular submerged bar, *J. Fluid Mech.*, 235, 453–79.

Rhee, J.P., 1997. On the transmission of water waves over a shelf, *Appl. Ocean Res.* 19, 161–9.

Rider, W.J. and Kothe, D.B., 1998. Reconstructing volume tracking, *J. Comput. Phys.*, 141, 112–52.

Ris, R.C., Booij, N., and Holthuijsen, L.H., 1999. A third-generation wave model for coastal regions, Part II, Verification, *J. Geophys. Res.*, C4, 104, 7667–81.

Rivero, F.J. and Arcilla, A.S., 1995. On the vertical distribution of $\langle \tilde{u}\tilde{w} \rangle$, *Coast. Eng.*, 25, 137–52.

Rivero, F.J., Arcilla, A.S., and Carci, E., 1997. An analysis of diffraction in spectral wave models, *Proc. 3rd Int. Symp. Ocean Wave Meas. and Anal.*, *WAVES'97*, ASCE, pp. 431–45.

Rodi, W., 1972. The prediction of free turbulent boundary layers using a two-equation model of turbulence, Ph.D. thesis, Imperial College, London.

Rodi, W., 1980. *Turbulence Models and Their Application in Hydraulics – A State-of-the-Art Review*, Delft, The Netherlands: IAHR publication.

Rogers, S.E., Kwak, D., and Kiris, C., 1991. Steady and unsteady solutions of the incompressible Navier–Stokes equations, *AIAA J.*, 29(4), 603–10.

Roseau, M., 1952. Contribution a la theorie des ondes liquides de gravite en profundeur variable, *Publ. Sci. Tech. du Ministerede Air*, No. 275, Paris.

Rotta, J.C., 1951. Statistische Theorie nichthomogener Turbulenz, *Z. Phys.*, 129, 547–72.

Saffman, P.G., 1970. A model for inhomogeneous turbulent flow, *Proc. R. Soc. Lond. Set. A*, 317(1529), 417–33.

Samet, H., 1990. *Applications of Spatial Data Structures: Computer Graphics, Image Processing, and GIS*, Reading, MA: Addison-Wesley.

Sander, J. and Hutter, K., 1992. Evolution of weakly nonlinear shallow-water waves generated by a moving boundary, *Acta Mech.*, 91(3–4), 119–55.

Sannasiraj, S.A., Sundar, V., and Sundaravadivelu, R., 1998. Mooring forces and motion responses of pontoon-type floating breakwaters, *Ocean Eng.*, 25(1), 27–48.

Sarpkaya, T., 1996. Vorticity, free surface and surfactants, *Annu. Rev. Fluid Mech.*, 28, 83–128.

Sarpkaya, T. and Isaacson, M., 1981. *Mechanics of Wave Forces on Offshore Structures*, New York, NY: Van Nostrand Reinhold.

Schäffer, H.A. and Madsen, P.A., 1995. Further enhancements of Boussinesq-type equations, *Coast. Eng.*, 26, 1–14.

Schultz, K. and Kallinderis, Y., 1998. Numerical prediction of vortex-induced vibrations, *Proc. 17th ASME Conf. OMAE*, paper No. 98-0362.

Schureman, P., 1958. *Manual of Harmonic Analysis of Tidal Observations*, Washington, DC: US Dept. of Com..

Scott, J.R., 1965. A sea spectrum for model tests and long-term ship prediction, *J. Ship Res.*, 9,145–52.

Seabra-Santos, F.J., Penouard, D.P., and Temperville, A.M., 1987. Numerical and experimental study of the transformation of a solitary wave over a shelf or isolated obstacle, *J. Fluid Mech.*, 176, 117–34.

Segur, H., Henderson, D., Carter, J., Hammack, J., Cong-Ming, L., Pheiff, D., and Socha, K., 2005. Stabilizing the Benjamin–Feir instability, *J. Fluid Mech.*, 539, 229–71.

Serre, F., 1953. Contribution 'a l' éude desecoulements permanents et variables dans les canaux, *Houille Blanche*, 8, 374–88.

Shankar, J., Cheong, H.-F. and Chan, C.-T., 1997. Boundary fitted grid models for tidal motions in Singapore coastal waters, *J. Hydraul. Res.*, 35(1), 3–20.

Shao, S.D., and Lo, E.Y.M., 2003. Incompressible SPH method for simulating Newtonian and non-Newtonian flows with a free surface, *Adv. Water Resour.*, 26(7), 787–800.

Shao, S.D. and Ji, C., 2006. SPH computation of plunging waves using a 2-D sub-particle scale (SPS) turbulence model, *Int. J. Numer. Meth. Fluids*, 51, 913–36.

Shen, C., 2000. Constituent Boussinesq equations for waves and currents, *J. Phys. Oceanogr.*, 31, 850–9.

Shen, Y.M., Ng, C.O., and Zheng, Y.H., 2004. Simulation of wave propagation over a submerged bar using the VOF method with a two-equation k-epsilon turbulence modeling, *Ocean Eng.*, 31(1), 87–95.

Shi, J.Z. and Lu, L.F., 2007. A model of the wave and current boundary layer structure, *Hydrol. Process.*, 21(13), 1780–1786.

Shih, T.H., Zhu, J., and Lumley, J.L., 1996. Calculation of wall-bounded complex flows and free shear flows, *Int. J. Numer. Meth. Fluids*, 23, 1133–44.

Shimizu, Y. and Tsujimoto, T., 1994. Numerical analysis of turbulent open-channel flow over a vegetation layer using a $k–\varepsilon$ turbulence model, *J. Hydrosci. Hydraul. Eng. JSCE*, 11, 57–67.

Shugan, I.V., Lee, K.J., and Sun, A.J., 2006. Kelvin wake in the presence of surface waves, *Phys. Lett. A*, 357, 232–5.

Shuto, N., 1973. Shoaling and deformation of nonlinear long waves, *Coast. Eng. Jpn*, 16, 1–12.

Shuto, N., 1974. Nonlinear long waves in a channel of variable section, *Coast. Eng. Jpn*, 17, 1–12.

Shuto, N., Goto, C., and Imamura, F., 1990. Numerical simulation as a means of warning for near-field tsunami, *Coast. Eng. Jpn*, 33(2), 173–93.

Silva, R., Losada, I.J., and Losada, M.A., 2000. Reflection and transmission of tsunami waves by coastal structures, *Appl. Ocean Res.*, 22(4), 215–23.

Sitanggang, K. and Lynett, P., 2005. Parallel computation of a highly nonlinear Boussinesq equation model through domain decomposition, *Int. J. Numer. Meth. Fluids*, 49(1), 57–74.

Skjelbreia, L. and Hendrickson, J., 1961. Fifth order gravity wave theory, *Proc. 7th Coast. Eng. Conf.*, 1, 184–96.

Skourup, J., Cheung, K.F., Bingham, H.B., and Buchmann, B., 2000. Loads on a 3D body due to second-order waves and a current, *Ocean Eng.*, 27, 707–27.

Slunyaev, A., Kharif, C., Pelinovsky, E., and Talipova, T., 2002. Nonlinear wave focusing on water of finite depth, *Physica D*, 173, 77–96.

Smagorinsky, J., 1963. General circulation experiments with the primitive equations: I. The basic equations, *Mon. Weather Rev.*, 91, 99–164.

Smith, G., 1991. Comparison of stationary and oscillatory flow through porous media, M.Sc. thesis, Queen's University, Canada.

Smith, R. and Sprinks, T., 1975. Scattering of surface waves by a conical island, *J. Fluid Mech.*, 72, 373–84.

Sobey, R.J., 1986. Wind-wave prediction, *Annu. Rev. Fluid Mech.*, 18, 149–72.

Sollitt, C.K. and Cross, R.H., 1972. Wave transmission through permeable break-waters, *Proc. 13th Int. Conf. Coast. Eng. ASCE*, New York, NY, 1827–46.

Sommerfeld, A., 1896. Mathematische theorie der diffraction, *Meth. Ann.*, 47, 317–74.

Sommerfeld, A., 1964. *Mechanics of Deformable Bodies*, Vol. 2 of *Lectures On Theoretical Physics*, New York: Academic Press.

Soomere, T., 2006. Nonlinear ship wake waves as a model of rogue waves and a source of danger to the coastal environment: a review, *Oceanologia*, 48(S), 185–202.

Sorensen, R.M., 1973. Ship-generated waves, *Adv. Hydrosci.*, 9, 49–83.

Spalart, P.R. and Allmaras, S. R., 1992. A one-equation turbulence model for aerodynamic flows, *AIAA Paper 92-0439*.

Spalding, D.B., 1980. Mathematical modeling of fluid mechanics, heat transfer and mass transfer processes, *Mech. Eng. Dept., Rep. HTS/80/1*, Imperial College of Science, Technology and Medicine, London.

Speziale, C.G., Abid, R., and Anderson, E.C., 1992. A critical evaluation of two-equation models of turbulence, *AIAA J.*, 30, 324–31.

Spring, B.H. and Monkmeyer, P.L., 1974. Interaction of plane waves with vertical cylinders, *Proc. 14th Int. Conf. Coast. Eng.*, pp. 1828–48.

Stansby, P.K. and Zhou, J.G., 1998. Shallow-water flow solver with non-hydrostatic pressure: 2D vertical plane problems, *Int. J. Numer. Meth. Fluids*, 28, 541–63.

Stelling, G.S. and Duinmeijer, S.P.A., 2003. A staggered conservative scheme for every Froude number in rapidly varied shallow water flows, *Int. J. Numer. Meth. Fluids*, 43(12), 1329–54.

Stelling, G.S. and Zijlema, M., 2003. An accurate and efficient finite-difference algorithm for non-hydrostatic free-surface flow with application to wave propagation, *Int. J. Numer. Meth. Fluids*, 43(1), 1–23.

Stive, M.J.F. and Wind, H.G., 1986. Cross-shore mean flow and in the surf zone, *Coast. Eng.*, 10, 325–40.

Stoker, J.J., 1957. *Water Waves*, New York: Interscience Publishers.

Stokes, G.G., 1880. On the theory of oscillatory waves, *Mathematical Physics Papers*, Cambridge: Cambridge University Press.

Streeter, V.L., Wylie, E.B., and Bedford, K.W., 1998. *Fluid Mechanics*, Singapore: McGraw-Hill.

Strelets, M., 2001. Detached-eddy simulation of massively separated flows, *AIAA Paper 01-0879*.

Strouboulis, T., Babuska, I., and Copps, K., 2000. The design and analysis of the generalized finite element method, *Comput. Meth. Appl. Mech. Eng.*, 181 (I-3), 43–69.

Su, M.Y., 1982a. Three-dimensional deep-water waves. Part 1. Experimental measurement of skew and symmetric wave patterns, *J. Fluid Mech.*, 124, 73–108.

Su, M.Y., 1982b. Evolution of groups of gravity waves with moderate to high steepness, *Phys. Fluids*, 25, 2167–74.

Su, X. and Lin, P., 2005. A hydrodynamic study on flow motion with vegetation, *Mod. Phys. Lett. B*, 19(28–29), 1659–62.

Sugimoto, N., Nakajima, N., and Kakutani, T., 1987. Edge-layer theory for shallow-water waves over a step – reflection and transmission of a soliton, *J. Phys. Soc. Jpn*, 56(5), 1717–30.

Suh, K.D., Dalrymple, R.A., and Kirby, J.T., 1990. An angular spectrum model for propagation of Stokes waves, *J. Fluid Mech.*, 221, 205–32.

Suh, K.D., Lee, C., and Woo, S.P., 1997. Time-dependent equations for wave propagation on rapidly varying topography, *Coast. Eng.*, 32, 91–117.

Suh, K.D., Jung, T.H., and Haller, M.C., 2005. Long waves propagating over a circular bowl pit, *Wave Motion*, 42(2), 143–54.

Sukumar, N., Moran, B., and Belytschko, T., 1998. The natural element method in solid mechanics, *Int. J. Numer. Meth. Eng.*, 43(5), 839–87.

Sulisz, W., 1985. Wave reflection and transmission at permeable breakwaters, *Proc. 13th Int. Conf. Coast. Eng. ASCE*, New York, NY, pp. 1827–46.

Sulisz, W., 2002. Diffraction of nonlinear waves by horizontal rectangular cylinder founded on low rubble base, *Appl. Ocean Res.*, 24(4), 235–45.

Summerfield, W., 1972. Circular islands as resonators of long-wave energy, *Philos. Trans. R. Soc. Lond.*, A272, 361–402.

Sussman, M., Smereka, P., and Osher, S., 1994. A level set approach for computing solutions to incompressible two-phase flow, *J. Comput. Phys.*, 114, 146–59.

Swift, M.R., Osborn, W.R., and Yeomans, J.M., 1995. Lattice Boltzmann simulation of nonideal fluids, *Phys. Rev. Lett.*, 75(5), 830–3.

Synolakis, C.E., 1987. The run-up of solitary waves, *J. Fluid Mech.*, 185, 523–45.

Tadepalli, S. and Synolakis, C.E., 1995. The run up of N-waves on sloping beaches, *Proc. R. Soc. A*, 445, 99–112.

Takano, K., 1960. Effects d'un obstacle parallelepipedique sur la propagation de la houle, *Houille blanche*, 15, 247–67.

Tanaka, M., 1986. The stability of solitary waves, *Phys. Fluids*, 29(3), 650–5.

Tang, C.J. and Chang, J.H., 1998. Flow separation during solitary wave passing over submerged obstacles, *J. Hydraul. Eng.*, 124(7), 742–9.

Tang, Y. and Ouellet, Y., 1997. A new kind of nonlinear mild-slope equation for combined refraction–diffraction of multifrequency waves, *Coast. Eng.*, 31, 3–36.

Tanioka, Y. and Satake, K., 2001. Detailed coseismic slip distribution of the 1944 Tonankai earthquake estimated from tsunami waveforms, *Geophys. Res. Lett.*, 28(6), 1075–8.

Tao, J.H., 1983, *Computation of Wave Run-up and Wave Breaking*, Internal Report, Danish Hydraulic Institute.

Tao, J.H., 2005. *Numerical Simulation of Water Waves*, Tianjin: Tianjin University Press (in Chinese).

Tao, J.H. and Han, G.., 2001. Numerical simulation of breaking wave based on higher-order mild slope equation, *Chin. Ocean Eng.*, 15(2), 269–80.

Tappert, F.D. and Zabusky, N.J., 1971. Gradient-induced fission of solitons, *Phys. Rev. Lett.*, 27(26), 1771–6.

Teng, B. and Taylor, R.E., 1995. Application of a higher order BEM in the calculation of wave run-up in a weak current, *Int. J. Offshore Polar Eng.*, 5, 219–24.

Teng, M.H. and Wu, T.Y., 1992. Nonlinear water waves in channels of arbitrary shape, *J. Fluid Mech.*, 242. 211–33.

Teng, M.H. and Wu, T.Y., 1997. Effects of channel cross-sectional geometry on long wave generation and propagation, *Phys. Fluids*, 9(11), 3368–77.

Teng, M.H. Feng, K., and Liao, T.I., 2000. Experimental study on long wave run-up on plane beaches, *Proc. 10th Int. Offshore Polar Eng. Conf.*, 3660–4.

Tennekes, H. and Lumley, J.L., 1972. *A First Course in Turbulence*, Cambridge: MIT Press.

Thompson, J.F., Thames, F.C., and Mastin, C.W., 1974, Automatic numerical generation of body-fitted curvilinear co-ordinate system for fields containing any number of arbitrary two dimensional bodies, *J. Comput. Phys.*, 15, 299–319.

Thompson, J.F., Warsi, Z.U.A., and Mastin, C.W., 1982. Boundary-fitted coordinate systems for numerical-solution of partial-differential equations – a review, *J. Comput. Phys.*, 47(1), 1–108.

Thompson, J.F., Warsi, Z.U.A., and Mastin, C.W., 1985. *Numerical Grid Generation: Foundations and Applications*, New York, NY: Elsevier Science Publ. Co. Inc..

Thomson, W. (Lord Kelvin), 1887, On ship waves, *Proc. Inst. Mech. Eng.*, pp. 409–33.

Thornton, E.B. and Guza, R.T., 1983. Transformation of wave height distribution, *J. Geophys. Res.*, 88(C10), 5925–38.

Tick, L.J., 1963. Nonlinear probability models of ocean waves, *Ocean Wave Spectra*, 163–9.

Ting, F.C.K. and Kim, Y.K., 1994. Vortex generation in water waves propagating over a submerged obstacle, *Coast. Eng.*, 24, 23–49.

Ting, F.C.K. and Kirby, J.T., 1995. Dynamics of surf-zone turbulence in a strong plunging breaker, *Coast. Eng.*, 24, 177–204.

Ting, F.C.K. and Kirby, J.T., 1996. Dynamics of surf-zone turbulence in a spilling breaker, *Coast. Eng.*, 27, 131–60.

Titov, V.V. and Synolakis, C.E., 1998. Numerical modeling of tidal wave runup, *J. Waterw. Port Coast. Ocean Eng. ASCE*, 124(4), 157–71.

Titov, V.V., Rabinovich, A.B., Mofjeld, H.O., Thomson, R.E., and González, F.I., 2005. The global reach of the 26 December 2004 Sumatra tsunami, *Science*, 309(5743), 2045–8.

Tolman, H.L., 1999. User manual and system documentation of WAVEWATCH-III version 1.18, *NOAA / NWS / NCEP / OMB Technical Note 166*.

Tome, M.F. and Mckee, S., 1994. Gensmac – a computational marker and cell method for free-surface flows in general domains, *J. Comput. Phys.*, 110(1), 171–86

Tome, M.F., Filho, A.C., Cuminato, J.A., Mangiavacchi, N., and McKee, S., 2001. GENSMAC3D: a numerical method for solving unsteady three-dimensional free surface flows, *Int. J. Numer. Meth. Fluids*, 37(7), 747–96.

Torrey, M.D., Cloutman, L.D., Mjolsness, R.C., and Hirt, C.W., 1985. *NASA-VOF2D: A Computer Program for Incompressible Flows with Free Surfaces*, Los Alamos National Laboratory, LA-10612-MS.

Torrey, M.D., Mjolsness, R.C., and Stein, R.L., 1987. *NASA-VOF3D: A Three-Dimensional Computer Program for Incompressible Flows with Free Surfaces*, Los Alamos National Laboratory, LA-11008-MS.

Troch, P., 1997. VOFbreak^2, a numerical model for simulation of wave interaction with rubble mound breakwaters, *Proc. 27th IAHR Congress*, San Francisco, CA, USA, pp. 1366–71.

Tsai, C.P., Chen, H.B., and Lee, F.C., 2006. Wave transformation over submerged permeable breakwater on porous bottom, *Ocean Eng.*, 33(11–12), 1623–43.

Tsai, W.T. and Yue, D.K.P., 1995. Effects of soluble and insoluble surfactant on laminar interactions of vortical flows with a free surface, *J. Fluid Mech.*, 289, 315–49.

Tsynkov, S.V., 1998. Numerical solution of problems on unbounded domains. A review, *Appl. Numer. Math.*, 27(4), 465–532.

Turnbull, M.S., Borthwick, A.G.L., and Taylor, R.E., 2003. Numerical wave tank based on a sigma-transformed finite element inviscid flow solver, *Int. J. Numer. Meth. Fluids*, 42(6), 641–63.

Twu, S.W. and Chieu, C.C., 2000. A highly wave dissipation offshore breakwater, *Ocean Eng.*, 27, 315–30.

Twu, S.W., Liu, C.C., and Twu, C.W., 2002. Wave damping characteristics of vertically stratified porous structures under oblique wave action, *Ocean Eng.*, 29, 1295–311.

Umeyama, M., 2005. Reynolds stresses and velocity distributions in a wave-current coexisting environment. *J. Waterw. Port Coast. Ocean Eng. ASCE*, 131, 203–12.

Ursell, F., 1952. Edge waves on a sloping beach, *Proc. R. Soc. Lond. Ser. A.*, 214, 79–97.

Ursell, F., 1953. The long-wave paradox in the theory of gravity waves, *Proc. Camb. Philos. Soc.*, 49, 685–94.

Vassberg, J.C., 2000. Multi-block mesh extrusion driven by a globally elliptic system, *Int. J. Numer. Meth. Eng.*, 49(1), 3–15.

Veeramony, J. and Svendsen, I.A., 2000. The flow in surf zone waves, *Coast. Eng.*, 39, 93–122.

Vidal, C., Losada, M.A., Medina, R., and Rubio, J., 1988. Solitary wave transmission through porous breakwaters, *Proc. 21st Int. Conf. Coast. Eng.*, ASCE, Costa del Sol-Malaga, Spain, pp. 1073–83.

Vreugdenhil, C.B., 1994. *Numerical Methods for Shallow-water Flow*, Dordrecht, The Netherlands: Kluwer Academic Publishers.

Walkley, M. and Berzins, M., 2002. A finite element method for the two-dimensional extended Boussinesq equations, *Int. J. Numer. Meth. Fluids*, 39(10), 865–85.

Wang, B.L. and Liu, H., 2006. Solving a fully nonlinear highly dispersive Boussinesq model with mesh-less least square-based finite difference method, *Int. J. Numer. Methods Fluids*, 52(2), 213–35.

Wang, K.H. and Ren, X.G., 1993. Water-waves on flexible and porous breakwaters, *J. Eng. Mech. ASCE*, 119(5), 1025–47.

Wang, K.H. and Ren, X.G., 1994. Wave interaction with a concentric porous cylinder system, *Ocean Eng.*, 21(4), 343–60.

Wang, S.K., Hsu, T.W., Tsai, L.H., and Chen, S.H., 2006. An application of Miles' theory to Bragg scattering of water waves by doubly composite artificial bars, *Ocean Eng.*, 33, 331–49.

Wang, Y.X. and Su, T.C., 1993. Computation of wave breaking on sloping beach by VOF method, *Proc. 3rd Int. Offshore Polar Eng. Conf.*, ISOPE, Golden CO, USA, pp. 96–101.

Wang, Y.Z. and Zheng, B., 2001. Vibrating-rocking motion of caisson breakwater under breaking wave impact, *Chin. Ocean Eng.*, 15(2), 205–16.

Watanabe, E., Wang, C.M., and Yang, X., 2003. Hydroelastic analysis of pontoon-type circular VLFS, *Proc 13th Int. Offshore Polar Eng*, Honolulu, HI, USA, pp. 93–9.

Watanabe, E., Utsunomiya, T., and Wang, C.M., 2004. Hydroelastic analysis of pontoon-type VLFS: a literature survey, *Eng. Struct.*, 26, 245–56.

Wehausen, J.V. and Laitone, E.V., 1960, Surface waves, in: S. Flügge (Ed.) *Encyclopedia of Physics, Vol. IX: Fluid Dynamics III*, Berlin: Springer-Verlag.

Wei, G. and Kirby, J.T., 1995. Time-dependent numerical code for extended Boussinesq equations, *J. Waterw. Port Coast. Ocean Eng. ASCE*, 121(5), 251–61.

Wei, G., Kirby, J.T., Grilli, S.T., and Subramanya, R., 1995. A fully nonlinear Boussinesq model for surface waves, Part 1. Highly nonlinear unsteady waves, *J. Fluid Mech.*, 294, 71–92.

Westerink, J.J., Luettich, R.A., Baptista, A.M., Scheffner, N.W., and Farrar, P., 1992. Tide and Storm Surge Predictions Using a finite element model, *J. Hydraul. Eng. ASCE*, 118, 1373–90.

Whalin, R.W., 1971. The limit of applicability of linear wave refraction theory in convergence zone, *Research Report H-71-3*, US Army Corps of Engineers, USA.

White, F.M., 2003. *Fluid Mechanics*, 5th ed., Singapore: McGraw-Hill.

Whitham, G.B., 1974. *Linear and Nonlinear Waves*, New York: Wiley.

Wiegel, R.L., 1960. A presentation of cnoidal wave theory for practical application, *J. Fluid Mech.*, 7, 273–86.

Wilcox, D., 1988. Multiscale model for turbulent flows, *AIAA J.*, 26, 1414–21.

Wilcox, D.C., 2004, *Turbulence Modeling for CFD*, 2nd ed., Palm Drive La Cañada, CA: DCW Industries, Inc.

Williams, A.N., 1996. Floating membrane breakwater, *J. Offshore Mech. Arctic Eng. Trans. ASME*, 118(1), 46–52.

Williams, A.N., Geiger, P.T., and Mcdougal, W.G., 1991. Flexible floating breakwater, *J. Waterw. Port Coast. Ocean Eng. ASCE*, 117(5), 429–50.

Witting, J.M., 1984. A unified model for the evolution of nonlinear water waves, *J. Comput. Phys.* 56, 203–36.

Woo, S.B. and Liu, P.L.F, 2004. Finite-element model for modified Boussinesq equations, II: Applications to nonlinear harbor oscillations, *J. Waterw. Port Coast. Ocean Eng. ASCE*, 130(1), 17–28.

Wu, C., Watanabe, E., and Utsunomiya, T., 1995. An eigenfunction expansion-matching method for analyzing the wave-induced responses of an elastic floating plate, *Appl. Ocean Res.*, 17(5), 301–10.

Wu, D.M. and Wu, T.Y., 1982. Three-dimensional nonlinear long waves due to moving surface pressure, *Proc. 14th Symp. on Naval Hydrodynamics*, Washington, DC, USA, pp. 103–25.

Wu, G.X., Ma, Q.W., and Taylor, R.E., 1998. Numerical simulation of sloshing waves in a 3D tank based on a finite element method, *Appl. Ocean Res*, 20(6), 337–55.

Wu, G.X., Taylor, R.E., and Greaves, D.M., 2001. The effect of viscosity on the transient free-surface waves in a two-dimensional tank, *J. Eng. Math.*, 40, 77–90.

Wu, J.C., 1982. Problems of general viscous flow, in: P.K. Banerjee and R.P. Shaw (Eds.), *Developments in Boundary Element Methods*, Vol. 2. London: Elsevier Applied Science Publishers.

Wu, N.J., Tsay, T.K., and Young, D.L, 2006. Meshless numerical simulation for fully nonlinear water waves, *Int. J. Numer. Meth. Fluids*, 50(2), 219–34.

Wu, T.Y., 1987. Generation of upstream advancing solitons by moving disturbances, *J. Fluid Mech.*, 184, 75–99.

Wu, T.Y., 1999. Modeling nonlinear dispersive water waves, *J. Eng. Mech.-ASCE*, 125(7), 747–55.

Wu, T.Y., 2000. A unified theory for modeling water waves, in: E. van der Giessen and T.Y. Wu (Eds.), *Advances in Applied Mechanics*, Vol. 37, New York: Academic Press, pp. 1–88.

Xiao, F., 1999. A computational model for suspended large rigid bodies in 3D unsteady viscous flow, *J. Comput. Phys.*, 155, 348–79.

Xie, L., Wu, K., Pietrafesa, L., and Zhang, C., 2001. A numerical study of wave–current interaction through surface and bottom stresses: wind driven circulation in the South Atlantic bight under uniform winds, *J. Geophys. Res.*, 106, 16841–52.

Xie, S., 1999. Wave forces on submerged semi-circular breakwater and similar structures, *Chin. Ocean Eng.*, 13(1), 63–72.

Xu, B.Y. and Panchang, V., 1993. Outgoing boundary conditions for finite-difference elliptic water-wave models, *Proc. R. Soc. Lond. A*, 441, 575–88.

Xu, B.Y., Panchang, V., and Demirbilek, Z., 1996. Exterior reflections in elliptic harbor wave models, *J. Waterw. Port Coast. Ocean Eng. ASCE*, 122(3), 118–26.

Yagawa, G. and Yamada, T., 1996. Free mesh method: a new meshless finite element method, *Comput. Mech.*, 18(5), 383–6.

Yakhot, V. and Orszag, S.A., 1986. Renormalization group analysis of turbulence. I. Basic theory, *J. Sci. Comput.*, 1(1), 3–51.

Yakhot, V., Orszag, S.A., Thangam, S., Gatski, T.B., and Speziale, C.G., 1992. Development of turbulence models for shear flows by a double expansion technique, *Phys. Fluids A*, 4 (7), 1510–20.

Yamasaki, J., Miyata, H., and Kanai, A., 2005. Finite-difference simulation of green water impact on fixed and moving bodies, *J. Mar. Sci. Technol.*, 10(1), 1–10.

Yan, G.W., 2000. A lattice Boltzmann equation for waves, *J. Comput. Phys.*, 161(1), 61–9.

Yan, H.M., Cui, W.C., and Liu, Y.Z., 2003. Hydroelastic analysis of very large floating structures using plate Green functions, *Chin. Ocean Eng.*, 17(2), 151–62.

Yan, S. and Ma, Q.W., 2007. Numerical simulation of fully nonlinear interaction between steep waves and 2D floating bodies using the QALE-FEM method, *J. Comput. Phys.*, 221, 666–92.

Yang, S.Q., Tan, S.K., Lim, S.Y., and Zhang, S.F., 2005. Velocity distribution in combined wave-current flows, *Adv. Water Resour.*, 29, 1196–208.

Yeung, R.W., 1981. Added mass and damping of a vertical cylinder in finite-depth waters, *Appl. Ocean Res.*, 3, 119–33.

Yilmaz, O. and Incecik, A., 1998, Analytical solutions of the diffraction problem of a group of truncated vertical cylinders, *Ocean Eng.*, 25(6), 385–94.

Yiu, K.F.C., Greaves, D.M., Leon, S.C., Saalehi, A., and Borthwick, A.G.L., 1996. Quadtree grid generation: Information handling, boundary fitting and CFD applications, *Comput. Fluids*, 25(8), 759–69.

Yoon, S.B. and Liu, P.L.F., 1987. Resonant reflection of shallow-water waves due to corrugated boundaries, *J. Fluid Mech.*, 180, 451–69.

Yoon, S.B. and Liu, P.L.F., 1989a, Stem waves along breakwater, *J. Waterw. Port Coast. Ocean Eng. ASCE*, 115(5), 635–48.

Yoon, S.B. and Liu, P.L.F., 1989b. Interaction of currents and weakly nonlinear water waves in shallow water, *J. Fluid Mech.*, 205, 397–419.

You, Z.J., 1996. The effect of wave-induced stress on current profiles, *Ocean Eng.*, 23, 619–28.

Younus, M. and Chaudhry, M.H., 1994. A depth-averaged k–ε turbulence model for the computation of free-surface flow, *J. Hydraulic Res.*, 32(3), 415–36.

Yu, X., 1995. Diffraction of water waves by porous breakwaters, *J. Waterw. Port Coast. Ocean Eng. ASCE*, 121(6), 275–82.

Yu, X., 2002. Functional performance of a submerged and essentially horizontal plate for offshore wave control: a review, *Coast. Eng. J.*, 44(2), 127–47.

Yu, X. and Chwang, A.T., 1994. Wave motion through porous structures, *J. Eng. Mech. ASCE*, 120(5), 989–1008.

Yu, X. and Zhang, B.Y., 2003. An extended analytic solution for combined refraction and diffraction of long waves over circular shoals, *Ocean Eng.*, 30(10), 1253–67.

Yue, D.K.P. and Mei, C.C., 1980. Forward diffraction of Stokes waves by a thin wedge, *J. Fluid Mech.*, 99(1), 33–52.

Yue, D.K.P., Chen, H.S., and Mei, C.C., 1978. A hybrid element method for diffraction of water waves by three dimensional bodies, *Int. J . Numer. Meth. Eng.*, 12, 245–66.

Yue, W.S., Lin, C.L., and Patel, V.C., 2003. Numerical simulation of unsteady multidimensional free surface motions by level set method, *Int. J. Numer. Meth. Fluids*, 42(8), 853–84.

Zakharov, V.E. and Shrira, V.I., 1990. On the formation of the directional spectrum of wind waves, *Sov. Phys. JETP* 71, 1091–100.

Zang, Y.L., Xue, S.T., and Kurita, S., 2000. A boundary element method and spectral analysis model for small-amplitude viscous fluid sloshing in couple with structural vibrations, *Int. J. Numer. Meth. Fluids*, 32(1), 79–96.

Zelt, J.A., 1991. The run-up of nonbreaking and breaking solitary waves, *Coast. Eng.*, 15(3), 205–46.

Zhang, D.H. and Chwang, A.T., 1996. Numerical study of nonlinear shallow water waves produced by a submerged moving disturbance in viscous flow, *Phys. Fluids*, 8, 147–56.

Zhang, D.H. and Chwang, A.T., 1999. On solitary waves forced by underwater moving objects, *J. Fluid Mech.*, 389, 119–35 .

Zhang, H., Zheng, L.L., Prasad, Y., and Hou, T.Y., 1998. A curvilinear level set formulation for highly deformable free surface problems with application to solidification, *Numer. Heat Transf. B Fundam.*, 34(1), 1–20.

Zhang, L, Kim, M.H., Zhang, J., and Edge, B.L., 1999. Hybrid model for Bragg scattering of water waves by steep multiply-sinusoidal bars, *J. Coast. Res.*, 15(2), 486–95.

Zhang, M.Y. and Li, Y.S., 1996. The synchronous coupling of a third-generation wave model and a two-dimensional storm surge model, *Ocean Eng.*, 23(6), 533–43.

Zhang, Y. and Zhu, S.P., 1994. New solutions for the propagation of long water waves over variable depth, *J. Fluid Mech.*, 278, 391–406.

Zhao, L., Panchang, V., Chen, W., Demirbilek, Z., and Chhabbra, N., 2001. Simulation of wave breaking effects in two-dimensional elliptic harbor wave models, *Coast. Eng.*, 42(4), 359–73.

Zhao, M., Teng, B., and Tan, L., 2004. A finite element solution of wave forces on submerged horizontal circular cylinders, *Chin. Ocean Eng.*, 18(3), 335–346.

Zhao, Y. and Anastasiou, K., 1996. Modelling of wave propagation in the nearshore region using the mild slope equation with GMRES-based iterative solvers, *Int. J. Numer. Meth. Fluids*, 23, 397–411.

Zhao, Z.D., Lian, J.J., and Shi, J.Z., 2006. Interactions among waves, current, and mud: numerical and laboratory studies, *Adv. Water Resour.*, 29(11), 1731–44.

Zheng, H.W., Shu, C., and Chew, Y.T., 2006. A lattice Boltzmann model for multiphase flows with large density ratio, *J. Comput. Phys.*, 218(1), 353–71.

Zheng, Y.H., Shen, Y.M., and Qiu, D.H, 2001. Numerical simulation of wave height and wave set-up in nearshore regions, *Chin. Ocean Eng.*, 15(1), 15–23.

Zhou, J.G. and Stansby, P.K., 1999. 2D shallow water flow model for the hydraulic jump, *Int. J. Numer. Meth. Fluids*, 29(4), 375–87.

Zhu, S.P., 1993. A new DRBEM model for wave refraction and diffraction, *Eng. Anal. Bound. Elem.*, 12(4), 261–74.

Zhu, S.P. and Zhang, Y., 1996. Scattering of long waves round a circular island mounted on a conical shoal, *Wave Motion*, 23, 353–62.

Zhuang, F. and Lee, J.J., 1996. A viscous rotational model for wave overtopping over marine structure, *Proc. 25th Int. Conf. Coast. Eng.*, ASCE, Orlando, FL, USA, pp. 2178–91.

Zienkiewicz, O.C., Taylor, R.L., 1989. *The Finite Element Method*, Vol. I, 4th ed., New York: McGraw-Hill.

Zilman, G. and Miloh, T., 2000. Hydroelastic buoyant circular plate in shallow water: a closed form solution, *Appl. Ocean Res.* 22, 191–8.

Zilman, G. and Miloh, T., 2001. Kelvin and V-like ship waves affected by surfactants, *J. Ship Res.*, 2, 150–63.

Zwillinger, D., 1997. *Handbook of Differential Equations*, 3rd ed., Boston, MA: Academic Press.

Subject index

Author index